Modern Engineering Statistics

LAWRENCE L. LAPIN
San Jose State University

An Alexander Kugushev Book

Duxbury Press
An Imprint of Wadsworth Publishing Company
I(T)P® An International Thomson Publishing Company

Belmont, CA • Albany, NY • Boston • Cincinnati • Detroit • Johannesburg
London • Madrid • Melbourne • Mexico City • New York
Paris • Singapore • Tokyo • Toronto • Washington

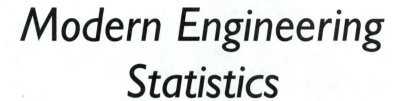

Assistant Editor: *Cynthia Mazow*
Editorial Assistant: *Martha O'Connor*
Marketing Manager: *Lauren Ward*
Production: *Ruth Cottrell*
Print Buyer: *Barbara Britton*
Permissions: *Peggy Meehan*
Copy Editor: *Charles Cox*
Cover Designer: *Ellen Pettengell*
Cover Image: *St. Louis Arch (Photonica/Mitsugu Hara)*
Compositor: *Interactive Composition Corporation*
Printer: *Phoenix Color*
Cover Printer: *Phoenix Color*

Printed in the United States of America
5 6 7 8 9 10

For more information, contact Duxbury Press at Wadsworth Publishing Company, 10 Davis Drive, Belmont, CA 94002, or electronically at http://www.thomson.com/duxbury.html

International Thomson Publishing Europe
Berkshire House 168-173
High Holborn
London, WC1V 7AA, England

Thomas Nelson Australia
102 Dodds Street
South Melbourne 3205
Victoria, Australia

Nelson Canada
1120 Birchmount Road
Scarborough, Ontario
Canada M1K 5G4

International Thomson Publishing GmbH
Königswinterer Strasse 418
53227 Bonn, Germany

International Thomson Editores
Campos Eliseos 385, Piso 7
Col. Polanco
11560 México D. F. México

International Thomson Publishing Asia
221 Henderson Road
#05-10 Henderson Building
Singapore 0315

International Thomson Publishing - Japan
Hirakawacho Kyowa Building, 3F
2-2-1 Hirakawacho
Chiyoda-ku, Tokyo 102, Japan

International Thomson Publishing Southern Africa
Building 18, Constantia Park
240 Old Pretoria Road
Halfway House, 1685 South Africa

Library of Congress Cataloging-in-Publication Data

Lapin, Lawrence L.
Modern engineering statistics / Lawrence Lapin,
p. cm.
Includes index.
ISBN 0-534-50883-9
1. Engineering—Statistical methods. 2. Probabilities.
I. Title.
TA340.L34 1996
620'.0072—dc20 96-35230

Contents

Preface

My Goals

This book has two distinguishing goals: it is designed specifically for a single-semester course and it intends to make statistics interesting and easy for engineering majors. While such students are among the most quantitatively skilled, they experience similar frustrations and anxieties to those of less mathematically sophisticated students. My goal is to make it easier for engineering students to assimilate statistics. This book should help them discover how the subject is relevant to their interests and immediate needs, because it has been "engineered" especially for them. It should provide the familiar feel of popular texts on purely engineering subjects. Like those books, this text is rich in examples that not only display the methodology but which also amply illustrate key theoretical concepts and carefully explain the underlying motivation. In doing this, my goal has been to provide strong explanations that should make learning statistics more relevant and meaningful.

The Coverage

Only essential statistical topics are covered, so that the needs of a single-semester course may be met without skipping half the book. I have made deliberate choices based on my teaching experience. One of these choices has been to reduce topical coverage to make room for greater practical depth. To that end, the organizational structure de-emphasizes mathematical statistics. Instead, *an early exposure is given to practical applications and key statistical concepts.*

The early chapters employ a maximum of descriptive statistics and data analysis, much heavier than in more traditional books. Thus statistics comes first and probability appears much later—turning tables on the conventional approach. A shorter-than-usual probability introduction is delayed until Chapter 6. An advantage of the early and substantial discussions of statistics (without probability) is that they strengthen the data analysis thrust. Two key applications, control charts and regression, are therefore meaningfully discussed *before* probability. Reliability applications are introduced alongside the early probability illustrations.

A small price is paid by deferring probability, in that inferential statistics does not really start until after Chapter 7. By that point, students will understand statistics and will have been using it in a manner meaningful to engineering majors. Inference frosts the cake—not just a way to illustrate mathematical statistics. This arrangement allows splitting some of the more challenging topics—such as regression—into two groups, with the more theoretical aspects appearing in post-probability chapters.

The data analysis thrust of this book continues all they way to the end, where an entire chapter on experimental design is provided. That topic should foster using statistics in achieving a key engineering function—*product* design. A second modern feature is statistical bootstrapping, introduced as an alternative to traditional statistical inference. Although the latter is emphasized, practitioners will appreciate the advantages of bootstrapping in avoiding unsubstantiated assumptions often implicit with traditional methods.

Applications

A paramount distinction of this book is that it is an order of magnitude more applied than usual. All examples and problems were specifically written with the particular needs of engineering students in mind. See, for instance, Noise-Induced Data Transmission Errors (p. 166), Satellite Ranging Errors (p. 220), Times Between Arrivals at Toll Booths (p. 235), Major Electrical Power Disruptions (p. 238), Survival of Satellite Power Systems (p. 248), etc. Whenever possible, real data are used in examples, problems, and illustrations. See, for instance, Space Shuttle Missions (p. 174 and p. 181), Isotope Production (p. 307), Airborne Radiometer Scans (p. 376), and Computer Evaluation of Algorithms (p. 488 and p. 493).

The Problems

Another distinguishing feature is the amount and variety of problem material. There are over 600 problems, most with several parts. Separate problem sets appear at the end of each major section within chapters. Most chapters also end with a review problem set. The computationally intensive problems should generally be solved with computer assistance. There is a balanced mix between shorter, conceptual exercises that will be solved by hand and longer, data-intensive problems that will be solved with computer assistance. Other problems involve interpreting or evaluating computer output. A few problems require students to make simple derivations.

Abbreviated answers to selected problems are provided in the back of this book. Very detailed solutions to such exercises can be found in the *Instructor's Manual*.

Use of the Computer

This book richly illustrates the popular software packages students will use in evaluating real-life applications. Included are illustrations featuring Minitab, SAS, and SPSS. Statistical bootstrapping employs the Resampling Stats package. The illustrations are accompanied by a short listing of the coding necessary to achieve the results shown. Computer output is sometimes annotated to facilitate explanatory discussions. Problem sets thoroughly integrate using computer. (See above.)

Emphasis and Outlook

The applied orientation favors insight over rigor, practicality over mathematical elegance, and simplicity over formalism. Although a few problems require them, few proofs are presented in the text proper, where just a few easy derivations are given. All of this is done to reinforce, not to impart unnecessary rigor. Throughout, the presentations maximize intuitive appeal and minimize mathematical derivation.

Instructor's Aids

Very detailed solutions of all problems may be found in the *Instructor's Manual.*

Acknowledgments

I wish to thank the many people who have assisted me in writing this book. The following colleagues provided comments and suggestions that were instrumental in the final product: William Astle, Colorado School of Mines; John Boyer, Kansas State University; Dave Cresap, University of Portland; Rebecca Elliot, University of Utah; James Halavin, Rochester Institute of Technology; Harry Posten, University of Connecticut; John Spurrier, University of South Carolina; James Swain, University of Alabama–Huntsville; Jean Weber, University of Arizona; and Bin Yu, University of California–Berkley.

Special mention goes to Janet Anaya, San Jose State University, who carefully checked the text and solutions for mathematical and computational accuracy. I am also indebted to my students who helped debug this book.

Lawrence L. Lapin

Modern Engineering Statistics

CHAPTER ONE

Introduction

Hardly anyone can get through a day without being exposed to statistics. Although most people envision masses of data when they hear the word, modern statistics is more concerned with interpreting a special kind of data to reach a decision that must otherwise be based on guesswork and hunch.

Today's engineers bear a great responsibility, unique among the professions, for creating works that are safe and reliable. In no other profession are the costs of failure greater, and the use of statistics is crucial to averting failure. Consider a major fault in the design of a large public project such as a dam or a nuclear power plant. Statistical methods are used extensively in evaluating design concepts at all levels and stages of such large projects. The probability of the various hazards that might cause nuclear-reactor accidents, such as tidal waves, earthquakes, or airplane crashes, is routinely evaluated.

Statistical analyses are involved in all the engineering disciplines. In many cases such analyses are helpful in making choices regarding designs, materials, procedures, technologies, or methods. A mechanical engineer might employ statistics to select materials strong enough to withstand anticipated forces. Electrical engineers need statistics to determine the reliability of subsystems. Civil engineers must design a roof to withstand the weight of snow that would accumulate in the worst imaginable storm. The list of statistical applications is indeed long.

1–1 The Meaning and Role of Statistics

Modern statistics has a very special meaning. The term *statistics* is a broad one and does indeed include the use of masses of numerical data. But the focus of this book is on *using numerical evidence in making decisions.* We especially need such aids when conclusions must be reached in the face of uncertainty. Thus, our notion of statistics must include those methods and procedures that allow us to translate numerical facts into action. Proper utilization of such tools requires some knowledge of the concepts and theory on which they are based.

A Working Definition of Statistics

To the nonengineer almost any collection of data constitutes statistics. Indeed, in the general usage of the word, statistics *are* numbers, as in the summaries of baseball results or in "vital statistics." But as an academic discipline, the term has a different, more precise meaning.

As a field of study, modern statistics encompasses a large body of methods and theory that is applied to numerical evidence to help make *decisions* or communicate information. (When statistics is referred to as an area of study, the word should be treated as a singular rather than a plural noun.) The solution of a partial differential equation and the inversion of a matrix are, therefore, not statistical exercises. Neither are the readings in a survey, the telemetry recordings from a space probe, or the digitized radar cross sections of ballistic objects. Those data are not statistical in the modern sense unless the numerical information is somehow used in the process of decision making.

Statistical decision making includes another dimension besides choice. This is *uncertainty.* Given a list of weighed items, we can easily pick the heaviest. Perusal of a topographic map can narrow the choice for a dam's location to a single spot. The target acquisition window for satellite-tracking radar can be chosen by extrapolating the satellite's last known trajectory. But none of these actions is a statistical decision unless the numerical evidence is somehow used in resolving uncertainties associated with the choices.

Definition

Statistics is the analysis of numerical data for the purpose of reaching a decision or communicating information in the face of uncertainty.

The Role of Modern Statistics

Our working definition of statistics is particularly appropriate to engineers, who, in their quantitative world, must make important decisions in the face of uncertainty. Many new engineers begin their careers as estimators, who establish the prices at which their firms should bid on new projects. From the start estimators must project future costs, choose vendors for parts and materials, and select appropriate methods and procedures for getting the job done. Wrong choices can lead to improperly low bids, resulting in losses to the firm, or to inordinately high bids and subsequent loss of potential work. Even the smallest projects must be priced under a great deal of uncertainty, and some form of statistics can play a major role in reaching many of the final decisions.

Engineering is a future-oriented profession that requires continuing research. The evaluation of new designs, concepts, procedures, and materials is at the forefront of engineering activity. Which show the most promise and should be implemented immediately? Which require further research? Statistics can be a very valuable tool in such evaluations, since it provides a scientific basis for choice. Through commonly accepted statistical procedures, scientists and engineers can communicate their findings to colleagues throughout the world. The statistical conclusions of one group may save another from pursuing dead-end paths or duplicating work already done.

Types of Statistics: Descriptive, Inferential, and Exploratory

The decision-making emphasis is a key element of modern statistics. Not many years back the primary focus of statistics was to summarize or describe numerical data.

Descriptive Statistics That portion of the field encompassing methods for summarizing is called **descriptive statistics**. A major element of descriptive statistics is displays of data, such as those in Figure 1–1. Although important in its own right, descriptive statistics provides necessary props for the larger class of statistical evaluations. Much of the next chapter is concerned with descriptive statistics, providing us with mortar and bricks to build the framework for the more esoteric statistics discussed next.

Inferential Statistics The main thrust of modern statistics is making generalizations about the whole (referred to as the *population*) by a thorough examination of the part (called the *sample*). Such conclusions are inferences rather than logical deductions, and this area may be referred to as **inferential statistics**.

Common sense tells us that parts might deviate, perhaps considerably, from the whole. Inferential statistics is therefore largely concerned with the quality of the generalizations made about populations using sample evidence. One need only reflect on the poor performance of many political polls to realize the dangers inherent in drawing conclusions from samples. No sample can be guaranteed to match its target population, but much of the statistical art is concerned with keeping such sampling error within reasonable bounds. The very act of sample selection may incorporate the scientific method so that the potential for *error* may itself be quantified.

Exploratory Data Analysis In recent years a third type of statistics has evolved. This is **exploratory data analysis**, a new statistical methodology employing descriptive statistics and extending procedures beyond traditional forms. Some procedures lack applicable probability theory, so that no universally acceptable methods yet exist for expressing the quality of their results.

Statistical Bootstrapping Some investigators include in the above grouping the newer computer-intensive procedures of **statistical bootstrapping**, introduced in Chapter 10. Investigators using these new methods can make use of smaller sample sizes and

Figure 1–1 *Graphical displays used in describing statistical data.*

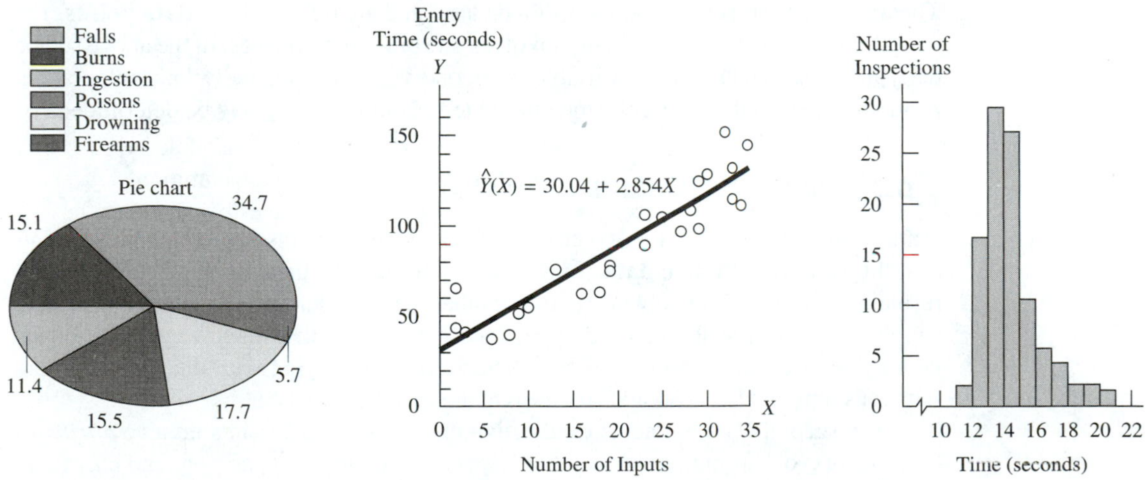

are not burdened by the restrictions that underlie many traditional methods, which sometimes require "well-behaved" data.

Problems

1–1 For each of the following decisions, give an example of potential numerical data that might be helpful in making the choice.
(a) A freshman is picking an engineering concentration.
(b) An electrical engineer is choosing alternatives for power supplies.
(c) A project engineer is selecting an electrical contractor.
(d) A computer scientist determines how much main memory to include in the primary circuitry.

1–2 Consider the following numerical data. Suggest for each a decision situation where the data might be used in reaching a choice.
(a) runs batted in by players on a baseball team
(b) mean times between failures of computer cooling-fan motors
(c) average starting salaries for engineers in various concentrations
(d) achievement scores from the students of two statistics classes taught differently

1–3 Pick a technical publication of interest. Identify four instances of numerical data that may be used for making a decision under uncertainty.
(a) List each item.
(b) Indicate for each example (1) the potential decision maker, (2) what choices might be considered, and (3) the uncertainties involved.

1–2 Statistical Data

A collection of data is referred to as a **data set**. To illustrate, consider the following daily temperatures (low, high) reported by Weather Service for 10 stations.

(48, 63)	(57, 85)	(65, 89)	(53, 71)	(49, 77)
(51, 74)	(51, 78)	(72, 81)	(56, 78)	(54, 73)

The above records are the elements of a data set and are referred to as **data points**.

A statistical evaluation usually involves one or more **variables**. In the above set the daily *high* temperature is one variable. A second variable is the daily *low* temperature. From those two a third variable, the diurnal temperature *range*, may be determined.

Classifications for Data and Variables

Statisticians divide data into two classes. When the data points involve *numerical values*, they are **quantitative data**. To illustrate, consider the data sets in Figure 1–2 representing Comp-u-Com's workstation motherboards obtained from three different suppliers, each of which involves a separate data set for that manufacturer's motherboard's status, with amounts and other characteristics varying with the unit itself and with its supplier. Unit cost, age, and weight are quantitative data.

The second class of data is **qualitative data**. Those data points involve *attributes* (such as sex, occupation, or some other category). Employee department and classifica-

Figure 1–2 *Data sets for Comp-u-Com workstation motherboards.*

Supplier	Serial Number	Country of Origin	Unit Cost	Age (days)	Weight (pounds)	Class
DanDee	9–15438	Korea	$147.88	182	5.4	XX
	.	.	.	.	.	.
	.	.	.	.	.	.
	.	.	.	.	.	.
Wysiwyg	8–004419	U.S.A.	$153.25	36	6.1	HH
	.	.	.	.	.	.
	.	.	.	.	.	.
	.	.	.	.	.	.
Pacific Rim	7–90837	Hong Kong	$155.16	17	5.8	XX
	.	.	.	.	.	.
	.	.	.	.	.	.
	.	.	.	.	.	.

tion are examples of qualitative data, which are ordinarily categorical in nature. The data set in Figure 1–2 shows for each motherboard two other subsets of qualitative data: the country of origin and the classification.

A variable need not be limited to a single data set. Following the same designations as the respective data type, a variable may be either quantitative (unit cost) or qualitative (country of origin). Levels for a quantitative variable are called **variates**. For the motherboards the unit costs for the three listed items are $147.88, $153.25, and $155.16. Levels for a qualitative variable are **attributes** or **categories**. The country of origin variable for the three motherboards has attributes Korea, U.S.A., and Hong Kong.

Types of Quantitative Data

Four types of quantitative data are encountered in statistical investigations.

1. **Nominal data** fall at the bottom of the spectrum. They are numbers that represent arbitrary *codes*. An engineering school, for example, might use numbers to denote undergraduate majors: 1 for electrical engineering, 2 for civil engineering, and so on.
2. **Ordinal data** convey *ranking* in terms of importance, strength, or severity. The *Beaufort* wind scale provides a good example; a value of 3 corresponds to a gentle sea breeze, while a 6 represents a strong breeze, and a 9 signifies a strong gale. The change in force between a gentle and a strong sea breeze is not equal to that between a strong gale and a strong breeze, as the numbers themselves would misleadingly indicate.
3. **Interval data** allow only addition and subtraction. The most common example of this type is *temperature*, for which scales are arbitrarily chosen. A 100°

Fahrenheit day is not twice as hot as a 50° day since $100°/50° = 2$ is not a meaningful ratio (and a different ratio applies under the *Celsius* scale).

4. **Ratio data** include times and many physical measurements of size, weight, or strength. The arithmetic operations of addition, subtraction, division, and multiplication are all valid with ratio data.

Most statistical investigations involve arithmetic operations, which limits them to interval or ratio data. Some statistical procedures can be meaningfully used with ordinal data. Although they involve numerical values, nominal data are generally treated as categories and evaluated in the same way as qualitative data.

Problems

1–4 Many publications of engineering interest contain data sets. Find an example of a data set. Then do the following.
(a) Identify the data points.
(b) Determine useful variables, indicating whether each is quantitative or qualitative. Identify for that data point the level or attribute for each variable found.

1–5 Give an example of each of the following data types.
(a) nominal (b) ordinal (c) interval (d) ratio

1–6 Indicate for each of the following situations which of the four types of quantitative data applies.
(a) survey questions that ask you to rate a car body design from 1 (ugly) to 5 (superb)
(b) area codes used in long distance dialing
(c) Richter scale used to express earthquake energy
(d) computer assembly times
(e) boiling temperatures for various liquids

1–3 The Population and the Sample

The basic statistical element is the **observation**, or single data point. Although we ordinarily associate observed data with numerical values such as volume or mass, they can also take the form of classifications such as "defective" or "nondefective." There are two basic accumulations of observations. The major grouping of observations is given the following definition.

Definition

A **statistical population** is the collection of all possible observations of a specified characteristic of interest.

Under the foregoing definition we limit the population to the observations. We do *not* include the objects, persons, or things being observed, which are referred to as **elementary units**. A single grouping of elementary units can give rise to any number of populations. Consider, for example, all the students attending a particular university. These persons could serve as the elementary units in populations of grade-point averages, earnings, sexes, majors, heights, ages, and so on. Consider the population of grade-point averages (GPAs): The collection of numerical GPA values, not the students themselves, constitutes the population. Do not confuse a statistical population with a demographic grouping of people (the size of which is often referred to as the "population").

A statistical population has an identity whether all, some, or even no observations are actually made. The population may already exist, as does the collection of starting salaries of all graduating engineers in the past year. Or it can lie in the future, such as the salaries of next year's graduates. Populations can be imaginary or hypothetical and may never actually come into existence. Consider the salaries of students who would have graduated from an experimental program that never got started or the salaries of those who would have worked on a major weapon system that has been cancelled.

For a variety of reasons the population itself often cannot be totally observed. Usually, the number of observations is limited by the availability of money or time. Whenever only a portion of the population is observed, the resulting observations constitute a sample. Theoretical or hypothetical populations that may never come into actual existence are sometimes referred to as **target populations**. They too can be partially observed by means of sampling. We make the following definition.

A **sample** is a collection of observations representing only a portion of the population.

Definition

Most statistical methods focus on the evaluation of samples. Since samples usually form the basis for drawing conclusions about the respective populations, care must be taken in selecting those elementary units to be observed for the sample. As we shall see, the elementary units are ordinarily chosen randomly.

The Role of Probability The key to scientific sampling is *random selection* of the units to be observed. Samples that are derived by such a procedure can be evaluated using the concepts of probability theory. Probability and frequency of occurrence provide a foundation for modern statistics. For that reason Chapters 6 and 7 are devoted to probability theory and applications.

Distinguishing among the Data Set, Population, and Sample

The statistical population and sample may encompass portions of several data sets. Figure 1–3 may be helpful in sorting things out. There we consider the population of unit costs for all workstation motherboards assembled by Comp-u-Com. Every motherboard supplier applies, not just the three shown, so that numerous data sets would be employed in defining the population data. Although the individual data points involve other types of information, only the unit costs from each applies to the population of interest. The number of elementary units in a population can involve fewer data points than the entire data set (which might be accomplished, for example, by removing all non-U.S. manufacturers or by including only the XX class of motherboards).

Since the population consists of all the possible observations that *might* be made, it exists whether or not those observations are actually made. (To avoid the expense of a complete enumeration, not all of the population data might be collected.) The sample contains just a subset of the population observations, but unlike the population, a level for the variable is actually found for each sample observation.

Although the population is usually not totally *known* (partially observed), it will generally be precisely *defined.* The sample results are the basis for making generalizations about the population.

Deductive and Inductive Statistics

An important dichotomy is fundamental to modern statistics.

Deductive Statistics In *logical deduction*, properties of the specific case can be ascribed from the general situation. **Deductive statistics** applies this process to *populations* whose characteristics are fully known, with a focus placed on the kinds of *samples* that might be obtained. To illustrate, consider the population in Figure 1–4 defined by six floppy disks. Each disk falls into exactly one of the following quality states: satisfactory (S) for receiving any data, unformatted (U) but good, incomplete (I) with some bad tracks, occupied (O) with some data, damaged (D) wafer, and broken (B) housing. Disks are selected at random using a sample of size $n = 2$. We may deduce that any one of the 15 outcomes pictured is an equally likely result for the sample denominations.

Deductive statistics is where analysis of the sampling process begins. It is used in developing statistical methodology. Only by thoroughly understanding how samples are generated from populations with *known* characteristics can we confidently deal with the reverse situation.

Figure 1–3 *Relationship between data sets, the population, and the sample.*

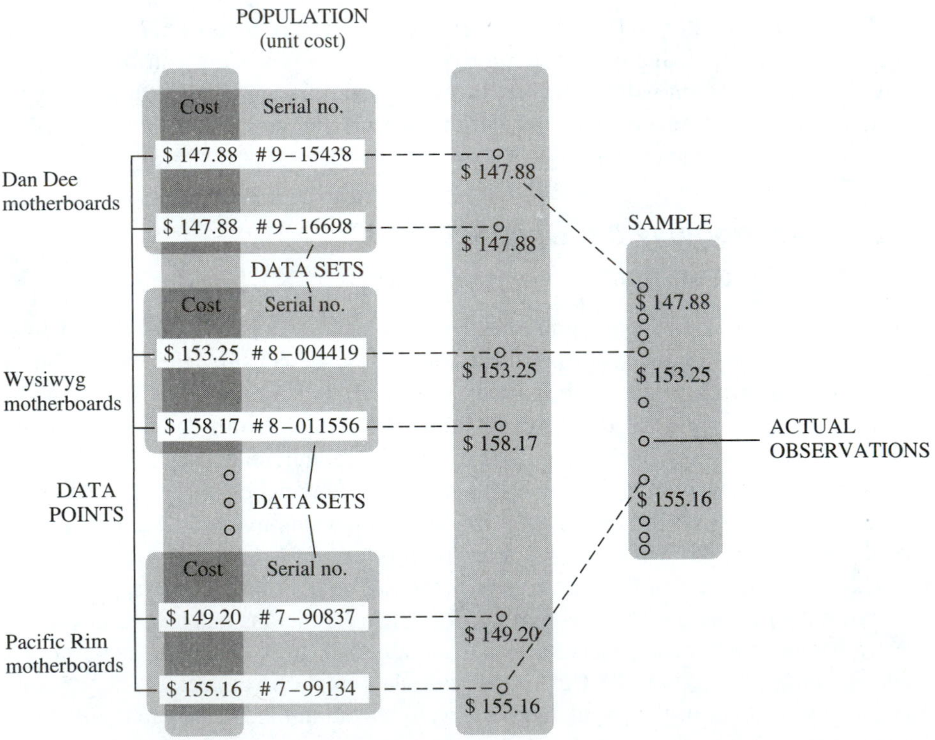

It is this second circumstance that is typically encountered in a statistical evaluation. This area of application involves *induction*, or generalizing about the whole from knowledge of just the part. Thus, it is called **inductive statistics** and involves making conclusions about the *unknown* population from the known sample.

Inductive Statistics Returning again to the selection of six floppy disks, Figure 1–5 illustrates this process. Except for the stipulation that the disk quality population consists of *some* mixture of possibilities, the population's characteristics are *unknown.* When we select the sample ($n = 2$), we get to see only a particular result—for example, (U, O). Numerous possibilities are plausible for the population, and some are more credible than the others. We can even imagine a population from which the sample result actually obtained is an impossibility.

The statistical art is largely concerned with using known sample information to draw conclusions, or make *inferences*, regarding the unknown population. The terms **inductive** and **inferential statistics** may be used interchangeably. An exposure to deductive statistics makes it easy to appreciate inductive statistics, the major concern of modern statistics.

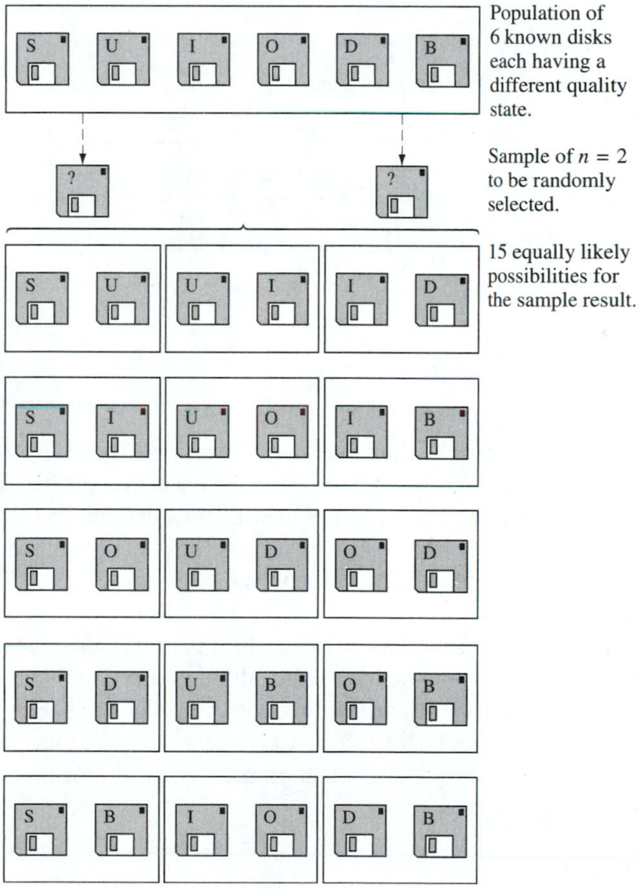

Figure 1–4 *Illustration of deductive statistics.*

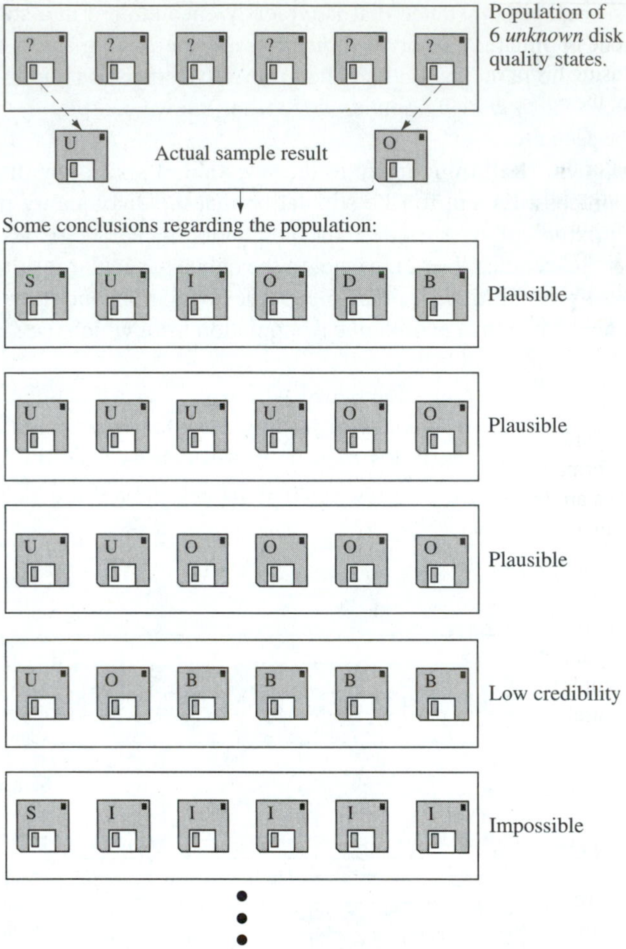

Figure 1–5 *Illustration of inductive statistics.*

Statistical Error

Statistics is partly art and partly science. It is an art because effective application depends so much on judgment and experience in selecting the best methods for a particular evaluation. The choice of procedures is considerable. Nevertheless, the scientific method dominates statistics, as can be seen by the emphasis that modern statistics places on the potential for error. The uncertainties involved in the sampling process justify this emphasis.

To a large extent sampling error may be quantified. As we shall see, this is achieved mainly through random selection. Also, the number of sample observations must be large enough to keep the chance of error at a controlled level. But scientific sampling includes many additional factors. For example, care must be taken to ensure that observations are not biased. We shall see that proper statistical procedure minimizes both error and bias.

Problems

1–7 Consider one of your acquaintances who is a practicing engineer. Comment on the suitability of using his or her earnings alone in attempting to make conclusions regarding the earnings of the entire group.

1–8 Refer to the deductive statistics illustration in Figure 1–4.
(a) List all of the possibilities when $n = 3$.
(b) What proportion of the possible sample outcomes involve either a damaged wafer or a broken housing?
(c) What proportion of the possible sample outcomes involve all ready-to-use diskettes?

1–9 Refer to the inductive statistics experiment in Figure 1–5 involving six floppy disks of unknown quality and for which the population is the quality states. A random sample of $n = 2$ is selected and a satisfactory and a broken disk are obtained. Indicate for each of the following population possibilities whether it is (1) plausible, (2) of low credibility, or (3) impossible.
(a) three satisfactory and three broken disks
(b) two unformatted, two satisfactory and two broken, disks.
(c) All disks are ready to receive data.
(d) There are one broken and five satisfactory disks.

1–10 For each of the following situations indicate whether deductive statistics or inductive statistics best categorizes the investigation.
(a) A listing is made of all the possible number outcomes from tossing a pair of dice.
(b) Twelve randomly chosen electrical engineering majors are tested to determine how well complex variables are understood by university students.
(c) A structural engineer uses very heavy stresses in measuring strains in metal rods.

1–4 The Need for Samples

There are several reasons why it is more desirable to seek incomplete or partial sample information than the complete data that might be obtained from a census. The need for partial observation rather than complete enumeration is so compelling in most investigations that sampling is taken for granted. Not every production avionics unit will be subjected to a vibration test, only a sample of jet-engine blades will be assessed for strength, and opinions regarding what options should be included in a computer software package are not sought from everybody.

Economic Advantages of Samples

A significant advantage of samples is that they are generally more economical than a census. Inspecting a portion of items being produced is obviously cheaper than assessing the entire run in detail. Even in cases where individual physical observations are cheap, we can usually achieve significant savings by relying on samples to trigger quality-control actions when huge quantities of units are involved. Moreover, in some situations the act of observation is very expensive, especially in materials testing. For example, metal-fatigue experiments may involve contorting each test unit thousands of times in special mechanical test stands. Investigations involving human subjects can be elaborate, too. Personal interviews for opinion surveys can cost upwards of $20 for one hour (during which as few as one or two responses might be obtained).

Thus, the economic advantages of sampling are compelling. Scarce research funds must not be squandered on an unnecessarily large number of observations. Not even the most affluent organization can afford to "bust its budget" in overzealous pursuit of total enumeration.

Although sampling can lead to significant savings, it does extract a price in addition to the data collection costs—because no sample is guaranteed to portray the underlying population accurately. For example, production problems will sometimes go undetected because telling evidence is lacking in the particular sample units observed. An entire production run might have to be scrapped later because a sample erroneously indicated that the proportion of defectives was lower than was actually the case. On the other hand, an unrepresentative sample could trigger costly remedial actions. Production might be halted for unneeded repairs. Such errors of omission and commission are especially insidious in basic or applied research, where unrepresentative samples may lead investigators down blind alleys or cause them prematurely to close off avenues that would have led to success.

Balancing the costs and benefits of sampling is an essential element of the statistical art. It is possible to control the incidence of sampling error so that an optimal overall solution is reached. As we shall see, this may be accomplished by judiciously determining the sample size and by establishing an efficient sampling design. The net gain from sampling rather than censusing can be dramatic.

The following example is not a traditional statistical application wherein a sample serves in drawing inferences about an uncertain population. Nevertheless, it graphically illustrates the economic advantages of the sampling concept.

Example:
Sampling Multiplies Telephone Line Capacity

Voice transmission via telephone is accomplished by converting sound into an electrical analog. The sound is transformed by a microphone/transmitter into an electronic mode that is relayed over the miles and then converted back into sound by a receiver/speaker. Early telephone equipment was designed entirely to accommodate voice sound, which we know to be quite slow compared with electronic speeds; one conversation required one line.

Essential voice characteristics of the original slow sound wave may be reconstructed from partial readings of the electronic waveform measured at intervals such as once every 100 microseconds. The reconstructed voice is analogous to the "continuous" motion perceived from the discrete frames of a motion picture in that only the signal for every hundredth microsecond of a conversation might be transmitted. In effect, then, a conversation may be transmitted and reconstructed into high-quality sound using only a *sample* of the electronic signal. Each of the 99 microseconds between samplings could be used to obtain a segment from each of 99 other conversations sharing the same line. In effect, sampling would allow for a hundredfold increase in the carrying capacity in a telephone line that otherwise could carry only one conversation at a time.

In this case transmitting voice samples provides a huge saving over a continuous transmission (really a census). As with any sampling scheme, however, there is a cost. Additional equipment is required to break down, merge, burst, and resynthesize the original voice. Here, however, the savings in transmission costs far outweigh equipment outlays.

Further Reasons for Sampling

There are a variety of reasons other than economic ones for sampling.

Timeliness A census may not be practical because the information thereby gained might take too long to acquire. It could take years to poll all potential users of a product regarding their design preferences, by which time the item will probably have become obsolete. Sample opinion data must be obtained for such an investigation. The exact incidence of defectives cannot be known until the end of a production run, but a plant superintendent cannot wait until that time before ordering necessary equipment adjustments. Instead, he will identify problems early by taking samples. Only a sample of items should be life-tested, for it could take years for an entire population to fail (after which all units would be useless for their intended purpose).

Characteristics of populations might shift over time so that later observations could represent a different entity than earlier ones. This is a common problem in assessing the preferences and attitudes of people, a fickle bunch who are capable of rapid shifts—as political pollsters know too well.

Large Populations As a practical matter, many populations needing measurement are too large for 100% observation to be achieved. Simply counting human beings is such a big job that the United States government conducts a census only once every 10 years, and the poorest nations don't even try. Imagine the problems faced by a food processor if the contents of each of the thousands of cans filled in one day were separately weighed. The census of a continuing process might never be completed, since *all* units will *never* be available for observation; such a population is theoretically infinite.

Destructive Nature of the Observation A large class of statistical investigations involves observations that destroy the units observed. Especially in engineering, test units must be damaged or destroyed to determine strength, durability, or lifetime. In these cases 100% observation is out of the question: We won't burn out all the light bulbs to find out what proportion were defective! Even surveys involving people might so change the observed individuals that they no longer reflect the population. In-depth interviews seeking attitudes about a product have been known to oversensitize test subjects so much that many of them wish never to encounter the item again—regardless of their initial predilections.

Inaccessible Populations Statisticians often measure populations that may never actually exist. Referred to as **target populations**, these are theoretical constructs that may occur only under laboratory conditions, so the only possible observations are those the investigator creates for the sample. For example, the only thin metal strips ever subjected to 10,000 quick twists on a test bench are those in the investigator's sample. Nevertheless, these strips represent the target population of all such objects that *might be* so twisted. In drug research only those patients who receive a particular dosage during evaluations might ever receive that therapy—but they are representatives of the target population of all people who would receive that same treatment.

Even when an actual population presently exists, some units may be inaccessible for observation. Consider the automotive engineer wishing to measure carbon deposits on valves of six-cylinder engines. Only engines from cars acquired by the researcher's firm would be available for disassembly.

Accuracy and Sampling

It is commonly believed that data acquired by a census are invariably more reliable than similar information based on sampling. Indeed, if the same data-gathering procedures were used for both the census and the sample, the census results would be superior.

However, the very act of observation can be a demanding one that must be performed under carefully controlled circumstances. The cost of a properly conducted census for even a small population can be so prohibitive that shortcuts might be mandatory. In recent times the observation quality of the U.S. Decennial Census has been downgraded; before 1970, census takers made personal visits to homes, whereas now the counting is done mainly on the basis of mailed questionnaires. A carefully conducted sampling study can yield higher-quality information than would be obtainable from a sloppy census.

Proper statistical procedure does not end with the making of observations. Data must then be gathered and evaluated. In 1990 a rash of lawsuits charged not only undercounting by the Census but also loss or destruction of data. And even when all census data can be accounted for, its mass can be so overwhelming that terrible mistakes occur.

Example: *Census Reports Huge Increase in Teenage Widowers*

In 1950 the U.S. Census reported that the number of teenage widowers had increased tenfold since 1940. Also, during that span of time the number of Indian divorcees grew by a like magnitude.

There was no sociological explanation for these anomalous findings. Dramatic changes in modern lifestyle had been reflected in marital status percentages that shifted by a few percent—not 1,000%, as in these findings. Two statisticians doubted the credibility of the census findings and proposed a scenario to explain what had happened.*

They purport that the oddball findings were the result of a data-processing snafu. In 1950 electronic data processing was used for the first time to tabulate the U.S. Census. Each individual was represented by a punched card. The reporters suggest that the holes in neighboring columns were reversed on a few dozen mispunched cards, so that most of the teenage widowers and Indian divorcees should have been counted as part of much larger groups.

---

*Ansley J. Coale and Frederick F. Stephan, "The Case of the Indians and Teen-age Widows," *Journal of the American Statistical Association*, LVII (June 1962), 338–437.

Problems

1–11 For the following investigations, explain why sample information might be preferred to that from a census.

(a) A personal computer manufacturer must determine the most preferred position of the monitor.

(b) Metal fatigue characteristics in the wing struts of a particular aircraft are to be evaluated.

(c) The mean lifetime of pressure seals must be established.

(d) Professional engineers' consulting fees are to be summarized.

1–12 Give an example of a type of investigation in which sampling is advantageous primarily for reasons of
(a) economy
(b) timeliness
(c) large population
(d) inaccessibility
(e) destructiveness of observations
(f) greater accuracy

1–13 For each of the following situations list the important reasons why sampling information would be sought rather than census data.
(a) Materials are being tested for strength.
(b) Customer opinions will be the basis for choosing chassis colors for a line of personal computers.
(c) A determination must be made as to whether the gas buildup inside CRTs is excessive.
(d) Comparisons will be made to determine how long various computers take to complete processing of benchmark programs.

1–5 Selecting the Sample

As we have seen, sampling error is a natural consequence in any statistical investigation. To a large extent it may be controlled by randomly selecting the sample units and may be made less significant by increasing the number of observations. Often troublesome is sampling bias, which is hard to predict; it may be minimized mainly by following good procedures in selecting sample units.

Sample Selection Using Random Numbers

We illustrate random sampling with the population of 100 Nobel physics laureates listed in Table 1–1. Ten names will be selected for a **simple random sample**, which gives each population unit (name) an equal chance of being selected.

A straightforward way of selecting our sample would be to write each name on a slip of paper. These could then be placed in a hat and thoroughly mixed, after which 10 slips of paper would be withdrawn, one at a time. Such a physical lottery would be cumbersome. And the randomness of the resulting sample might be doubted by some (as was the randomness of the 1970 U.S. Draft Lottery).

Rather than conducting physical lotteries, it is accepted statistical procedure to select samples using **random numbers**. These values are a succession of digits that were themselves generated by a lottery in such a way that every possible integer is equally likely. The following is a list of 5-digit random numbers taken from the first two columns of Appendix Table F:

12651	**61**646
81769	**74**436
36737	**98**863
82861	**54**371
21325	**15**732

These random numbers were generated at the RAND Corporation by an electromechanical device. Because any value between 0 and 9 had the same chance of appearing in each digit position, all 5-digit integers had an equal chance of appearing as the successive entries on RAND's list. In addition to equal likelihood, random numbers exhibit no sequencing patterns or serial correlation—each successive value on the list is independent of the preceding ones.

The usual procedure for selecting a sample by random numbers is first to assign each population unit an identification number. The Nobel laureates in Table 1–1 are numbered in chronological sequence. A random number is read from the table, and the population unit with the matching identification number is selected for inclusion in the sample. If the unit has been previously selected, or if there is no match, the next random number on the

Table 1–1 *Winners of Nobel Prize in Physics*

01 Hess (1936, Austria)	35 Jensen (1963, Germany)	68 Wilson (1978, U.S.)
02 Davisson (1937, U.S.)	36 Wigner (1963, U.S.)	69 Glashow (1979, U.S.)
03 Thomson (1937, Gr. Brit.)	37 Basov (1964, U.S.S.R.)	70 Salam (1979, Pakistan)
04 Fermi (1938, U.S.)	38 Prochorov (1964, U.S.S.R.)	71 Weinberg (1979, U.S.)
05 Lawrence (1939, U.S.)	39 Townes (1964, U.S.)	72 Cronin (1980, U.S.)
06 Stern (1943, U.S.)	40 Feynman (1965, U.S.)	73 Fitch (1980, U.S.)
07 Rabi (1944, U.S.)	41 Schwinger (1965, U.S.)	74 Bloembergen (1981, U.S.)
08 Pauli (1945, U.S.)	42 Tomonaga (1965, Japan)	75 Schaalow (1981, U.S.)
09 Bridgman (1946, U.S.)	43 Kastler (1966, France)	76 Siegbahn (1981, Sweden)
10 Appleton (1947, Gr. Brit.)	44 Bethe (1967, U.S.)	77 Wilson (1982, U.S.)
11 Blackett (1948, Gr. Brit.)	45 Alvarez (1968, U.S.)	78 Chandrasekhar (1983, U.S.)
12 Yukawa (1949, Japan)	46 Gell-Mann (1969, U.S.)	79 Fowler (1983, U.S.)
13 Powell (1950, Gr. Brit.)	47 Neel (1970, France)	80 Rubbia (1984, Italy)
14 Cockcroft (1951, Gr. Brit.)	48 Alfvén (1970, Sweden)	81 van der Meer (1984, Netherlands)
15 Walton (1951, Ireland)	49 Gabor (1971, Gr. Brit.)	82 von Klitzing (1985, Germany)
16 Bloch (1952, U.S.)	50 Bardeen (1972, U.S.)	83 Binnig (1986, Germany)
17 Purcell (1952, U.S.)	51 Cooper (1972, U.S.)	84 Rohrer (1986, Switzerland)
18 Zernike (1953, Neth.)	52 Schrieffer (1972, U.S.)	85 Ruska (1986, Germany)
19 Born (1954, Gr. Brit.)	53 Giaever (1973, U.S.)	86 Bednorz (1987, Germany)
20 Bothe (1954, Germany)	54 Esaki (1973, Japan)	87 Muller (1987, Switzerland)
21 Kusch (1955, U.S.)	55 Josephson (1973, Gr. Brit.)	88 Lederman (1988, U.S.)
22 Bardeen (1956, U.S.)	56 Ryle (1974, Gr. Brit.)	89 Schwartz (1988, U.S.)
23 Brattain (1956, U.S.)	57 Hewish (1974, Gr. Brit.)	90 Steinberger (1988, U.S.)
24 Shockley (1956, U.S.)	58 Rainwater (1975, U.S.)	91 Dehmelt (1989, Germany, U.S.)
25 Lee (1957, U.S.)	59 Mottelson (1975, Denmark)	92 Paul (1989, Germany)
26 Yang (1957, U.S.)	60 Bohr (1975, Denmark)	93 Ramsey (1989, U.S.)
27 Cherenkov (1958, U.S.S.R.)	61 Richter (1976, U.S.)	94 Friedman (1990, U.S.)
28 Frank (1958, U.S.S.R.)	62 Ting (1976, U.S.)	95 Kendall, (1990, U.S.)
29 Chamberlain (1959, U.S.)	63 Anderson (1977, U.S.)	96 Taylor (1990, Canada)
30 Segrè (1959, U.S.)	64 Mott (1977, Gr. Brit.)	97 de Gennes (1991, France)
31 Glaser (1960, U.S.)	65 Van Vleck (1977, U.S.)	98 Charpak (1992, Poland, France)
32 Hofstadter (1961, U.S.)	66 Kapitsa (1978, U.S.S.R.)	99 Hulse (1993, U.S.)
33 Landau (1962, U.S.S.R.)	67 Penzias (1978, U.S.)	100 Taylor (1993, U.S.)
34 Goeppert-Mayer (1963, U.S.)		

list is used. In our example the identification numbers have only two digits (except for the number 100, which will be treated as 00); we will achieve our matches by using only the first two digits of the random numbers (shown as boldface in the above list).

The following names constitute the random sample of physics prize winners

12 Yukawa	61 Richter
81 van der Meer	74 Bloembergen
36 Wigner	98 Charpak
82 von Klitzing	54 Esaki
21 Kusch	15 Walton

Of course, a different 10 names would be obtained for a different set of random numbers. The starting point on the random number list should itself be selected at random, and it should be decided in advance which digit positions (if any) should be discarded.

Presumed Randomness and Computer-Generated Random Numbers

Statistical methodology often presumes that sample units have been randomly selected. The key issues are representativeness of the sample and independence amongst the sample units in terms of key factors. Random selection has proven to be an effective way of achieving those features in samples. But the randomness of a sample is often simply assumed—even though random numbers have not been explicitly used in the manner illustrated with the Nobel laureates. In getting sample readings for statistical process control, sample units may be selected at staggered times and randomness may simply be assumed. Although the risks may be judged low, there is always the potential for hidden danger in merely assuming randomness.

Even when random numbers are used as illustrated, it may be inconvenient to accomplish this with a preexisting list. Another source of the numbers is computer generation, so that a random number is created as needed. Usually, computer-generated values are properly considered as pseudorandom, since they start with a prescribed seed value and are created by a mathematical algorithm, the last number being used to create the next number. There is nothing random at all about the string of values thereby obtained. Nevertheless, pseudorandom numbers mimic true random numbers, generally having the essential properties ascribed to physically generated numbers: any digit is followed by a 0, 1, . . . , or 9 each about ten percent of the time (uniform distribution), and regardless of level, any digit will be followed just as often by a number < 5 as one ≥ 5 (no serial correlation). These characteristics ordinarily allow us to use computer-generated pseudorandom numbers without compromising the validity of an investigation. Computer-generated values are especially useful in statistical procedures involving simulations, where repeated sample selection is done inside the computer itself.

Problems

1–14 Using the leading 2 digits in the *second* column of Appendix Table F, select a random sample of 10 Nobel laureates from Table 1–1.

1–15 The RAND Corp. generated random numbers using an electromechanical device. Ideal random numbers give an equal chance for any integer between 0 and 9 appearing in any particular digit position and in such a way that successive values are not in any way influenced by what digits have been obtained in prior positions. Comment on the suitability of using as random numbers values created by the following means.

(a) A wheel of fortune has 10 nails along its outer edge, each positioned over an arc of 36 degrees. The wheel is set in motion by hand, revolving at least 5 times before the friction of a rubber strap stops the wheel so that it stops with the strap lying inside one of the wedges defined by the two nails. The wedges are numbered 0, 1, . . . , 8, and 9. Spin the wheel 5 times to generate a 5-digit random number.

(b) A pair of 6-sided die cubes (each having one face with 1, 2, 3, 4, 5, or 6 dots) is tossed and the number of dots on the top faces are summed. A sum of 10 is used as zero, while an 11 is read as a one. Sums of 12 are ignored. The values obtained serve as random digits, so that 5 rolls of the dice generate a 5-digit random number.

(c) Four-digit random numbers are generated from a telephone book by reading the last four digits of successive telephone numbers.

(d) The constant π is expanded to thousands of places. Successive nonoverlapping 5-digit sequences are used as random numbers.

(e) A computer program generates random numbers as needed by taking a large seed number, squaring it, and using the middle digits as the random number. That value is then the seed for the next random number, computed in the same fashion. The process continues as long as necessary.

(f) Three-digit base 10 random numbers are generated by tossing a coin nine times. The first coin toss determines the leading binary digit: 0 if tail and 1 if head. The second binary digit is determined in the same way from the second toss, and so on. The resulting binary number is converted to base 10. Any value > 999 is ignored.

1–6 Engineering Applications of Statistics

This book is organized to foster your understanding and basic competency in modern statistical methodology as that applies to engineers. Statistics is especially important to engineers, who (much more than people in most disciplines) are faced with the need to analyze data.

Statistics is a very broad field, hosting a variety of methodologies, many intended for particular disciplines and areas of application. Although many of its statistical procedures overlap with those of other fields, engineering statistics embraces certain techniques and orientations that serve as its distinguishing features. Several of the more prominent statistical applications associated with engineering are briefly described below.

Statistical Process Control

Statistical process control is a major concern for those engineers who are faced with continuing processes, most notably industrial and chemical engineers. These engineers must cope with destabilizing forces that may give rise to quality problems, such as excessive wear during finishing, higher-than-normal impurities in chemical products, or nonuniform surfaces on bearing parts. Provision must be made for continuing monitorship of such processes, which must be designed to facilitate corrections and improvements. When maintainability is a key issue, good engineering does not end with the functional aspects of the design. Operators of production equipment or facilities need the capability to fix problems as they arise.

Chapter 3 illustrates the role that statistics plays in rapidly identifying the presence of a quality problem. Often there is no obvious clue about the existence of such problems, but the statistical evaluation signals the need for a search for the cause, allowing for an *early* investigation that may lead to a solution before too much damage has been done.

Quality Assessment

Although statistical process control is useful for monitoring or directing operations under the engineer's purview, there is another dimension to statistical quality control covering outside vendors and contractors who supply finished components required for assembly into more complex products. Acceptance sampling procedures, based on the statistical testing methods described in Chapter 11, are useful for helping decide when incoming lots are good enough to use or are so bad that they must be rejected.

Model Building and Predicting

A key part of the scientific method as applied to engineering is the ability to make predictions of one variable when the level of one or more other variables is given. This is done by means of a mathematical *model*, such as the familiar physical law $f = ma$, which allows us to calculate the force f by knowing the mass m and the acceleration a. Such laws seem today like obvious facts, but they are rooted in empirical testing. Translating the observed experimental data into a model is in large measure a statistical exercise.

In Chapter 4 we will see how statistical data can be used to determine such an empirical relationship, as illustrated by the following equation relating the strain E that will be experienced for a particular level of stress S applied to a metal rod having new metallurgical properties.

$$S = -5,000 + 10^7 E \quad \text{and} \quad E = .0005 + 10^{-7} S$$

where E denotes strain and S expresses the stress, with

$$S = \text{load/area} \quad \text{and} \quad E = \text{elongation/length}$$

Using this model allows us to make *predictions* of what the strain will be on a two-inch-diameter rod subjected to stress by a dead weight of 10,000 pounds:

$$E = .0005 + 10^{-7}[10,000/(\pi(1)^2] = .00082$$

Chapter 4 describes the statistical procedure called regression analysis, which can help us find an appropriate set of constants for relationships like those shown above and can be used to fit appropriate functions to the empirical data. In Chapter 5 we will see how several predictor variables may be employed using a multiple regression analysis.

Communicating with and Acting on Experimental Results

Practically all experiments involve a need to summarize the data and use the results to make an estimate. *Statistical estimation* is crucial to any engineering application. A key issue of statistics is the acknowledgment of sampling error, since the sampling process involves uncertainty. Much of the last half of this book deals with statistical estimation.

The estimate is usually just a part of the statistical evaluation. Ordinarily some kind of action will be triggered by the results obtained. This will involve *statistical testing*, in which an action is taken consistent with some hypothesis regarding the unknown population. For example, a sample can be used in determining whether or not to accept or reject a shipment of components. It can tell us when to adjust control valves. But sampling error can lead to the wrong choice. Statistics can help us control the incidence of such errors. Chapter 10 investigates a variety of testing procedures.

Assessing Design Reliability

We can all empathize with the civil engineer who must build a bridge that won't fail. Similar challenges are faced by nuclear engineers who must avoid catastrophic accidents and by aeronautical engineers who must design crash-proof aircraft. Those designers must contend with obvious reliability issues. But reliability should be the concern of any product designer. A company that produces fans that fail in the first month is not likely to succeed.

Engineers are usually the ultimate guarantors of the safety and performance of their designs over proper lifespans. Statistical procedures may be used to assess product or system reliability on a trial basis early in the design process. Much of that methodology rests on a foundation of probability, the basic tools of which are presented in Chapter 6. Chapter 11 lays the foundation for many of the statistical testing procedures for measuring how reliable a product or component actually is.

Experimental Design

A very modern application of statistics, described in Chapter 14, may be helpful in product design or for establishing parameters for a process. As an example, consider the design of a structural member for which there is a choice of materials, fabrication routine, geometry, porosity, plus many other factors. Statistical methods may be used to direct design experiments in which various combinations of factor levels or attributes are explored in an efficient investigation aimed at discovering the optimal configuration. Engineers can confidently let statistics lead the way to the design finalization, even though empirical evaluations may be made with only a small subset of all possibilities.

Problems

1–16 Give an example of a circumstance of interest to you in which statistics might be used for each of the following:

(a) process control
(b) model building and predicting
(c) estimating
(d) testing
(e) assessing designs

1–17 Using stress as the abscissa and strain as the ordinate, plot the following data points on a graph.

	Load (lbs)	Area (sq. in.)	Elongation (inches)	Length (inches)
(1)	5,000	.5	.001	20
(2)	7,000	.4	.002	18
(3)	4,000	.6	.0005	15
(4)	1,000	.2	.001	10
(5)	1,500	.1	.002	20

(a) Visually fit a line through your points and determine its intercept and slope.
(b) Comment on the suitability of using your equation to make strain predictions for given levels of stress.

Review Problems

1–18 You are assisting a project engineer to determine specifications for a computer keyboard.
(a) List four questions for which numerical data might be helpful in making the choice.
(b) For each question in (a), provide (1) the source of data and (2) the type of data.

1–19 Indicate for each of the following situations whether the type of statistics involved is primarily inductive or deductive.
(a) A player assesses the chance that a particular card will win a bingo game.
(b) A quality assurance manager rejects shipments of parts when more than 3 items in a sample of 100 are defective, since the evidence then strongly indicates that the shipment is of poor quality.
(c) A professor knows the SAT scores of 25 students in her class. If she selects 3 students at random, she wants to know the possible levels of the average score and how likely each possibility is.
(d) The government uses experimental results in deciding whether or not a new fuel injection system would have excessive exhaust emissions.

1–20 For each of the following statistical investigations identify a potential source of error and suggest how it might be remedied.
(a) While shaking hands with potential voters, a city council candidate asks how he or she will vote.
(b) A software company asks persons to evaluate Hyper-Flow against competing packages. The questionnaire is inside the Hyper-Flow box.
(c) To minimize costs, the attractiveness of a new design is determined by asking opinions of people who pass by the company

CHAPTER TWO

Describing, Displaying, and Exploring Statistical Data

In our daily lives we have all become familiar with displays and summaries such as pie charts, percentages, and averages computed for every imaginable numerical quantity. These are elements of descriptive statistics, which organizes the data collected in statistical investigations and provides a structural framework for describing and summarizing the results.

2–1 The Frequency Distribution

Data organization is a basic need in any kind of statistical evaluation. Consider the values in Table 2–1. These quantities represent the observed times required to complete the calibration inspection of a particular test device. Each number represents a value obtained from a series of stopwatch observations of various laboratory assistants performing the same task. They were recorded in a log and are listed in their original sequence. Since the data points have not been arranged in any meaningful fashion, they are referred to as **raw data**. Descriptive statistics is largely concerned with arranging and manipulating raw data so that they may be interpreted.

One helpful arrangement is achieved by grouping the raw data into categories. Since the observations are numerical values, they may be placed into a series of contigu-

Table 2–1
Sample times (in seconds) to inspect test devices for calibration

12.8	15.6	13.5	15.7	15.3	15.2	20.1	14.2	12.9	14.0
16.9	14.3	15.5	14.6	13.0	14.7	19.0	13.0	11.3	14.2
14.5	14.8	14.2	13.0	13.1	12.5	16.1	19.1	16.7	13.2
15.0	12.7	13.6	13.3	13.2	14.7	12.9	13.1	17.3	15.4
17.9	13.0	14.3	14.2	15.7	15.6	13.0	13.9	14.2	16.0
12.9	13.1	13.3	12.3	13.1	13.6	13.2	18.5	13.2	13.7
12.6	14.4	14.5	13.9	17.0	13.7	12.7	16.8	13.3	14.7
14.2	13.0	14.6	14.0	12.9	14.7	12.8	12.0	14.2	12.8
13.7	15.2	14.8	13.0	11.7	12.2	13.3	13.8	14.2	14.3
14.7	12.6	18.9	14.3	14.4	15.5	16.8	17.0	13.2	12.9

Table 2–2 *Sample frequency distribution of inspection times*

Time (seconds)	Tally	Number of Inspections
11.0–under 12.0	//	2
12.0–under 13.0	𝍸 𝍸 𝍸 /	16
13.0–under 14.0	𝍸 𝍸 𝍸 𝍸 𝍸 ////	29
14.0–under 15.0	𝍸 𝍸 𝍸 𝍸 𝍸 //	27
15.0–under 16.0	𝍸 𝍸 /	11
16.0–under 17.0	𝍸 /	6
17.0–under 18.0	////	4
18.0–under 19.0	//	2
19.0–under 20.0	//	2
20.0–under 21.0	/	1
		Total 100

ous blocks called **class intervals**. The number of observations falling into each interval is then determined. The respective count provides the **class frequency**. Table 2–2 shows the resulting summary for the inspection times, which is referred to as the **sample frequency distribution**. The first class interval is 11.0–under 12.0, the second 12.0–under 13.0, and so forth. Each class interval has a width of 1.0, with the class limits chosen so there is no ambiguity regarding where to place an observation; thus, the upper limit of the first interval is under 12.0, whereas the lower limit of the second is 12.0 exactly. Both the number of classes and the interval widths must be chosen by the investigator.

The frequency distribution for inspection times applies to sample data. Should the data represent the entire population, then the resulting table would provide the population frequency distribution. Ordinarily, only samples are available, and the exact population frequency distribution remains unknown.

The Histogram and Frequency Curve

It is often easier to draw conclusions about the sample data when the frequency distribution has been portrayed graphically. Figure 2–1 shows a **histogram** in which the inspection times are represented by a bar chart having frequency as the vertical axis and the observed values as the horizontal axis. Frequency distributions may be categorized by the basic shapes of their histograms, with various distribution families encountered in certain statistical applications.

An alternative display is the **frequency polygon**, shown for the same data in Figure 2–2. Here a dot is plotted above the midpoint of each interval at a height matching the class frequency. These dots are then connected, with outside line segments touching the horizontal axis one-half of an interval width below and above the lowest and highest intervals, respectively. Notice that the area circumscribed by the frequency polygon is about the same shape and size as that presented by the collective bars of the histogram.

Although the population frequency distribution is usually not available, the sample counterpart suggests the shape it would take when graphed. Figure 2–3 shows the **frequency curve** for calibration inspection times suggested by the sample data, as plotted by either the histogram in Figure 2–1 or the frequency polygon in Figure 2–2. Since the

Figure 2–1
Histogram for the frequency distribution of inspection times.

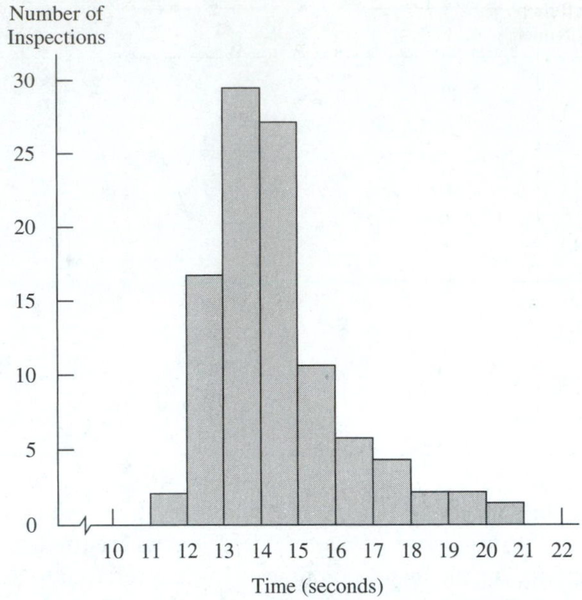

size of this particular population is arbitrarily large, the ordinate is given in terms of **relative frequency**. Since the entire population will be very large in relation to its sample, the frequency curve will be smoother than its sample counterparts. (Imagine times measured to the nearest 100 microseconds and a histogram portraying millions of observations graphed with intervals of width .001.)

Figure 2–2
Frequency polygon for the frequency distribution of inspection times.

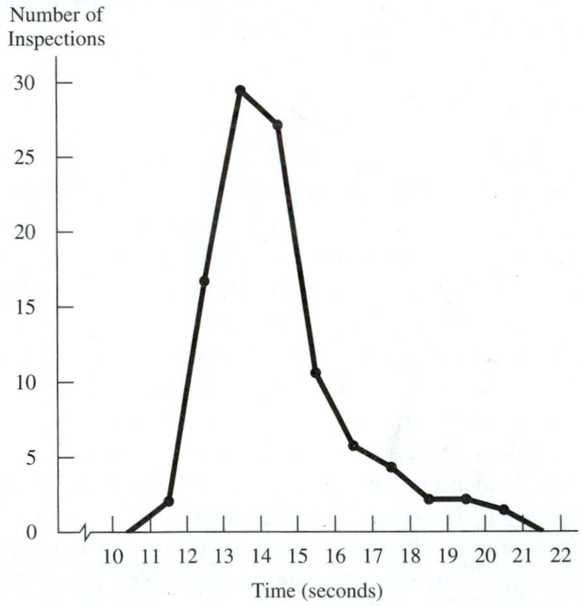

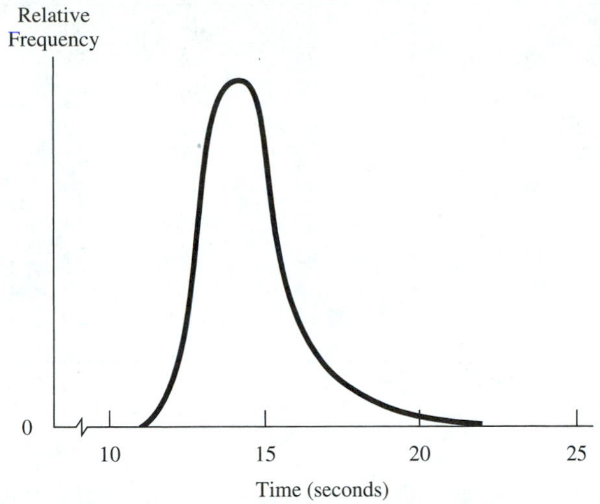

Figure 2–3
Suggested shape of smoothed frequency curve for the entire population of inspection times.

Stem-and-Leaf Plots

One summary that is useful for arranging experimental data is the **stem-and-leaf plot**. Here the raw data are arranged tabularly by locating each observation on a "tree." This is done by separating the values into a stem digit and a leaf digit. To illustrate, consider the temperature data in Table 2–3. Scanning these values, we see that the first digits range from a low of 1 to a high of 8. Listing these integers in the first column, the following table is constructed:

Stem (1st digit)	Leaf (2nd digit)
1	8
2	
3	5 7 8 0 7 8 9
4	7 9 9 6 6 5 7 2 2 3 9 6 9 6 3 9 8 8 3 5 9 7 6 5 6 9
5	1 0 8 5 3 6 6 4 5 0 6 0 2 3 1 2 7 0 5 2 4 3 1 9 2 8 5
6	0 8 8 9 2 0 2 6 2 1 9 0 5 3 9 8 3 6 6 0
7	6 0 2 5
8	0

The second digit for each temperature is entered to the right of the line in the row corresponding to its first digit. For example, the first temperature, 47, has a value of 7 (the leaf) in the 4 row (the stem); the second value, 49, has a leaf with a value of 9, also on the row with a stem of 4. The third value, 51, becomes the first leaf, 1, on the row with a stem of 5.

One advantage of the stem-and-leaf plot is that all the original raw data are portrayed. This is useful for later computation of summary statistics. As we shall see, two such statistics—the median and the mode—can be quickly found from the plot itself.

Table 2–3
April normal temperatures (°F) for selected U.S. cities

47	49	51	49	60	46	50	58	46
55	45	47	42	42	68	53	56	56
35	43	54	76	55	50	68	49	46
56	37	38	69	62	60	50	70	72
62	66	49	46	62	52	43	61	53
51	49	30	52	57	69	50	55	52
54	48	60	65	37	53	48	80	
63	51	69	68	63	18	59	38	
43	66	52	39	75	58	45	66	
49	47	46	55	45	60	46	49	

Source: *The World Almanac and Book of Facts,* 1988, p. 181.

The main advantage, however, is that it provides the essential features of the histogram; you can see these easily by rotating the plot 90 degrees. The disadvantage is that the stem-and-leaf plot becomes cumbersome if the number of observations is large or if the data are expressed to more than two significant figures.

Computer-Generated Displays

It can be very time consuming to generate histograms by hand. Computer assistance can not only eliminate that burden but also avoid potential errors. Several commercial statistical software packages are available. Especially valuable to engineers are Minitab, SAS, and SPSS. These are illustrated throughout this book.

Figure 2–4 shows a histogram for the preceding temperature data that was generated with SPSS.

Figure 2–4
Computer-generated histogram for April normal temperature data, obtained using SPSS.

```
ENG89      APRIL NORMAL TEMPERATURE

     COUNT    MIDPOINT     ONE SYMBOL EQUALS APPROX. 1.00 OCCURRENCES

         1       15.00  *
         0       25.00
         7       35.00  *******
        26       45.00  **************************
        27       55.00  ***************************
        20       65.00  ********************
         4       75.00  ****
         1       85.00  *
                        I....+....I....+....I....+....I....+....I....+....I
                        0         10        20        30        40        50
                                  HISTOGRAM FREQUENCY

VALID  CASES      86        MISSING CASES      0
```

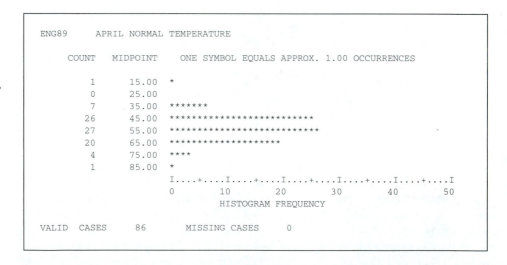

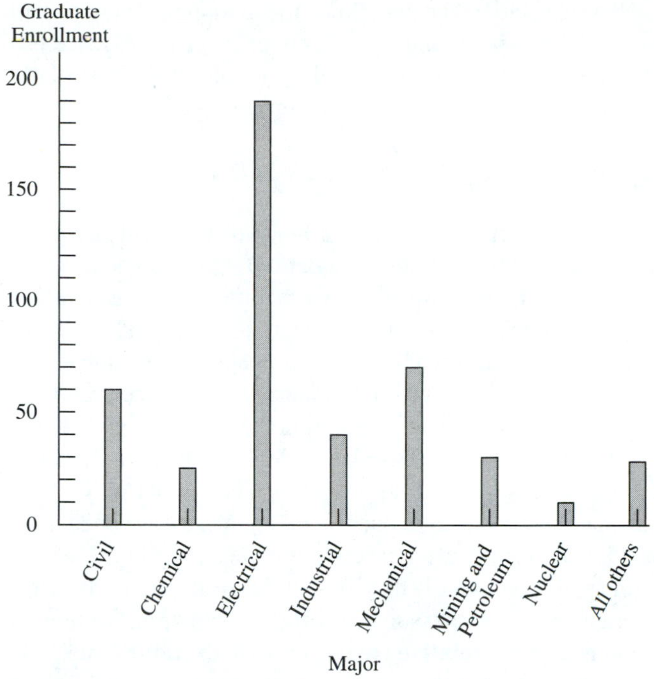

Figure 2–5 *Chart for qualitative frequency distribution of engineering majors.*

Frequency Distributions for Qualitative Variables

There is less choice in portraying qualitative data. Each category or attribute occurs with some frequency, which may be summarized in a table or bar chart. For example, consider Figure 2–5, which represents the frequency distribution for the majors of a randomly selected group of engineering students. Notice that the bars do not touch, which reflects the fact that the chart is a graphical representation of discrete categories. Similar graphs might be appropriate for portraying the frequency distributions of discrete quantitative variables such as number of patents held, number of degrees, or family size.

The **pie chart** is a popular graphical display for qualitative data. Figure 2–6 shows a pie chart generated by a computer using StatGraphics. It represents the frequency

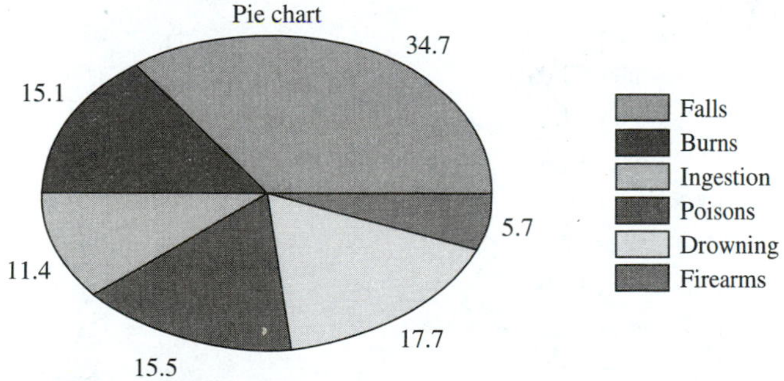

Source: *The World Almanac and Book of Facts,* 1988, p. 810.

Figure 2–6 *Computer-generated pie chart for cause of accidental death (nonvehicular), obtained using StatGraphics.*

distribution for type of accidental death. Since the more frequent observation categories have larger wedges, this type of display quickly conveys the essence of the frequency distribution. The piece of pie is sized according to the category's **relative frequency**, with the angle of the slice corresponding to its proportion of 360 degrees.

Relative and Cumulative Frequency Distributions

Frequency of occurrence is the foundation for much of statistical analysis because it provides a meaningful arrangement of observed data. But there are some drawbacks to basing evaluations directly on the number of data points found in each class interval. It is clumsy to compare groups of different size in terms of straight tallies. For example, knowing that there are 254 electrical engineering majors at one university and only 154 at a second says little about the relative importance of that concentration in the two engineering schools unless the respective total enrollments are also included in the comparison. The first school has 1,547 engineering students, so that the proportion, or relative frequency, of electrical engineers is $\frac{254}{1,547} = .164$. The second university has 655 engineering students, and the relative frequency of the electrical concentration is a much higher $\frac{154}{655} = .235$. Thus, electrical engineering is more dominant in the second institution.

The foregoing suggests that it may be helpful to divide each of the original frequencies by the sample size, expressing the distribution in terms of relative frequencies. The resulting summary is the **relative frequency distribution**. Table 2–4 shows the relative frequency distribution for the sample results given earlier for the calibration inspection times. After the original frequencies have been converted to relative frequencies, it is easier to compare samples of various sizes taken at different times, in other places, or for other inspection tasks.

It sometimes is advantageous to make another transformation of class frequencies to obtain another form of the frequency distribution. This is the **cumulative frequency distribution**, which is constructed by adding the frequency of each class to the sum of the frequencies for the lower classes. Table 2–5 provides the cumulative frequency distribution for the sample of calibration inspection times. Cumulative frequencies are use-

Table 2–4
Relative frequency distribution of sample inspection times

Time (seconds)	Number of Times (Frequency)	Relative Frequency
11.0–under 12.0	2	$\frac{2}{100} = .02$
12.0–under 13.0	16	$\frac{16}{100} = .16$
13.0–under 14.0	29	$\frac{29}{100} = .29$
14.0–under 15.0	27	$\frac{27}{100} = .27$
15.0–under 16.0	11	$\frac{11}{100} = .11$
16.0–under 17.0	6	$\frac{6}{100} = .06$
17.0–under 18.0	4	$\frac{4}{100} = .04$
18.0–under 19.0	2	$\frac{2}{100} = .02$
19.0–under 20.0	2	$\frac{2}{100} = .02$
20.0–under 21.0	1	$\frac{1}{100} = .01$
		Total 1.00

Table 2–5 *Cumulative frequency distribution of sample inspection times*

Time (seconds)	Frequency (number of times)	Cumulative Frequency	Cumulative Relative Frequency
11.0–under 12.0	2	2	.02
12.0–under 13.0	16	2 + 16 = 18	.18
13.0–under 14.0	29	18 + 29 = 47	.47
14.0–under 15.0	27	47 + 27 = 74	.74
15.0–under 16.0	11	74 + 11 = 85	.85
16.0–under 17.0	6	85 + 6 = 91	.91
17.0–under 18.0	4	91 + 4 = 95	.95
18.0–under 19.0	2	95 + 2 = 97	.97
19.0–under 20.0	2	97 + 2 = 99	.99
20.0–under 21.0	1	99 + 1 = 100	1.00

ful in describing different levels of the variable of interest. We see that 95 out of the 100 times fall below 18.0 seconds, whereas only 2 out of 100 fall below 12.0 seconds.

When plotted on a graph such as Figure 2–7, the cumulative frequency distribution gives another visual summary of the sample. Each dot is plotted directly above the

Figure 2–7 *Cumulative frequency distributions of inspection times.*

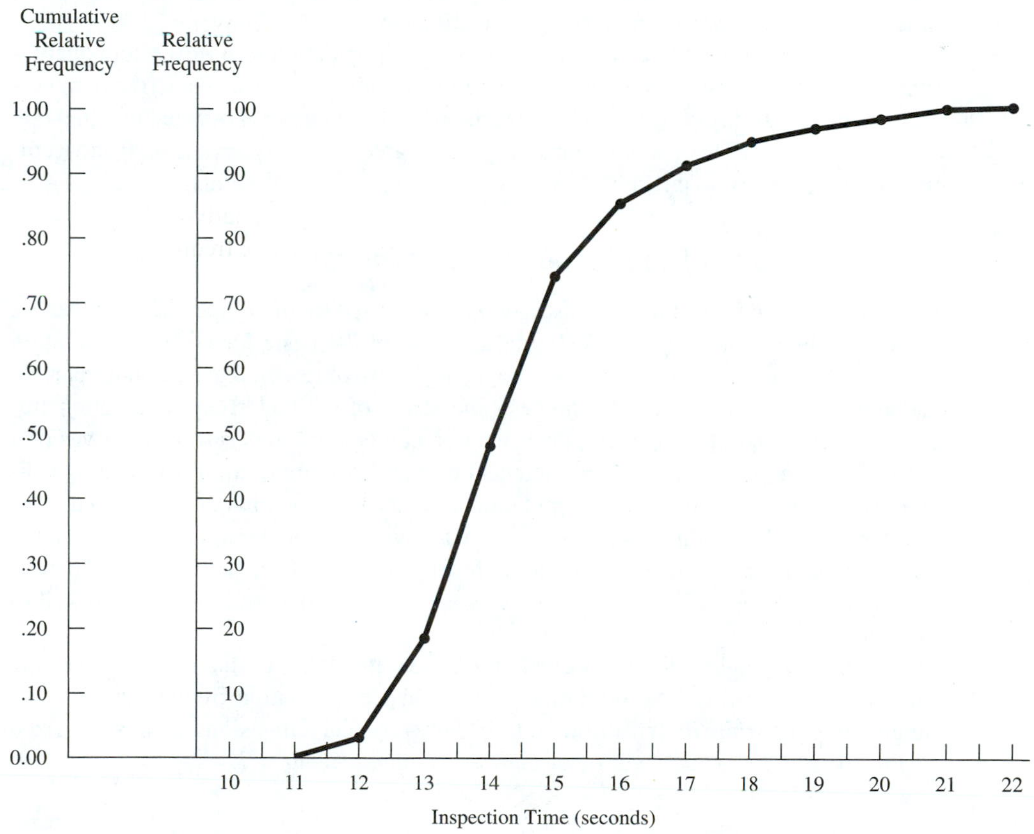

upper class limit at a height equal to the cumulative frequency for that interval. These are then connected by line segments, with the lowest line touching the horizontal axis at the lower limit of the smallest class. Such a graph can be useful when the original data are unavailable. The interpolated cumulative frequency of any particular observation value may be read directly from the graph, and the approximate observation level corresponding to any cumulative frequency may be obtained similarly, by reading the graph in reverse. Thus, about 88 out of 100 observations fall below 16.5 seconds, whereas 14.1 seconds is the time below which about half (50) of the observations fall. Often the ordinate is cumulative *relative* frequency, so that the proportion of the observations falling below a particular level may be easily determined from the graph.

Multidimensional Data Displays

Some of the most interesting aspects of statistical investigations arise when there are two or more dimensions to a statistical population, such as chemical yield, processing temperature, and pressure setting; stress and strain; or college major and starting salary. Statistical investigations attempt to uncover relationships between two or more variables, such as how yield responds to different settings of temperature and pressure or how majors differ in salary level. A starting point is often a data display.

Figure 2–8 shows a data display for two quantitative population variables. A computer-generated **scatter diagram** relates concrete strength Y to pulse velocity X. A visual inspection indicates that Y should increase with X. Later chapters show how a line or a curve can be used to summarize that relationship.

A second two-dimensional display arises when population factors are both qualitative. Figure 2–9 shows a computer-generated **crosstabulation** for the active membership of an engineering college alumni association. Each member is counted into the total of a particular cell. There is one row for highest degree received (three levels) and a column for major field (five levels).

Common Forms of the Frequency Distribution

Quantitative samples or populations may be categorized by the shapes of their relative frequency distributions. Figure 2–10 shows some of the basic forms. The population may be represented either by a relative frequency curve or its cumulative counterpart. In each case the sample histogram suggests the shape of the underlying population frequency curve. The same mathematical function can be used to describe all curves of a particular shape so that individual members of one distribution family are distinguishable by the level of one or more population parameters, which may be estimated from the sample results. Often, all variables associated with a particular application will have frequency distributions of the same basic form. As we shall see, a common statistical procedure might be used to evaluate all samples that have data fitting a single distribution family.

Part (a) of Figure 2–10 represents the observations of rod diameters, where the sample histogram closely approximates a bell-shaped population frequency curve belonging to the **normal distribution** family. Many physical dimensions are characterized by this distribution, which will be discussed in detail in Chapter 7.

```
DATA LIST /STRENGTH 1-5(2) VELOCITY 7-11(3).
BEGIN DATA.
36.77 4.293
43.33 4.256
44.09 4.337
      :
27.51 4.035
END DATA.
SET PRINTER ON.
PLOT SYMBOLS='*'
/VSIZE=30 /HSIZE=70
/FORMAT=DEFAULT
/TITLE='SCATTER DIAGRAM'
/VERTICAL='COMPRESSIVE STRENGTH (MPa)'
/HORIZONTAL='PULSE VELOCITY (mm/us)'
/PLOT=STRENGTH WITH VELOCITY.
SET PRINTER OFF.
```

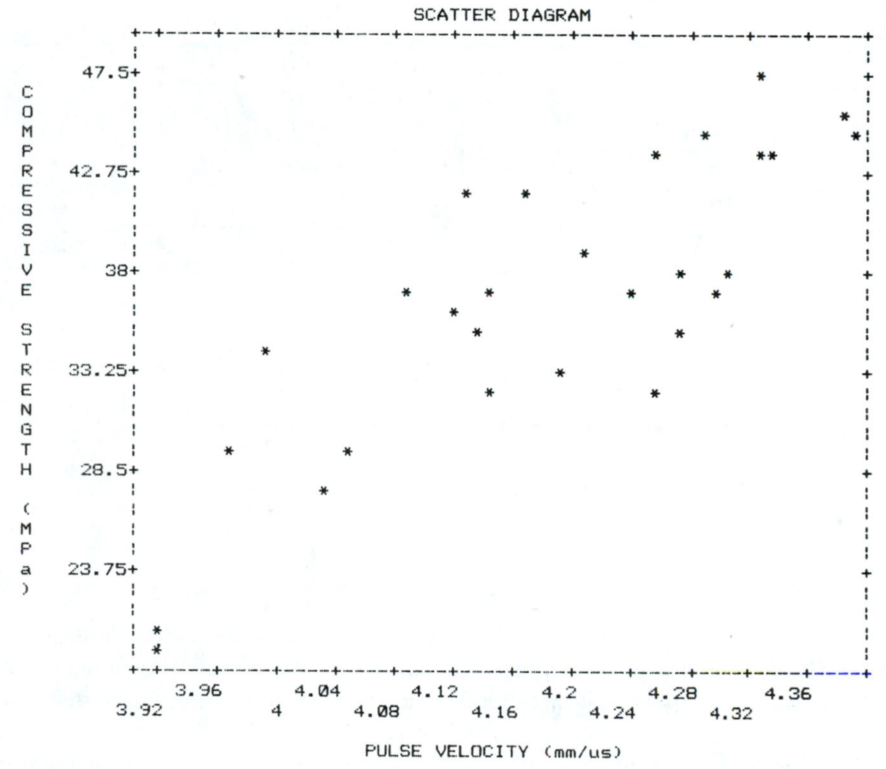

Figure 2–8
SPSS scatter diagram for concrete strength vs. pulse velocity.

Source: Scanlon and Mikhailovsky, "Strength Evaluation of an Existing Concrete Bridge," *Canadian Journal of Civil Engineering,* Vol. 14 (1987): 150.

Figure 2–9
SPSS crosstabulation for engineering college active alumni association membership.

```
DATA LIST FIXED / FREQ 1-5 DEGREE 7 MAJOR 9.
WEIGHT BY FREQ.
VALUE LABELS
  DEGREE  1 'BACHELOR' 2 'MASTER'/
  MAJOR  1 'ELECT' 2 'MECH' 3 'CIVIL' 4 'CHEM' 5 'INDUS'.
BEGIN DATA.
   47 1 1
  113 1 2
   45 1 3
   12 1 4
   24 1 5
   28 2 1
    9 2 2
    3 2 3
    5 2 4
   15 2 5
END DATA.
CROSSTABS  TABLES=DEGREE BY MAJOR
 /OPTIONS=3,4,5
 /STATISTIC=1
FINISH.
```

Crosstabulation: DEGREE
By MAJOR

DEGREE	MAJOR-> Count / Row Pct / Col Pct / Tot Pct	ELECT 1	MECH 2	CIVIL 3	CHEM 4	INDUS 5	Row Total
1 BACHELOR	Count	47	113	45	12	24	241
	Row Pct	19.5	46.9	18.7	5.0	10.0	80.1
	Col Pct	62.7	92.6	93.8	70.6	61.5	
	Tot Pct	15.6	37.5	15.0	4.0	8.0	
2 MASTER	Count	28	9	3	5	15	60
	Row Pct	46.7	15.0	5.0	8.3	25.0	19.9
	Col Pct	37.3	7.4	6.3	29.4	38.5	
	Tot Pct	9.3	3.0	1.0	1.7	5.0	
	Column Total	75	122	48	17	39	301
		24.9	40.5	15.9	5.6	13.0	100.0

In (b) the times between failures of electronics components closely fit an **exponential distribution**, the frequency curve of which has a single "tail" extending indefinitely to the right. Exponential distributions are common in reliability and queueing evaluations.

The frequency distribution for the random decimals in (c) is described by the **uniform distribution**. Such a rectangular shape applies to variables that are equally likely to fall anywhere within a fixed interval.

A common characterization for a wide variety of data is an unsymmetrical hill-shaped curve. Such skewed curves are often encountered when there is an upper or lower limit on the possible values. They take two forms. The gasoline mileage data in (d) suggest a **negatively skewed distribution**, which has the longer tail pointing left. This configuration is similar to the **positively skewed distribution** with the longer tail pointing right that was found earlier for the calibration inspection times.

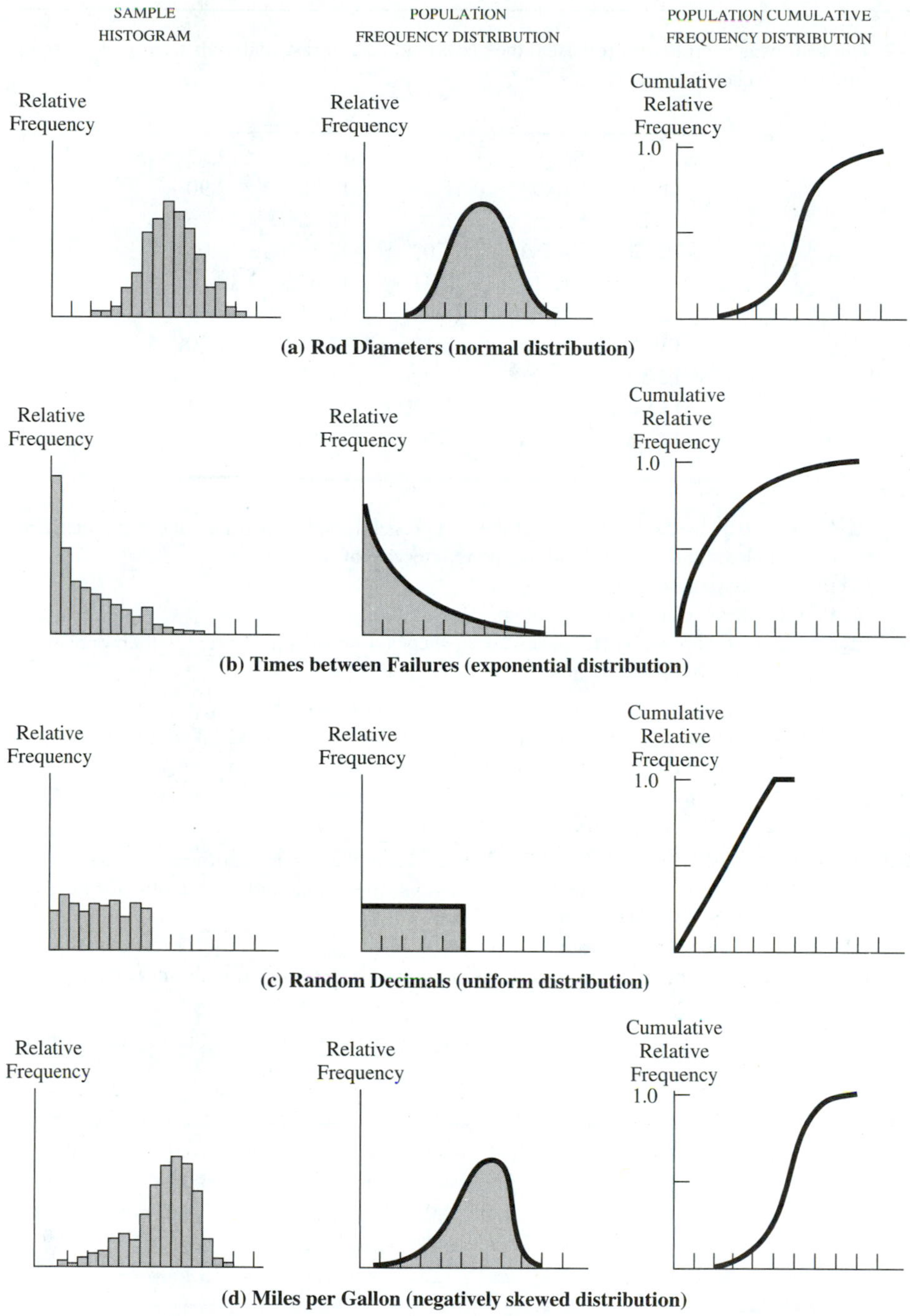

Figure 2–10
Frequency distributions in commonly encountered forms.

Problems

2–1 The following sample data represent the per pound cost of raw materials used in processing batches of a chemical feedstock:

\$12.01	\$14.90	\$11.24	\$14.40	\$12.87
11.30	16.98	15.23	12.38	12.90
12.29	11.51	12.06	15.06	11.16
10.12	14.82	17.02	12.56	13.95
11.13	12.14	13.41	15.67	12.17
12.15	12.57	14.21	13.21	16.50
13.81	15.58	11.10	12.03	11.20
12.27	17.05	12.57	13.11	13.31
13.84	11.62	12.33	16.31	12.74
14.25	18.63	13.34	13.43	14.78

(a) Using \$10.00 as the lower limit of the first class interval, construct a table summarizing the sample frequency distribution having intervals of width \$1.00.
(b) Plot the above as a histogram.
(c) Plot the above as a frequency polygon.
(d) Of the basic population frequency curve shapes described in the text, which one do the above data most closely match?

2–2 Using the raw materials costs from Problem 2–1, complete the following:
(a) Construct a table for the sample frequency distribution having \$2.00 widths for the class intervals, with \$10.00 as the first lower limit.
(b) Plot the above as a histogram.
(c) Plot the above as a frequency polygon.

2–3 Using the raw materials costs from Problem 2–1, complete or answer the following:
(a) Construct a table for the sample frequency distribution having \$.50 widths for the class intervals, with \$10.00 as the first lower limit.
(b) Plot the above as a histogram.
(c) Do you think a more accurate data summary might be obtained with intervals narrower or wider than \$.50?

2–4 The following precipitation levels (inches) apply for selected cities during the month of April:

2.9	3.7	3.2	4.0	3.9	2.1	2.9	2.9	1.1
0.4	3.0	3.2	1.0	2.2	5.4	3.5	3.6	4.0
0.7	2.8	1.8	1.5	4.0	4.0	0.3	2.2	3.3
3.8	2.6	2.2	4.2	5.4	4.8	3.3	2.7	1.8
4.4	2.6	2.9	2.0	1.2	3.6	3.9	0.8	3.1
3.1	3.7	0.3	3.7	4.1	4.5	2.3	1.5	3.4
3.4	3.3	1.2	5.9	3.6	3.8	4.0	3.6	
5.0	3.4	2.6	3.3	5.8	0.6	2.9	2.4	
1.6	3.6	0.7	2.9	3.1	2.9	2.0	3.2	
1.2	1.8	3.6	2.7	3.4	2.9	0.5	2.4	

Source: *The World Almanac and Book of Facts*, 1988, p. 181.

(a) Construct a stem-and-leaf plot.
(b) Using intervals of width 1″ and starting with 0″ as the lower limit of the first, construct a table for the frequency distribution.
(c) Plot the distribution from (b) as (1) a histogram and (2) a frequency polygon.

2–5 Using standard widths of 5.0 years, plot a histogram for the following age frequency distribution:

Age (years)	Frequency
20–under 25	5
25–under 30	13
30–under 35	17
35–under 40	8
40–under 80	12

2–6 Using standard widths of 5.0 mpg, plot a histogram for the following gasoline mileage frequency distribution:

Mileage (mpg)	Frequency
5–under 20	9
20–under 25	12
25–under 30	14
30–under 35	22
35–under 40	7
40–under 60	4

2–7 A sample of engineering students has been categorized by concentration, level, and financial aid. The number of persons in each category is summarized here:

Concentration	Level	Financial Aid	Number
Electrical	Undergraduate	None	23
Electrical	Undergraduate	Some	6
Electrical	Graduate	None	12
Electrical	Graduate	Some	15
Mechanical	Undergraduate	None	14
Mechanical	Undergraduate	Some	4
Mechanical	Graduate	None	5
Mechanical	Graduate	Some	12
Civil	Undergraduate	None	34
Civil	Undergraduate	Some	10
Civil	Graduate	None	7
Civil	Graduate	Some	24

Construct a table and a bar chart for the sample frequency distribution characterized by (a) concentration, (b) level, and (c) financial aid.

2–8 Referring to the data in Problem 2–7, obtain a crosstabulation of concentration vs. level.

2–9 The following data provide the number of professional U.S. women employed (thousands) in 1986:

A. Engineering/Computer science	347
B. Health care	1937
C. Education	2833
D. Social/Legal	698
E. Arts/Athletics/Entertainment	901
F. All others	355

Source: *The World Almanac and Book of Facts*, 1988, p. 86.

(a) Construct a bar chart.
(b) Construct a pie chart.

2–10 The following sample data represent the gasoline mileages (in miles per gallon) determined for cars in a particular weight class:

25.1	29.0	34.5	35.7	37.9
34.9	24.3	26.6	27.3	32.0
30.0	34.5	35.3	33.5	36.6
34.8	16.2	13.1	24.5	33.6
28.0	33.9	30.7	32.0	37.7
21.1	31.2	35.6	34.4	25.2
35.9	18.3	29.4	29.5	34.8
29.4	26.4	38.8	36.0	28.7
23.4	35.3	33.7	38.1	28.6
34.2	34.8	39.2	39.9	36.8

(a) Using 10 mpg as the lower limit of the first class interval, construct a table summarizing the sample frequency distribution having intervals of width 5 mpg.
(b) Plot the above as a histogram.
(c) Plot the above as a frequency polygon.
(d) Of the basic population frequency curve shapes described in the text, which one does the above data most closely match?

2–11 Using the gasoline mileages from Problem 2–10, answer the following:
(a) Construct a table for the sample frequency distribution having 2 mpg widths for the class intervals, with 10 mpg as the first lower limit.
(b) Plot the above as a histogram.
(c) Do you think a more accurate data summary might be obtained with intervals narrower or wider than 2 mpg?

2–12 The following data apply to automobiles driven in the United States:

Year	Average Distance (1,000 miles) X	Average Fuel Consumption (gallons) Y
1960	9,450	661
1965	9,390	667
1970	9,980	735
1975	9,630	712
1976	9,760	711
1978	10,050	715
1979	9,480	664
1980	9,140	603
1981	9,000	579
1982	9,530	587
1983	9,650	578
1984	9,790	553
1985	9,830	549

Source: *The World Almanac and Book of Facts*, 1988, p. 127.

Construct a scatter diagram.

2–13 The frequency distribution of times between arriving teleport messages at the central processing unit of a time-sharing network is as follows:

Time (milliseconds)	Number of Messages
0.0–under 5.0	152
5.0–under 10.0	84
10.0–under 15.0	56
15.0–under 20.0	31
20.0–under 25.0	14
25.0–under 30.0	6
30.0–under 35.0	2

(a) Plot the above as a histogram.

(b) Which one of the basic population shapes described in the text does the above most closely approximate?

2–14 Referring to the interarrival time data in Problem 2–13:

(a) Construct a table showing the relative frequencies. Then find the cumulative relative frequencies for each class interval.

(b) Plot the cumulative relative frequency distribution.

(c) Use your graph to determine approximately the durations below which each of the following percentages of interarrival times falls: (1) 90%, (2) 25%, (3) 50%, and (4) 75%.

2–15 The following is the frequency distribution of unscheduled downtime (hours) experienced by a computer mainframe over a sample period:

Downtime	Frequency
0–under 1	35
1–under 2	19
2–under 3	12
3–under 4	7
4–under 5	3
5–under 6	2

(a) Construct a table showing the cumulative number of occurrences for each class interval. From these determine the cumulative relative frequencies.
(b) Plot the cumulative relative frequency distribution.
(c) Use your graph to determine approximately the proportion of downtime (1) below 1.5 hours, (2) below 3.6 hours, (3) at or above 2.3 hours, and (4) between 1.8 and 4.3 hours.

2–16 The cumulative relative frequency distribution for the depth (in feet) of oil well shafts in a particular region is given here:

Depth of Well	Cumulative Proportion
0–under 1,000	.09
1,000–under 2,000	.35
2,000–under 3,000	.72
3,000–under 4,000	.88
4,000–under 5,000	.95
5,000–under 6,000	.98
6,000–under 7,000	.99
7,000–under 8,000	1.00

Altogether there are 700 wells in the region.
(a) Make a table for the relative frequency distribution.
(b) Using your answer to part (a), determine the original frequency distribution.
(c) Using your answer to part (b), determine the cumulative frequency distribution.

2–17 For each of the following populations, sketch an appropriate shape for the frequency distribution and explain the reasons for your choice.
(a) content volumes in bottles leaving a filling machine
(b) millisecond portion of a computer's clock when processing begins on each new job
(c) the number of engineers employed in each of several private companies
(d) times required to install car bumpers

2–2 Summary Statistical Measures: Location

Although the frequency distribution arranges the raw data into a meaningful pattern, that summary cannot by itself answer many important statistical questions. For example, an industrial engineer wishing to select the faster of two production methods might obtain sample completion times from pilot runs and then try to reach a decision by comparing the two resulting sample frequency distributions. But unless the sample results are summarized further, beyond frequency tables or histograms, such an evaluation would be cumbersome and might actually cloud the issue. The faster procedure ought to be more clearly indicated by the "average" completion times under the two production methods.

Averages are one class of **statistical measures**. These quantities express various properties of the statistical data. Our discussion begins with measures of **location**. There are two types of location measures. One group expresses **central tendency**. Such quantities indicate the central point around which observations tend to cluster, thus giving a capsule summary of their magnitude. Another group of location measures is concerned with **positions** other than the center and are placed according to frequency of occurrence. Later in this chapter we will describe **variability** or **dispersion**. These quantities express the degree to which observations differ. Even a set of qualitative observations has a summary measure, the **proportion**.

Statistics and Parameters

Summary data measures fall into two major groupings, depending on whether the observations they describe are a population or a sample. When the data constitute a population, each summary measure is referred to as a **population parameter**. But ordinarily not all possible population observations are made, so usually the only available observations are those in a representative sample. A measure that summarizes sample data is called a **sample statistic**. Ordinarily, it is the statistic that is computed from those observations actually made. Important population parameters have counterpart sample statistics that measure the same characteristic, and the latter will often serve to estimate the unknown parameter values.

The Arithmetic Mean

The **arithmetic mean** is the most commonly used and best understood measure of central tendency. Consider the quantitative scholastic aptitude test (SAT) scores achieved by five engineering school applicants: 755, 613, 584, 693, 622. The mean 653.4 is calculated by adding these values and dividing by the number of students (5):

$$\frac{755 + 613 + 584 + 693 + 622}{5} = \frac{3,267}{5} = 653.4$$

If the five students represent only a sample from a population of all applicants to a particular engineering school, this calculation is the **sample mean**. If the observations are instead the scores of all the engineering applicants graduating from a particular high school, then they comprise an entire population, and 653.4 would then be the **population mean**.

It is convenient to express this procedure symbolically. Consider first the sample mean. Denoting the ith observation in the collection of raw data as X_i and representing the number of observations by n, also referred to as the **sample size**, we denote the sample mean as $\bar{X}$ (X-bar) and compute it as follows:

SAMPLE MEAN

For the engineering applicants we used $n = 5$, with $X_1 = 755$, $X_2 = 613$, $X_3 = 584$, $X_4 = 693$, and $X_5 = 622$ to compute the value of the sample mean, $\overline{X} = 653.4$.

A similar computation provides the **population mean**, which is denoted by the symbol μ (lowercase Greek *mu*). When all of the observations are available, μ is computed in exactly the same way as $\overline{X}$. We reserve the uppercase N to denote the population size. It is rare, however, for μ actually to be computed for most populations encountered in engineering—because of their size, most must be represented by samples. Thus, μ will ordinarily remain an unknown quantity, whereas only $\overline{X}$ actually is computed and serves as an estimator of μ. We may compute the sample mean for the raw calibration and inspection time data from Table 2–1 by first summing all the entries and then dividing by $n = 100$: $\overline{X} = \frac{1{,}434.2}{100} = 14.342$ seconds.

Sometimes, however, the raw data may be unavailable, and the investigator may have access only to the summary provided by the frequency distribution. It is possible in this case to use the grouped data to approximate the value of the sample mean that might be computed were all the raw data accessible. This requires manipulating the quantities describing each class of the frequency distribution. We denote the frequency of the kth class interval as f_k and the **midpoint** as X_k. The following expression may then be used.

SAMPLE MEAN APPROXIMATED FROM GROUPED DATA

$$\overline{X} = \frac{\sum f_k X_k}{n}$$

where $n = \sum f_k$.

The above is equivalent to summing the products of each class midpoint X_k with the relative frequency, f_k/n. Such a calculation is a **weighted average**. Weighted averages play an important role in statistics.

Table 2–6 illustrates the sample mean calculated by using grouped data from the sample frequency distribution found earlier for the calibration inspection times. The value 14.42 computed from the grouped data is only an *approximation* to the sample mean computed directly from the raw data ($\overline{X} = 14.342$). Ordinarily $\overline{X}$ will be computed using just one of the two procedures.

The Median

Another measure of central tendency is the **median**. This quantity often serves as a preferred location for the central value when the observed data involve a skewed frequency distribution. Ordinarily, only the **sample median**, denoted by *m*, is obtained from those observations actually made. It is located by first ordering the data by increasing magnitude and then finding *that value above and below which an equal number of observations lie*. When there is an odd number of observations, the sample median will be the value of middle size.

For example, the SAT scores given earlier may be listed in increasing magnitude:

584 613 622 693 755

The sample median is $m = 622$.

When there is an even number of observations, the sample median is found by averaging the middle two values. As an example, consider the following partial order se-

Table 2–6
Calculation of the mean calibration inspection time using grouped data

Time (seconds)	Frequency (Number of Times) f_k	Class Interval Midpoint X_k	$f_k X_k$
11.0–under 12.0	2	11.5	23.0
12.0–under 13.0	16	12.5	200.0
13.0–under 14.0	29	13.5	391.5
14.0–under 15.0	27	14.5	391.5
15.0–under 16.0	11	15.5	170.5
16.0–under 17.0	6	16.5	99.0
17.0–under 18.0	4	17.5	70.0
18.0–under 19.0	2	18.5	37.0
19.0–under 20.0	2	19.5	39.0
20.0–under 21.0	1	20.5	20.5
		Total	1,442.0

$$\bar{X} = \frac{\sum f_k X_k}{n} = \frac{1{,}442.0}{100} = 14.42 \text{ seconds}$$

quence listing prepared from the original raw data from the calibration inspection times:

47th	48th	49th	50th	51st	52nd	53rd	54th
13.9	14.0	14.0	14.2	14.2	14.2	14.2	14.2

The sample median for the 100 observations is the average of the 50th and 51st values, $m = (14.2 + 14.2)/2 = 14.2$. Since both quantities are the same, the median is the same, too. When there is an even number of observations, the middle two typically will differ. For example, suppose that one more applicant having an SAT score of 648 is added to the original group of engineering applicants, raising n to 6. The middle two values in the order sequence are then 622 and 648, and the sample median is $m = (622 + 648)/2 = 635$.

The procedure for finding m can be very time consuming when n is large, and $\bar{X}$ is ordinarily easier to compute. But the median is an average of **position**, making it often the better representative value. This is especially true with economic data, such as family income. Such data are ordinarily summarized in terms of the median income rather than the mean. The influence of one very rich family could so affect the arithmetic mean that the resulting summary might poorly describe the entire group. But the wealthy family would have "one vote" in placing the median, making it the more democratic measure of central tendency. The median is also useful in life testing, where the mean might never be calculated within an experiment's duration because it ends before the last items have failed.

The Mode

A third measure of location is the **mode**. Although rarely used alone, this quantity is an important descriptive measure and has some conceptual significance. The mode is *the most frequently occurring value*. For example, suppose that the number of patents held by the members of the mechanical engineering faculty at a certain university are as follows:

0 0 3 7 1 0 2 10 27 15 0 0 1 2

The mode is 0, since that value occurs five times, with 1 and 2 tied for second place with two professors for each.

When the raw data represent values on a continuum, often no single value occurs more than once. In those cases the mode is taken to be *the midpoint of the class interval with the highest frequency*. For example, consider the following grouped sample data:

Class Interval	Frequency
95.0–under 100.0	7
100.0–under 105.0	23
105.0–under 110.0	22
110.0–under 115.0	17
115.0–under 120.0	4
	$n = \overline{73}$

The mode is the midpoint of the second class interval, $(100.0 + 105.0)/2 = 102.5$.

When the observations are represented by a frequency curve, the usual descriptive display for entire populations, the mode is the point of highest relative frequency, which is the location of maximum clustering. This feature makes the mode useful in comparing the other measures of central tendency.

Many descriptive statistics are computed as standard output by statistical software. Consider the following example.

Example: *Microwave remote sensing data*

The following brightness temperatures (°K) were obtained from a radiometer sensor on board an aircraft flown on a mission to measure the soil moisture of the ground.

228 230 238 245 246 260

The above readings apply to soils known to have 13% moisture. The Minitab printout in Figure 2–11 provides several descriptive summaries.

Figure 2–11 *Minitab printout summarizing the brightness temperature radiometer data.*

```
MTB > Describe c1.

               N       MEAN      MEDIAN      TRMEAN      STDEV      SEMEAN
Temp.          6     241.17      241.50      241.17      11.84        4.83

             MIN        MAX          Q1          Q3
Temp.     228.00     260.00      229.50      249.50
```

Source: LeVine et al., "ESTAR: A Synthetic Aperture Microwave Radiometer for Remote Sensing Applications," *Proceedings of the IEEE*, December 1994, p. 1795.

Finding the Median and Mode with a Stem-and-Leaf Plot

When the raw data are arranged in a stem-and-leaf plot, the median and mode can be easily found by sorting the leaf digits in ascending value. The following *ordered* stem-and-leaf plot is obtained for the temperature data of Table 2–3:

Stem (1st digit)	Leaf (2nd digit)
1	8
2	
3	0 5 7 7 8 8 9
4	2 2 3 3 3 5 5 5 6 6 6 6 6 6 7 7 7 8 8 9 9 9 9 9 9 9
5	0 0 0 0 1 1 1 2 2 2 2 3 3 3 4 4 5 5 5 5 6 6 6 7 8 8 9
6	0 0 0 0 1 2 2 2 3 3 5 6 6 6 8 8 8 9 9 9
7	0 2 5 6
8	0

The median of the 86 observations is the average of the 43rd and 44th values (the leaf digits for these are shaded):

$$\frac{52 + 52}{2} = 52$$

The mode is 49, since the 4-stem has seven 9-leaves, the most frequently occurring of all the leaves on a single stem.

Frequency Distribution Forms and Summary Measures

Figure 2–12 shows three cases ordinarily encountered in statistical investigations that measure physical characteristics or summarize the times for completing an operation. The **symmetrical** frequency curve in (a) represents a population shape often encountered in measurements of dimension. Here the mean, median, and mode all are a common value. In the **skewed** distributions in (b) and (c), the mean is the more extreme population measure, with the median lying between the mode and mean. The direction of skew is expressed by the sign of the difference, mean – median.

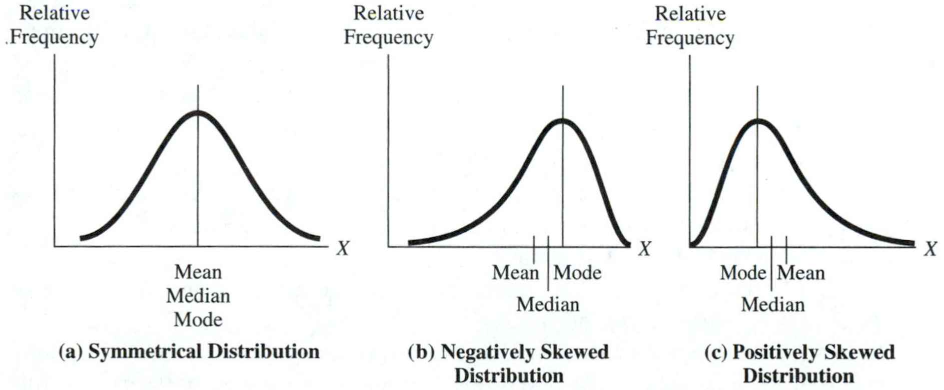

Figure 2–12 *Positional comparisons of the three measures of central tendency for symmetrical and skewed frequency distributions.*

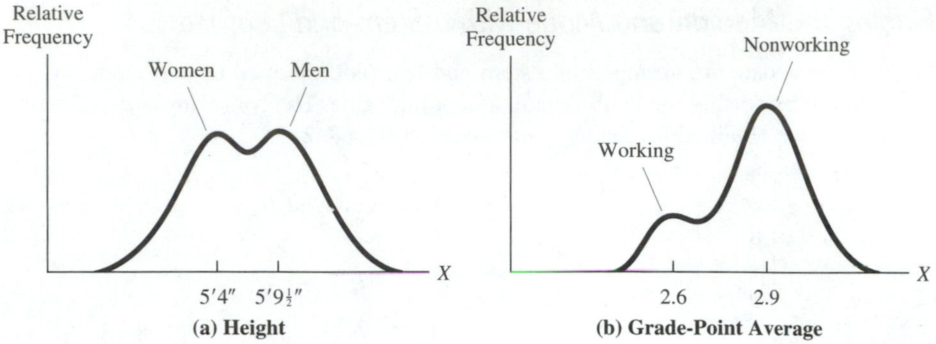

Figure 2–13
Bimodal frequency distributions.

The mode concept is used to characterize a population type whose frequency distribution has a double-humped shape, such as those in Figure 2–13. These frequency curves each represent a **bimodal distribution**. A frequency distribution like this is usually obtained when *anthropometric data* (human dimensions) are collected from measurements of men and women. The curve in (a) represents the heights of students at a particular university where the sexes are represented equally. It is generally good statistical practice to make separate evaluations of men's and women's heights, treating the populations individually by sex.

Indeed, discovery of a bimodal frequency distribution should signal an investigator that some underlying nonhomogeneity exists in the raw data and that a better description might be obtained by finding the cause and splitting the data into two groups. For example, curve (b) was found for the GPAs of the engineering students in a particular school. Investigation showed that most students whose grades are reflected by the lower hump held jobs, whereas those in the major grouping were largely free of work commitments. Separate evaluations were then made of working and nonworking students. [The lower mode in curve (b) is only a "local" one since it pertains to a small subset of observations. But the frequency curve is still classified as bimodal.]

Percentiles, Fractiles, and Quartiles

The final group of location measures includes those concerned with frequency information. Most familiar is the **percentile**, which is a point below which a stated percentage of the observations lie. The percentile will be a value in the same units as the observations themselves, but it need not be unique. For example, an industrial engineer might find that 17 of 20 units involve assembly times below 23 seconds. The percentage of items having such times is $100(\frac{17}{20}) = 85$, and 23 seconds is the 85th percentile. If that engineer finds that 31 seconds is the 95th percentile, then 95% of the 20 test units—19 of them—have assembly times falling below 31 seconds.

An alternative way of conveying this information is to use fractiles. A **fractile** is that point below which a stated fraction of the values lie. Thus, 23 seconds is the .85-fractile and 31 seconds is the .95-fractile.

We have so far encountered one important percentile: the median. *The median is both the 50th percentile and the .50-fractile.*

Frequency information is also expressed in **quartiles**, which divide raw data into four groups of equal frequency. The **first quartile** is the same point as the 25th percentile

(.25-fractile), the **second quartile** is the 50th percentile (.50-fractile and also the median), and the **third quartile** is equal to the 75th percentile (.75-fractile).

Example: *Class standing in engineering school*

The 347 graduates in Bob Blackwell's class were ranked according to their grade-point averages (4.0 being the maximum). Bob's GPA of 3.57 seemed quite high. But that placed him only in the 77th percentile, since 267 (77% of 347) classmates had lower GPAs. Eighty classmates (including Bob) achieved GPAs of at least 3.57.

It is interesting that the 90th percentile corresponds to a slightly higher GPA of 3.65. Bob is remorseful that just 15 more points on a thermodynamics midterm would have given him an A instead of a B and would have raised his GPA beyond 3.65. That small difference in GPA would have raised him to the 90th percentile and placed him ahead of 312 classmates.

Finding Percentiles from Ungrouped Data When all raw data are available, the percentiles may be established by ranking the values from lowest to highest. For example, the following ten values are sample weights (in grams) of coating materials used in a masking process:

Value	5.3	5.4	5.7	6.0	6.1	6.1	6.2	6.4	6.5	6.6
Position	1	2	3	4	5	6	7	8	9	10

The 10th percentile is any point above 5.3 but not exceeding 5.4 grams. We might use 5.31, 5.33, 5.39, or 5.40 as the 10th percentile. The 20th percentile may be any point greater than 5.4 but not exceeding 5.7 grams.

There is considerable leeway in selecting percentiles directly from raw data. To avoid ambiguity, we adopt the following procedure.

Interpolation Procedure for Locating Percentiles

1. **Sort the raw data in ascending order.** Denote as X_1 the first sorted value, X_2 the second, and so on up to X_n, with n representing the total number of observations. The subscripts represent the *position* of the data value.
2. **Establish the decimal equivalent.** Denote by d the decimal equivalent of the desired percentage point, which must meet the following restriction:

$$\frac{1}{n} \leq d \leq \frac{n-1}{n}$$

   Denote by Q_d the corresponding percentile value that is to be found.
3. **Find the relative position of the desired percentile.** This is

$$(n+1)d$$

   Then let k be the largest integer such that

$$k \leq (n+1)d$$

   The desired percentile will lie between X_k and X_{k+1}.
4. **Compute the percentile value.** The following expression applies:

$$Q_d = X_k + [(n+1)d - k](X_{k+1} - X_k)$$

We illustrate this procedure using the above $n = 10$ sample weights. First we find the 10th percentile ($d = .10$). The relative position of $Q_{.10}$ is $(10+1)(.10) = 1.1$. Using $k = 1$, we compute

$$\begin{aligned} Q_{.10} &= 5.3 + [(10+1)(.1) - 1](5.4 - 5.3) \\ &= 5.31 \end{aligned}$$

Next we locate the 25th percentile. The 25th percentile lies between positions 2 and 3, so we have

$$\begin{aligned} Q_{.25} &= 5.4 + [(10+1)(.25) - 2](5.7 - 5.4) \\ &= 5.63 \end{aligned}$$

The 75th percentile lies between positions 8 and 9. We obtain the following:

$$\begin{aligned} Q_{.75} &= 6.4 + [(10+1)(.75) - 8](6.5 - 6.4) \\ &= 6.43 \end{aligned}$$

The median is the 50th percentile:

$$\text{Median} = Q_{.50} = 6.1 + [(10+1)(.50) - 5](6.1 - 6.1) = 6.1$$

It can be a time-consuming and error-prone task to sort a large set of raw data by hand. It may be easier first to construct an ordered stem-and-leaf plot. Returning to the temperature data given earlier, we can easily count out the positions for any of the $n = 86$ observations.

```
                                   Leaf (2nd digit)
             1 | 8
             2 |                          21st    22nd
             3 | 0 5 7 7 8 8 9
Stem         4 | 2 2 3 3 3 5 5 5 6 6 6 6 6 6 7 7 7 8 8 9 9 9 9 9 9 9
(1st digit)  5 | 0 0 0 0 1 1 1 2 2 2 2 3 3 3 4 4 5 5 5 5 6 6 6 7 8 8 9
             6 | 0 0 0 0 1 2 2 2 3 3 5 6 6 6 8 8 8 9 9 9
             7 | 0 2 5 6
             8 | 0
                        65th  66th
```

The 25th percentile has relative position $(86+1)(.25) = 21.75$, falling above the 21st observation by .75 times the distance between that point and the 22nd observation, both of which have value 46, so that $Q_{.25} = 46$. The 75th percentile is positioned at $(86+1)(.75) = 65.25$, so

$$\begin{aligned} Q_{.75} &= X_{65} + [65.25 - 65](X_{66} - X_{65}) \\ &= 60 + .25(61 - 60) = 60.25 \end{aligned}$$

Finding Percentiles from Grouped Data When the raw data are unavailable, but they have already been grouped into a frequency distribution, it is simple to read the percentile directly from the graph for the cumulative relative frequency distribution. Figure 2–14 shows the cumulative relative frequency plot for the inspection times used earlier. The

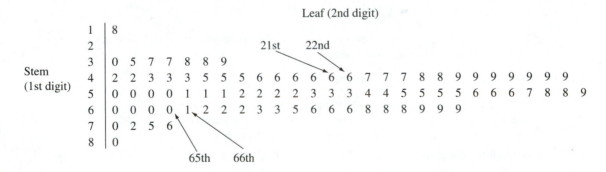

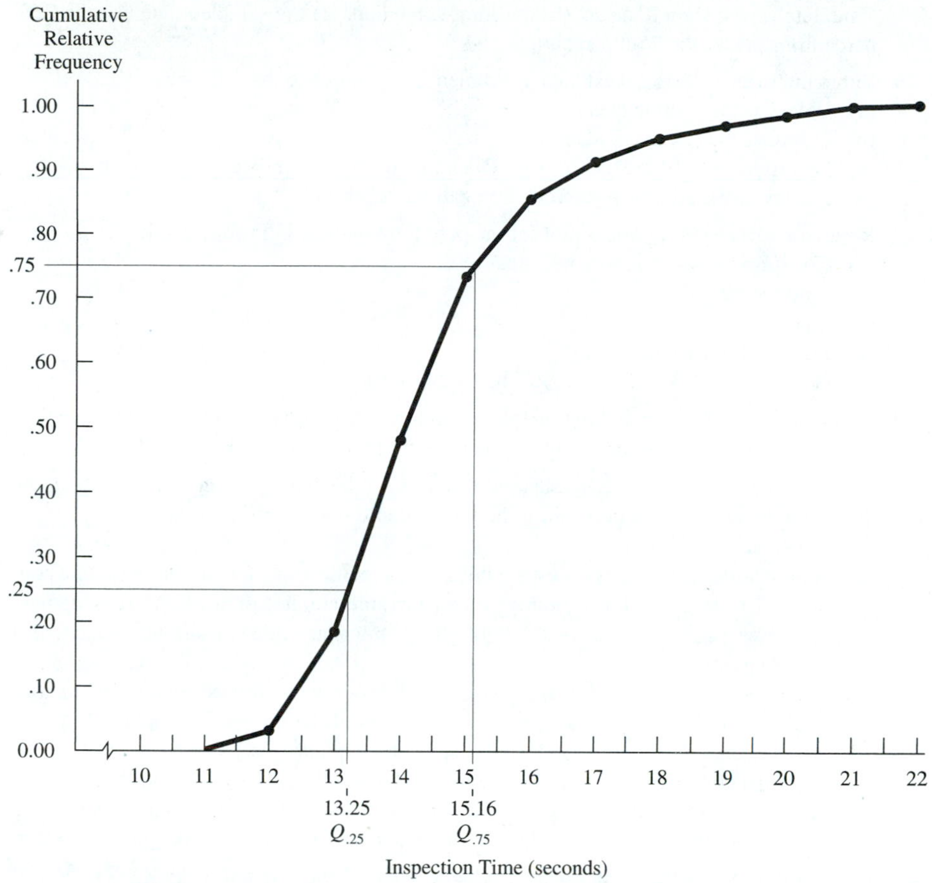

Figure 2–14
Finding percentiles graphically.

ordinate now expresses cumulative *relative* frequencies. The 25th percentile is the horizontal coordinate of that point on the curve at height .25:

$$Q_{.25} = 13.25 \text{ seconds}$$

and the 75th percentile is that time where the curve has height .75:

$$Q_{.75} = 15.16 \text{ seconds}$$

The above values should agree closely with those that could be computed from grouped data.

Problems

2–18 The following numbers of positions have been held by a random sample of aerospace engineers during the 10 to 15 years since their graduation:

1	2	3	3	2	5	4	4	3	1
2	1	4	6	5	5	4	2	3	2

Calculate (a) the sample mean, (b) the sample median, (c) the sample mode, (d) the 25th percentile, and (e) the 75th percentile.

2–19 Referring to the materials cost data in Problem 2–1, complete the following:

(a) Calculate the sample mean.

(b) Calculate the sample median.

(c) Determine the difference $\bar{X} - m$. Does this difference indicate that the sample frequency distribution is positively or negatively skewed?

2–20 Referring to the stem-and-leaf plot for the precipitation data in Problem 2–4:

(a) Construct the ordered stem-and-leaf plot.

(b) Find the median.

(c) Find the mode.

(d) Find the following percentiles:

(1) 10th (2) 25th (3) 75th (4) 90th

2–21 Referring to the interarrival time sample frequency distribution in Problem 2–13, calculate (a) the sample mean, (b) the sample median, and (c) the sample mode.

2–22 For each of the following situations, indicate a possible source of nonhomogeneity in the data, and discuss why the population should or should not be split to serve the purposes of the statistical investigation.

(a) A human-factors engineer is establishing new specifications for aircraft hydraulic controls. Actuating one lever requires a considerable amount of strength. The engineer chooses a maximum force that 95% of sample pilots are able to exert while seated in a normal position.

(b) Six months ago a new accident prevention program was initiated by a plant superintendent. For the past year a record has been kept of the number of weekly accidents. An investigation is being conducted to investigate plant safety standards.

(c) A computer software engineer has obtained data on the processing times taken to run payrolls at two identically equipped computer facilities. One center serves New York City clients, the other is located in Nebraska and services clients in neighboring states.

2–23 Referring to the gasoline mileages in Problem 2–10, answer the following:

(a) Calculate the sample mean.

(b) Calculate the sample median.

(c) Determine the difference $\bar{X} - m$. Does this difference indicate that the sample frequency distribution is positively or negatively skewed?

2–24 Referring to the equipment downtime sample frequency distribution in Problem 2–15, calculate (a) the sample mean, (b) the sample median, and (c) the sample mode.

2–25 The following sample data apply to the average yield of a final product (in grams) from each liter of chemical feedstock:

29.84	31.19	31.91	33.78	29.68
32.23	26.23	30.97	31.92	29.19
26.83	31.55	32.00	27.90	27.93
26.59	30.05	26.25	31.23	27.47
31.00	30.02	32.35	29.15	28.82
29.94	30.21	30.62	29.03	31.23
29.81	31.59	30.52	31.01	28.35
27.94	26.68	29.84	30.60	28.10
29.48	27.76	27.59	31.04	30.81
32.22	32.22	27.02	29.03	32.65

Compute (a) the sample mean and (b) the sample median.

2–26 Refer to the data in Problem 2–25. The following frequency distribution applies.

Yield	Number of Batches
25.5–under 26.5	2
26.5–under 27.5	5
27.5–under 28.5	7
28.5–under 29.5	6
29.5–under 30.5	8
30.5–under 31.5	11
31.5–under 32.5	9
32.5–under 33.5	1
33.5–under 34.5	1

(a) Using the grouped data, calculate the sample mean.
(b) Plot the cumulative relative frequency distribution. Then determine the following percentiles:
(1) 10th (2) 25th (3) 50th (4) 75th (5) 90th
(c) What is the value of the sample median?
(d) The above statistics are only approximations to the respective values found in Problem 2–25. Determine how much the grouped approximation value lies above or below its counterpart statistic computed from the raw data directly for (1) the sample mean and (2) the sample median.

2–27 The following hourly labor costs were computed for a random sample of small construction projects:

$21.52	$20.76	$21.87	$18.54	$21.36
22.56	22.35	19.81	19.85	24.87
19.83	20.38	21.95	21.33	19.42
23.11	20.73	20.93	20.84	21.23
18.75	23.17	19.05	22.13	19.91
20.50	20.12	21.00	21.92	25.12
22.48	21.75	23.39	20.38	20.13
19.61	19.37	21.05	20.74	20.58
21.11	22.74	20.31	22.05	20.62
20.65	20.84	22.87	22.98	21.75
20.43	24.36	22.17	21.47	21.94
21.61	19.72	21.24	20.26	22.45
19.24	22.25	24.10	23.60	23.05
19.51	20.37	19.73	20.15	19.95
20.48	19.62	20.26	19.84	20.04

Compute (a) the sample mean and (b) the sample median.

2–28 Refer to the data in Problem 2–27. The following frequency distribution applies:

Hourly Cost	Number of Projects
\$18.50–under 19.50	6
\$19.50–under 20.50	23
\$20.50–under 21.50	18
\$21.50–under 22.50	15
\$22.50–under 23.50	8
\$23.50–under 24.50	3
\$24.50–under 25.50	2

Repeat (a)–(d) as in Problem 2–26.

2–29 The following brightness temperatures (°K) were obtained by ESTAR in aircraft from scans of the ground having about 16% moisture.*

215 256 222 233 229 232

Repeat (a)–(e) as in Problem 2–18.

2–3 Summary Statistical Measures: Variability

We have seen several measures of central tendency (location), each of which expresses the center of the observed values in a different way. Often, comparisons of means will be sufficient for evaluating two or more groups of data. However, other summary measures can be used for making comparisons. After location, **measures of variability** provide the next most important descriptive summaries. Such a quantity expresses the degree to which individual observation values differ from each other.

The Importance of Variability

Variability and **dispersion** are the synonymous terms used in statistics to characterize individual differences. The greater the variability between observations, the more they will be spread out. Populations or samples having high variability will have a frequency distribution involving wider class intervals or more classes than a low-variability group measured on the same scale.

The following example will help explain the concept of variability and demonstrates its worth in statistical evaluations.

*LeVine et al., "ESTAR: A Synthetic Aperture Microwave Radiometer for Remote Sensing Applications," *Proceedings of the IEEE,* December 1994, p. 1795.

Example:* *Comparing queueing systems for computer workstations

Figure 2–15 shows two frequency polygons constructed from the waiting times experienced by two sample groups of students seeking access to workstations at a university's central equipment room. Two different "queue disciplines" were employed. Under method A, arriving students would stand behind the chair of a student using a terminal, waiting until that or a nearby "unclaimed" workstation was free. Method B is a managed system, whereby arriving students are placed in a single queue in order of their arrival. These students are fed one at a time to workstations as they become free.

Figure 2–15 *Frequency distributions for waiting time for computer workstations.*

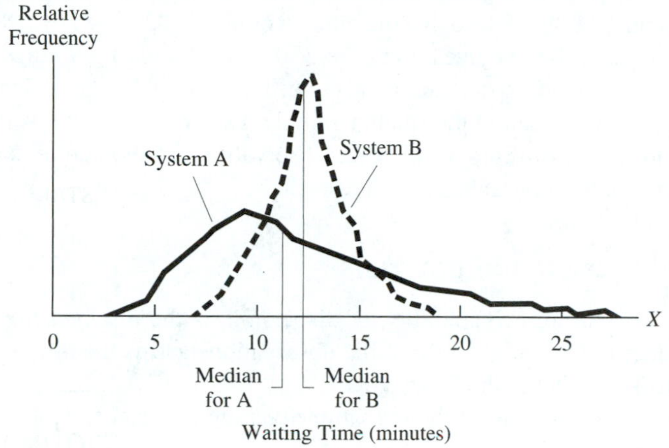

Notice that sample group B experienced somewhat higher times, as indicated by the level of the sample median. This reflects the "overhead" imposed by the queue manager in determining when workstations are free and in moving the newest student into the work area. But the waiting times under B are less varied and cluster more tightly around the center of the distribution. The anarchy of method A often results in extraordinarily long waiting times. When asked to state their preference for queueing method, the great majority favored method B, which was more predictable, even though it involved longer average waits.

This example shows that variability can provide a second "dimension" for statistical evaluations in which central tendency by itself can often be inadequate for making decisions. There are two main summary measures of variability, one of which takes two forms.

The Range

The simplest measure of variability is the **sample range**, which may be computed by subtracting the smallest observation from the largest. As an example, using the original SAT scores for the five engineering applicants, the highest score was 755 and the lowest was 584. The sample range is

$$\text{Range} = 755 - 584 = 171$$

This value indicates that there is a 171-point spread in SAT scores achieved by the applicants in the sample.

The range is the crudest measure of variability, but it nevertheless can be quite useful. Consider the daily temperature cycle for two locales, both of which experience a mean of 85°F. One cycle might occur in a very pleasant subtropical region where the diurnal temperature ranges from a low of 75° to a high of 95°, with a range of 20°. The other might be in a far less hospitable desert region, where the diurnal temperature can swing from 50° to 120° in the summer and from –20° to 50° in the winter—a 70° range in both cases. An engineer must obviously design different facilities for housing personnel and equipment in the two regions.

One practical difficulty with using the range is that it tends to become larger with increasing sample size. In a group of 100 engineering applicants, the low SAT score might be 485 and the high 783, so that the range would be 783 − 485 = 298. In a group of 10,000 applicants, the extremes might be 350 and 800, giving a range of 450, yet there may be no reason to expect individuals to differ from each other much more in the 10,000-applicant group than in the smaller sample. One way around this distorting effect is to eliminate the **outliers**, or extreme observation values, before computing the range. Doing so can be very subjective, however.

Interquartile Range and Box Plots

A related useful measure of variability, or dispersion, is the *interquartile range*. This represents scatter in the middle 50% of the observations and is the difference between the third quartile ($Q_{.75}$) and the first ($Q_{.25}$).

To illustrate, consider again the ten sample coating material weights (grams):

5.3 5.4 5.7 6.0 6.1 6.1 6.2 6.4 6.5 6.6

Recall that in Section 2–2 we found the following quartile values for these weights:

$$Q_{.25} = 5.63 \quad Q_{.50} = 6.10 \quad Q_{.75} = 6.43$$

The interquartile range then is

$$Q_{.75} - Q_{.25} = 6.43 - 5.63 = .80$$

The range and interquartile range may be combined in a **box plot**. Figure 2–16 shows the box plot for the above data. The plot begins with a line segment starting at the minimum observed level and ends with a line segment ending at the maximum value. The box begins at the first quartile, ends at the third, and is divided at the median. The overall length of the plot gives the range, while the length of the box provides the interquartile range.

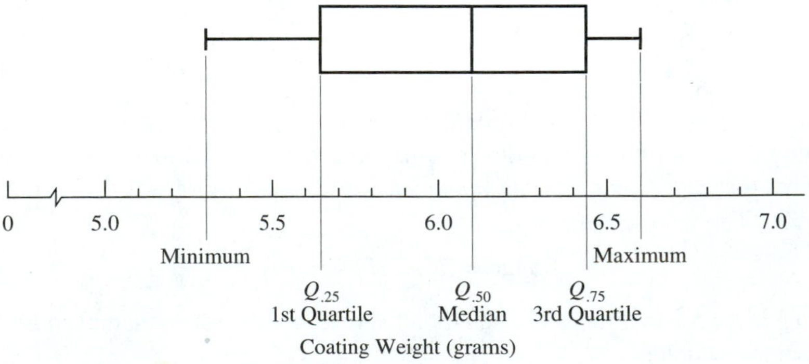

Figure 2–16
Box plot for sample of coating material weights.

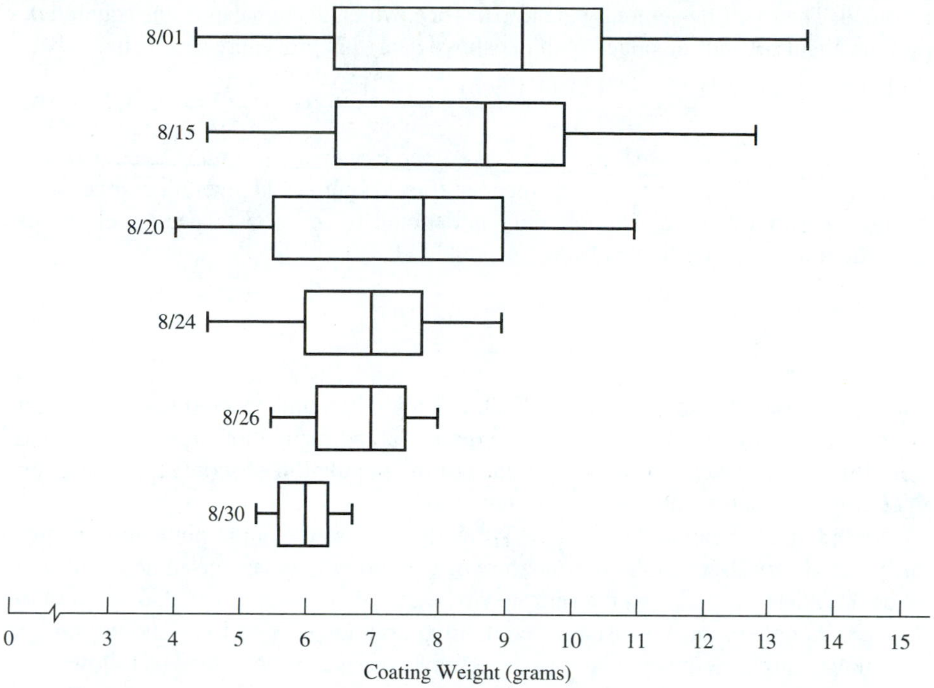

Figure 2–17 *Box plots from successive samples of coating weights.*

Box plots can be useful when comparing data groups. Figure 2–17 shows box plots from successive samples of coating weights. Notice that as the coating process was refined over time, the sample units received less material, as reflected by progressively lower medians. Concurrently, the weights for individual units became less varied and more uniform, demonstrated by progressively shorter plots and boxes.

The Variance and Standard Deviation

The most important measure of variability is based on the **deviations** of individual observations about the central value. For this purpose the mean usually serves as the center. For example, we determined the mean of 653.4 for the five SAT scores given in Section 2–2. The deviation for each score may be computed by subtracting that value from each:

$$\begin{aligned} 755 - 653.4 &= 101.6 \\ 613 - 653.4 &= -40.4 \\ 584 - 653.4 &= -69.4 \\ 693 - 653.4 &= 39.6 \\ 622 - 653.4 &= -31.4 \end{aligned}$$

These deviations may be summarized by a collective measure that considers each deviation. Obviously, they cannot be averaged directly, since their mean must be zero (which you can verify as an exercise). One measure of the average deviation is achieved by first taking the absolute values. Unfortunately, absolute values are mathematically cumbersome. Instead, the deviations are first squared. Then the squared deviations are averaged. The resulting summary is the **variance**, which is the mean of the squared deviations about the central value. With the above data, this procedure results in

$$\frac{(101.6)^2 + (-40.4)^2 + (-69.4)^2 + (39.6)^2 + (-31.4)^2}{5} = \frac{19,325.2}{5} = 3,865.04$$

When all possible observations are taken, the raw data constitute the entire population. The quantity calculated in this manner after a complete enumeration or census is the **population variance**. This quantity is denoted by σ^2 (lowercase Greek sigma, squared) and may be computed from

$$\sigma^2 = \frac{\sum_{i=1}^{N}(X_i - \mu)^2}{N}$$

where X_i denotes the ith observation. Each deviation is found by subtracting the population mean μ from the observation. The divisor is the population size N. In the SAT score illustration we computed the variance of the population of scores for $N = 5$ engineering applicants from a particular high school.

In practice σ^2 cannot be computed directly because the entire population is ordinarily not observed. An analogous measure of variability may be determined with sample data. This is referred to as the **sample variance**, which is represented as s^2. In place of μ, the sample mean $\overline{X}$ is used for computing individual deviations, and the sample size minus 1 replaces the population size. Sample variance is computed as follows:

SAMPLE VARIANCE

$$s^2 = \frac{\sum_{i=1}^{n}(X_i - \overline{X})^2}{n - 1}$$

We illustrate this equation using the sample times from Table 2–1 for completing the calibration inspections. Subtracting the sample mean $\overline{X} = 14.342$ from each observed value, squaring the differences, and summing, the total of 292.484 is obtained. Thus,

$$s^2 = \frac{292.484}{100 - 1} = 2.9544$$

The expression has a slightly different form than that for σ^2 because its divisor is $n - 1$ rather than n. Reducing the divisor by 1 in computing the sample variance results in a better estimator of σ^2. The theoretical reason for this is described in Chapter 10.

The variance is analogous to the second moment in physics. It not only provides a useful index for gauging individual differences among the observations, but σ is a parameter in certain mathematical functions that are used to represent frequency curves for some populations. The variance also has important properties that are useful to statistical theory and can facilitate statistical evaluations.

As a practical matter, though, the variance is in units different from the observations themselves. In our SAT score illustration the variance is 3,865.04 *squared* points. Likewise, the sample variance for the calibration inspection times is in seconds squared. Partly because of the confusion of different units and partly because the variance can be orders of magnitude above (or below) the size of the observed values, statistical evaluations are conducted with the variance's positive square root. The resulting quantity is called the **standard deviation**. The population and sample standard deviations are denoted, respectively, as σ and s.

The standard deviations for our two illustrations are as shown:

$$\sigma = \sqrt{\sigma^2} = \sqrt{3{,}865.04} = 62.17 \qquad \text{(population of SAT scores)}$$

$$s = \sqrt{s^2} = \sqrt{2.9544} = 1.72 \qquad \text{(sample of inspection times)}$$

These values are in the same units as the observations themselves and provide more convenient summaries than the variances. Since one measure can be obtained directly from the other, the variance and standard deviation are used interchangeably.

When computing s^2 (or s) by hand, it is usually convenient to use the following equivalent expression for the sample variance:

$$s^2 = \frac{\sum_{i=1}^{n} X_i^2 - n\overline{X}^2}{n-1}$$

One disadvantage of the above is the possibility of roundoff errors. (As an exercise you may verify that this expression is equivalent to the earlier one for s^2.)

As with the sample mean, when the raw data are no longer available, the sample variance may be approximated from grouped data provided by the frequency distribution. This is done using the following expression.

SAMPLE VARIANCE APPROXIMATED FROM GROUPED DATA

$$s^2 = \frac{\sum f_k X_k^2 - n\overline{X}^2}{n-1}$$

Here the summation is taken over all classes and X_k is the midpoint of the kth class. The sample mean will ordinarily also be approximated from grouped data. As an illustration, the grouped approximation is applied in Table 2–7 with the frequency distribution for calibration inspection times. Notice that the variance computed under this procedure is slightly larger than the value for s^2 computed earlier directly from the raw data. Keep in mind that statistics computed from grouped data are only approximate values.

The Meaning of the Standard Deviation

The population standard deviation is an important index of variability that conveys a great deal of information. It is a fundamental parameter of the mathematical function describing the normal curve that fits the frequency distribution of so many populations. But σ is useful in describing many other frequency distributions as well.

Empirical Rule

Together the mean and standard deviation may give a rich description of the underlying data that neither measure alone can provide. Many physical measurements are consistent with the following.

> For many data sets approximately 95 percent of all observed values will lie within a distance of two standard deviations (or standard errors) from the mean. Furthermore, about 99.7 percent of those values will lie within a distance of three standard deviations.

Empirical Rule

Table 2–7
Calculation of the sample variance for calibration inspection time using grouped data

Time (seconds)	Frequency (Number of Times) f_k	Class Interval Midpoint X_k	X_k^2	$f_k X_k$	$f_k X_k^2$
11.0–under 12.0	2	11.5	132.25	23.0	264.50
12.0–under 13.0	16	12.5	156.25	200.0	2,500.00
13.0–under 14.0	29	13.5	182.25	391.5	5,285.25
14.0–under 15.0	27	14.5	210.25	391.5	5,676.75
15.0–under 16.0	11	15.5	240.25	170.5	2,642.75
16.0–under 17.0	6	16.5	272.25	99.0	1,633.50
17.0–under 18.0	4	17.5	306.25	70.0	1,225.00
18.0–under 19.0	2	18.5	342.25	37.0	684.50
19.0–under 20.0	2	19.5	380.25	39.0	760.50
20.0–under 21.0	1	20.5	420.25	20.5	420.25
			Totals	1,442.0	21,093.00

$$\bar{X} = \frac{\sum f_k X_k}{n} = \frac{1,442.0}{100} = 14.42 \text{ seconds}$$

$$s^2 = \frac{\sum f_k X_k^2 - n\bar{X}^2}{n-1} = \frac{21,093.00 - (100)(14.42)^2}{100 - 1} = 3.0238 \text{ seconds}^2$$

This rule works best when the data have a two-tailed frequency distribution of a bell shape—like the one in Figure 2–10(a), although it will be approximately correct even when there is a modest skew. As we shall see in Chapter 7, this rule derives from a property of the normal distribution. Experience has shown that a great many physical phenomena generate values that are consistent with the empirical rule. It will also apply to possible levels of sample means computed from populations having a variety of shapes.

To illustrate, we consider the calibration inspection times from Table 2–7, for which the computed mean is $\bar{X} = 14.342$ and the computed standard deviation is $s = 1.72$. The empirical rule tells us that approximately 95 percent of the inspection times should lie between

$$\begin{aligned} &\bar{X} - 2s && \text{and} && \bar{X} + 2s \\ &= 14.34 - 2(1.72) && && = 14.34 + 2(1.72) \\ &= 10.90 && && = 17.78 \text{ seconds} \end{aligned}$$

Limits computed in this manner will not generally be exact for any particular data. But an examination of the original raw calibration inspection time data listed in Table 2–1 shows that precisely 94% of the actual observed times fell between 10.90 and 17.78 seconds. In this instance the empirical rule and actual results agree very closely.

Of course, there are some data sets where the empirical rule would give a poor approximation. It works best with data whose histogram resembles the symmetrical bell shape shown in Figure 2–10(a), depicting the frequency curve for a normal distribution. Fortunately, frequency distributions for a great many physical measurements fall into that family.

Of course, the empirical rule would not really be needed if the complete data set were available, as it is with the calibration times. In such cases we could simply count observations, getting for any percentage the precise limits for the corresponding range of values. But often we must conjecture about a data set *yet-to-be* measured or not completely observed. In those cases having good estimates of the mean and standard deviation and some idea regarding the general shape of the underlying frequency distribution can be very useful information indeed.

Quality control provides an interesting application. Consider the plight of an industrial engineer as she searches for proper control settings for a new rolling machine used to make metal sheets. The nominal thickness is .0625″, a target that will only occasionally be met exactly because of naturally occurring noise. Five pilot runs have established the following average sheet thickness levels:

$$.064'' \quad .062'' \quad .065'' \quad .069'' \quad .061''$$

From these the sample mean and standard deviation are computed.

$$\bar{X} = \frac{.064 + .062 + .065 + .069 + .061}{5} = .0642''$$

$$s^2 = \frac{(.064 - .0642)^2 + (.062 - .0642)^2 + (.065 - .0642)^2 + (.069 - .0642)^2 + (.061 - .0642)^2}{5 - 1}$$

$$= \frac{.00000004 + .00000484 + .00000064 + .00002304 + .00001024}{4}$$

$$= .00000970$$

$$s = \sqrt{.00000970} = .0031''$$

The empirical rule indicates that if the process is on target with a true mean of .0625″, then approximately 95 percent of the sheets will fall between

$$.0625'' - 2(.0031) = .0563'' \qquad \text{and} \qquad .0625'' + 2(.0031) = .0687''$$

The above may be used as control limits for deciding when to adjust the control settings:

Adjust whenever the average measured sheet thickness $\leq .0563''$ (LCL)

or $\geq .0687''$ (UCL)

The empirical rule tells us that the engineer will make adjustments only about 5% of the time when the process is actually on target.

She might think the above limits are too wide for consistent quality. To accomplish that, she must find a way to reduce the variability in thickness, getting a lower standard deviation. Otherwise, tighter limits will just lead to excessive adjustments.

Composite Summary Measures

The various summary statistical measures can be combined to provide meaningful information. Two of these are especially useful in data evaluations.

The **coefficient of variation** relates variability in the sample or population to the mean. It is calculated by dividing the standard deviation by the sample mean:

SAMPLE COEFFICIENT OF VARIATION

$$v = \frac{s}{\bar{X}}$$

The population coefficient of variation is computed by the analogous ratio, σ/μ.

To illustrate, consider again the inspection times from Table 2–1. We have

$$v = \frac{s}{\overline{X}} = \frac{1.72}{14.342} = .120$$

This indicates that the sample standard deviation for the inspection times is only 12% as large as the mean level.

By itself, the standard deviation does not convey the *relative* degree of variability. For example, consider the mean and standard deviation in men's hand length and reach:*

Hand Length	**Reach**
$\mu = 7.44''$	$\mu = 32.33''$
$\sigma = .34''$	$\sigma = 1.63''$
$\sigma/\mu = .045$	$\sigma/\mu = .050$

The respective standard deviations suggest that reach has about five times the variability of hand length. But the relative variability, as expressed by the coefficient of variation, is nearly identical for the two populations.

The coefficient of variation might be used as the basis for sorting statistical distributions into like clusters. Table 2–8 shows the results of an experiment on work-time measurement. Each entry represents sample observations of the time to complete a work element. The coefficients of variation range from .215 to .572. Unfortunately, the researchers found no consistent pattern relating type of work element to coefficient of variation.

Their data illustrate a second composite measure, the **coefficient of skewness**. This quantity expresses the direction of and the degree to which a frequency distribution is skewed; it involves the mean, median, and standard deviation. The following expression is used to compute it:

SAMPLE COEFFICIENT OF SKEWNESS

$$\text{SK} = \frac{3(\overline{X} - m)}{s}$$

To illustrate, consider the inspection times from Table 2–1. We compute

$$\text{SK} = \frac{3(\overline{X} - m)}{s} = \frac{3(14.342 - 14.2)}{1.72} = 0.25$$

which exceeds zero, indicating a slight *positive skew*.

The research data in Table 2–8 show coefficients of skewness ranging from 0.298 to 3.224. The researcher concluded that such wide-ranging skewness reflects a lack of homogeneity in the parent work-time distributions.

*Source: H. T. E. Hertzberg, G. S. Daniels, and E. Churchill, *Anthropometry of Flying Personnel,* WADC Technical Report 52–321, U.S.A.F. (September 1954).

Table 2–8 *Sample data from work-time measurement experiment*

Work Element	Sample Mean (0.01 min) $\bar{X}$	Sample Median (0.01 min) m	Sample Standard Deviation (0.01 min) s	Sample Coefficient of Variation v	Sample Coefficient of Skewness SK
1	11.139	9.833	3.338	0.346	1.174
2	5.604	4.613	2.354	0.420	1.263
3	2.540	1.908	0.588	0.232	3.224
4	4.229	3.133	1.068	0.253	3.079
5	9.957	9.081	2.141	0.215	1.227
6	2.913	2.068	1.665	0.572	1.523
7	2.576	1.858	1.451	0.563	1.484
8	5.990	5.070	2.021	0.337	1.366
9	4.467	3.295	2.435	0.545	1.444
10	2.969	2.432	0.881	0.297	1.829
11	3.532	2.790	1.039	0.294	2.142
12	4.465	3.698	1.446	0.324	1.591
13	5.903	4.736	1.832	0.310	1.908
14	3.305	3.197	1.087	0.329	0.298
15	3.210	2.520	1.446	0.452	1.432
16	5.984	5.362	1.789	0.299	1.043
17	4.126	3.378	1.678	0.219	1.337
18	7.270	6.461	2.034	0.263	1.193
19	2.770	2.133	1.063	0.384	1.797
20	6.673	5.914	1.968	0.295	1.157
21	8.530	7.833	1.845	0.216	1.052
22	3.928	3.325	1.688	0.430	1.072
23	5.247	4.604	1.469	0.280	1.313
24	5.094	4.402	1.234	0.242	1.682
25	5.452	4.750	2.079	0.381	1.013
26	3.653	2.958	0.982	0.269	2.123

Source: Knott, Kenneth, and Roy J. Sury, "A Study of Work-Time Distributions on Unpaced Tasks," *IIE Transactions* (March 1987): 50–55.

Problems

2–30 The following sample data represent the observed number of busy workstations in a computer network:

10	4	15	17	6	12	9	13	15	5

(a) Calculate the sample range.
(b) Calculate the sample variance and the sample standard deviation.

(c) Find the following fractiles:
(1) .25 (2) .50 (3) .75
(d) Calculate the interquartile range.
(e) Construct the box plot.

2–31 The following grade-point averages apply to a random sample of graduating seniors.

3.65	2.73	2.35	3.09	3.28
3.51	2.86	2.59	3.13	3.24

Repeat (a)–(e) as in Problem 2–30.

2–32 Using the ESTAR brightness temperature data in Problem 2–29, repeat (a)–(e) as in Problem 2–30.

2–33 Refer to the data in Problem 2–30.
(a) Compute the sample coefficient of variation.
(b) Find the sample median. Then compute the sample coefficient of skewness.

2–34 Repeat Problem 2–33 using the data from Problem 2–31 instead.

2–35 Refer to the chemical yield data in Problem 2–25. Compute (a) the sample range, (b) the sample variance, and (c) the sample standard deviation.

2–36 Using the grouped data from the sample frequency distribution for interarrival times in Problem 2–13, compute the sample variance and standard deviation.

2–37 For each of the following evaluations discuss why a measure of central tendency may not be a wholly adequate summary by itself.
(a) A human-factors engineer is using a sample of past temperatures experienced in a clean room to determine specifications for a revised air-conditioning system.
(b) An industrial engineer wants to sequence assembly tasks to achieve balance among workstations so that excessive buffer stocks are avoided and worker idle time is minimized. Completion times for each task are used in the evaluation.
(c) In planning for facilities expansion, a construction maintenance superintendent requires data on the repair times of equipment.

2–38 Refer to the hourly labor cost data in Problem 2–27. Compute (a) the sample range, (b) the sample variance, and (c) the sample standard deviation.

2–39 Using the grouped data from the sample frequency distribution for computer downtimes in Problem 2–15, compute the sample variance and standard deviation.

2–40 Using your answers to Problem 2–30 and assuming that the sample mean and standard deviation are good approximations to the population values, apply the empirical rule to determine the limits for the range that contains
(a) 95 percent of the possible levels for the number of busy workstations.
(b) 99.7 percent of the possible levels.

2–41 You know that when a process for filling jars is operating as intended, it will yield mean contents of 16.0 ounces per jar. The standard deviation of .2 ounce is assumed always to apply. A sample jar is removed from the line every 10 minutes, and its contents are precisely measured.
(a) Find the limits between which approximately 95 percent of the sample jar content weights should lie.
(b) If the process is readjusted whenever a sample jar's content weight falls outside the limits found in (a), indicate for each of the following results whether or not an adjustment will be made: (1) 15.3 oz, (2) 16.1 oz, (3) 16.3 oz, (4) 16.5 oz, (5) 15.8 oz.

2–4 Summary Statistical Measures: The Proportion

Qualitative observations form an important class of statistical data. The primary summary measure for such data is the **proportion**. When the observations constitute the complete population, this quantity is the **population proportion**, a parameter we denote by the symbol π (lowercase Greek *pi*, which represents the first letter in the word *proportion* and is a quantity between 0 and 1—*not* 3.1416). The analogous measure where the observations are sample data and only representative of some population is the **sample proportion**, which we denote by P. Ordinarily, only this sample statistic is computed, so that P serves as an estimator of π. The following expression is used to calculate P.

SAMPLE PROPORTION

$$P = \frac{\text{Number of observations in category}}{\text{Sample size}}$$

If a sample of $n = 100$ engineers contains 16 civil engineers, then the sample proportion of persons in that discipline is $P = \frac{16}{100} = .16$. A sample of 37 silicon chips might contain three unusable ones, so the sample proportion unusable is $P = \frac{3}{37} = .08$. The manufacturing process for the same chips might consistently produce unusable chips at a lower rate, so the population proportion of unusables might be only $\pi = .05$.

The proportion is an important decision parameter in a wide variety of statistical applications. It is the central focus of many quality-control decisions, where equipment must be adjusted if the proportion of defectives is too high. A major concern in chemical processing is the proportion of volume containing impurities (often expressed as a percentage or in parts per million). The proportion also is an important descriptor of various systems in which the proportion of time in a particular state is crucial.

Example: *Evaluating equipment maintenance policies*

The proportion played a key role in one engineer's evaluation of maintenance policies for papermaking equipment. Under the present policy, random observations provided the following:

Equipment State	Number of Observations	Proportion
Satisfactory	78	$\frac{78}{135} = .578$
Erratic	42	$\frac{42}{135} = .311$
Inoperable	15	$\frac{15}{135} = .111$
	135	1.000

The engineer combined the above proportions with cost information to determine the net cost of the present maintenance schedule, under which a complete adjustment was done whenever the equipment was found erratic. A second maintenance policy of "benign neglect," with repairs made only when the equipment was inoperable, was similarly evaluated. The proportion of time in the latter state was found to be so high that the proposed plan would have been more costly than the present one.

Problems

2–42 The following sample data apply to positions held by recent civil engineering graduates:

	Government (1)	Construction (2)	Design Firm (3)	Graduate School (4)	Other (5)
Men	28	42	17	15	21
Women	17	8	9	17	15

(a) Calculate the sample proportion of women in each position category.
(b) Which category has proportionally the lowest concentration of women?
(c) Which category has proportionally the highest concentration of women?

2–43 Refer to the completion time data in the calibration inspection illustration in Table 2–1. Determine the sample proportion of times (a) below 14.0 seconds, (b) below 16.0 seconds, (c) greater than or equal to 15.0 seconds, and (d) greater than or equal to 18.0 seconds.

2–44 An inspector scraps any production batch yielding a sample proportion of defectives exceeding .10. For each of the following batches (a) determine the sample proportion defective, and (b) indicate whether the batch will be saved or scrapped.

	Batch				
	(1)	(2)	(3)	(4)	(5)
Items inspected	50	75	100	150	200
Items defective	8	3	21	7	15

2–45 Refer to the student data in Problem 2–7. Determine the sample proportion of students who are (a) electrical engineers, (b) graduates, (c) receiving financial aid, (d) undergraduate electrical engineers, and (e) electrical engineers receiving financial aid.

2–46 Refer to the student data in Problem 2–7. Determine the following:
(a) The sample proportion of undergraduate students who receive financial aid.
(b) The sample proportion of graduate students who receive financial aid.
(c) Does the sample evidence indicate that graduates involve a higher or lower proportion of students on financial aid than do undergraduates?

2–47 A plant initiates a policy that the machinery must be adjusted whenever the proportion of broken parts in any hour exceeds 2% of total production during that span. In a particular hour the following findings were achieved:

	Machine			
	(1)	(2)	(3)	(4)
Number of broken parts	5	23	12	3
Number of satisfactory parts	512	624	387	432

(a) Determine the population proportion of broken parts for each machine.
(b) Which machines will be adjusted?

Review Problems

2–48 The number of lost days per month of drilling experienced by a sample of offshore platforms during 20 months are given below:

0	5	7	4	4	10	5	2	1	0
2	3	3	6	5	4	4	2	0	1

Determine values for the following sample statistics: (a) mean, (b) median, (c) mode, (d) range, (e) variance, (f) standard deviation, (g) coefficient of variation, and (h) coefficient of skewness.

2–49 The following sample data for completing the framing of a standard tract house are incomplete:

Time (days)	Number of Houses	Relative Frequency	Cumulative Frequency
5–under 6	—	—	—
6–under 7	19	—	27
7–under 8	—	.35	—
8–under 9	15	.15	—
9–under 10	—	—	91
10–under 11	—	—	—
Totals	100	1.00	

(a) Determine the missing values.
(b) Plot the cumulative frequency distribution for these data.

2–50 Suppose that your statistics instructor plans to give five quizzes during the term and that your grade will be based on the computed measure of central tendency, which you get to select now. Would you prefer the mean or the median? Explain.

2–51 The following sample data represent the annual maintenance costs (in thousands of dollars) for heavy earth-moving equipment:

24	21	20	31	25
41	43	32	17	25
32	30	30	26	34
36	39	27	28	31
21	19	25	26	31

(a) Construct a table for the sample frequency distribution using intervals of width 5 thousand dollars, with 15 thousand dollars as the lower limit of the first interval.
(b) Plot a histogram and frequency polygon for these data.

2–52 Refer to the data in Problem 2–51:
(a) Construct a stem-and-leaf plot.
(b) Construct an ordered stem-and-leaf plot.
(c) Using the original ungrouped data, find the following sample statistics: (1) mean, (2) standard deviation, and (3) mode.
(d) Using the original ungrouped data, find the following percentiles: (1) 10th, (2) 25th, (3) 75th.

(e) What is the value of the sample median?
(f) Compute the following sample statistics: (1) range, (2) interquartile range.
(g) Construct the box plot.

2–53 Refer to the data presented in Problem 2–51. Answer the following:
(a) From your grouped data, determine the value of (1) the sample mean, (2) the sample mode.
(b) From your grouped data, determine the value of (1) the sample variance and (2) the sample standard deviation.

2–54 Refer to the data presented in Problem 2–51. Answer the following:
(a) Determine the relative frequencies for each class interval.
(b) Determine the cumulative relative frequency for each class interval.
(c) Plot your results from (b). Then determine approximately from your graph the proportion of maintenance costs falling below (1) 23 thousand dollars and (2) 32 thousand dollars.
(d) Determine from your graph the approximate costs below which the proportion of observed values is (1) .25, (2) .50, (3) .75, and (4) .90.

2–55 A sample of projects has been categorized by size and by duration. The following data apply:

Size	Duration	Number of Projects
Small	Short	12
Medium	Short	23
Large	Short	8
Small	Long	18
Medium	Long	15
Large	Long	5

Construct a table for the frequency distribution for the samples characterized by (a) size and (b) duration.

2–56 Refer to the sample data in Problem 2–55. Determine the sample proportion of projects that are (a) small, (b) large, (c) small and short, and (d) large and long.

2–57 A chemical engineer fine-tuned a chemical process by experimenting with control settings. The following completion times (in minutes) were obtained for sample runs, with all settings unchanged during a single day.

Day 1	Day 2	Day 3	Day 4	Day 5
33	34	38	35	38
51	37	30	39	39
22	44	35	37	37
26	28	33	37	38
	27	42		
		32		

The engineer has two objectives: (1) to minimize variability in completion times and (2) to minimize average completion time.

(a) Determine for each day the following sample statistics: (1) mean, (2) standard deviation, (3) first quartile, (4) median, (5) third quartile.
(b) Compute for each day the sample (1) coefficient of variation and (2) coefficient of skewness.
(c) Construct on the same graph a box plot for each day's completion times.
(d) Comment on how near the engineer has come to finding settings that meet his goals.

CHAPTER THREE

Statistical Process Control

One of the most important statistical applications for engineers is **statistical quality control**. This importance is due to the nature of engineering itself, which is largely concerned with the creation of things, the operation of processes, or the design of public works and facilities. End products must adhere to the engineer's specifications, as should the components and materials from which they are fabricated. The design engineer depends on others to create the final product, and high quality is expected.

Quality control is ordinarily associated with manufacturing, and one of the main responsibilities of industrial engineers is to ensure that proper procedures are being employed to monitor production and to remedy any unusual problems that might prevent the final product from being satisfactory. Although the industrial engineer plays the terminal role, all engineers involved with the creation of any final product should be concerned with its quality.

Statistical process control is used to monitor a production process or an operation to determine whether or not it is operating as designed, that is "in control." For this purpose **control charts** are used to monitor current production on a "real time" basis. This chapter presents the methodology of statistical process control.

3–1 The Control Chart

Historically, statistical quality-control procedures have been developed mainly for the manufacturing environment, especially when production volumes are high. These methods permit early detection and speedy correction of trouble. Because uncorrected problems could result in the scrapping of thousands of finished units, the potential cost savings from preventive measures can be huge. In manufacturing applications of statistical quality control the primary focus is the production *process* itself. The disposition of those items already produced is an important but secondary consideration.

Problems are detected indirectly after a sample establishes at a given time that the process either is **in control** or is not. This is true as long as the observed units fall within reasonable bounds, which are predetermined in accordance with what would be expected from a process that is functioning as intended. They allow for the same kinds of chance variations that would arise in any sampling situation.

Successive samples falling within the limits are expected to differ from each other, and any variation is assumed to be caused by *chance*. Should a sample fall outside those limits, it is then presumed that there is some **assignable cause**. Such an occurrence is a signal for action, and a rigorous search should be made for the actual cause. It could be traced to a variety of sources, such as malfunctioning machinery, an inattentive operator, a poor batch of raw materials, or an environmental disturbance.

Statistical Preliminaries

Figure 3–1 sketches the theoretical underpinnings of statistical process control. The pictured case applies when using the mean to signal whether the process is under control. It depicts a series of samples taken at different times from the underlying population, whose frequency curve is shown at the top of Figure 3–1. The mean $\bar{X}$ will be computed for each of these. Those computed values are treated as a distinct data set, apart from the population. That collection of sample means will have its own frequency distribution, pictured as the bottom curve in Figure 3–1. The frequency curve for the $\bar{X}$s will generally have a bell shape (regardless of the shape of the parent population's frequency

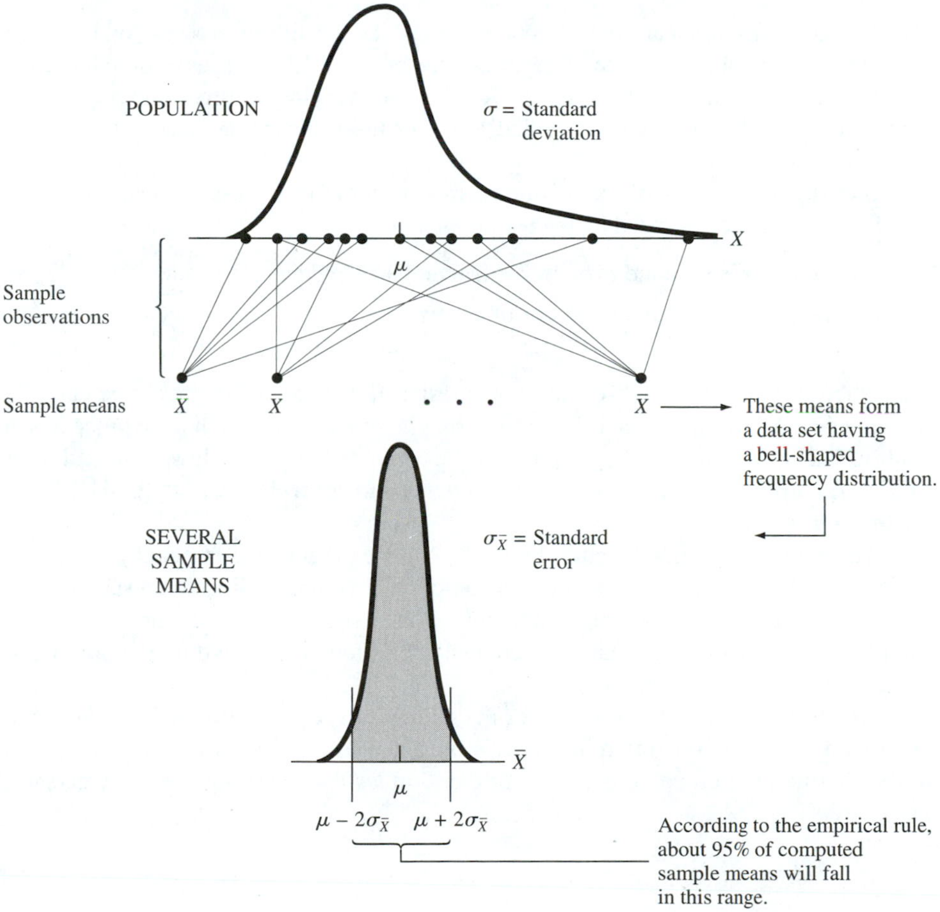

Figure 3–1 *Theoretical underpinnings of statistical processs control with the mean.*

curve). Its central value μ coincides with that of the population, but the $\bar{X}$ data set has a smaller standard deviation.

The smaller standard deviation for the $\bar{X}$s should be no surprise, since we would expect a set of means to show less variability than the individual population values from which they were generated. To distinguish it from the population standard deviation σ, the summary measure of $\bar{X}$ variability is called the *standard error* of the sample mean. To avoid confusing that quantity with σ itself and the standard deviation of the sample (s), it is represented symbolically as $\sigma_{\bar{X}}$. Later in this book we will see that $\sigma_{\bar{X}}$ can be expressed as a mathematical function of σ and the sample size n. Like the population parameters μ and σ, whose true values remain *unknown* throughout the statistical control process, $\sigma_{\bar{X}}$ can only be estimated from sample results by making simple calculations.

The empirical rule tells us that about 95% of the computed $\bar{X}$s should lie within $\mu \pm 2\sigma_{\bar{X}}$ and that over 99% will fall inside $\mu \pm 3\sigma_{\bar{X}}$. As long as the underlying population is stable, very few computed means will ever occur outside these limits. Remedial action is triggered when such an unusually large or small $\bar{X}$ is obtained.

Concept of Statistical Control

To illustrate the statistical control concept, consider the filling process for beverage cans. The process objective is to place an average of $\mu_0 = 12.0$ ounces of fluid into each can. It is inherent for such a process to yield cans of varying volume. The process design recognizes this through a specified level for the standard deviation of $\sigma_0 = .10$ ounce per can.

Statistical control involves a periodic assessment of the process by means of a *sample*. Two actions are triggered by the result obtained:

1. The process is found to be **in control** and no remedies are taken.
2. The process is found to be **out of control** and a search is conducted for assignable cause, with necessary steps taken to remedy the problem.

The action depends on the observed sample result. In the case of the beverage cans the *sample mean* $\bar{X}$ is used to establish, at any particular time, whether or not the process is in control. That is accomplished by comparing the computed mean to a **lower control limit** LCL and an **upper control limit** UCL. The process is **in control** if LCL $\leq \bar{X} \leq$ UCL and out of control otherwise.

The concepts are illustrated in Figure 3–2. Case (a) shows a frequency curve that *might* apply for the current population of can-content volumes. Since only sample cans are precisely measured, the true population characteristics remain unknown. Using a sample of $n = 6$ cans, the mean $\bar{X}$ is computed and found to lie within the prescribed control limits.

At another time a sticky valve results in an average overfilling of the cans, and μ shifts upward, as illustrated in case (b). But the presence of a problem is unknown to the foreman until a value is found for the $\bar{X}$ in a subsequent sample. A large sam-

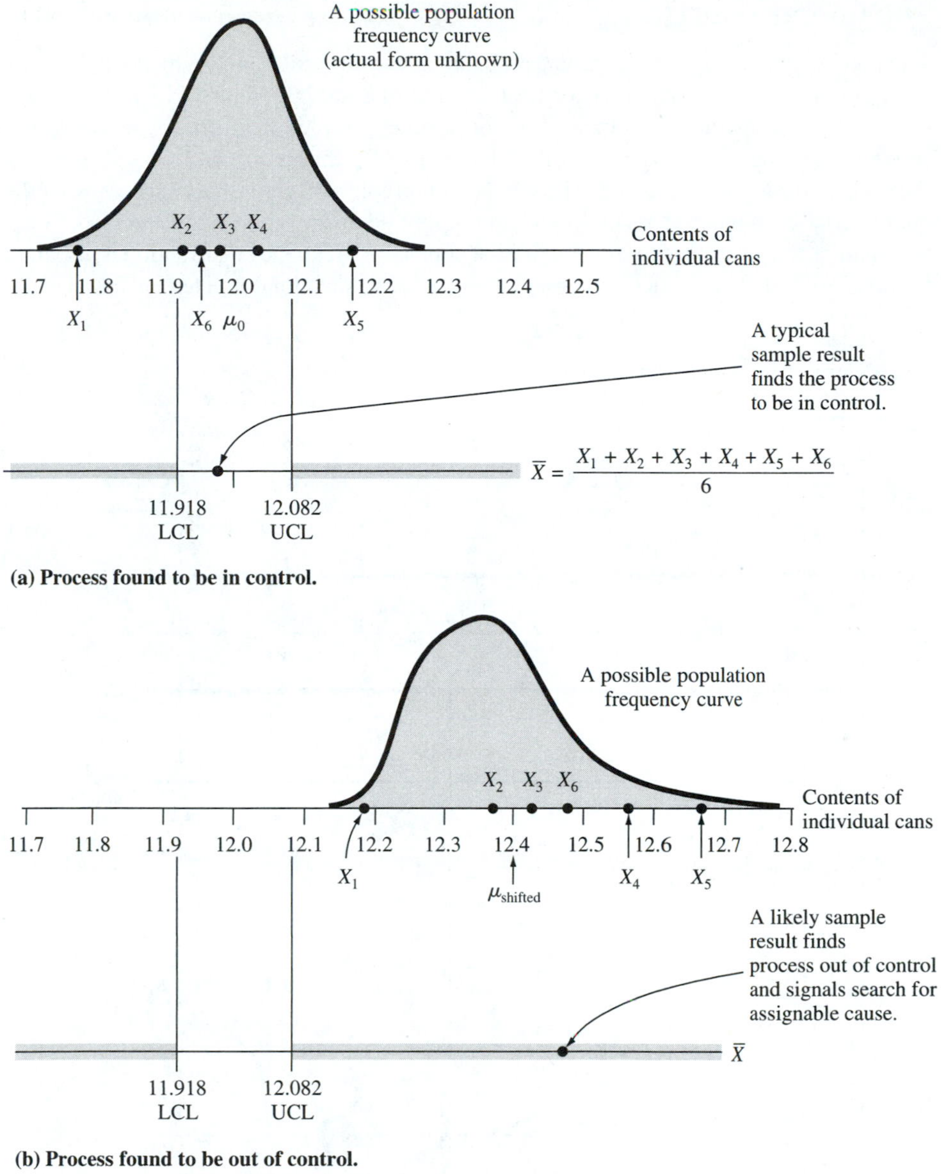

Figure 3–2
Two cases for statistical process control for filling beverage cans.

ple mean is consistent with a process that is overfilling, and since the computed $\bar{X} > \text{UCL}$, a search for cause is launched. The foreman finds and corrects the problem. But the process mean μ remains unknown—both before and after the repair. A subsequent sample finds the process to be in control.

The Control Chart

The mechanism for detecting quality problems is the **control chart**, which summarizes on a graph the results obtained for a succession of samples. Figure 3–3(a) shows the control chart constructed from a series of several samples from the beverage-filling process. Horizontal lines represent the LCL and UCL. In Section 3–2 we will see how the control limits are computed. The companion graph in Figure 3–3(b) shows what the characteristics of the underlying population might be during each sample period.

Notice that μ shifts over time, while the population frequency distribution is also fluid in form. Such shifts are often minor and may be due to such random elements as changes

Figure 3–3
Control chart illustrating unknown true population characteristics.

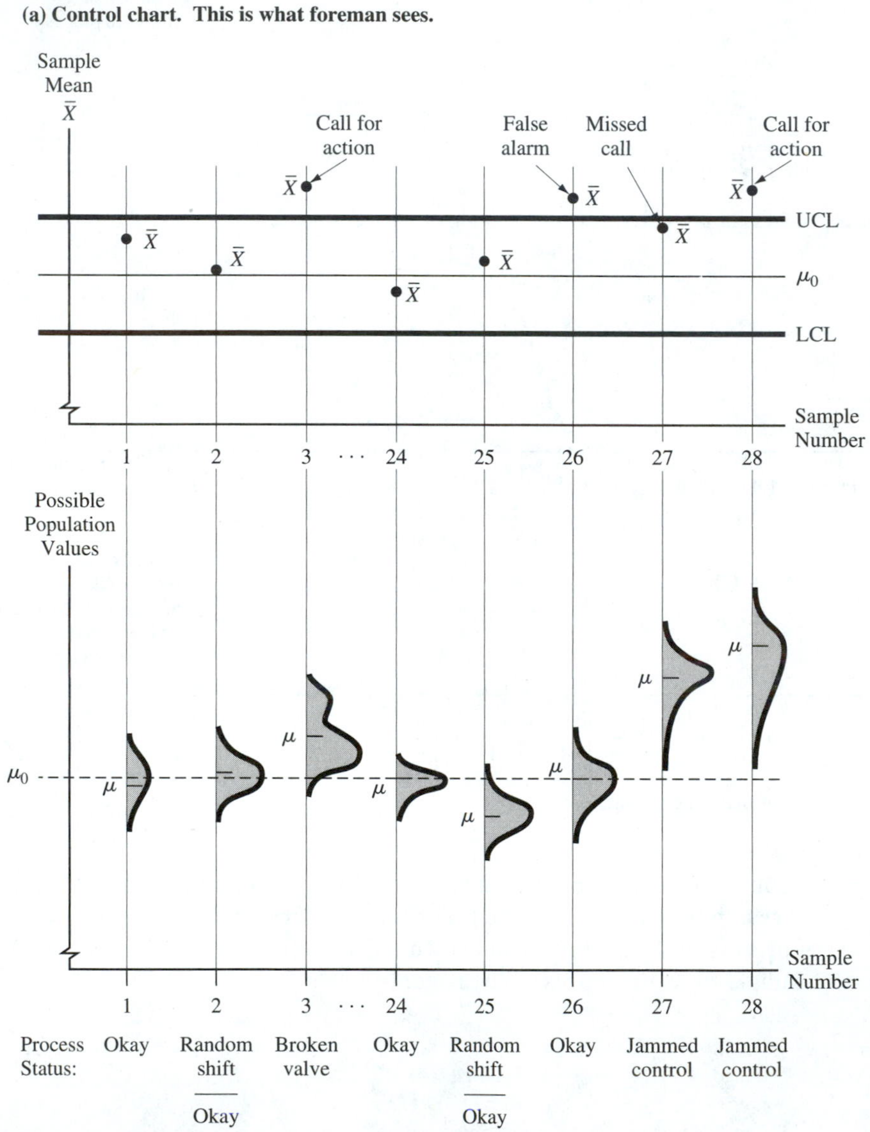

(b) Possible populations at time of sample. This is what may actually be happening.

in temperature and liquid viscosity. Chart (b) is not part of the statistical control procedure and appears here only to explain what *might* give rise to what we actually see. (Remember, the population details remain *unknown*, and we get to see only the *sample* results.)

During Period 1 the process is operating exactly as intended. Chart (b) sketches the population frequency curve indicating how individual can contents will vary due to expected random causes. The sample mean is likely (but not certain) to fall between the LCL and UCL. In Period 2 μ shifts upward, possibly because the beverage batch has slightly different sugar levels. The true level of μ remains unknown to the foreman, who sees only the computed $\overline{X}$, the level of which places the process in control at that time. In Period 3 there is a broken control valve (causing the population frequency distribution to be bimodal). After a search for cause, triggered by the high level of $\overline{X}$ signaling the process was out of control, the valve failure was discovered and a new one installed.

Figure 3–3 shows the continuation of the control chart at Periods 24 and 25, both of which were found to be in control. In Period 26 the process was also operating satisfactorily, but the sample mean was large enough to find the process out of control. The search for assignable cause found nothing to remedy during that time, and the foreman concluded that the sample had triggered a *false alarm*. In Period 27 μ shifts dramatically up. But the sample yields an untypically low value (for the population as it existed *then*), and $\overline{X}$ fell within (LCL, UCL), signaling that the process was in control. This was a *missed call* for needed action. Period 28 yielded a large $\overline{X}$ that finally found the process to be out of control, triggering a search for assignable cause, a jammed control.

The selection of control limits is motivated by a need to balance the cost of monitorship with the incidence of false alarms and missed calls. Figure 3–4 shows two extreme cases. If the limits are too far apart, there will be a long lag between the onset and discovery of a problem, and the number of missed calls will be excessive, as shown in (a). If the limits are too close, the incidence of false alarms, shown in (b), will be unacceptably high. A combination of judgment and statistical theory is used in selecting acceptable control limits that meet cost limitations.

Types of Control Charts A variety of control charts may be constructed. Section 3–2 describes in detail control charts for quantitative data. There we will see how to compute the LCL and UCL for the mean, used when the focus is placed on *central tendency*. But *variability* is an important element of quality control as well. We will see how control limits may be found for the range and standard deviation. Section 3–3 describes control charts for qualitative data using the sample proportion.

Two major approaches are described for constructing control charts. The first approach is based on *given specifications*. The second finds limits using only *historical data*, when no predetermined standards are in place. That approach would be suited to a new operation whose standards are evolving. Once standards are determined, they may be used directly in a first approach to finding consistent control limits.

Problems

3–1 A process for filling jars is designed to operate with a mean of 16.0 ounces and standard deviation of .5 ounce. The control limits for the sample mean have been established at LCL $=$ 15.59 and UCL $= 16.41$ for samples of size $n = 6$. Indicate for each of the following sample results whether or not the process should be found in control.

(a) 16.5, 16.2, 16.1, 17.0, 16.6, 16.4
(b) 15.5, 15.9, 16.1, 16.0, 15.6, 15.4
(c) 17.0, 16.2, 15.9, 16.3, 16.4, 14.9
(d) 16.4, 15.5, 15.8, 16.1, 16.8, 15.2

Figure 3–4
Two extreme cases for control limits.

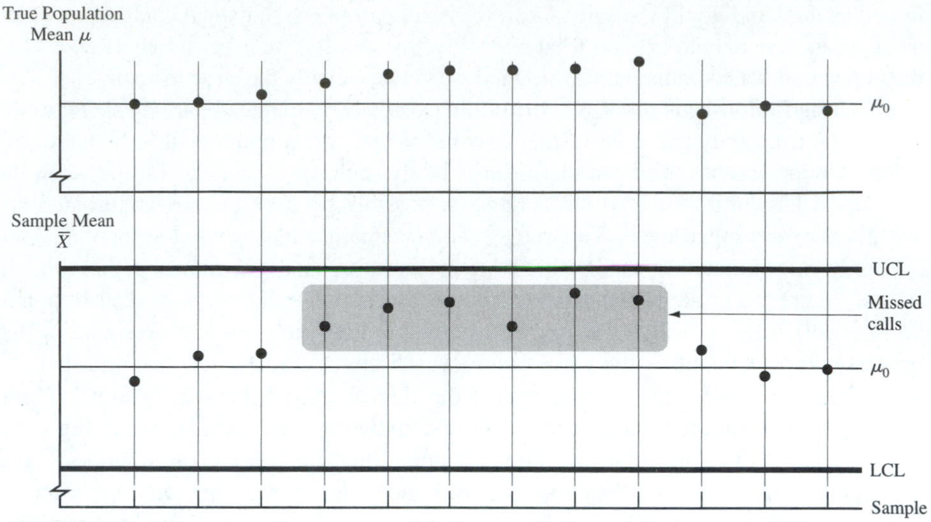

(a) Control limits that are too far apart.

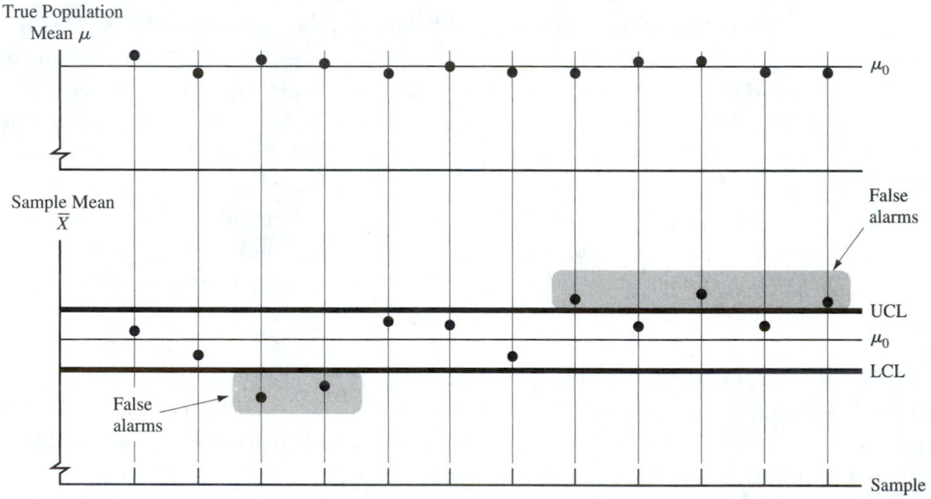

(b) Control limits that are too close together.

3–2 Consider the process in Problem 3–1. A second test focuses on process variability. In this case the sample standard deviation s is used, with LCL $= .168$ and UCL $= .784$. Indicate for samples (a)–(d) whether or not the process should be found in control.

3–3 Consider the following situations. Briefly describe how statistical process control might be implemented to improve operations.

(a) The local post office advertises: "Nobody must wait more than five minutes." The postmaster wants to keep waiting time small, but needs a more realistic slogan.

(b) The file server at a university computer laboratory needs to be purged of inactive storage whenever turnaround times are excessive.

(c) An instrumentation maintenance foreman needs to know when to overhaul transits.

(d) A taxi fleet is maintained on an as-needed basis. Suggest how miles per gallon might be used to determine when to tune an engine.

3–2 Control Charts for Quantitative Data

Control charts based on the *sample mean* are universal. Two types of control charts may be used in monitoring variability. One of these is based on the *sample standard deviation*. Although the standard deviation is very familiar to people with any formal statistical education, it presents conceptual problems to people with little or no statistical background, and it is cumbersome to compute. For those reasons, an alternative control chart that is based on the *sample range* is frequently used.

Computing Control Limits for the Mean Using Specifications

The least complicated case for computing control limits applies when process standards are known. The following expressions are used in that case.

CONTROL LIMITS FOR THE MEAN USING SPECIFICATIONS

$$\text{LCL} = \mu_0 - SL\frac{\sigma_0}{\sqrt{n}} \qquad \text{UCL} = \mu_0 + SL\frac{\sigma_0}{\sqrt{n}}$$

where the following parameters apply:

$\mu_0 =$ specified level for population mean

$\sigma_0 =$ specification for population standard deviation

$SL =$ sigma limit (2 or 3)

The specified level μ_0 for the population mean serves as the central value. The lower control limit lies a fixed distance below the center, with the upper control limit falling the same distance above that value. The distance used depends upon (1) the number of sigma limit SL and (2) the standard error for the sample mean, here $\sigma_{\bar{X}} = \sigma_0/\sqrt{n}$. The commonly used sigma limits are $SL = 2$ or 3, which—according to the empirical rule—will provide limits within which 95 percent or 99.7 percent, respectively, of all $\bar{X}$s will lie.

We illustrate the computation of these limits with the contents of beverage cans discussed earlier, for which the specified population mean is $\mu_0 = 12.0$ and $\sigma_0 = .10$ ounce per can. Using $SL = 2$ we have

$$\text{LCL} = 12.0 - 2(.10/\sqrt{6}) = 11.918$$

$$\text{UCL} = 12.0 + 2(.10/\sqrt{6}) = 12.082$$

When the beverage-filling process is operating at the standard, about 95.5% of the computed $\bar{X}$s will fall between the control limits. But about 4.5% of them will fall either below the LCL or above the UCL, triggering false alarms. The incidence of false alarms can be controlled by using wider intervals, which may be achieved by using $SL = 3$ instead.

But the incidence of missed calls, occurring when μ_0 shifts, tends to be higher when the control limit range is greater. Judgment must be used in determining which level of *SL* brings the better balance. In either case the incidence of false alarms and missed calls both can be reduced by increasing the sample size *n*, but that will increase cost.

Control Limits for the Range and Standard Deviation Using Specifications

The following expressions apply for computing the control limits for the range and standard deviation.

CONTROL LIMITS USING SPECIFICATIONS

Range:

$$\text{LCL} = d_2\sigma_0 - SLd_3\sigma_0 \qquad \text{UCL} = d_2\sigma_0 + SLd_3\sigma_0$$

Standard Deviation:

$$\text{LCL} = c_4\sigma_0 - SLc_5\sigma_0 \qquad \text{UCL} = c_4\sigma_0 + SLc_5\sigma_0$$

where the following apply:

$\mu_0 =$ specification for population mean

$\sigma_0 =$ specification for population standard deviation

$SL =$ sigma limit (2 or 3)

with constants of proportionality: d_2 d_3 c_4 c_5

The central values for both sets of control limits involve σ_0 and constants of proportionality: d_2 and d_3 for the range; c_4 and c_5 for the standard deviation. Appendix Table K lists values for these constants, all of which depend on the sample size used. The same considerations noted with the mean apply in selecting *SL* and the sample size *n*.

Notice that the central value for the *R*-chart is $d_2\sigma_0$, reflecting the need to convert σ_0 into a range measure while also incorporating the role of the sample size *n* through the constant d_2. Analogously, the central value for the *s*-chart $c_4\sigma_0$, the average level for individual *s* values, varies with the sample size, as reflected in c_4.

Control limits for the range are computed for the beverage-filling process. With $n = 6$, Table K provides $d_2 = 2.534$ and $d_3 = .848$. Using the same population parameter standard deviation as before and $SL = 2$, we compute the control limits for the *R*:

$$\text{LCL} = 2.534(.10) - 2(.848)(.10) = .084$$
$$\text{UCL} = 2.534(.10) + 2(.848)(.10) = .423$$

Figure 3–5(a) shows the control chart for the same observations made earlier. The *R* values were computed from the same sample data as the $\overline{X}$s used in the earlier control chart. The process was found to be out of control due to excessive variability in Period 3 (caused by a broken valve) and in Period 28 (caused by a jammed control).

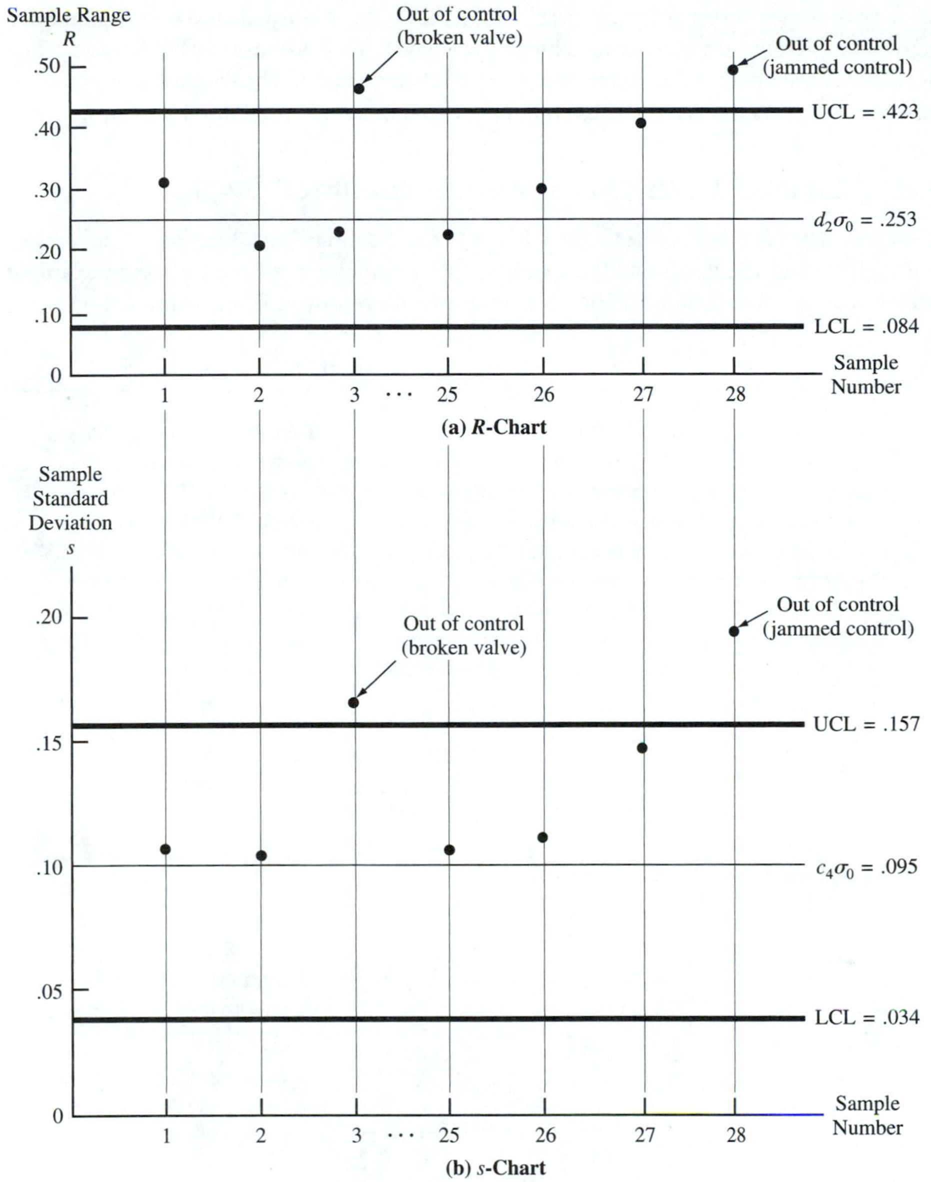

Figure 3–5
Range and standard deviation control charts for beverage-filling process.

Although these findings coincide with those signaled by the $\overline{X}$-chart, the two procedures may each uncover problems not identified by the other, since central tendency specifications can be met when variability is in violation, and vice versa.

Control limits for the standard deviation are also computed for the beverage-filling process. With $n = 6$, Table K provides $c_4 = .952$ and $c_5 = .308$, so that with $SL = 2$, we compute the control limits for s:

$$\text{LCL} = .952(.10) - 2(.308)(.10) = .034$$

$$\text{UCL} = .952(.10) + 2(.308)(.10) = .157$$

Using the same data as before, Figure 3–5(b) shows the control chart for s. The findings agree completely with those reached with the R-chart. That will usually be the case, since R and s are alternative measures of the same characteristic. The R-chart is favored by many because ranges are easier to compute by hand than are standard deviations.

Using Control Charts to Uncover Process Instability

Control charts may be helpful even when no specifications have been set. Indeed, useful applications are not limited to manufacturing situations. As the following example suggests, these tools can be helpful for processes far removed from traditional ones.

Example: *Control Charts and Bribery in Jai Alai*

One interesting application of control charts took place in a Florida trial involving bribery charges against jai alai bettors and players. Some of those players had been suspected of taking bribes as enticements to purposefully lose their matches—especially during the earlier matches in a session. A corrupted player's influence most directly affects combination bets, such as the quiniela (won by picking the top two players, in any order). Treating the total number of quiniela bets placed as the process variable, the $\bar{X}$ and R control charts in Figure 3–6 were constructed. The investigators examined the raw data and judgmentally selected those sessions beginning with the 30th as a period of stability. The control limits were established using only the data from those later sessions.

Figure 3–6 *Control charts for number of jai alai quiniela bets used as evidence in bribery trial.*

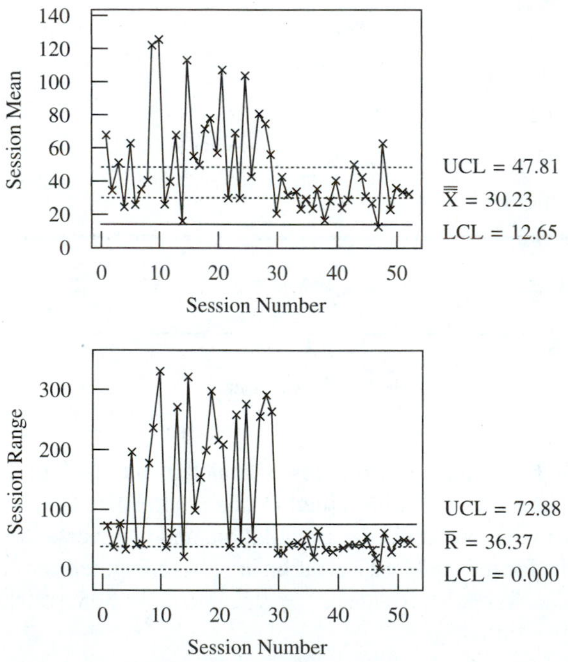

The jury and other key persons in the trial of a bribery suspect were struck by the strength of the evidence of unusually strong and erratic betting in the earlier sessions. Unbeknownst to the statistical investigators, the defendant had been arrested after the 29th jai alai session, so that his influence stopped then. The game-fixing conspirator was convicted of bribery soon thereafter.

Source: John M. Charnes and Howard S. Gitlow, "Using Control Charts to Corroborate Bribery in Jai Alai," *The American Statistician*, November 1995, pp. 386–89.

Control Charts for the Mean When No Specifications Are Given

New processes are often launched before specifications have been established for monitoring quality. In those cases control charts may still be used to uncover problems, but the sample data themselves are used in establishing the control limits. The following expression is used in finding these for the mean.

CONTROL CHART LIMITS FOR THE MEAN
(NO SPECIFICATIONS)

$$\text{LCL} = \bar{\bar{X}} - SL\frac{\hat{\sigma}}{\sqrt{n}} \qquad \text{UCL} = \bar{\bar{X}} + SL\frac{\hat{\sigma}}{\sqrt{n}}$$

where $\bar{\bar{X}}$ = grand sample mean

SL = sigma limit (2 or 3)

with the estimated standard deviation

$$\hat{\sigma} = \frac{\bar{R}}{d_2}$$

computed using

$\bar{R}$ = mean of sample ranges

d_2 = range conversion constant

Illustration: *Controlling Thickness in Rolled Sheet Metal*

A detailed illustration will help us explain the concepts and procedures. An industrial engineer for a steel mill has implemented a procedure for controlling the quality of rolled sheet metal. A major product is sheets one-twentieth of an inch thick. Although individual thicknesses will vary within the same sheet and from sheet to sheet, no sheet should deviate much from the target dimension. The true population mean μ for individual sheet thickness is unknown and may be estimated from the sample data. The standard deviation σ of that population may also be estimated from the sample.

In initiating the quality-control procedure the baseline data in Table 3–1 were collected. Every hour, a sample of $n = 5$ sheets is selected and measured for thickness. The mean $\bar{X}$, range R, and standard deviation s are computed for each sample.

Table 3–1 *Thickness of Rolled Steel Sheets*

Sample Number	Average Thickness of Each Sheet in Sample (inches)					Mean $\overline{X}$	Range R	Standard Deviation s
1	.0494	.0489	.0508	.0485	.0501	.0495	.0023	.0009
2	.0514	.0520	.0510	.0469	.0485	.0500	.0051	.0022
3	.0481	.0504	.0486	.0496	.0500	.0493	.0023	.0010
4	.0508	.0477	.0481	.0487	.0458	.0482	.0050	.0018
5	.0491	.0443	.0533	.0498	.0480	.0489	.0090	.0032
6	.0458	.0508	.0492	.0478	.0526	.0492	.0068	.0026
7	.0473	.0527	.0494	.0511	.0521	.0505	.0054	.0022
8	.0525	.0492	.0492	.0475	.0495	.0496	.0050	.0018
9	.0533	.0524	.0516	.0490	.0487	.0510	.0046	.0021
10	.0502	.0502	.0513	.0497	.0514	.0506	.0017	.0008
11	.0507	.0485	.0456	.0475	.0485	.0482	.0051	.0018
12	.0509	.0477	.0520	.0486	.0479	.0494	.0043	.0019
13	.0467	.0516	.0510	.0498	.0507	.0500	.0049	.0019
14	.0503	.0471	.0490	.0515	.0490	.0494	.0044	.0016
15	.0506	.0485	.0515	.0498	.0494	.0500	.0030	.0011
16	.0501	.0495	.0475	.0475	.0487	.0487	.0026	.0012
17	.0518	.0499	.0499	.0471	.0521	.0502	.0050	.0020
18	.0479	.0506	.0520	.0506	.0493	.0501	.0041	.0015
19	.0481	.0480	.0517	.0512	.0503	.0499	.0037	.0017
20	.0479	.0506	.0492	.0523	.0502	.0500	.0044	.0016
21	.0521	.0497	.0523	.0516	.0521	.0516	.0026	.0011
22	.0475	.0478	.0494	.0524	.0517	.0498	.0049	.0022
23	.0499	.0488	.0511	.0473	.0456	.0485	.0055	.0022
24	.0500	.0500	.0492	.0502	.0509	.0501	.0017	.0006
25	.0483	.0502	.0476	.0468	.0489	.0484	.0034	.0013
						1.2411	.1068	.0423

Using $SL = 2$, the control chart for the mean is plotted in Figure 3–7, after 25 periods of sample rolled steel thickness data were obtained. Those data were used to find the control limits.

To estimate μ, the sample results are pooled, and the **grand mean** $\overline{\overline{X}}$ is computed by averaging the individual sample means. Since the grand mean estimates the central value about which individual $\overline{X}$s will cluster, $\overline{\overline{X}}$ provides the centerline of the $\overline{X}$-chart. For the rolled sheets we have

$$\overline{\overline{X}} = \frac{1.2411}{25} = .04964 \text{ inch}$$

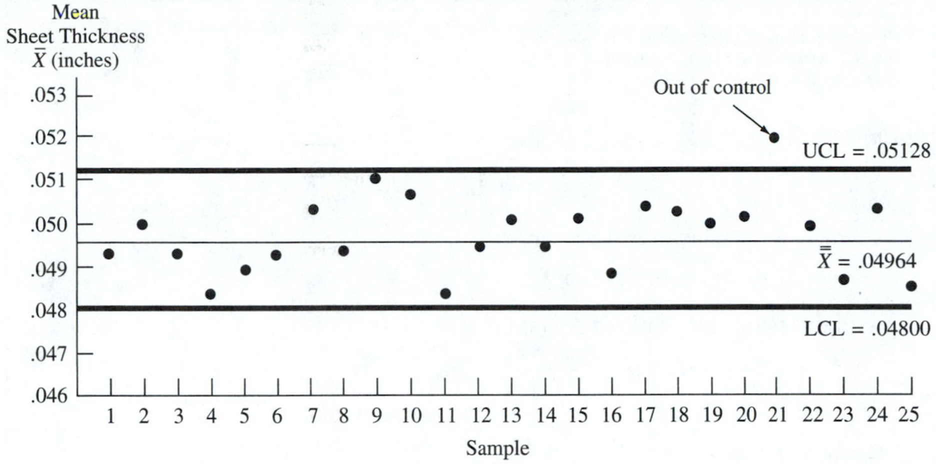

Figure 3–7
Control chart for thickness of rolled steel sheets obtained using the mean ($SL = 2$).

Thus, the process mean μ is estimated to be .04964 inch, and the centerline for the $\bar{X}$-chart in Figure 3–7 is placed at a height of .04964. The estimated standard deviation for X, denoted as $\hat{\sigma}$, is determined from the samples by first averaging the individual sample ranges and using the constant of proportionality d_2 seen earlier. Values for d_2 are provided in Appendix Table K. For $n = 5$, $d_2 = 2.326$.

For the rolled sheet thickness data we have

$$\bar{R} = \frac{.1068}{25} = .00427 \text{ inch}$$

so that

$$\hat{\sigma} = \frac{.00427}{2.326} = .001836 \text{ inch}$$

Using these values we compute the control limits.

$$\text{LCL} = .04964 - 2(.001836/\sqrt{5}) = .04964 - .00164 = .04800$$

$$\text{UCL} = .04964 + 2(.001836/\sqrt{5}) = .04964 + .00164 = .05128$$

Sample 21 exceeds the upper control limit, and the assignable cause proved to be a control that needed to be recalibrated.

Control Limits for the Range and Standard Deviation When No Specifications Are Given

Analogous expressions provide the control limits for the range and standard deviation.

CONTROL CHART LIMITS FOR THE RANGE AND STANDARD DEVIATION (NO SPECIFICATIONS)

Range:

$$\text{LCL} = \bar{R} - SL\hat{\sigma}_R \qquad \text{UCL} = \bar{R} + SL\hat{\sigma}_R$$

where $\bar{R}$ = mean of sample ranges
SL = sigma limit (2 or 3)

with the estimated standard deviation

$$\hat{\sigma}_R = \bar{R}\left(\frac{d_3}{d_2}\right)$$

Standard Deviation:

$$\text{LCL} = \bar{s} - SL\hat{\sigma}_s \qquad \text{UCL} = \bar{s} + SL\hat{\sigma}_s$$

where $\bar{s}$ = mean of sample standard deviations
SL = sigma limit (2 or 3)
with the estimated standard deviation

$$\hat{\sigma}_s = \bar{s}\left(\frac{c_5}{c_4}\right)$$

computed using constants of proportionality d_2, d_3, c_4, c_5

Using the sample data from Table 3–1 for rolled steel thickness, the control chart in Figure 3–8(a) for the sample range was constructed. The control limits are computed from the above expression using the sample data. Appendix Table K provides $d_2 = 2.326$, as before, and $d_3 = .864$. Again using $SL = 2$, we have for the R-chart

$$\hat{\sigma}_R = .00427(.864/2.326) = .00159$$
$$\text{LCL} = .00427 - 2(.00159) = .00109$$
$$\text{UCL} = .00427 + 2(.00159) = .00745$$

Sample 5 was found to be out of control (caused by an erratic sensor switch).

The control chart for the sample standard deviation is constructed in Figure 3–8(b) using the same sample results. The control limits are based on a central value computed by averaging the individual sample standard deviations:

$$\bar{s} = .0423/25 = .00169$$

Appendix Table K provides $c_4 = .940$ and $c_5 = .341$. Again using $SL = 2$, we have for the s-chart

$$\hat{\sigma}_s = .00169(.341/.940) = .00061$$
$$\text{LCL} = .00169 - 2(.00061) = .00047$$
$$\text{UCL} = .00169 + 2(.00061) = .00291$$

As we would expect, exactly the same sample results are identified as out of control as found with the R-chart.

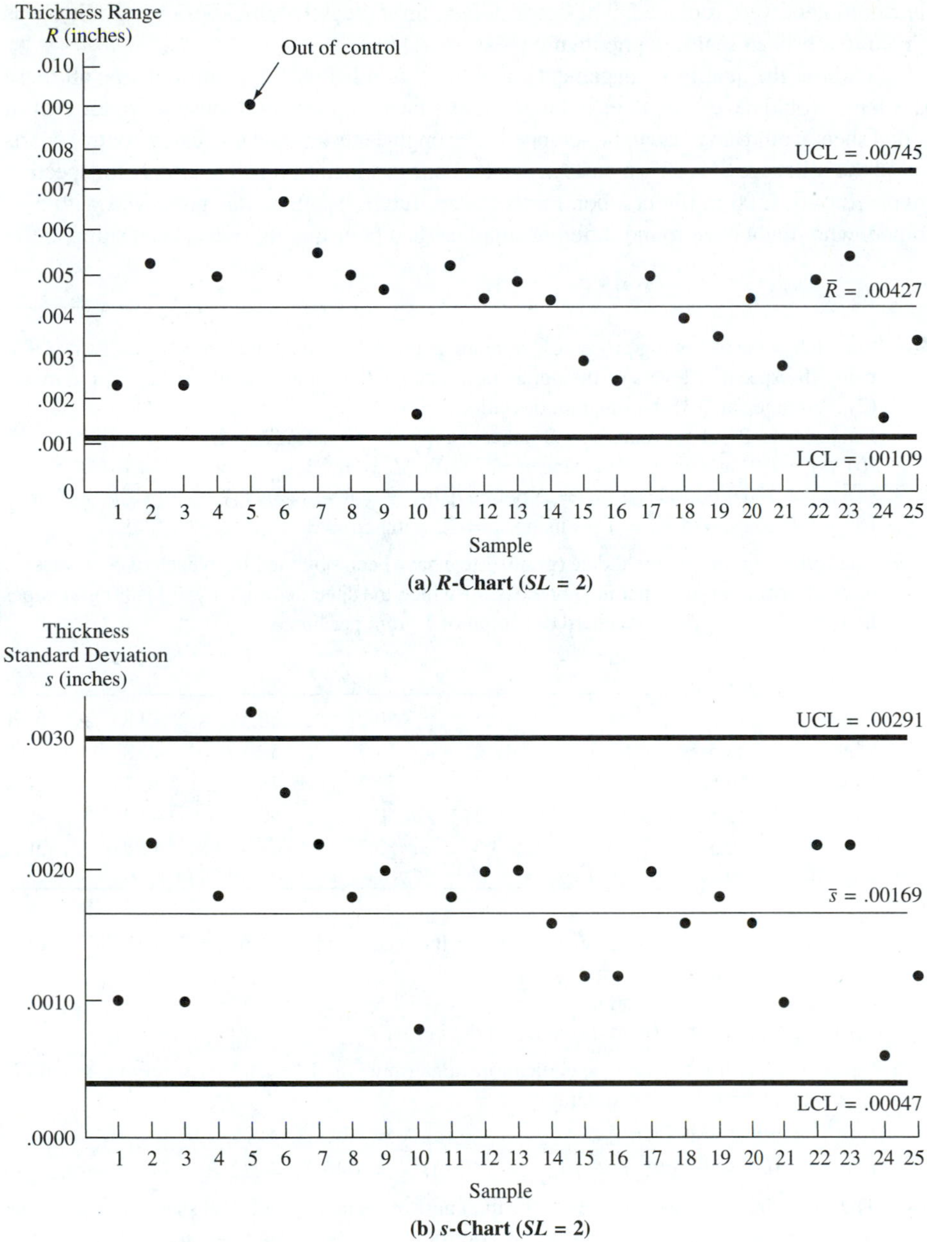

Figure 3–8 *Control charts for thickness of rolled steel sheets obtained using R and s.*

Implementing Statistical Control

The industrial engineer kept a log of the quality-control findings during a 200-hour run with $\frac{1}{20}$-inch rolled steel sheets. In that time $\overline{X}$ exceeded the UCL twice and fell below the LCL once. The problems were promptly remedied by adjusting roller settings and replacing a malfunctioning scanner. In two separate samples R exceeded its UCL. In one case many of the starting ingots were too cool, a condition corrected when holding-oven thermometers were replaced. The other defect, an increase in thickness variability, was traced to a broken seal in a pneumatic press.

Without the quality management capability provided by the control charts, all these problems would have been detected many hours later, and several thousand square feet of rolled sheet would have had to be scrapped. The engineer estimated that using control charts saved the mill over $50,000 in costs in just this one production run. The costs of inspection involved only $500 in direct labor. Furthermore, future business was preserved with customers who might have found different suppliers had poor-quality sheets been shipped.

Problems

3–4 For each of the following processes, compute at the indicated sigma limits the LCL and UCL using the specified levels of the population mean and standard deviation for (1) the mean, (2) the range, and (3) the standard deviation.

(a) $\mu_0 = 16.0$ ounces and $\sigma_0 = .2$ ounce using $n = 10$ with $SL = 2$

(b) $\mu_0 = 1.0''$ and $\sigma_0 = .01''$ using $n = 7$ with $SL = 3$

(c) $\mu_0 = 100.0$ pounds and $\sigma = .5$ pound using $n = 4$ with $SL = 2$

(d) $\mu_0 = 30$ psi and $\sigma_0 = 1$ psi using $n = 12$ with $SL = 3$

3–5 The following sample yield data (grams/liter) have been obtained from each of 5 successive days of operation of a chemical process for which the specified mean yield is 50 grams per liter of feedstock with a standard deviation of 1 gram per liter.

Day 1:	49.5	49.9	50.5	50.2	50.5	49.8	51.1
Day 2:	48.5	52.3	48.2	51.2	50.1	49.3	50.0
Day 3:	50.5	51.7	49.5	51.2	48.3	50.2	50.4
Day 4:	49.8	49.7	50.2	50.6	50.3	49.4	49.3
Day 5:	50.5	50.9	49.5	50.2	49.8	49.8	50.3

(a) Construct the 2-sigma control limits for the mean. Indicate on which days (if any) the process was out of control.

(b) Repeat (a) using instead the range.

(c) Repeat (a) using instead the standard deviation.

3–6 Using 3-sigma limits and no specifications, determine the LCL and UCL for (1) $\bar{X}$ and (2) R in each of the following situations:

(a) $\bar{\bar{X}} = .10$, $\bar{R} = .005$, $n = 5$

(b) $\bar{\bar{X}} = .005$, $\bar{R} = .001$, $n = 10$

(c) $\bar{\bar{X}} = 5.62$, $\bar{R} = .13$, $n = 4$

(d) $\bar{\bar{X}} = 1.46$, $\bar{R} = .22$, $n = 7$

3–7 The following sample data apply to the thickness of $\frac{1}{20}$-inch rolled steel sheets. Compute the sample mean for each. Then, using the control limits from Figure 3-7, indicate whether or not $\bar{X}$ places the process in control.

(a) .0505	.0475	.0516	.0452	.0555
(b) .0504	.0485	.0492	.0515	.0502
(c) .0523	.0515	.0531	.0528	.0541
(d) .0488	.0435	.0505	.0562	.0512
(e) .0453	.0479	.0462	.0474	.0458

3–8 Using the data in Problem 3–7, compute the range for each sample. Then, using the control limits from Figure 3-8(a), indicate whether or not R places the process in control.

3–9 Using the data in Problem 3–7, compute the standard deviation for each sample. Then using the control limits from Figure 3-8(b), indicate whether or not s places the process in control.

3–10 Refer to the data for the thickness of rolled steel sheets in Table 3–1. Suppose that the two out-of-control samples are removed and the control limits recalculated. Determine the new 2-sigma control limits for (a) the mean, (b) the range, and (c) the standard deviation.

3–11 The following sample data apply to diameters of $\frac{1}{4}$-inch ball bearings:

(1) .254	.262	.245	.266	.259
(2) .241	.235	.241	.231	.242
(3) .255	.245	.253	.249	.252
(4) .255	.263	.265	.271	.255
(5) .241	.249	.250	.252	.253
(6) .255	.253	.247	.244	.248
(7) .235	.260	.253	.242	.259
(8) .254	.246	.246	.252	.260
(9) .243	.254	.256	.252	.246
(10) .259	.257	.253	.246	.249

(a) Compute the mean and range for each sample. Then find the 3-sigma control limits for $\bar{X}$ and construct the control chart, including the data point for each sample. Indicate for which samples (if any) the process is out of control.

(b) Find the control limits for R and construct the control chart, including the data point for each sample. Indicate for which samples (if any) the process is out of control.

(c) Eliminating the samples where the process was found to be out of control in parts (a) and (b), determine the revised control limits for (1) $\bar{X}$ and (2) R.

3–3 Control Charts for Qualitative Data Using the Proportion

The underlying principles of control charts apply to qualitative data, placing an item or event into one of two categories, such as defective or nondefective, late or on time, and operable or inoperable. At any time the process output is governed by the *population proportion* π of undesirable items or events. Like μ or σ with quantitative data, the level of π remains unknown throughout the process. Statistical control is determined in these applications by the *sample proportion P*.

Statistical control procedures using P parallel those based on $\bar{X}$. There are two ways for computing control limits: (1) when the level for π is specified and (2) when no specification is given.

Control Limits for the Proportion with Given Specifications

When the level for π is specified, the following expressions apply for computing the control limits.

> CONTROL LIMITS FOR THE PROPORTION WITH GIVEN SPECIFICATIONS
>
> $$\text{LCL} = \pi_0 - SL\sqrt{\frac{\pi_0(1-\pi_0)}{n}} \qquad \text{UCL} = \pi_0 + SL\sqrt{\frac{\pi_0(1-\pi_0)}{n}}$$
>
> where the following apply:
>
> SL = sigma limit (2 or 3)
>
> π_0 = specified level for population proportion

Analogously to the earlier expressions, the central value is the specified level π_0 for the population proportion. The lower control limit is that value minus SL times an expression involving π_0 and n. The latter quantity is the standard deviation for P. (We shall see in later chapters why that particular expression applies.)

To illustrate, suppose that an electroplating facility checks finished products for significant blemishes. The specifications are that only 5% of production be allowed to have blemishes, so that $\pi_0 = .05$ is the specified level for the population proportion blemished. Samples of $n = 25$ items are removed periodically from production and categorized as blemished or satisfactory. Using $SL = 2$, the following control limits are computed.

$$\text{LCL} = .05 - 2\sqrt{.05(1-.05)/25} = .05 - .087 = -.037 \text{ or } 0$$
$$\text{UCL} = .05 + 2\sqrt{.05(1-.05)/25} = .05 + .087 = .137$$

The sample proportion P is compared to these limits, and the process will be found in control whenever $\text{LCL} \leq P \leq \text{UCL}$.

It may seem a bit odd that there is a lower control limit. After all, why would it be a problem if there are too few blemished units in the sample? In many cases there is no problem if P is small, and in those situations the process is found out of control only when $P > \text{UCL}$. However, abnormally low levels for P could signal that there is a problem with the inspection process itself, so that defective items are going undetected, or that production staff are trying too hard to avoid defectives and are being careless about cost control. And of course, there is always the serendipitous possibility that the low-P sample will lead to the discovery of a factor whose presence or absence might actually improve the process.

Another consideration may be the large range defined by the control limits. Tighter limits can be achieved, but only by paying the price of using larger sample sizes. For example, if the sample size is raised to $n = 100$, the following limits would instead apply.

$$\text{LCL} = .05 - 2\sqrt{.05(1-.05)/100} = .05 - .044 = .006$$
$$\text{UCL} = .05 + 2\sqrt{.05(1-.05)/100} = .05 + .044 = .094$$

Control Limits for the Proportion When No Specifications Are Given

A more elaborate procedure applies when no specifications are given. The control limits are computed from the following expression.

CONTROL LIMITS FOR THE PROPORTION
WITH NO SPECIFICATIONS GIVEN

$$\text{LCL} = \bar{P} - SL\sqrt{\frac{\bar{P}(1-\bar{P})}{n}} \qquad \text{UCL} = \bar{P} + SL\sqrt{\frac{\bar{P}(1-\bar{P})}{n}}$$

where

$$\bar{P} = \frac{\text{Combined number unsatisfactory}}{\text{Combined sample size}}$$

$SL =$ sigma limit (2 or 3)

The above replaces π_0 by an estimate of π computed from the sample. That quantity is the **combined sample proportion** $\bar{P}$.

Illustration: *Unsatisfactory Memory Chips*

An industrial engineer at a microelectronics manufacturing plant is setting up statistical quality-control procedures for a new high-density computer memory chip. Production begins September 1 on a limited scale, with volume gradually increasing thereafter. The engineer wants to minimize the incidence of defective chips. There are many ways for a chip to be unsatisfactory—fuzzy lines, uneven layering, breaks, poor texturing, and so on. Rather than accounting separately for each possibility, the testers will simply classify each chip as satisfactory or defective.

The engineer selects a random sample of $n = 1,000$ chips daily for the entire month. Table 3–2 shows the results obtained. These data were used in constructing the control chart for the proportion of defective chips, shown in Figure 3–9.

The abscissa of the control chart represents successive samples. In the present illustration, there is one sample in each successive day. The ordinate shows possible levels for the sample proportion defective. The value of this statistic computed for each sample is plotted as a dot.

Notice that the dots fluctuate around a central value, the combined proportion defective $\bar{P}$. This value is computed from the data in Table 3–2, treating the 30 days' observations as a single sample of size 30,000:

$$\bar{P} = \frac{\text{Combined number of defectives}}{\text{Combined sample size}} = \frac{1{,}907}{30{,}000} = .0636$$

Table 3–2 *Sample Data from Production of Silicon Memory Chips*

September Date	Number of Defectives	Proportion Defective *P*	September Date	Number of Defectives	Proportion Defective *P*
1	41	.041	16	49	.049
2	75	.075	17	66	.066
3	77	.077	18	45	.045
4	56	.056	19	54	.054
5	55	.055	20	42	.042
6	108	.108	21	74	.074
7	127	.127	22	49	.049
8	57	.057	23	70	.070
9	46	.046	24	45	.045
10	135	.135	25	44	.044
11	47	.047	26	72	.072
12	72	.072	27	51	.051
13	76	.076	28	43	.043
14	61	.061	29	57	.057
15	59	.059	30	54	.054
				1,907	

For the silicon memory chip illustration, the following control limits apply:

$$\begin{aligned}\text{LCL} &= .0636 - 3\sqrt{\frac{.0636(1 - .0636)}{1{,}000}} \\ &= .0636 - 3(.0077) \\ &= .0405\end{aligned}$$

and

$$\begin{aligned}\text{UCL} &= .0636 + 3(.0077) \\ &= .0867\end{aligned}$$

These limits are represented by the boldface horizontal lines shown on the control chart in Figure 3–9.

Using the Control Chart

In our memory chips example three points lie above the UCL—.108, .127, and .135—and it is assumed that the process was out of control on the corresponding dates—September 6, 7, and 10. Since no control limits had yet been established on those dates, the underlying production problems went undetected. (A later investigation showed that several key personnel were ill on the three days.) These three out-of-control data points must be eliminated from the database before the control chart can be used in actual production decisions.

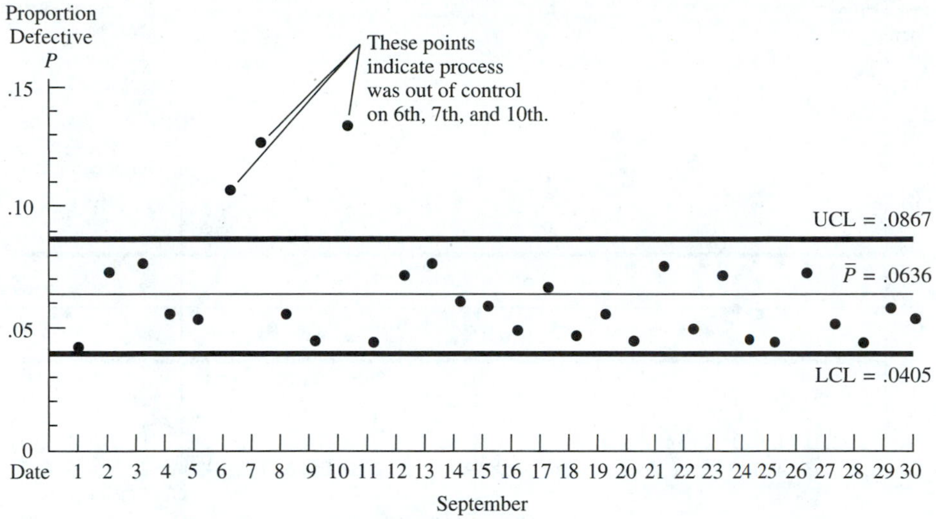

Figure 3–9
Initial control chart for proportion of defective memory chips.

Figure 3–10 shows the revised control chart, which was then used to monitor October production. The combined proportion defective is recomputed to be $\bar{P} = \frac{1{,}537}{27{,}000} = .0569$. This proportion necessitates revised control limits: LCL $= .0349$ and UCL $= .0789$.

During the first week in October, the following observations were made:

Date	Proportion Defective P	Assumed Status
10/01	.055	In control
10/02	.067	In control
10/03	.075	In control
10/04	.093	Out of control
10/05	.053	In control
10/06	.044	In control
10/07	.049	In control

The sample results for October 4 indicated that an inordinately high proportion of defective chips was being produced. A thorough investigation of the chips in the sample for that day showed a very high incidence of cracked layers. The cause was traced to a malfunctioning thermostat in one of the baking ovens, which was replaced.

Problems

3–12 Using 3-sigma limits, determine the UCL and LCL for P in each of the following situations with the specified level:

(a) $\pi_0 = .10, \quad n = 100$
(b) $\pi_0 = .05, \quad n = 1{,}000$
(c) $\pi_0 = .07, \quad n = 500$
(d) $\pi_0 = .08, \quad n = 200$

Figure 3–10
Final control chart for proportion of defective memory chips.

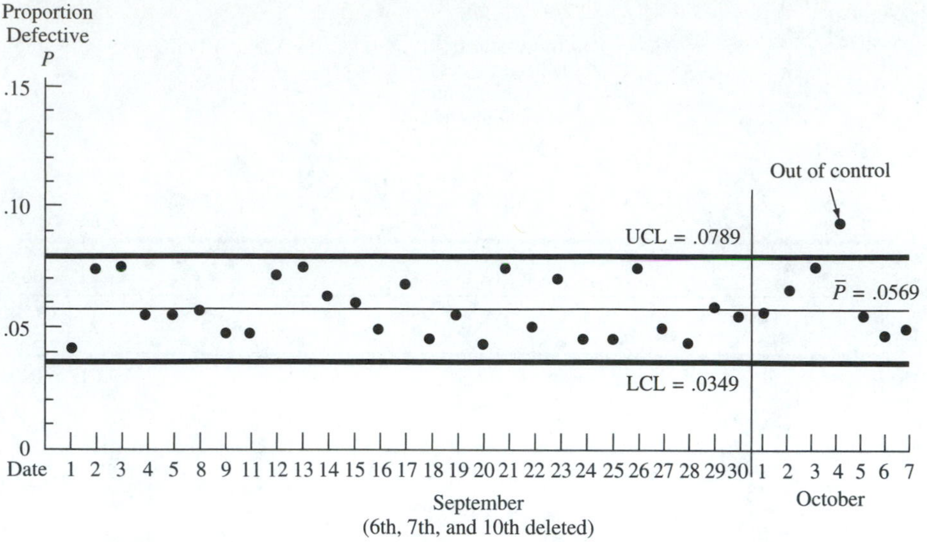

3–13 The following numbers of defectives were found in successive samples of 1,000 items taken from a production process:

7/01	75	7/06	109	7/11	75	7/16	52
02	64	07	111	12	74	17	54
03	75	08	63	13	63	18	61
04	48	09	71	14	58	19	63
05	49	10	48	15	53	20	59

(a) Determine the control limits for the sample proportion. Construct the 3-sigma control chart and plot the points for these data.
(b) Identify those dates when the process was out of control.
(c) Determine the new control limits applicable after removing the out-of-control data points. Indicate for each of the following numbers of defectives found in samples of size 1,000 whether the process was or was not in control:
(1) 95 (2) 35 (3) 55 (4) 63 (5) 64

3–14 Determine the 2-sigma control limits for the sample proportion, using the data in Table 3–2. Then indicate on which dates the process was out of control.

3–15 The following numbers of defectives were found in successive samples of 100 items taken from a production process:

8/06	8	8/11	12	8/16	22	8/21	11
07	15	12	3	17	13	22	7
08	12	13	9	18	10	23	15
09	19	14	14	19	15	24	24
10	7	15	10	20	18	25	2

Answer parts (a)–(b) as in Problem 3–13.

3–16 Determine the 2-sigma control limits for the sample proportion, using the data in Problem 3–13. Then indicate on which dates the process was out of control.

3–4 Further Issues in Statistical Quality Control

Statistical process control is but one aspect of the broad topic of statistical quality control. With manufacturers everywhere needing to compete in a global economy, product quality has become an issue of paramount importance like never before. Quality applications have become one of the most dynamic areas of statistics. Two of these statistical quality control procedures are described here.

Acceptance Sampling

The objective of acceptance sampling is determining how to dispose of items that have already been produced or that have been provided by a supplier or vendor. The main consideration is the lot, the characteristics of which are unknown and cannot be completely observed. Whether the lot is accepted or rejected will depend on the results found from a sample. The key issue is that some of the time a good lot will be rejected or a bad lot will be accepted. Those errors cannot be totally avoided when the final action is based on a sample. Chapter 11 discusses procedures for achieving a proper balance between these two types of errors while achieving overall efficiency in the resources expended.

Optimization of Quality Loss Function: Taguchi Method

The control chart comes into play during production, long after a product has been designed. New thinking criticizes manufacturers who rely exclusively on such procedures because the quality emphasis is thereby placed at the very end of the product cycle rather than at the beginning. Modern thinking moves some of the quality control emphasis to the product design phase, where attention is given to making the final product robust to potential forces that might detract from final quality. One man, Genichi Taguchi, has so eloquently spoken on behalf of that approach, that a key statistical procedure for doing this is often referred to as the Taguchi method.

Taguchi advocates using a quality loss function to help direct the product design effort. A good design is one that optimizes this function by minimizing deviations from a targeted ideal. All of this is accomplished by using sample data taken from experiments in which various levels of design parameters are systematically tried out. His method is a special case of experimental design, which is presented in Chapter 14 as a statistical tool that can be used not only in enhancing final product quality but for basic performance criteria as well.

CHAPTER FOUR

Making Predictions: Regression Analysis

SO FAR OUR sampling applications have focused on a single population, with observed values denoted by X. In this chapter we extend sampling to incorporate a second population, represented by Y. Our present concern is predicting values for one population on the basis of observations taken from the other. Such predictions are important in engineering when Y cannot be observed directly; in such cases its value must be forecast from a known value for X. For example, the unknown time Y to process a raw chemical might be expressed in terms of the known density X of the starting solution. By measuring and timing sample batches of the chemical, a relationship may be established between these two quantities, such as $Y = 9.5 + .45X$. The process is called *regression analysis*.

Engineers encounter some of the elements of regression analysis whenever they fit a curve to experimental data. But regression analysis has a wider scope than "curve-fitting," primarily because it is a rigorous procedure that controls prediction errors. A considerable refinement to the freehand fairing of a curve, the *method of least squares* is the central regression technique that minimizes the collective deviations of the individual observations from the resulting regression surface. An accompanying mathematical function, or *regression equation*, completely specifies the resulting line or curve. This feature ensures that any two analysts can achieve identical results from the same given data and assumed shape—permitting greater objectivity and making the curve-fitting more of a science than an art.

4–1 Linear Regression Using Least Squares

The basic application is **linear regression**, wherein Y is related to X by a straight line, $Y = a + bX$. After collecting a set of sample data points (X, Y), an investigator arithmetically combines those observations using the method of least squares, which finds the slope b and Y-intercept a providing the best fit. Because it is just a sample summary, the resulting expression is referred to as the **estimated regression equation**. The regression line may then be used to predict the level for the **dependent variable** Y that best corresponds to a particular setting of the **independent variable** X. The computed Y-intercept and slope are only estimates of the true line parameters. Since a and b will vary from one sample to the next, they are referred to as **estimated regression coefficients**.

Illustration:* *Workstation Data Entry

Figure 4–1 shows how a line might provide a satisfactory basis for predicting the unknown time Y required to complete a workstation entry involving a known number of inputs X. The plotted points are **raw data** obtained from a random sample of 25 entries made at computer terminals by materials-control clerks in an electronics assembly facility. Such a graphical display of sample data is sometimes called a **scatter diagram**. The method of least squares was used to determine the regression line fitting the observed data.

The Method of Least Squares

The **method of least squares** finds that particular line where the aggregate deviation of the data points above or below it is minimized. Rather than measuring each point's separation in terms of the physical distance, the procedure is instead based upon *vertical* deviations, which are then squared. This not only eliminates difficulties in measuring perpendicular line segments, but it provides summary statistics having desirable properties.

To distinguish the actual and predicted values of the dependent variable, we use Y_i to denote the ith observation and $\hat{Y}(X)$ to denote a predicted value for a given level of X. For the ith observation the independent variable value is X_i, and

$$\hat{Y}(X_i) = a + bX_i$$

The goal of least-squares regression is to find the values of a and b that minimize the following:

$$\sum_{i=1}^{n}[Y_i - \hat{Y}(X_i)]^2 = \sum_{i=1}^{n}[Y_i - (a + bX_i)]^2$$

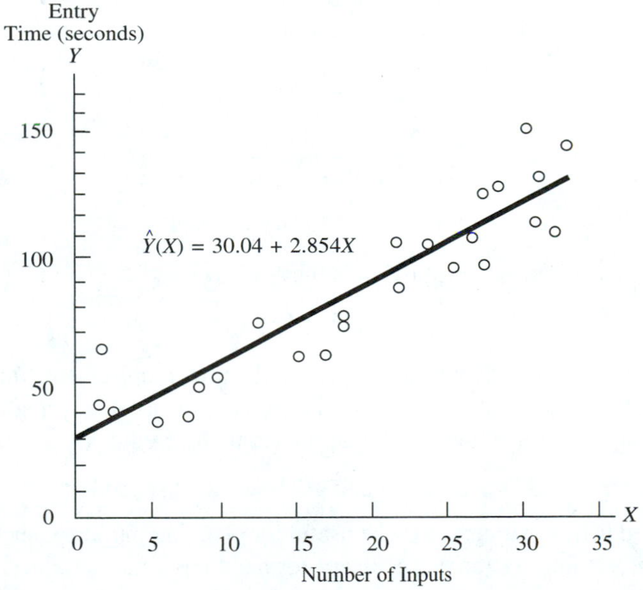

Figure 4–1
Scatter diagram with regression line for workstation data entry.

Taking the partial derivatives of this expression with respect to a and to b, and then setting the resulting terms equal to 0 and simplifying, we obtain the following.

NORMAL EQUATIONS FOR LINEAR REGRESSION

$$\sum_{i=1}^{n} Y_i = na + b \sum_{i=1}^{n} X_i$$

$$\sum_{i=1}^{n} X_i Y_i = a \sum_{i=1}^{n} X_i + b \sum_{i=1}^{n} X_i^2$$

A simultaneous solution of the normal equations provides us with the following expression for the slope b. (For simplicity the subscripts are dropped.)

$$b = \frac{n \sum XY - \sum X \sum Y}{n \sum X^2 - (\sum X)^2}$$

It is easiest to express a in terms of the value found for b by transforming the first normal equation, getting

$$a = \frac{1}{n}\left(\sum Y - b \sum X\right)$$

We may simplify this by using the fact that $\bar{X} = (\sum X)/n$ and $\bar{Y} = (\sum Y)/n$ and use the resulting expressions to compute the following.

ESTIMATED LINEAR REGRESSION COEFFICIENTS

$$b = \frac{n \sum XY - \sum X \sum Y}{n \sum X^2 - (\sum X)^2}$$

$$a = \bar{Y} - b\bar{X}$$

To illustrate, consider the raw workstation data in Table 4–1. Substituting the intermediate calculations into the above, we find the following estimated regression coefficients:

$$b = \frac{25(52{,}670) - (506)(2{,}195)}{25(13{,}130) - (506)^2} = 2.854$$

$$a = 87.80 - 2.854(20.24) = 30.04$$

The estimated linear regression equation is thus

$$\hat{Y}(X) = 30.04 + 2.854X$$

This regression line may be used to predict the time requirements for any particular entry for which the number of inputs X is known. For instance, should $X = 32$ inputs be necessary for controlling a particular material, the predicted entry time would be

$$\hat{Y}(32) = 30.04 + 2.854(32) = 121.4 \text{ seconds}$$

Values obtained in this manner would be useful for planning purposes, such as forecasting the data processing resources necessary for handling a new product line or an increase in orders.

Table 4–1
Raw Data for a Sample of Data Entries and Intermediate Calculations for Finding Estimated Regression Coefficients

Observation i	Number of Inputs X	Entry Time (seconds) Y	XY	X^2
1	2	66	132	14
2	19	77	1,463	361
3	6	37	222	36
4	23	106	2,438	529
5	10	55	550	100
6	23	89	2,047	529
7	9	52	468	81
8	30	128	3,840	900
9	18	63	1,134	324
10	25	104	2,600	625
11	19	76	1,444	361
12	2	44	88	4
13	27	97	2,619	729
14	28	109	3,052	784
15	8	40	320	64
16	29	124	3,596	841
17	29	98	2,842	841
18	16	63	1,008	256
19	33	131	4,323	1,089
20	3	41	123	9
21	34	111	3,774	1,156
22	32	151	4,832	1,024
23	13	76	988	169
24	33	114	3,762	1,089
25	35	143	5,005	1,225
	506 $= \sum X$	2,195 $= \sum Y$	52,670 $= \sum XY$	13,130 $= \sum X^2$

$\bar{X} = \frac{506}{25} = 20.24$ $\quad \bar{Y} = \frac{2,195}{25} = 87.80$ $\quad s_X = 10.97$ $\quad s_Y = 33.99$

Rationale and Meaning of Least Squares

Figure 4–2 will help us understand the underlying rationale for the method of least squares. Each observed data point lies above or below the regression line. The vertical deviation for the ith observation is denoted by the difference $Y_i - \hat{Y}(X_i)$. For example, the 22nd observation has coordinates (32, 151) and lies above the estimated regression line by

$$Y_{22} - \hat{Y}(32) = 151 - 121.4 = 29.6 \text{ seconds}$$

This deviation is portrayed by the heavy line segment connecting the observed point to the regression line. Each vertical deviation represents the amount of *error* associated with using the regression line to predict a new entry's time. The values $a = 30.04$ and $b = 2.854$ minimize this error.

Figure 4–2
Fitting a regression line to the workstation data.

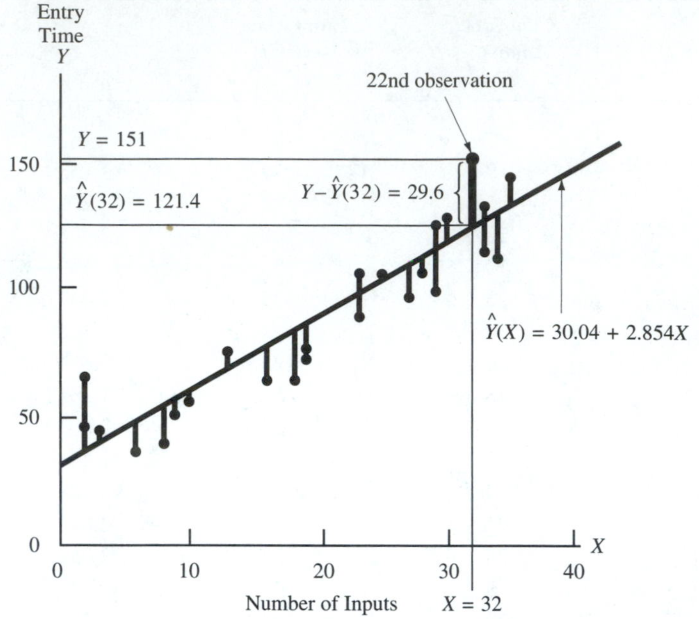

Some of the observation points fall below the regression line and provide negative deviations. It is easy to establish that

$$\sum[Y - \hat{Y}(X)] = 0$$

so the positive and negative deviations cancel. This is one reason why the procedure is based on *squared* deviations. The estimated regression line obtained for the workstation sample minimizes the sum of the squared deviations for all the data points.

The value of the slope, $b = 2.854$, indicates that the time needed for any particular entry is estimated to increase by 2.854 seconds for every additional data input required. The Y-intercept, $a = 30.04$, tells us that regardless of the number of inputs, any entry is estimated to require 30.04 seconds of setup time—including accessing the operating system, logging in and out, and other chores that do not depend on the complexity of the entry.

Knowing that a particular entry involves $X = 30$ inputs, $\hat{Y}(30) = 115.7$ seconds may be used to estimate the workstation time requirements. But the proper interpretation of that time is that *on the average* all entries involving the same number of inputs will take 115.7 seconds. This quantity is a point estimate, and individual times will vary from entry to entry, even when they appear to have identical input requirements.

Measuring the Variability of Results

The fundamental measure of variability in regression analysis is the degree of scatter exhibited by the data points about the regression surface. In seeking a summary value for this scatter, or dispersion, we may take guidance from our treatment of the single population. We have seen that the variance is a satisfactory measure of dispersion. Recall that the variance is the mean of the squared deviations of individual values about the central value. This suggests that a regression analog may be found by averaging the squared deviations about the regression *line:*

$$\frac{\sum[Y - \hat{Y}(X)]^2}{n}$$

The square root of the mean squared deviations is referred to as the **standard error of the estimate** about the regression line. We will slightly modify this expression before taking the square root, obtaining as the standard error of the estimate

$$s_{Y \cdot X} = \sqrt{\frac{\sum[Y - \hat{Y}(X)]^2}{n-2}}$$

For consistency with our past notation, we use the letter s to indicate that the calculations are based on sample data. (For theoretical reasons the divisor is n minus the number of

Table 4–2 *Calculation of the Standard Error of the Estimate for Workstation Regression*

Obs. i	Number of Inputs X	Entry Time Y	$\hat{Y}(X) = 30.04 + 2.854X$	$Y - \hat{Y}(X)$	$[Y - \hat{Y}(X)]^2$
1	2	66	35.7	30.3	918.09
2	19	77	84.3	−7.3	53.29
3	6	37	47.2	−10.2	104.04
4	23	106	95.7	10.3	106.09
5	10	55	58.6	−3.6	12.96
6	23	89	95.7	−6.7	44.89
7	9	52	55.7	−3.7	13.69
8	30	128	115.7	12.3	151.29
9	18	63	81.4	−18.4	338.56
10	25	104	101.4	2.6	6.76
11	19	76	84.3	−8.3	68.89
12	2	44	35.7	8.3	68.89
13	27	97	107.1	−10.1	102.01
14	28	109	110.0	−1.0	1.00
15	8	40	52.9	−12.9	166.41
16	29	124	112.8	11.2	125.44
17	29	98	112.8	−14.8	219.04
18	16	63	75.7	−12.7	161.29
19	33	131	124.2	6.8	46.24
20	3	41	38.6	2.4	5.76
21	34	111	127.1	−16.1	259.21
22	32	151	121.4	29.6	876.16
23	13	76	67.1	8.9	79.21
24	33	114	124.2	−10.2	104.04
25	35	143	129.9	13.1	171.61
				−0.2	4,204.86

$$s_{Y \cdot X} = \sqrt{\frac{4{,}204.86}{25-2}} = 13.52$$

regression coefficients.) The subscript $Y \cdot X$ indicates that the deviations are about the regression line that provides predicted levels for Y given X. Table 4–2 shows the calculations used to determine the value $s_{Y \cdot X} = 13.52$ for the workstation regression results.

Notice the resemblance of the foregoing expression to that for the sample standard deviation of the individual Y observations, which we designate by s_Y,

$$s_Y = \sqrt{\frac{\sum(Y - \bar{Y})^2}{n-1}} = \sqrt{\frac{\sum Y^2 - \frac{1}{n}(\sum Y)^2}{n-1}}$$

The computed sample standard deviation for the workstation data entry times is $s_Y = 33.99$. This is calculated without reference to the independent variable X. The sample standard deviation is the mean of the squared deviations about $\bar{Y}$, which is the center of the observed data. We may view s_Y as a measure of the **total variability** in Y. Ordinarily, deviations about Y will be greater than those about the regression line, so the level of s_Y will be greater than that for $s_{Y \cdot X}$. The regression will therefore provide a more reliable prediction of the mean level for Y based on a known X than what would be achieved using Y alone as the estimator for the population mean.

This feature is illustrated in Figure 4–3, in which two frequency curves are sketched. The curve on the left represents the distribution of individual Y s and has variability estimated by s_Y. On the right is a more compact frequency curve representing the vertical deviations about the regression line, with variability represented by $s_{Y \cdot X}$.

We must treat any prediction $\hat{Y}(X)$ made from the estimated regression equation as a *conditional mean* given that level for X, and not an exact value. The computed $\hat{Y}(X)$ is only an estimate of the true mean, and actual Ys may vary about $\hat{Y}(X)$. The standard error $s_{Y \cdot X}$ helps to measure that variability.

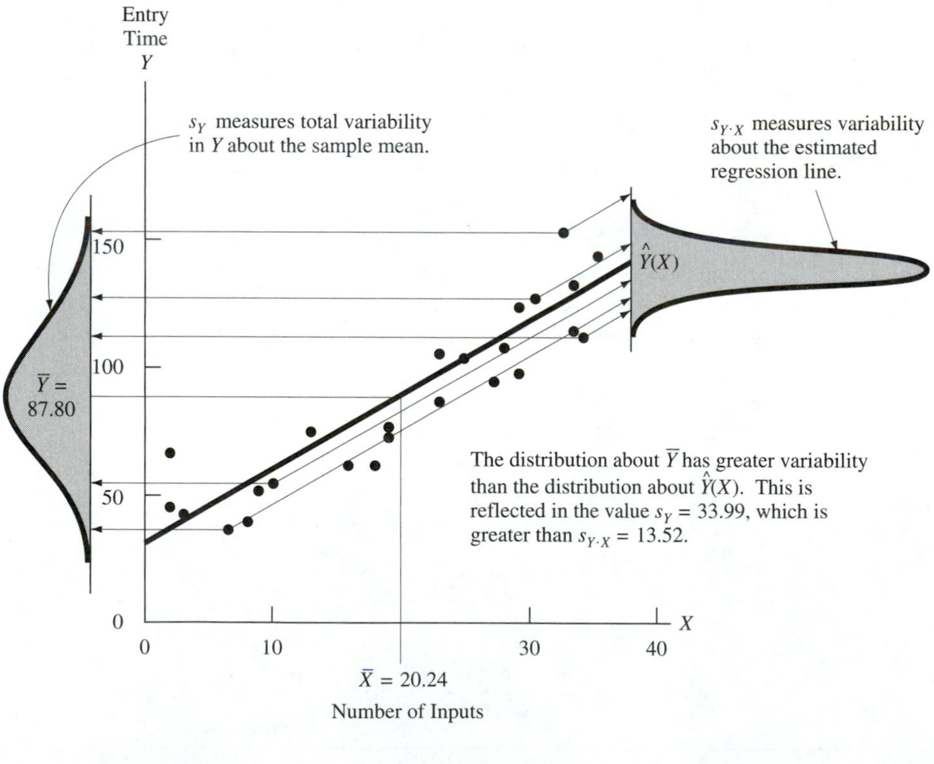

Entry Time Y
s_Y measures total variability in Y about the sample mean.
$s_{Y \cdot X}$ measures variability about the estimated regression line.
150
100
50
0
$\bar{Y}$ = 87.80
$\hat{Y}(X)$
The distribution about $\bar{Y}$ has greater variability than the distribution about $\hat{Y}(X)$. This is reflected in the value s_Y = 33.99, which is greater than $s_{Y \cdot X}$ = 13.52.
0
10
30
40
X
$\bar{X}$ = 20.24
Number of Inputs

Figure 4–3
Illustration of total variability and variability about the estimated regression line.

Computer-Assisted Regression Analysis

Hand computation of regression coefficients can be a very time-consuming chore, with a great potential for error. Plotting a scatter diagram by hand can also be tedious. Regression evaluations are therefore usually performed with computer assistance. Most statistical software packages and many general purpose data-handling programs will compute the coefficients of the estimated regression line. Figure 4–4 shows the Minitab printout for a regression run with the workstation data. These results differ slightly from the earlier computations, due to rounding errors.

```
MTB > Regress c2 1 c1.

The regression equation is
Time = 30.0 + 2.85 Inputs

Predictor        Coef       Stdev     t-ratio        p
Constant       30.040       5.762        5.21    0.000
Inputs         2.8537      0.2514       11.35    0.000

s = 13.51        R-sq = 84.9%      R-sq(adj) = 84.2%
```

Figure 4–4 *Minitab printout for regression with workstation data.*

The statistical software will even plot scatter diagrams. Figure 4–5 shows a scatter diagram for the results of a regression analysis made with the SPSS software package. The experiment was performed in conjunction with experiments concerning the design and operation of coal conversion plants. In this run the natural logarithm of viscosity is the dependent variable, with the reciprocal of temperature X as the independent variable. The raw data are provided in Table 4–3. Although the regression line is not plotted on the graph, the computer run provided the following estimated regression equation:

$$\hat{Y}(X) = -4.275 + 1{,}282.8X$$

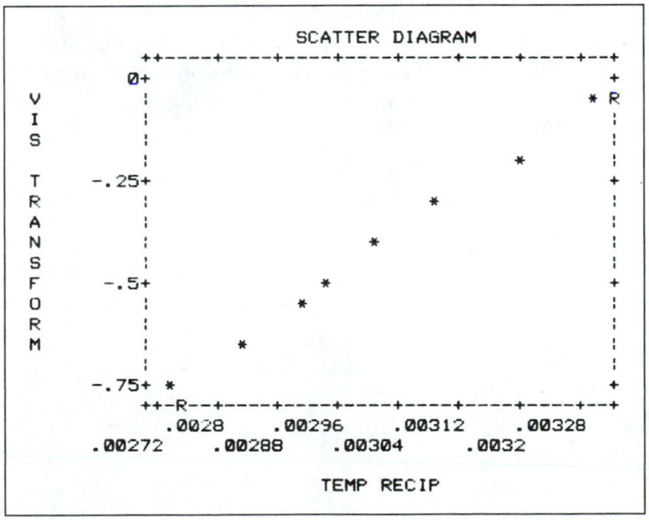

Figure 4–5 *SPSS scatter diagram for coal liquid viscosity data.*

Table 4–3
Viscosity Data for a Coal Liquid

Average Temperature (K)	Viscosity (mPa·s)	Temperature Reciprocal X (1/K)	Viscosity Transform Y (ln[mPa·s])
303.2	.9625	.0032982	−.0382212
313.1	.8373	.0031939	−.1775728
323.5	.7260	.0030912	−.3202052
323.1	.7376	.0030905	−.3043536
333.1	.6558	.0030021	−.4218994
340.1	.6009	.0029430	−.5093268
343.1	.5875	.0029146	−.5318791
353.7	.5234	.0028273	−.6474093
353.1	.5300	.0028321	−.6348783
364.6	.4685	.0027427	−.7582192

Source: Byers and Williams, *Journal of Chem. Eng. Data,* Vol. 32, 1987, pp. 349–354.

Problems

4–1 Consider the following sample results, where the number of data points X is used to predict computer processing time Y (in seconds):

X	Y	X	Y	X	Y
105	44	211	112	55	34
511	214	332	155	128	73
401	193	322	131	97	52
622	299	435	208	187	103
330	143	275	138	266	110

(a) Plot the above on a scatter diagram.
(b) Use the method of least squares to determine the expression for the estimated regression line. Then plot the line on your scatter diagram.
(c) Compute the sample standard deviation for computer processing time.
(d) Compute the standard error of the estimate for computer processing time.
(e) Determine the predicted processing time when the number of data points is
(1) 200 (2) 300 (3) 400 (4) 500

4–2 The following sample observations have been obtained by a chemical engineer investigating the relationship between weight of final product Y (in pounds) and volume of raw materials X (in gallons):

X	Y	X	Y
14	68	22	95
23	105	5	31
9	40	12	72
17	79	6	45
10	81	16	93

(a) Plot the given data on a scatter diagram.
(b) Use the method of least squares to determine the expression for the estimated regression line. Then plot the line on your scatter diagram.
(c) Compute the sample standard deviation for final product weight.
(d) Compute the standard error of the estimate for final product weight.
(e) Determine the predicted final product weight when the volume of raw material in gallons is (1) 5, (2) 10, (3) 15, and (4) 20.

4–3 The following data were obtained from a stress test on rods fabricated from an experimental alloy:

Lateral Strain ϵ_Y	Longitudinal Strain ϵ_X	Lateral Strain ϵ_Y	Longitudinal Strain ϵ_X
.0006	.002	.0035	.010
.0011	.003	.0046	.015
.0014	.004	.0069	.020
.0016	.005	.0086	.025
.0022	.006	.0102	.030

(a) Use the method of least squares to determine the estimated regression equation for predicting lateral strain from longitudinal strain.
(b) **Poisson's ratio** is the lateral strain divided by the longitudinal strain. If the true Y-intercept is zero, what is the estimated value of this quantity provided by your regression results?

4–4 An industrial engineer wishes to establish a relationship between cost per production batch of laminated wafers and size of run. The following data have been obtained from previous runs:

Size X	Cost Y	Size X	Cost Y
1,550	$17,224	786	$10,536
2,175	24,095	1,234	14,444
852	11,314	1,505	15,888
1,213	13,474	1,616	18,949
2,120	22,186	1,264	13,055
3,050	29,349	3,089	31,237
1,128	15,982	1,963	22,215
1,215	14,459	2,033	21,384
1,518	16,497	1,414	17,510
2,207	23,483	1,467	18,012

(a) Use the method of least squares to find the estimated regression equation.
(b) The *fixed cost* of a production run is the component of total expense that does not vary with quantity produced, whereas the *variable cost* is the added cost per incremental unit.

Use your answer from (a) to find for any production run (1) the fixed cost and (2) the variable cost.

(c) For a run of 2,000 units use your regression line to predict (1) the total production cost and (2) the average cost per unit.

4–5 The following data apply for a sample of engineering students:

Grade-Point Average (GPA)	Graduate Achievement Examination Score	Grade-Point Average (GPA)	Graduate Achievement Examination Score
3.3	550	3.7	650
3.9	670	2.8	450
2.7	510	2.9	520
3.1	480	3.9	710
3.2	450	2.6	420
3.8	710	3.2	570

(a) Using a student's GPA as the independent variable, find the estimated regression equation for predicting that person's achievement score.

(b) Using a student's achievement score as the independent variable, find the estimated regression equation for predicting that person's GPA.

(c) Plot the equations from (a) and (b) on a graph. Should the results be collinear? Explain.

4–6 A stress-strain relationship is to be established from the following sample data collected for steel rods of a particular composition:

Load P (pounds)	Diameter D (inches)	Elongation δ (inches)	Length L (inches)
5,000	.50	.06	22
1,000	.25	.07	31
1,500	.25	.13	32
2,000	.25	.11	30
10,000	.50	.12	24
15,000	.50	.14	18
10,000	1.00	.02	15
20,000	1.00	.05	15
30,000	1.00	.07	12
12,000	.50	.13	24
4,000	.25	.20	36
3,000	.25	.25	36
14,000	.50	.17	24
16,000	.50	.18	25
25,000	1.00	.06	13

(a) Compute the stress (σ) and strain (ϵ) for each sample point using the following relationships:

$$\sigma = \frac{P}{A} \text{ psi} \quad \epsilon = \frac{\delta}{L}$$

where A is the *area* of the right section. Then plot the resulting points on a scatter diagram with σ as the ordinate and ϵ as the abscissa.

(b) Treating stress as the dependent variable and strain as the independent variable, find the estimated regression line and plot it on your graph.

(c) The **modulus of elasticity** E may be estimated by the slope of the regression line. What is the point estimate for this quantity?

4–7 The following viscosity data were obtained from an experiment involving a toluene and tetralin mixture:

Average Temperature (K)	Viscosity (mPa · s)	Average Temperature (K)	Viscosity (mPa · s)
298.2	.7036	368.2	.3383
308.2	.6177	308.2	.6155
318.2	.5530	328.2	.4902
328.2	.4909	348.2	.4031
338.2	.4404	358.2	.3676
348.2	.4004	368.2	.3369
358.2	.3655		

Source: Byers and Williams, *Journal of Chem. Eng. Data*, Vol. 32, 1987, pp. 349–354.

(a) Using the reciprocal of temperature as the independent variable X, find the equation for the estimated regression line for predicting the natural logarithm Y of viscosity.

(b) Using the fact that viscosity $= e^Y$, find the estimated viscosity when the temperature is (1) 300 K, (2) 320 K, and (3) 350 K.

4–8 The following data were obtained from an experiment involving measurements of infrared energy levels for a chemical substance at various temperatures.

Temperature (K)	Temperature Reciprocal X (1/K)	Libration Bandwidth Y (cm^{-1})
17.9	.056	43
16.9	.059	43
16.1	.062	34
14.9	.067	35
14.5	.069	38
14.1	.071	31
13.0	.077	29

Source: Baciocco et al., *Journal of Chem. Physics*, Vol. 87, 15 August 1987, pp. 1913–1916.

(a) Find the estimated regression equation using the temperature reciprocal as the independent variable and libration bandwidth as the dependent variable.
(b) Determine the estimated libration bandwidth when the temperature is (1) 14 K, (2) 15 K, and (3) 16 K.

4–2 Correlation and Regression Analysis

Regression analysis is largely concerned with predicting the level of the dependent variable Y for a known level of the independent variable X. It therefore focuses on finding an appropriate function expressing the relationship between X and Y.

Correlation measures the *strength* of the relationship between X and Y. It gives a valuable perspective of bivariate relationships, and thus explains many of the concepts that underlie the regression model, making correlation analysis a helpful complement to regression analysis.

The Correlation Coefficient

The **correlation coefficient** r expresses the strength of the *linear* relationship between X and Y. Figure 4–6 illustrates several cases. The scatter diagrams in (a) and (b) portray sample data where all the observed points fall on a single line. These ideal situations indicate that X and Y are *perfectly correlated*, so the computed values for the sample correlation coefficient are precisely *one*, in absolute value. The slope of the lines indicates the direction of the relationship. In (a), where $r = +1$, X and Y are *positively correlated*. The data in (b) indicate that the variables are *negatively correlated*.

More typical sample results are shown on the scatter diagrams in (c) and (d), where the observed data points cluster about the respective fitting lines. Notice that the data cluster more tightly about the line in (c), where $r = +.97$, than they do in (d), where $r = -.61$. The variables are more highly correlated in (c), indicating a stronger relationship between X and Y. The closer r is in absolute value to 1, the greater the degree of correlation, and the data will be less scattered about the fitting line. The diagrams in (e) and (f) exhibit *zero correlation* between X and Y. In both cases, $r = 0$, and a horizontal line provides the best fit to the data. Although the data in diagram (f) indicate a pronounced nonlinear relationship, that is not reflected in r, which expresses the strength and direction of *linear* relationships only.

Computing the Correlation Coefficient

The correlation coefficient is a collective measure conveying the distances that separate individual data points from the two sample means, $\overline{X}$ and $\overline{Y}$. To illustrate, we will use the original scatter diagram for workstation data entry, as shown in Figure 4–7. Consider the 19th observation in the data set. The distances of that point, with coordinates (X_{19}, Y_{19}), from the respective means are the differences $(X_{19} - \overline{X})$ and $(Y_{19} - \overline{Y})$. A collective distance measure is the product of these:

$$(X_{19} - \overline{X})(Y_{19} - \overline{Y})$$

Figure 4–6 Scatter diagrams for various degrees of correlation.

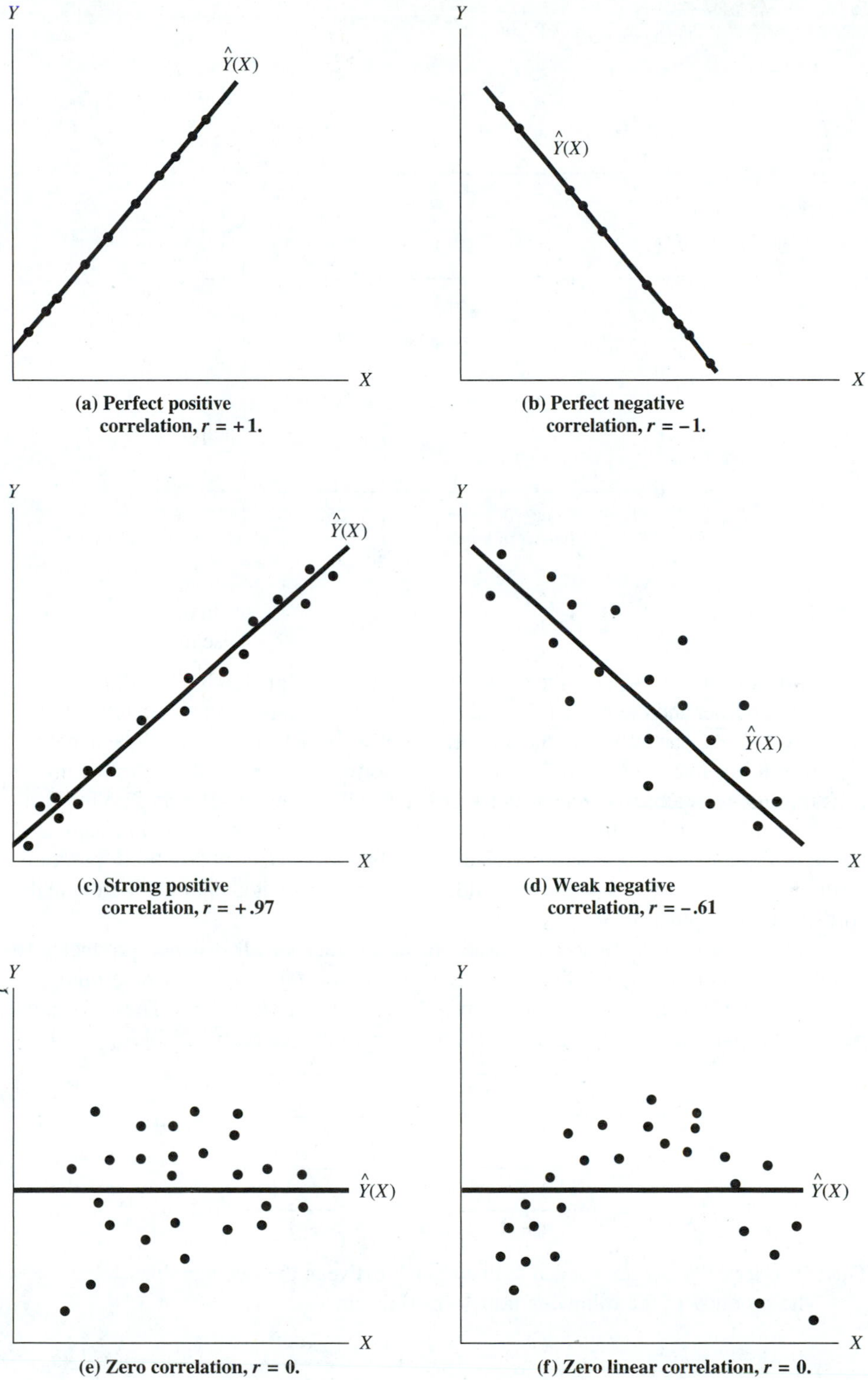

Figure 4–7
Scatter diagram for workstation data entry.

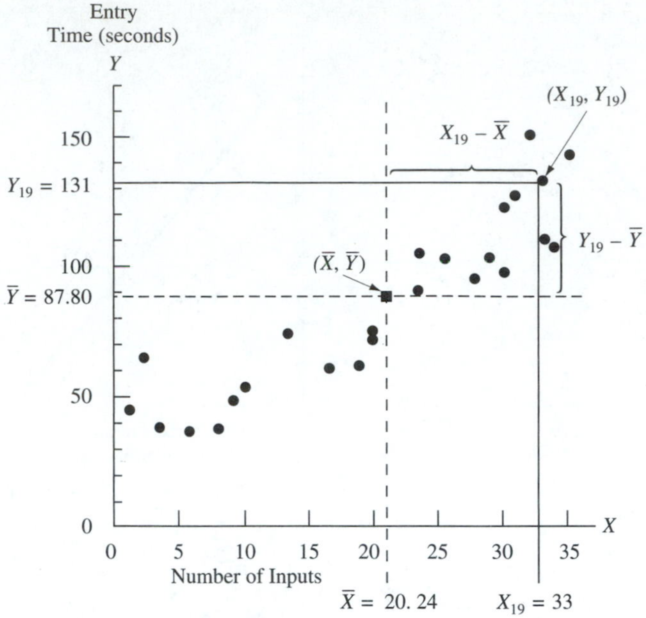

The above distance product is the signed area of the rectangle having the 19th transaction at one corner and the point $(\overline{X}, \overline{Y})$ diagonal to it. For data points lying to the left of or below $(\overline{X}, \overline{Y})$, one or both of the differences $(X - \overline{X})$ and $(Y - \overline{Y})$ will be negative.

If you imagine the dashed line at $\overline{X}$ as the north-south axis and the dashed line at $\overline{Y}$ as the east-west axis, then the distance products will be negative for data points lying in the northwest and southeast quadrants and positive for points in the northeast and southwest quadrants. Positively correlated variables have data points lying predominantly in the northeast and southwest, while negatively correlated ones fall mainly in the northwest and southeast.

The correlation coefficient is based on the average of all distance products. To avoid scale distortions, the distances themselves must be adjusted before determining r. That is accomplished by dividing by the respective *standard deviations*. These are computed in the usual fashion from the data values separately for each variable.

$$s_X = \sqrt{\frac{\sum(X - \overline{X})^2}{n - 1}} = \sqrt{\frac{\sum X^2 - (\sum X)^2/n}{n - 1}}$$

and

$$s_Y = \sqrt{\frac{\sum(Y - \overline{Y})^2}{n - 1}} = \sqrt{\frac{\sum Y^2 - (\sum Y)^2/n}{n - 1}}$$

The subscripts X and Y are needed to distinguish between the two variables.

The products of the following transformed distances

$$\frac{(X - \overline{X})}{s_X} \quad \text{and} \quad \frac{(Y - \overline{Y})}{s_Y}$$

are then averaged (using the $n - 1$ divisor) to compute the correlation coefficient.

$$r = \frac{\sum \left[\frac{(X - \overline{X})}{s_X}\right]\left[\frac{(Y - \overline{Y})}{s_Y}\right]}{n - 1}$$

The following equivalent expression may be used to compute the correlation coefficient.

$$r = \frac{\sum(X - \overline{X})(Y - \overline{Y})}{(n - 1)s_X s_Y}$$

Since the numerator is equal to

$$\sum(X - \overline{X})(Y - \overline{Y}) = \sum XY - n\overline{X}\overline{Y}$$

the following equivalent expression serves as the main expression for computing r.

CORRELATION COEFFICIENT

$$r = \frac{\sum XY - n\overline{X}\overline{Y}}{(n - 1)s_X s_Y}$$

When the sample standard deviations are not computed separately, the following expression can also be used.

$$r = \frac{\sum XY - n\overline{X}\overline{Y}}{\sqrt{(\sum X^2 - n\overline{X}^2)(\sum Y^2 - n\overline{Y}^2)}}$$

(As we shall see, an alternative procedure is sometimes used to find r using information obtained during regression analysis.)

To illustrate, we have from Table 4–1 the sample means

$$\overline{X} = 20.24 \qquad \text{and} \qquad \overline{Y} = 87.80$$

The standard deviations for the number of entries X and their times Y are

$$s_X = 10.97 \qquad \text{and} \qquad s_Y = 33.99$$

The product sum from Table 4–1 is $\sum XY = 52{,}670$. Using the above values in the main expression for r, the correlation coefficient is computed.

$$r = \frac{52{,}670 - 25(20.24)(87.80)}{(25 - 1)(10.97)(33.99)} = .921$$

The above indicates that there is a fairly strong positive correlation between number of inputs and entry time.

Problems

4–9 Refer to the data in Problem 4–1 for the number of data points X and processing time Y. Calculate the sample correlation coefficient.

4–10 Refer to the data in Problem 4–2 for the weight of the final product Y and volume of raw materials X. Calculate the sample correlation coefficient.

4–11 The following data pertain to drying time and solvent content of sample batches of an experimental coating substance:

Solvent Content (%) X	Drying Time (hours) Y	Solvent Content (%) X	Drying Time (hours) Y
2.5	2.3	3.4	1.9
2.7	2.1	3.5	1.9
3.0	2.2	4.0	1.6
3.1	2.1	4.2	1.6
3.2	2.0	4.5	1.5

Calculate the correlation coefficient.

4–12 The following data pertain to seasonal rainfall and augmented reservoir storage:

Seasonal Rainfall (inches) X	Reservoir Storage (acre-feet) Y	Seasonal Rainfall (inches) X	Reservoir Storage (acre-feet) Y
25.3	45,200	25.1	46,700
17.2	37,750	28.3	52,450
19.4	39,480	20.5	39,700
22.1	41,150	15.2	32,250
23.4	42,650	18.6	37,420

Calculate the sample correlation coefficient.

4–13 Artificial intelligence researchers have studied algorithms for solving constraint-satisfaction problems. Using the same problem on each, they obtained the following approximate numbers of consistency checks:

RBT X	ABT Y	RBT X	ABT Y
850	550	2,400	700
1,200	1,950	2,450	1,750
1,275	700	2,600	1,750
1,600	2,250	2,800	700
1,800	700	2,900	1,750
1,900	1,950	5,000	1,750
2,150	3,600	5,250	3,575
2,200	2,250		

Source: Dichter and Pearl, *Artificial Intelligence*, Vol. 34, No. 1, December 1987, p. 31.

(a) Plot these data on a scatter diagram.
(b) Compute the correlation coefficient.

4–3 Multiple Regression Analysis

We have seen how the value of one variable may be used in predicting another variable. We can do this by using sample data to fit an equation expressing the dependent variable Y in terms of an independent variable X. Since there is a single independent variable, such a procedure is often called *simple regression*. The underlying concepts and procedures extend to more than one predictor, so *multiple regression* incorporates several independent variables in making predictions of Y. Multiple regression ordinarily explains more of the total variation in Y.

By using several independent variables in multiple regression, an investigator may make greater use of information. For example, an industrial engineer can better predict the cost of making a particular production run if, in addition to the quantity of raw materials, he knows how much labor will be available. Similarly, a chemical engineer ought to make a better yield prediction if she knows not only the temperature and pressure settings but also impurity levels for the raw feedstocks. A software engineer can more finely gauge a job's processing time if, besides knowing the number of input data points, he also knows memory requirements and the amount of output.

Multiple regression analysis extends simple linear regression by considering two or more independent variables. Extending our earlier notation, the independent variables will be denoted as $X_1, X_2, X_3, \ldots$. In the case of two independent variables we use the following.

ESTIMATED MULTIPLE REGRESSION EQUATION

$$\hat{Y} = a + b_1X_1 + b_2X_2$$

As in simple regression, $\hat{Y}$ represents the value for Y computed from the estimated regression equation. With the dependent variable Y and the two independent variables X_1 and X_2, a total of three variables is considered. A graphical representation of the scatter diagram will be three dimensional.

Regression in Three Dimensions

Figure 4–8 represents a three-dimensional scatter diagram. The estimated regression equation is represented by a **regression plane**, which, like the regression line, is positioned to fit the observed data points. As in simple regression, the method of least squares establishes the particular plane providing the best fit.

The Y-intercept a is the level of Y when both X_1 and X_2 are zero. The values of b_1 and b_2 are referred to as the **estimated partial regression coefficients**. These provide the slope of the respective lines obtained by slicing the regression plane with cuts parallel to the respective axes. The constant b_1 expresses the net change in Y for a unit increase in X_1 when X_2 is held at a constant value. Likewise, b_2 represents the response of Y to a unit increase in X_2 when X_1 is fixed at some level.

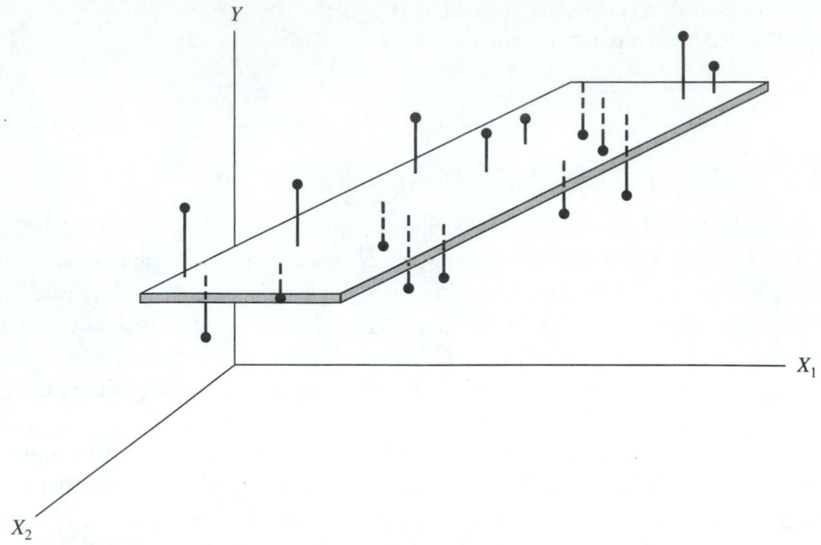

Figure 4–8
Scatter diagram and regression plane for multiple regression with three variables.

Example:
Hydrological Survey: Predicting River Flow

Multiple regression has been used to assist in a hydrological survey for selecting a dam site. The seasonal mean river flow Y (in cubic feet per second, or cfps) past a particular point is estimated by the equation

$$\hat{Y} = 2{,}500 + 4{,}900X_1 + .23X_2$$

where X_1 represents the annual rainfall (inches) and X_2 the average release (also in cubic feet per second) from upstream dams.

Here $a = 2{,}500$ cfps represents the predicted flow that might be achieved if there were no rainfall and no releases (so that all river flow resulted from aquifer seepage). The partial regression coefficient $b_1 = 4{,}900$ indicates that the mean river flow would increase by 4,900 cfps for each additional average inch of rainfall in the drainage basin, regardless of releases from upstream dams. The constant $b_2 = .23$ signifies that only 23% of the water volume released from dams ever reaches the downstream observation location, with the rest lost to evaporation, seepage, and earlier withdrawal; this quantity applies regardless of what the rainfall happens to be. Thus, we predict the river's flow in a year having a mean rainfall of $X_1 = 18.4$ inches and when the average dam release is $X_2 = 150{,}000$ cfps:

$$\hat{Y} = 2{,}500 + 4{,}900(18.4) + .23(150{,}000) = 127{,}160 \text{ cfps}$$

Of course, the actual river flow under the stipulated conditions might differ from this amount, perhaps considerably.

A regression equation ordinarily does not explain all the variation in Y. Some of it might be explained in terms of factors not included. In the hydrological survey example, mean temperature might serve as an additional independent variable, and its inclusion in the regression evaluation might further explain variation in river flow and thereby improve predictions of Y. Furthermore, a regression evaluation is ordinarily based on sample data only, which makes it subject to sampling error. Finally, even if the true regression equation were known, individual Ys involving exactly the same Xs would still exhibit some variability.

As in simple regression analysis, the regression coefficients are found by fitting a plane through the observed data. Here too, the vertical deviations $Y - \hat{Y}$ are the basis for the evaluation. Again, the method of least squares minimizes

$$\sum(Y - \hat{Y})^2 = \sum[Y - (a + b_1X_1 + b_2X_2)]^2$$

Mathematically, the regression parameters are found by taking the respective partial derivatives of the above and setting each resulting expression equal to zero. That provides three equations in three unknowns. As in simple regression, these are referred to as **normal equations**.

NORMAL EQUATIONS FOR MULTIPLE REGRESSION

$$\sum Y = na + b_1\sum X_1 + b_2\sum X_2$$
$$\sum X_1Y = a\sum X_1 + b_1\sum X_1^2 + b_2\sum X_1X_2$$
$$\sum X_2Y = a\sum X_2 + b_1\sum X_1X_2 + b_2\sum X_2^2$$

In finding the estimated regression coefficients the simplest procedure is to directly solve these equations simultaneously rather than to derive further expressions.

Illustration:* *Predicting Gasoline Mileage

To illustrate finding the multiple regression equation, we consider the problem of predicting gasoline mileage Y (in miles per gallon). The independent variables are fuel octane rating X_1 and average speed X_2 (miles per hour). The sample data in Table 4–4 were obtained from 20 test runs with cars of the same make driven with different fuels and at various speeds.

The intermediate calculations that provide the inputs for expressing the normal equations are shown in Table 4–5. Substituting the appropriate column totals, the following normal equations are obtained:

$$604.5 = 20a + 1{,}819b_1 + 1{,}088b_2$$
$$55{,}066.5 = 1{,}819a + 165{,}535b_1 + 99{,}063b_2$$
$$32{,}905.2 = 1{,}088a + 99{,}063b_1 + 59{,}626b_2$$

Solving these equations simultaneously, the following estimated regression coefficients are determined:

$$a = -63.535$$
$$b_1 = 1.1789$$
$$b_2 = -.24743$$

Ordinarily, multiple regression results will be obtained with the assistance of a digital computer.

The computed values for a, b_1, and b_2 provide the estimated multiple regression equation

$$\hat{Y} = -63.535 + 1.1789X_1 - .24743X_2$$

Table 4–4 Mileages, Octanes, and Speeds for 20 Automobile Test Runs

Run	Gasoline Mileage (mpg) Y	Octane X_1	Average Speed (mph) X_2
1	24.8	88	52
2	30.6	93	60
3	31.1	91	58
4	28.2	90	52
5	31.6	90	55
6	29.9	89	46
7	31.5	92	58
8	27.2	87	46
9	33.3	94	55
10	32.6	95	62
11	30.6	88	47
12	28.1	89	58
13	25.2	90	63
14	35.0	93	54
15	29.2	91	53
16	31.9	92	52
17	27.7	89	52
18	31.7	94	53
19	34.2	93	54
20	30.1	91	58

We may use this equation to predict the gasoline mileage from any particular test run. Suppose that the fuel is rated at 90 octane and that the test car will be driven at an average speed of 60 miles per hour. With $X_1 = 90$ and $X_2 = 60$, the predicted mileage is

$$\hat{Y} = -63.535 + 1.1789(90) - .24743(60) = 27.72 \text{ mpg}$$

The proper interpretation of $b_1 = 1.1789$ is that each unit increase in fuel octane rating brings about an estimated increase in gasoline efficiency of 1.1789 miles per gallon. Furthermore, this mileage increase is estimated to be the same no matter what the average speed happens to be. Likewise, $b_2 = -.24743$ signifies that gasoline consumption increases with speed, so that mileage is estimated to decline by .24743 mpg for each additional mile per hour—regardless of the fuel octane rating.

The Y-intercept $a = -63.535$ by itself has little meaning since the origin represents zero octane and zero speed. It will help if we shift the origin to minimum practical levels for the independent variables. The lowest fuel octane rating in the sample was 87, and a minimal speed for the test car is assumed to be 45 mph. Considering the transformations

$$X_1' = X_1 - 87$$
$$X_2' = X_2 - 45$$

Table 4–5 *Intermediate Calculations for Obtaining Regression Coefficients*

Y	X_1	X_2	X_1Y	X_2Y	$X_1\ X_2$	X_1^2	X_2^2	Y^2
24.8	88	52	2,182.4	1,289.6	4,576	7,744	2,704	615.04
30.6	93	60	2,845.8	1,836.0	5,580	8,649	3,600	936.36
31.1	91	58	2,830.1	1,803.8	5,278	8,281	3,364	967.21
28.2	90	52	2,538.0	1,466.4	4,680	8,100	2,704	795.24
31.6	90	55	2,844.0	1,738.0	4,950	8,100	3,025	998.56
29.9	89	46	2,661.1	1,375.4	4,094	7,921	2,116	894.01
31.5	92	58	2,898.0	1,827.0	5,336	8,464	3,364	992.25
27.2	87	46	2,366.4	1,251.2	4,002	7,569	2,116	739.84
33.3	94	55	3,130.2	1,831.5	5,170	8,836	3,025	1,108.89
32.6	95	62	3,097.0	2,021.2	5,890	9,025	3,844	1,062.76
30.6	88	47	2,692.8	1,438.2	4,136	7,744	2,209	936.36
28.1	89	58	2,500.9	1,629.8	5,162	7,921	3,364	789.61
25.2	90	63	2,268.0	1,587.6	5,670	8,100	3,969	635.04
35.0	93	54	3,255.0	1,890.0	5,022	8,649	2,916	1,225.00
29.2	91	53	2,657.2	1,547.6	4,823	8,281	2,809	852.64
31.9	92	52	2,934.8	1,658.8	4,784	8,464	2,704	1,017.61
27.7	89	52	2,465.3	1,440.4	4,628	7,921	2,704	767.29
31.7	94	53	2,979.8	1,680.1	4,982	8,836	2,809	1,004.89
34.2	93	54	3,180.6	1,846.8	5,022	8,649	2,916	1,169.64
30.1	91	58	2,739.1	1,745.8	5,278	8,281	3,364	906.01
604.5	1,819	1,088	55,066.5	32,905.2	99,063	165,535	59,626	18,414.25
$=\sum Y$	$=\sum X_1$	$=\sum X_2$	$=\sum X_1Y$	$=\sum X_2Y$	$=\sum X_1X_2$	$=\sum X_1^2$	$=\sum X_2^2$	$=\sum Y^2$

the estimated regression equation may be expressed as

$$\begin{aligned}\hat{Y} &= -63.535 + 1.1789(X_1' + 87) - .24743(X_2' + 45)\\ &= 27.895 + 1.1789X_1' - .24743X_2'\end{aligned}$$

Under the shift in origin, the Y-intercept is 27.895 mpg, which may be interpreted as the estimated mileage when octane and speed are at $(X_1', X_2') = (0, 0)$, or their minimal practical levels $(X_1, X_2) = (87, 45)$.

Advantages of Multiple Regression

Let's compare the multiple regression results to what would be achieved with separate simple regression evaluations. This will highlight the advantages of simultaneous consideration of the two independent variables.

Figure 4–9 shows three two-dimensional scatter diagrams for the gasoline mileage data from Table 4–4. In Figure 4–9(a) mileage is plotted against octane. Notice the fairly high sample correlation coefficient value of .74 for these two variables, indicating that octane rating X_1 by itself might provide adequate predictions of gasoline mileage Y. It will be convenient if we use double subscripts to denote which variables are involved, so Y_1 indicates that the association between Y and X_1 is being represented by $r_{Y1} = .74$.

A similar figure applies to each variable pair. In relating gasoline mileage Y to average speed X_2, the sample correlation coefficient is $r_{Y2} = .081$. This second relationship is not a very strong one, as reflected by the scatter diagram in (b), which exhibits a pattern of almost zero correlation. This seems to indicate that average speed will have little effect on gasoline mileage, so by itself X_2 will provide poor predictions of Y. Furthermore, this appears to contradict our earlier multiple regression finding that Y is estimated to decline by nearly .25 mpg for each additional mph of average speed. How can we resolve this paradox?

Our answer lies in the scatter diagram in Figure 4–9(c), where octane rating X_1 is plotted against average speed X_2. There is a fairly strong positive relationship between these variables, confirmed by the sample correlation coefficient value of $r_{12} = .53$. For the particular sample test runs in the investigation, the higher speeds tended to occur when higher-octane fuel was used, while low octanes prevailed for the lower speeds. The effect of speed on gasoline mileage is camouflaged by this strong interaction between speed and octane.

Figure 4–9

Scatter diagrams for variable pairs in gasoline mileage illustration.

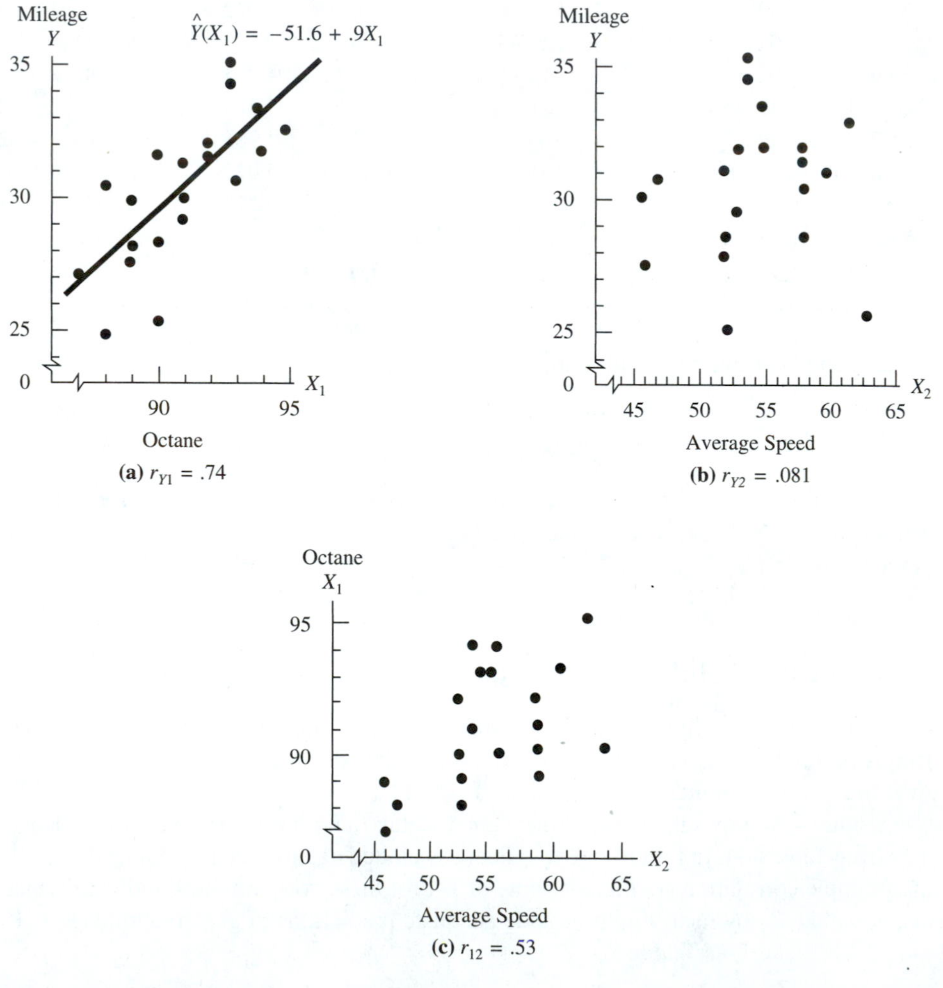

Our example illustrates a general inadequacy in using separate two-variable evaluations to establish how the dependent variable relates to several independent variables. Although the sample correlation between mileage Y and speed X_2 is only $r_{Y2} = .081$, it would be a blunder to discard X_2 from the analysis because of that small correlation alone. This is because X_2 exhibits a significant correlation with X_1, $r_{12} = .53$, so that the influence of average speed X_2 on mileage Y can be explained only by considering how Y relates to *both* X_1 and X_2 through multiple regression.

Residuals and the Standard Error of the Estimate

The least squares procedure fits a regression surface Y through the observed data points so that the sum of squared vertical deviations, $\sum(Y - \hat{Y})^2$, is minimized. As with simple regression, we use the square root of the average of the squared deviations about the regression plane to define the standard error of the estimate for values of Y:

$$S_{Y\cdot 12} = \sqrt{\frac{\sum(Y - \hat{Y})^2}{n - 3}}$$

which summarizes variability in Y about the regression plane. The subscript $Y\cdot 12$ shows that two independent variables X_1 and X_2 are being used to predict Y.

The vertical deviations, $Y - \hat{Y}$, are referred to as *residuals*. This terminology is often used because $\sum(Y - \hat{Y})^2$ represents the variation in Y left unexplained by the regression analysis. This unexplained variation is what is left over, or the "residual," from the evaluation. Including a further independent variable in the regression analysis might bring about a reduction in unexplained variation. The residuals resulting from an expanded regression would then be smaller than the current ones.

The predicted values and residuals computed for the gasoline mileage investigation are shown in Table 4–6. Notice that, except for rounding errors, the residuals (vertical deviations) sum to zero—a consequence of the least squares procedure. The standard error of the estimate for the gasoline mileage multiple regression is

$$S_{Y\cdot 12} = \sqrt{\frac{45.4751}{20 - 3}} = 1.64$$

This summarizes the degree to which the gasoline mileage data points are scattered about the regression *plane*.

One demonstration of multiple regression's power is provided by comparing the standard error to its counterpart in a simple regression analysis involving mileage and octane only. Using the computed values from Table 4–5, the following estimated regression equation is obtained:

$$\hat{Y}(X_1) = -51.602 + .8997X_1$$

The scatter of data about this regression *line* is summarized by the standard error of the estimate

$$s_{Y\cdot X_1} = 1.8968 \text{ mpg}$$

The standard error for the regression plane is smaller, which indicates that a multiple regression analysis gives the better interpretation of the data. Including X_2(speed) as the second variable actually increases the explained variation in Y and permits more accurate predictions of gasoline mileage than would be possible using octane as the only predictor.

Table 4–6
Actual, Predicted, and Residual Values from Three-Variable Regression Analysis in Gasoline Mileage Investigation

Actual Mileage Y	Predicted Value $\hat{Y}$	Residual Value $Y - \hat{Y}$	Squared Deviation $(Y - \hat{Y})^2$
24.8	27.34	−2.54	6.4516
30.6	31.26	−.66	.4356
31.1	29.39	1.71	2.9241
28.2	29.70	−1.50	2.2500
31.6	28.96	2.64	6.9696
29.9	30.00	−.10	.0100
31.5	30.57	.93	.8649
27.2	27.65	−.45	.2025
33.3	33.67	−.37	.1369
32.6	33.12	−.52	.2704
30.6	28.58	2.02	4.0804
28.1	27.04	1.06	1.1236
25.2	26.98	−1.78	3.1684
35.0	32.74	2.26	5.1076
29.2	30.63	−1.43	2.0449
31.9	32.06	−.16	.0256
27.7	28.52	−.82	.6724
31.7	34.17	−2.47	6.1009
34.2	32.74	1.46	2.1316
30.1	29.39	.71	.5041
604.5	604.51	−0.01	45.4751

Regression with Many Variables

Multiple regression may involve more than two predictor variables. For example, in addition to rainfall X_1 and dam releases X_2, the mean daily temperature X_3 in its watershed area might be used in predicting a river's flow Y. With three independent variables, the estimated regression equation may be expressed as

$$\hat{Y} = a + b_1 X_1 + b_2 X_2 + b_3 X_3$$

This expression describes a four-dimensional linear surface referred to as a regression **hyperplane**. This surface may be fitted to the observed data points in the same way as in the case of two independent variables. In establishing values for the estimated regression coefficients, additional intermediate calculations must be made to find the constants in *four* normal equations.

Ordinarily, higher-dimensional regression analyses will be performed with the assistance of a digital computer. It is therefore not necessary to give a detailed discussion of the algebraic expressions that would be needed for solving such a problem by hand.

All the earlier multiple regression concepts extend to the case of three independent variables. The amount of variation in Y explained by the regression may be gauged by the standard error of the estimate, denoted by $S_{Y \cdot 123}$, calculated in the same way as before with divisor $n - 4$. Additional independent variables X_4, X_5, and so on, might be included in a multiple regression analysis.

We will now extend our gasoline mileage example one more dimension and illustrate computer applications of multiple regression with three independent variables. To demonstrate both computer applications and higher-dimensional regression analysis at the same time, we will expand the gasoline mileage illustration. Table 4–7 shows the data for an augmented regression analysis that now includes the load X_3 carried during each test run. Figure 4–10 shows a printout for a computer run using the SAS package to perform a multiple regression with those expanded gasoline mileage data. The SAS report is broken down into three parts. Shown first is the analysis of variance (which will be described in Chapter 13). That is followed by the regression results and the predicted values—with residuals—computed for all observations.

Although computer packages will vary in presentation, most provide the same information as shown in Figure 4–10, where the more important values are identified with annotated symbols added to the printout. (SAS uses a synonym "root MSE" for the standard error of the estimate $S_{Y\cdot 123}$.)

The results indicate that gasoline mileage may be predicted from the estimated regression equation

$$\hat{Y} = -59.005 + 1.2097X_1 - .2970X_2 - .0098X_3$$

The estimated partial regression coefficient $b_1 = 1.2097$ indicates the estimated increase in gasoline mileage that might be achieved for each additional point in fuel octane rating.

Table 4–7
Sample Gasoline Mileage Data from 20 Automobile Test Runs, Using Octane, Speed, and Load as Independent Variables

Run	Gasoline Mileage (mpg) Y	Octane X_1	Average Speed (mph) X_2	Load (pounds) X_3
1	24.8	88	52	646
2	30.6	93	60	465
3	31.1	91	58	359
4	28.2	90	52	665
5	31.6	90	55	214
6	29.9	89	46	606
7	31.5	92	58	458
8	27.2	87	46	557
9	33.3	94	55	605
10	32.6	95	62	407
11	30.6	88	47	259
12	28.1	89	58	423
13	25.2	90	63	596
14	35.0	93	54	286
15	29.2	91	53	612
16	31.9	92	52	399
17	27.7	89	52	444
18	31.7	94	53	697
19	34.2	93	54	376
20	30.1	91	58	363

Figure 4–10
SAS printout for multiple regression with gasoline mileage data using $m = 4$ variables.

GASOLINE MILEAGE DATA

Model: MODEL1
Dep Variable: MILEAGE

Analysis of Variance

Source	DF	Sum of Squares	Mean Square	F Value	Prob>F
Model	3	134.59603	44.86534	83.070	0.0001
Error	16	8.64147	0.54009		
C Total	19	143.23750			

Root MSE	$S_{Y\cdot123}$	0.73491	R-Square	0.9397
Dep Mean	$\bar{Y}$	30.22500	Adj R-Sq	0.9284
C.V.		2.43146		

Parameter Estimates

Variable	DF	Parameter Estimate		Standard Error	T for H0: Parameter=0	Prob > |T|
INTERCEP	1	-59.004933	a	7.09881835	-8.312	0.0001
OCTANE	1	1.209731	b_1	0.08812242	13.728	0.0001
SPEED	1	-0.296972	b_2	0.04181721	-7.102	0.0001
LOAD	1	-0.009833	b_3	0.00119102	-8.256	0.0001

Obs	MILEAGE Y	Predict Value $\hat{Y}$	Std Err Predict	Residual	Std Err Residual	Student Residual
1	24.8000	25.6566	0.344	-0.8566	0.649	-1.319
2	30.6000	31.1093	0.263	-0.5093	0.686	-0.742
3	31.1000	30.3261	0.248	0.7739	0.692	1.118
4	28.2000	27.8892	0.288	0.3108	0.676	0.460
5	31.6000	31.4331	0.356	0.1669	0.643	0.260
6	29.9000	29.0415	0.357	0.8585	0.642	1.337
7	31.5000	30.5623	0.208	0.9377	0.705	1.330
8	27.2000	27.1038	0.381	0.0962	0.629	0.153
9	33.3000	32.4272	0.340	0.8728	0.652	1.339
10	32.6000	33.5051	0.369	-0.9051	0.635	-1.424
11	30.6000	30.9469	0.431	-0.3469	0.595	-0.583
12	28.1000	27.2773	0.327	0.8227	0.658	1.250
13	25.2000	25.3010	0.483	-0.1010	0.554	-0.182
14	35.0000	34.6512	0.341	0.3488	0.651	0.536
15	29.2000	29.3231	0.236	-0.1231	0.696	-0.177
16	31.9000	32.9243	0.257	-1.0243	0.688	-1.488
17	27.7000	28.8526	0.223	-1.1526	0.700	-1.646
18	31.7000	32.1165	0.425	-0.4165	0.600	-0.695
19	34.2000	33.7663	0.280	0.4337	0.679	0.638
20	30.1000	30.2867	0.245	-0.1867	0.693	-0.270

Sum of Residuals	-2.59348E-13
Sum of Squared Residuals	8.6415
Predicted Resid SS (Press)	12.7560

This is very close to the counterpart coefficient computed for the earlier two-independent-variable regression. There has also been modest change in the coefficient for speed, $b_2 = -.2970$, which gives .2970 as the estimated drop in mileage attributed to a 1-mph increase in speed. Both of these apply regardless of the levels for the other independent variables. The partial regression coefficient is $b_3 = -.0098$ for the load carried. This indicates an estimated decrease of almost .01 mile per gallon for each pound of load—again regardless of octane or speed. Stated differently, each 100 pounds of load is estimated to decrease gasoline mileage by about 1 mph. The Y-intercept for the new regression hyperplane is $a = -59.005$, about 4.4 mpg higher than before.

We can see that including X_3 as an additional variable will improve predictions made using the resulting estimated regression equation. This is because the standard error, $S_{Y \cdot 123} = .7349$, is smaller than when X_1 and X_2 were the only predictors (and $S_{Y \cdot 12} = 1.6352$, as we found in the preceding section). Including X_3 in the regression brings about a substantial reduction in standard error. This indicates that the latest estimated multiple regression equation should considerably improve the precision of prediction intervals for Y.

Adding more dimensions to the regression analysis by including further independent predictor variables might lead to even more precision. But adding more variables could just as well raise the standard error and actually cloud the multiple regression results. Consider the observations in Table 4–8 for the stops per mile observed for each car during its test run. These data were incorporated into a five-dimensional regression computer run, the SAS output for which is shown in Figure 4–11. The estimated multiple regression equation is

$$\hat{Y} = -56.526 + 1.1763X_1 - .2814X_2 - .0093X_3 - .9555X_4$$

Table 4–8 *Observations of Stops per Mile as a Fourth Independent Variable in Gasoline Mileage Investigation*

Run	Mileage (mpg) Y	Stops per Mile X_4	Run	Mileage (mpg) Y	Stops per Mile X_4
1	24.8	1.2	11	30.6	.2
2	30.6	.7	12	28.1	.7
3	31.1	.3	13	25.2	.6
4	28.2	.6	14	35.0	.3
5	31.6	1.2	15	29.2	.8
6	29.9	.1	16	31.9	.7
7	31.5	.2	17	27.7	1.4
8	27.2	.1	18	31.7	1.1
9	33.3	.3	19	34.2	.2
10	32.6	.4	20	30.1	.1

Notice that the previous regression coefficients change very little. The value $b_4 = -.9555$ signifies that gasoline mileage is estimated to drop by nearly 1 mpg for each stop per mile.

The standard error of the estimate for the 5-variable regression is $S_{Y \cdot 1234} = .6281$, which is smaller than before. This indicates that inclusion of X_4 (stops per mile) should further improve the quality of multiple regression predictions, although the improvement is much less substantial than that achieved by prior inclusion of X_3 (load).

Figure 4–11
SAS printout for multiple regression with gasoline mileage data using $m = 5$ variables.

```
                              GASOLINE MILEAGE DATA

Model: MODEL1
Dep Variable: MILEAGE
                              Analysis of Variance

                              Sum of          Mean
          Source       DF     Squares        Square        F Value        Prob>F

          Model         4   137.32040      34.33010         87.028        0.0001
          Error        15     5.91710       0.39447
          C Total      19   143.23750

              Root MSE       0.62807      R-Square         0.9587
              Dep Mean      30.22500      Adj R-Sq         0.9477
              C.V.           2.07799

                              Parameter Estimates

                         Parameter        Standard      T for H0:
Variable      DF          Estimate           Error    Parameter=0     Prob > |T|

INTERCEP       1        -56.525667      6.13973412         -9.207         0.0001
OCTANE         1          1.176293      0.07637879         15.401         0.0001
SPEED          1         -0.281352      0.03622883         -7.766         0.0001
LOAD           1         -0.009309      0.00103723         -8.975         0.0001
STOPS          1         -0.955497      0.36358473         -2.628         0.0190

                            Predict    Std Err                 Std Err      Student
    Obs      MILEAGE         Value    Predict    Residual     Residual     Residual

      1      24.8000       25.1975      0.342     -0.3975        0.527       -0.755
      2      30.6000       30.9908      0.229     -0.3908        0.585       -0.668
      3      31.1000       30.5699      0.231      0.5301        0.584        0.908
      4      28.2000       27.9465      0.247      0.2535        0.577        0.439
      5      31.6000       30.7276      0.406      0.8724        0.479        1.820
      6      29.9000       29.4853      0.349      0.4147        0.522        0.794
      7      31.5000       30.9202      0.224      0.5798        0.587        0.988
      8      27.2000       27.5889      0.374     -0.3889        0.505       -0.771
      9      33.3000       32.6528      0.303      0.6472        0.550        1.176
     10      32.6000       33.6073      0.318     -1.0073        0.542       -1.860
     11      30.6000       31.1624      0.378     -0.5624        0.502       -1.121
     12      28.1000       27.2393      0.279      0.8607        0.562        1.530
     13      25.2000       25.4939      0.419     -0.2939        0.468       -0.629
     14      35.0000       34.7275      0.293      0.2725        0.555        0.491
     15      29.2000       29.1437      0.213      0.0563        0.591        0.095
     16      31.9000       32.6798      0.239     -0.7798        0.581       -1.342
     17      27.7000       28.0631      0.356     -0.3631        0.517       -0.702
     18      31.7000       31.5947      0.414      0.1053        0.472        0.223
     19      34.2000       33.9852      0.253      0.2148        0.575        0.374
     20      30.1000       30.7238      0.268     -0.6238        0.568       -1.098

Sum of Residuals                -2.70006E-13
Sum of Squared Residuals              5.9171
Predicted Resid SS (Press)           11.0738
```

Problems

4–14 In preparing a hydrological survey, a civil engineer has recorded the following sample data:

Seasonal Mean River Flow (cfps)	Watershed Rainfall (inches)	Upstream Dam Releases (cfps)	Average Daily High Temperature (°F)
240,000	25	90,000	62
210,000	19	100,000	62
220,000	23	65,000	61
175,000	16	120,000	64
200,000	20	135,000	64
215,000	17	100,000	60
150,000	14	120,000	62
175,000	15	125,000	64
220,000	21	75,000	61
230,000	16	95,000	60
235,000	24	65,000	60
190,000	19	95,000	64
210,000	21	75,000	63
210,000	20	80,000	62
230,000	23	65,000	62

(a) Using watershed rainfall and upstream dam releases as the independent variables and seasonal mean river flow as the dependent variable, determine the equation for the estimated regression plane.
(b) Compute the standard error of the estimate for Y.
(c) Predict the seasonal mean river flow for the following cases.

	Watershed Rainfall (in.)	Upstream Dam Releases (cu. ft./sec.)
(1)	20	100,000
(2)	20	110,000
(3)	21	120,000
(4)	21	90,000

4–15 Refer to the hydrological data in Problem 4–14.
(a) Using watershed rainfall and average daily high temperature as the independent variables and seasonal mean river flow as the dependent variable, determine the equation for the estimated regression plane.

(b) Compute the standard error of the estimate for Y.
(c) Predict the seasonal mean river flow for the following cases.

	Watershed Rainfall (inches)	Average Daily High Temperature (°F)
(1)	20	60
(2)	25	62
(3)	15	62
(4)	15	60

4–16 The following data provide the final product yield from 15 pilot batches tested at a chemical plant:

Final Product Yield (pounds)	Settling Time (hours)	Solid Catalyst (pounds)	Liquid Catalyst (gallons)
400	20	150	210
420	25	230	220
380	15	140	210
440	20	310	170
340	8	50	90
420	15	200	230
330	20	110	150
290	9	70	120
340	16	150	110
330	11	120	150
310	14	90	100
260	8	80	70
290	9	70	110
250	5	40	60
400	12	170	220

(a) Using settling time and weight of solid catalyst as the independent variables and final product yield as the dependent variable, determine the equation for the estimated regression plane.
(b) Compute the standard error of the estimate for Y.
(c) Predict the final product yield for the following cases.

	Settling Time (hours)	Solid Catalyst (pounds)
(1)	10	100
(2)	15	150
(3)	20	200
(4)	25	300

4–17 Refer to the chemical processing data in Problem 4–16.

(a) Find the estimated regression equation using settling time and volume of liquid catalyst as the independent variables used to predict final product yield. Then compute the standard error of the estimate.

(b) Find the estimated regression equation using weight of solid catalyst and volume of liquid catalyst as the independent variables used to predict final product yield. Then compute the standard error of the estimate.

(c) Which of the above multiple regressions would provide the better predictions of the final product yield? Explain.

4–18 Refer to the hydrological survey data in Problem 4–14.

(a) Find the estimated multiple regression equation for the hyperplane, incorporating all three predictors to forecast seasonal mean river flow.

(b) Predict the seasonal mean river flow for the following cases.

	Watershed Rainfall (inches)	Upstream Dam Releases (cfps)	Average Daily High Temperature (°F)
(1)	20	100,000	60
(2)	15	110,000	61
(3)	25	120,000	62
(4)	20	100,000	60

4–19 Refer to the chemical processing data in Problem 4–16.

(a) Find the estimated multiple regression equation for the hyperplane, incorporating all three predictors to forecast final product yield.

(b) Predict the final product yield for the following cases.

	Settling Time (hours)	Solid Catalyst (pounds)	Liquid Catalyst (gallons)
(1)	10	100	200
(2)	20	150	100
(3)	15	250	150
(4)	25	200	100

4–20 The following readings may be incorporated into the hydrological survey results given in Problem 4–14. (a) Find the multiple regression equation for the hyperplane, incorporating irrigation withdrawals as the fourth independent variable.

Seasonal Mean River Flow (cfps)	Irrigation Withdrawals (cfps)
240,000	78,000
210,000	81,000
220,000	152,000
175,000	103,000
200,000	82,000
215,000	88,000
150,000	275,000
175,000	141,000

Seasonal Mean River Flow (cfps)	Irrigation Withdrawals (cfps)
220,000	100,000
230,000	139,000
235,000	95,000
190,000	81,000
210,000	101,000
210,000	64,000
230,000	96,000

(b) Predict the seasonal mean river flow for the following cases.

	Watershed Rainfall (inches)	Upstream Dam Releases (cfps)	Average Daily High Temperature (°F)	Irrigation Withdrawals (cfps)
(1)	20	100,000	60	75,000
(2)	15	110,000	61	100,000
(3)	25	120,000	62	50,000
(4)	20	100,000	60	90,000

4–21 The following readings may be incorporated into the chemical processing data given in Problem 4–16.

Final Product Yield (pounds)	Chamber Temperature (°F)	Chamber Pressure (psi)
400	280	150
420	320	175
380	310	155
440	290	160
340	265	190
420	295	145
330	325	160
290	340	155

Final Product Yield (pounds)	Chamber Temperature (°F)	Chamber Pressure (psi)
340	290	145
330	340	170
310	285	140
260	355	170
290	315	130
250	325	135
400	305	150

(a) Find the multiple regression equation for the original variables plus chamber temperature as the fourth independent variable. Then predict the final product yield for the following cases.

	Settling Time (hours)	Solid Catalyst (pounds)	Liquid Catalyst (gallons)	Chamber Temperature (°F)
(1)	10	100	200	300
(2)	20	150	100	310
(3)	15	250	150	320
(4)	25	200	100	340

(b) Find the multiple regression equation for the hyperplane based on the original variables and the incorporation of chamber pressure as the fourth independent variable. Then predict the final product yield for the following cases.

	Settling Time (hours)	Solid Catalyst (pounds)	Liquid Catalyst (gallons)	Chamber Pressure (psi)
(1)	10	100	200	130
(2)	20	150	100	140
(3)	15	250	150	150
(4)	25	200	100	160

(c) Find the multiple regression equation for the hyperplane based on all five independent variables. Then predict the final product yield for the following cases.

	Settling Time (hours)	Solid Catalyst (pounds)	Liquid Catalyst (gallons)	Chamber Temperature (°F)	Chamber Pressure (psi)
(1)	10	100	200	300	130
(2)	20	150	100	310	140
(3)	15	250	150	320	150
(4)	25	200	100	340	160

4–22 The following data apply to a random sample of 25 engineers:

Earnings Y	Professional Experience (years) X_1	Engineering Education (years) X_2	Positions Held X_3	Number of Employees Supervised X_4	Patents Held X_5
$36,000	10	4.0	1	0	0
58,700	4	5.0	2	50	0
52,400	7	5.5	4	30	0
153,200	25	4.0	3	500	2
53,100	9	4.5	4	2	15
38,200	9	4.0	5	0	0
41,400	8	4.0	3	0	5
40,800	12	5.0	5	0	1
43,700	11	6.0	4	12	0
42,500	3	5.5	2	6	0
47,300	5	4.5	1	5	8
56,300	6	8.5	2	0	23
34,500	5	4.0	3	0	0
35,900	5	7.0	3	2	0
42,200	8	9.0	5	3	0
63,400	13	6.5	7	0	15
56,800	21	10.5	12	4	8
45,700	12	4.5	5	8	4
61,300	8	5.0	12	25	3
55,900	23	5.0	10	47	0
97,500	14	4.0	5	123	0
42,400	15	6.0	4	0	0
43,800	8	4.5	3	5	1
37,900	3	5.0	2	2	2
43,500	2	5.0	1	10	3

(a) Determine the estimated multiple regression equation for earnings as a function of the remaining five independent variables.
(b) Holding all other variables constant, determine the estimated worth (increase in earnings) from: (1) one year of experience, (2) one year of engineering education, (3) one more position held, (4) an additional employee being supervised, (5) one patent.
(c) Calculate the standard error of the estimate.
(d) Predict the annual earnings for each of the following engineers.

	Professional Experience (years)	Engineering Education (years)	Positions Held	Number of Employees Supervised	Patents Held
(1)	20	4	1	0	10
(2)	10	5	5	20	0
(3)	5	6	2	1	1
(4)	3	8	1	0	3

4–23 Table 4–9 shows the results obtained by two researchers studying methods for decontaminating stainless steel canisters used in disposing of radioactive waste. A corrosive concentration of varying strength was applied to sample canisters and the weight loss was measured.

(a) Using loss of weight as the dependent variable *Y*, with cerium concentration X_1, temperature X_2, and contact time X_3 as the independent variables, determine the estimated regression equation.

(b) Find point estimates for each of the following cases:
(1) 0.01 M Ce, 25°C, 3 hours
(2) 0.05 M Ce, 45°C, 6 hours
(3) 0.10 M Ce, 50°C, 1 hour
Do you notice anything strange about the estimated values?

(c) Holding temperature fixed at 65°C and time at 6 hours, plot a scatter diagram of weight loss versus cerium concentration. What does the graph suggest regarding using a *linear* regression model in evaluating these data?

Table 4–9
Results of Cerium (IV) Contact with Stainless Steel

Temp. X_2, °C	Contact Time X_3, h	Loss of Weight *Y*, mg for Cerium Concentration X_1, M				
		0.115 M	0.057 M	0.028 M	0.01 M	0.0025 M
25	1	3.85	1.89	0.5	—	—
	3	13.75	7.1	3.0	—	—
	6	34.6	15.4	6.4	—	—
	12	88.4	36.8	15.9	—	—
45	1	19.2	11.7	4.4	2.6	1.5
	3	76.9	39.7	16.9	10.8	3.2
	6	152.2	77.1	36.4	14.6	5.2
	12	220.2	100.2	49.3	19.3	5.1
65	1	101.5	41.0	15.4	7.1	2.9
	3	139.8	113.0	44.0	9.4	4.0
	6	228.0	100.0	54.7	17.0	4.6
	12	276.7	100.0	54.1	26.0	4.7
78	1	112.6	49.7	27.5	—	—
	3	211.1	86.9	43.9	—	—
	6	265.9	112.9	48.0	—	—
	12	280.0	122.0	65.8	—	—

Source: Bray and Thomas. *Transactions of American Nuclear Society,* Vol. 55, November 15–19, 1987, pp. 230–231.

Review Problems

4–24 The following sample data provide the reaction time and retort chamber temperature for refining a chemical feedstock:

Temperature (°F) X	Time (hours) Y	Temperature (°F) X	Time (hours) Y
200	3.3	240	2.8
205	3.3	245	2.7
210	3.0	250	2.8
215	3.1	255	2.8
220	3.0	260	2.6
225	2.9	265	2.6
230	2.9	270	2.5
235	2.8		

(a) Plot these data on a scatter diagram. Find the estimated regression equation and plot this on your graph.

(b) Compute (1) the sample standard deviation in reaction time and (2) the standard error of the estimate from regression. Can a significant amount of the variation in temperature be explained by the regression analysis? Explain.

4–25 A government transportation engineer has collected the following data for an experimental synthetic fuel used in test track runs with a benchmark automobile:

Speed (miles/hour) X	Mileage (miles/gallon) Y	Speed (miles/hour) X	Mileage (miles/gallon) Y
40	30.0	60	28.3
42	29.7	62	27.9
45	29.3	63	28.1
48	29.1	65	27.7
50	29.2	68	27.3
53	28.7	70	27.0
55	28.6	72	26.9
57	28.7		

(a) Plot the data in the table on a scatter diagram. Find the estimated regression equation and plot this on your graph.

(b) Compute (1) the sample standard deviation in mileage and (2) the standard error of the estimate from regression. Can a significant amount of the variation in mileage be explained by the regression analysis? Explain.

4–26 A software engineer wishes to predict processing times for thermodynamic evaluations made by an experimental program run on a particular computer. The following data resulted from 20 test runs:

Processing Time (minutes)	Required RAM (k bytes)	Amount of Input (k bytes)	Amount of Output (k lines)
5.2	19	5	1
17.3	105	10	2
15.5	70	15	5
23.4	80	20	8
15.4	24	12	10
9.5	15	2	5
6.2	22	3	4
10.0	35	10	3
7.7	42	5	2
6.3	15	2	2
7.2	8	4	5
8.5	7	5	6
8.9	12	10	3
5.6	15	7	2
4.1	17	4	1
9.7	18	3	6
13.4	24	8	5
11.7	25	8	4
8.4	32	10	3
12.1	79	12	2

(a) Using required RAM (random access memory) and amount of input as the independent variables and processing time as the dependent variable, determine the equation for the estimated regression plane.
(b) Complete a table showing actual, predicted, and residual values for processing time.
(c) Compute the standard error of the estimate for Y.

4–27 Refer to the software test data in Problem 4–26.
(a) Find the estimated regression equation using required RAM and output as the independent variables used to predict processing time. Then compute the standard error of the estimate.
(b) Find the estimated regression equation using amount of input and output as the independent variables used to predict processing time. Then compute the standard error of the estimate.
(c) Which of the foregoing multiple regressions would provide the better predictions of final processing time? Explain.

4–28 Using the computer test run data in Problem 4–26, find the estimated multiple regression equation for the hyperplane, incorporating all three predictors to forecast processing time.

4–29 The following readings may be incorporated into the computer test run data given in Problem 4–26:

Processing Time (minutes)	Number of Nonzero Parameters	Required Disk Storage (m bytes)	Processing Time (minutes)	Number of Nonzero Parameters	Required Disk Storage (m bytes)
5.2	50	1.4	7.2	38	2.1
17.3	60	2.5	8.5	33	4.4
15.5	40	1.8	8.9	49	3.7
23.4	58	2.2	5.6	27	4.8
15.4	43	1.7	4.1	35	3.0
9.5	57	2.2	9.7	48	1.8
6.2	31	3.1	13.4	65	4.5
10.0	52	1.9	11.7	63	2.8
7.7	45	1.3	8.4	39	1.5
6.3	59	1.7	12.1	27	2.5

(a) Find the multiple regression equation for the original variables plus the number of nonzero parameters as the fourth independent variable.

(b) Find the multiple regression equation for the hyperplane based on the original variables and the incorporation of required disk storage as the fourth independent variable.

(c) Find the multiple regression equation for the hyperplane based on all five independent variables.

CHAPTER FIVE

Statistical Analysis in Model Building

MODEL BUILDING IS a key element in modern engineering, where there is great need for data analysis that provides a summary of the relationship between two or more variables. Such knowledge is needed by engineers when building a mathematical model that may be later used in follow-on research or by practitioners. Even the most basic models have an empirical component. For example, every high school physics student encounters physical laws, such as

$$y = \frac{1}{2}gt^2$$

which tells us how far a falling object travels toward the earth by time t. We have learned that $g = 32$ feet/second^2 is the gravitational constant that makes it work. The constant g has been estimated through tests conducted by Galileo and other pioneering physicists and has been refined over the years.

But we could perform experiments to gather the data necessary to make our own estimate of g. This might be done by precisely measuring how far objects have fallen at various times t. Those data could then be plotted onto a scatter diagram like the one in Figure 5–1.

The linear regression methods of Chapter 4 must be adapted for finding how y relates to t, because a *nonlinear* relationship applies.

5–1 Nonlinear Regression

This chapter focuses on using statistical data to help explain nonlinear relationships in terms of a mathematical model. This is accomplished through a variety of regression analyses, which find the parameters giving the least squares fit to the specified functional form. Its role in model formulation makes regression a very helpful tool in making the building blocks for constructing a theory.

But regression analysis alone cannot indicate which functional form is best. Its shortcomings would be readily apparent if we were to use x to represent the distance and then attempted to find the least-squares fit for the following:

$$x = ge^{\alpha t}$$

Although we might obtain a very good fit, the result would be inadequate because we know it could not jibe with the underlying theory of falling objects. For instance, the

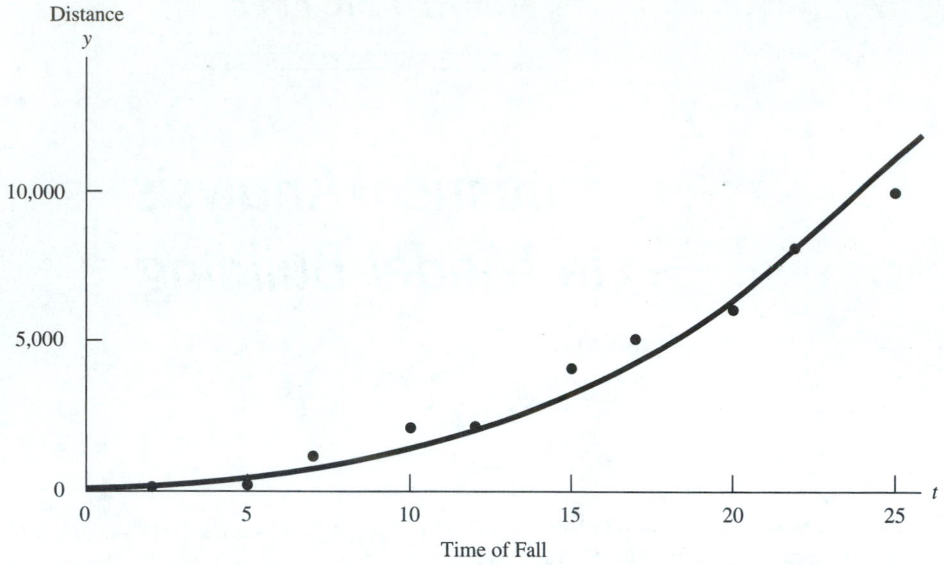

Figure 5–1
Scatter diagram that might be obtained in gravitational measurement experiment.

first derivative of the above, dx/dt, does not provide a meaningful expression for velocity as does

$$v = dy/dt = gt$$

found from the original relation

$$y = \frac{1}{2}gt^2$$

Using a Linear Surrogate

The velocity expression suggests how we might utilize the underlying theory to obtain an easier model to work with. In this case by using the relationship between velocity v and time t we can fit a *line* to observed data values. (We suppose that velocity can be readily measured with Doppler radar at any time t from start of fall. Our present concern is not with any higher instrumentation costs of this second experiment.) Figure 5–2 shows a scatter diagram that might be obtained. The slope of the regression line fitting these data provides the estimate of g.

The linear regression model can also be employed by first transforming the original data.

Transforming Variables to Get a Linear Relationship

Some underlying relationships are suited to a transformation of variables so that the new variables may be related by a linear function. Consider the scatter diagram in Figure 5–3, where the fractions Y of light bulbs still burning in chandeliers after X hours are plotted. Here the data are consistent with the exponential function of the form

$$Y = e^{-\lambda X}$$

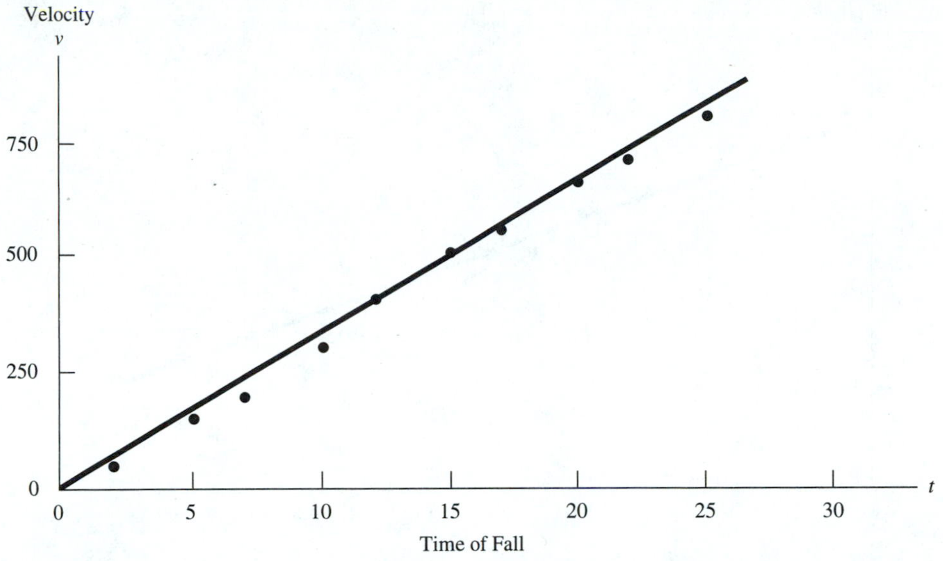

Figure 5–2
Scatter diagram that might be obtained in second gravitational measurement experiment.

The symbol λ (lowercase Greek lambda) denotes the mean failure rate (burnouts per hour), which, following a transformation of variables, may be estimated from a linear regression analysis.

A transformation of Y to natural logarithms provides

$$\ln Y = \ln(e^{-\lambda X}) = -\lambda X$$

which expresses a linear relationship between $\ln Y$ and X. The value of λ may be estimated by the slope of the regression line fitting the data in Figure 5–4. There $\ln Y$ is the dependent variable and time X is the independent variable.

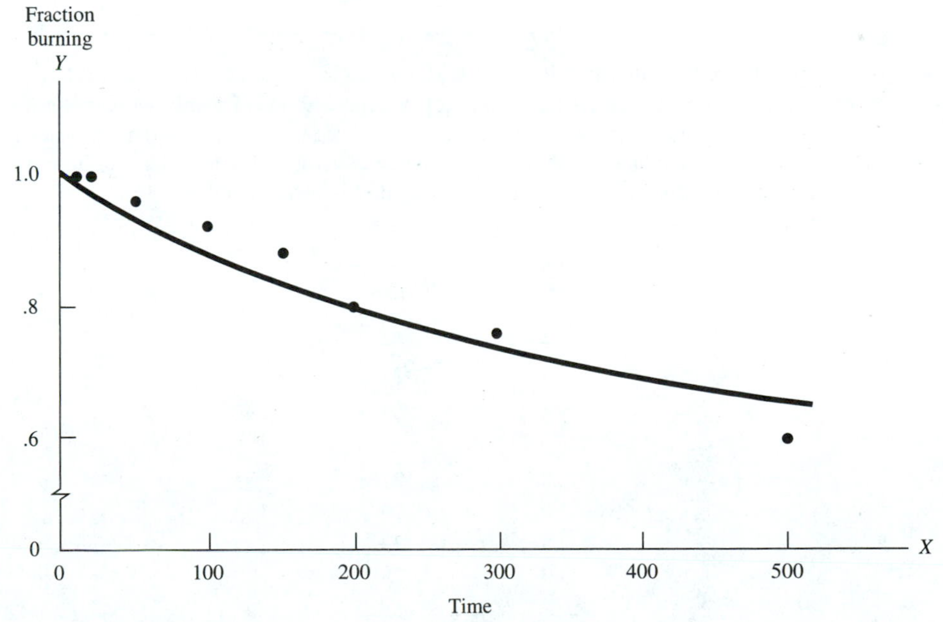

Figure 5–3
Scatter diagram for fraction of light bulbs still burning versus time.

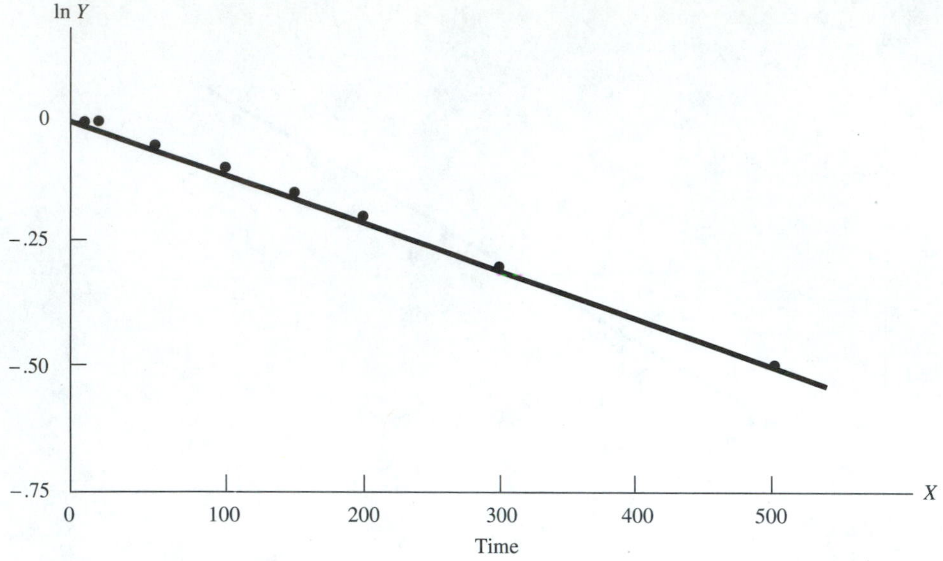

Figure 5–4 *Scatter diagram for natural logarithm of the fraction of light bulbs still burning versus time.*

Fitting a Polynomial and Other Multiple Regression Procedures

Later in this chapter we will extend the procedures of *multiple* regression to fitting a polynomial to the empirical data. Thus, we could work directly with the original falling-object model and use distance of fall as the dependent variable, finding directly an expression for the curve sketched in Figure 5–1. We will also see how to use indicator variables as additional independent variables used in building models where data are obtained under differing conditions.

Problems

5–1 To help her son, Eddie, with a physics project, Major Edwina Rickenbacher performs a gravity experiment while flying a jet fighter aircraft through a series of nearly identical inside loops. During these maneuvers she activates a device designed by her son. The instrument includes a yard-long vacuum tube that on command releases an object and measures the time of fall for different distance settings. The following data were obtained.

Distance (feet)	Time (seconds)
1.0	.14
1.5	.17
1.5	.18
2.0	.21
2.0	.20
2.5	.22
3.0	.25

(a) Using time as the abscissa, plot the above data on a scatter diagram.
(b) Free-falling objects on earth travel a distance of $\frac{1}{2}gt^2$. For the observed times, plot on your graph the distance predicted by that model.
(c) Do the Rickenbacher data agree with what would be expected with free-falling objects? Explain.

5–2 Suggest how you might collect data to perform experiments for measuring the following.
(a) velocity of light c in a vacuum
(b) electron charge e
(c) Planck's constant h

5–3 Consider the following data sets. In each case (1) plot a scatter diagram and (2) suggest a possible functional form relating Y to X.

(a)		(b)		(c)	
Y	X	Y	X	Y	X
1.1	1	1.1	0	3.2	.5
3.7	2	1.6	1	9.8	1.0
26.0	5	7.0	2	31.8	1.5
50.5	7	20.0	3	98.8	2.0
37.3	6	52.5	4	321.5	2.5
87.4	9			995.0	3.0

5–2 Curvilinear Regression

We now consider **curvilinear regression**, where the dependent variable Y may best be described in terms of X by a nonlinear function. A broad class of nonlinear regression functions may be established using the procedures described so far. This is accomplished by transforming X or Y so that the underlying curve relating Y to X may be expressed as a line involving the transformed variables.

Although a variety of two-variable relationships may be treated as essentially linear, an alternative approach is to fit a polynomial to the raw data directly. That more general procedure, **polynomial regression**, will be described in Section 5–3.

Logarithmic Transformations

In some applications the relationship between X and Y might best be described by the following multiplicative model:

$$\hat{Y}(X) = ab^x$$

Taking the natural logarithm of both sides, we may express this equivalently as

$$\ln \hat{Y}(X) = \ln a + X \ln b$$

which is the expression for a **logarithmic regression line** relating the transformed dependent variable ln Y to X. The logarithmic regression line is found by using the method of least squares exactly as before.

ESTIMATED REGRESSION COEFFICIENTS FOR THE LOGARITHMIC REGRESSION LINE

$$\ln b = \frac{n\sum X \ln Y - \sum X \sum \ln Y}{n\sum X^2 - \left(\sum X\right)^2}$$

$$\ln a = \frac{\sum \ln Y}{n} - \bar{X} \ln b$$

(These computations may be performed using logarithms to the base 10. Natural logarithms are computationally more convenient, however.)

Illustration: *Artificial Intelligence and Computer Chess*

Consider the data in Table 5–1 resulting from an artificial intelligence investigation involving computer chess games. The skill index levels Y have been established for a particular heuristic chess algorithm after several successive games X played against an established program. Each data point resulted from a reinitialized computer run with new program parameters; computer time limitations determined the number of games played in each run. In relating Y to X, it is plausible to assume exponential improvement in chess-playing skill as the number of games is increased. This is demonstrated by the scatter diagram in Figure 5–5 (a), where the data closely fit a regression curve having appropriate form.

Table 5–1 *Computer Chess Results for Artificial Intelligence Study*

Number of Games X	**Skill Index** Y	$\ln Y$	$X \ln Y$	X^2
3	16	2.773	8.318	9
7	34	3.526	24.682	49
8	40	3.689	29.512	64
12	93	4.533	54.396	144
14	125	4.828	67.592	196
9	50	3.912	35.208	81
9	48	3.871	34.839	81
4	18	2.890	11.560	16
11	70	4.248	46.728	121
13	110	4.700	61.100	169
12	85	4.443	53.316	144
8	45	3.807	30.456	64
7	40	3.689	25.823	49
6	33	3.497	20.982	36
3	20	2.996	8.988	9
126 $=\sum X$		57.402 $=\sum \ln Y$	513.500 $=\sum \ln Y$	1,232 $=\sum X^2$

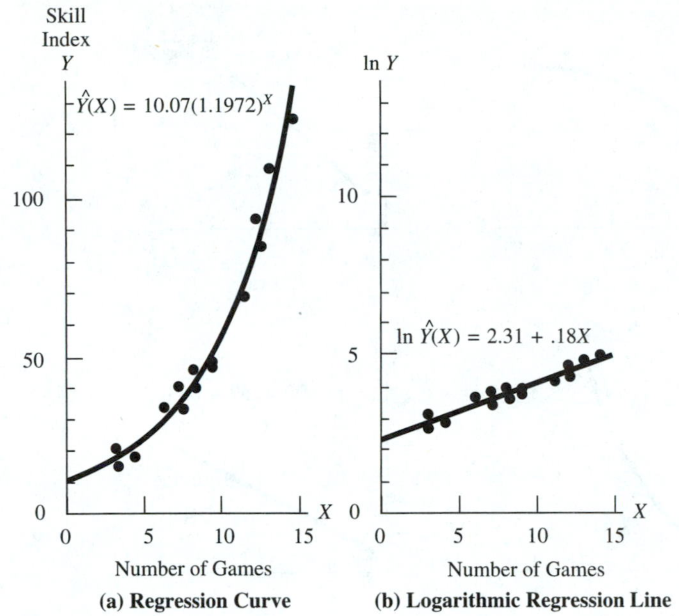

Figure 5–5
Using least squares to find estimated regression curve (a) by first finding logarithmic regression line (b).

The regression curve was found using the computed values from Table 5–1.

$$\ln b = \frac{15(513.50) - 126(57.402)}{15(1{,}232) - (126)^2} = .18$$

$$\ln a = \frac{57.402}{15} - 8.40(.18) = 2.31$$

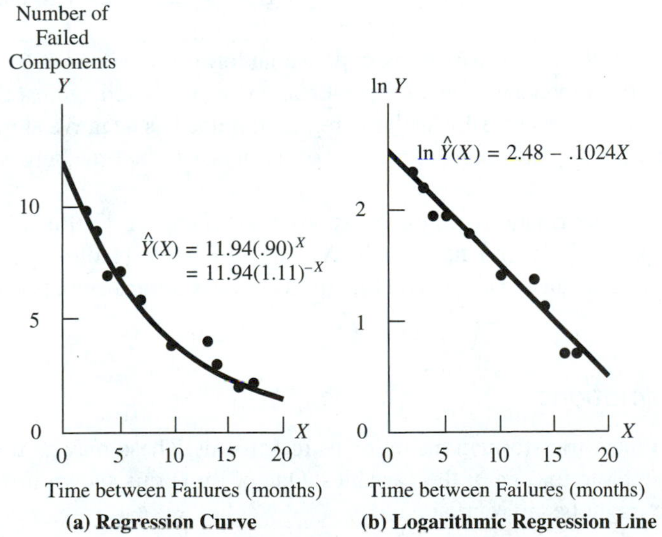

Figure 5–6
Estimated regression curve (a) and estimated logarithmic regression line (b) for components in environmental tests.

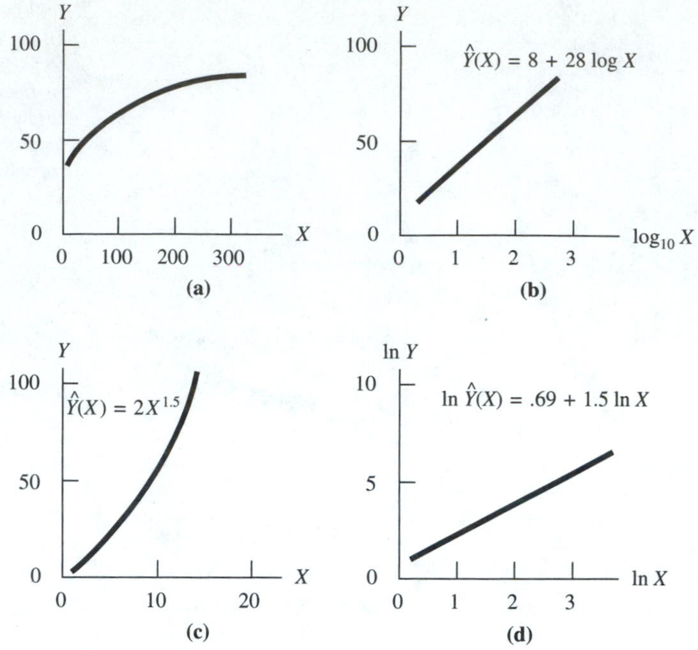

Figure 5–7
Further logarithmic transformations for obtaining linear regression relationships.

so that the logarithmic regression line is

$$\ln \hat{Y}(X) = 2.31 + .18X$$

This line is plotted in Figure 5–5 (b). If we raise e to the power of the preceding expression, the corresponding regression curve is

$$\hat{Y}(X) = e^{2.31+.18X} = 10.07(1.1972)^X$$

which indicates that there is a 19.72% increase in the program's skill level with each chess game played.

The regression curve might assume a *negative* exponential form, as in Figure 5–6 (a), where the number of failed components Y in environmental testing is plotted against the time X between failures. Notice that the logarithmic regression line has negative slope, reflecting the inverse relationship between the number of failures and the time between failures.

A variety of other transformations might be appropriate. Figure 5–7 illustrates transformations involving the independent variable X. The line in (b) results from a change in scale using log X as the abscissa. The line in (d) reflects a transformation of both regression variables.

Reciprocal Transformations

Another family of curvilinear transformations involves reciprocals. These may be useful when there is an asymptote for one of the variables. One of the forms shown in the table on the facing page might be appropriate.

Original Relationship	Transformation	Linear Relationship
$\hat{Y} = a + \frac{b}{X}$	$X' = \frac{1}{X}$	$\hat{Y}(X') = a + bX'$
$\frac{1}{\hat{Y}} = a + bX$	$\hat{Y}' = \frac{1}{\hat{Y}}$	$\hat{Y}'(X) = a + bX$
$\frac{1}{\hat{Y}} = a + \frac{b}{X}$	$X' = \frac{1}{X}, \quad \hat{Y}' = \frac{1}{\hat{Y}}$	$\hat{Y}'(X') = a + bX'$

Problems

5–4 Consider the following data relating the area of aquifer contamination Y (acres) to the time X (years) after release of toxic chemical:

X	Y	X	Y
.5	1.5	5.1	35.8
1.3	4.8	5.6	44.5
2.4	5.3	6.2	68.7
3.6	10.1	7.3	165.6
4.4	19.7	8.1	253.4

(a) Plot these data on a scatter diagram.
(b) Use curvilinear regression in fitting a curve of form $\hat{Y} = ab^X$. Plot your curve on the diagram found in (a).

5–5 Using the artificial intelligence sample data in Table 5–1, determine the coefficients for a regression equation of the form $\hat{Y} = aX^b$.

5–6 Using the data in Problem 5–4, determine the coefficients for a regression curve of the form $\hat{Y} = aX^b$.

5–7 An FAA engineer has collected the following aircraft-landing-accident data:

Number of Crashes	Nearest Distance from Runway (feet)
32	100
15	200
6	1,000
2	2,000
1	5,000

Determine the logarithmic regression line using distance as the independent variable.

5–8 The following data apply to a test performed on a sample of bitumen removed from a tar sand deposit:

Temperature X **(°F)**	**Viscosity** Y **(poise)**	**Temperature** X **(°F)**	**Viscosity** Y **(poise)**
750	50	620	820
800	15	650	400
700	100	680	150
850	9	710	110
590	950	550	1,200

(a) Determine the coefficients to the following regression equation:

$$\hat{Y} = ab^{(1{,}000/X)}$$

(b) Plot the corresponding curvilinear regression line and the transformed data points, using

$$X' = \frac{1{,}000}{X} \qquad \text{and} \qquad Y' = \ln Y$$

5–9 An evaluation was made of a method for decontaminating stainless steel canisters used in disposing of radioactive waste. A corrosive concentration of varying strength was applied to sample canisters and the weight loss was measured. The following data were obtained:

Cerium Concentration M	**Canister Weight Loss (mg)**	
	One Hour	*Three Hours*
.0025	2.9	4.0
.0100	7.1	9.4
.0280	15.4	44.0
.0570	41.0	113.0
.1150	101.5	139.8

Source: Bray and Thomas, *Transactions of American Nuclear Society*, Vol. 55, November 15–19, 1987, pp. 230–231.

(a) For the one-hour results, find the estimated curvilinear regression equation for estimating canister weight loss from cerium concentration. Use a natural logarithmic transformation on both the independent and dependent variables.

(b) Plot the scatter points and sketch the regression surface using (1) arithmetic scales and (2) log-log scales.

5–3 Polynomial Regression

Section 5–2 introduced curvilinear regression, in which various nonlinear relationships between X and Y could be treated as essentially linear. This was made possible by transforming one or both variables, using an inverse or logarithm to a convenient base. But such a procedure is limited to a narrow class of functional shapes. More generally, it is possible to fit Y to a polynomial in the independent variable X. As long as the polynomial involves high enough powers of X, virtually any curve form can be selected to fit the data. This may be accomplished by extending the least-squares multiple regression procedure.

Parabolic Regression

The simplest such polynomial is an estimated regression equation of the form

$$\hat{Y} = a + b_1 X + b_2 X^2$$

which gives a parabola. This equation is equivalent to that of the estimated regression plane when $X_1 = X$ and $X_2 = X^2$. Making these variable substitutions, we can perform a multiple regression to find the values of a, b_1, and b_2.

Illustration: *Finding a Stress-Strain Curve*

Consider the stress-strain data in Table 5–2 obtained from testing an experimental alloy. The above least-squares procedure provides the stress-strain curve shown in Figure 5–8. A parabola was determined to provide the best fit over the middle-strength range.

Table 5–2 *Test Data for Experimental Alloy Tension Member*

Test	Stress (ksi) Y	Strain (in./in.) X	Test	Stress (ksi) Y	Strain (in./in.) X
1	91	.001	6	110	.006
2	97	.002	7	112	.009
3	108	.003	8	105	.011
4	111	.005	9	98	.016
5	114	.006	10	91	.017

The intermediate calculations are given in Table 5–3. Substituting the appropriate values into the expressions on page 109, we obtain the following normal equations:

$$1{,}037 = 10a + .076b_1 + 858 \times 10^{-6}b_2$$
$$7.786 = .076a + 858 \times 10^{-6}b_1 + 11{,}662 \times 10^{-9}b_2$$
$$.085454 = 858 \times 10^{-6}a + 11{,}662 \times 10^{-9}b_1 + 173{,}574 \times 10^{-12}b_2$$

Solving these equations simultaneously, we obtain the following estimated regression coefficients:

$$a = 89.6 \qquad b_1 = 5{,}378 \qquad b_2 = -311{,}829$$

The estimated parabolic regression equation for the stress-strain curve is

$$\hat{Y} = 89.6 + 5{,}378X_1 - 311{,}829X_2$$

or

$$\hat{Y} = 89.6 + 5{,}378X - 311{,}829X^2$$

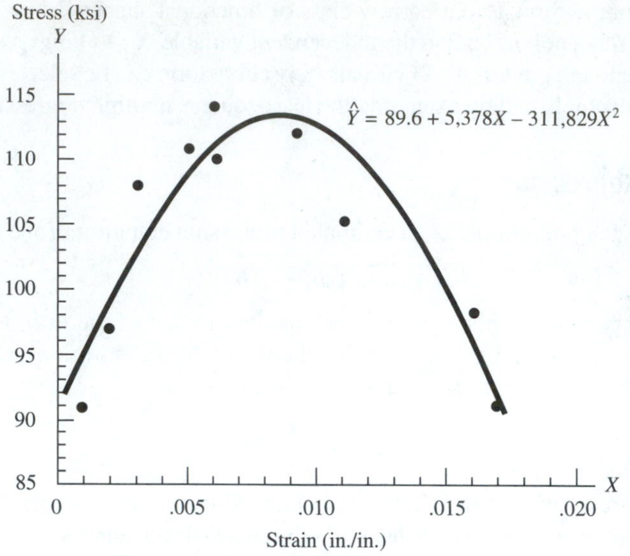

Figure 5–8 *Scatter diagram and parabolic stress-strain regression curve for experimental alloy tension member.*

Regression with Higher-Power Polynomials

The parabolic regression procedure may be extended by incorporating higher powers of X as additional independent variables. In the case of a third-degree polynomial, the estimated regression equation will be of the form

$$\hat{Y} = a + b_1X + b_2X^2 + b_3X^3$$

A perfect fit can be achieved for any data when the highest power of X matches the number of data points. A close fit to almost any data might be achieved using a polynomial involving modest powers. However, unless this is carefully done, predictions of Y made from a higher-power regression equation may substantially miss subsequent actual levels. The greatest challenge in such a curve-fitting exercise is providing a rationale that

Table 5–3 *Intermediate Calculations in Support of Parabolic Regression to Find Stress-Strain Curve*

Y	X_1 (X)	X_1Y	X_2 (X^2)	X_2Y	X_1X_2	X_2^2
91	.001	.091	1×10^{-6}	.000091	1×10^{-9}	1×10^{-12}
97	.002	.194	4×10^{-6}	.000388	8×10^{-9}	16×10^{-12}
108	.003	.324	9×10^{-6}	.000972	27×10^{-9}	81×10^{-12}
111	.005	.555	25×10^{-6}	.002775	125×10^{-9}	625×10^{-12}
114	.006	.684	36×10^{-6}	.004104	216×10^{-9}	$1{,}296 \times 10^{-12}$
110	.006	.660	36×10^{-6}	.003960	216×10^{-9}	$1{,}296 \times 10^{-12}$
112	.009	1.008	81×10^{-6}	.009072	729×10^{-9}	$6{,}561 \times 10^{-12}$
105	.011	1.155	121×10^{-6}	.012705	$1{,}331 \times 10^{-9}$	$14{,}641 \times 10^{-12}$
98	.016	1.568	256×10^{-6}	.025088	$4{,}096 \times 10^{-9}$	$65{,}536 \times 10^{-12}$
91	.017	1.547	289×10^{-6}	.026299	$4{,}913 \times 10^{-9}$	$83{,}521 \times 10^{-12}$
1,037	.076	7.786	858×10^{-6}	.085454	$11{,}662 \times 10^{-9}$	$173{,}574 \times 10^{-12}$
$= \sum Y$	$= \sum X_1$	$= \sum X_1Y$	$= \sum X_2$	$= \sum X_2Y$	$= \sum X_1X_2$	$= \sum X_2^2$

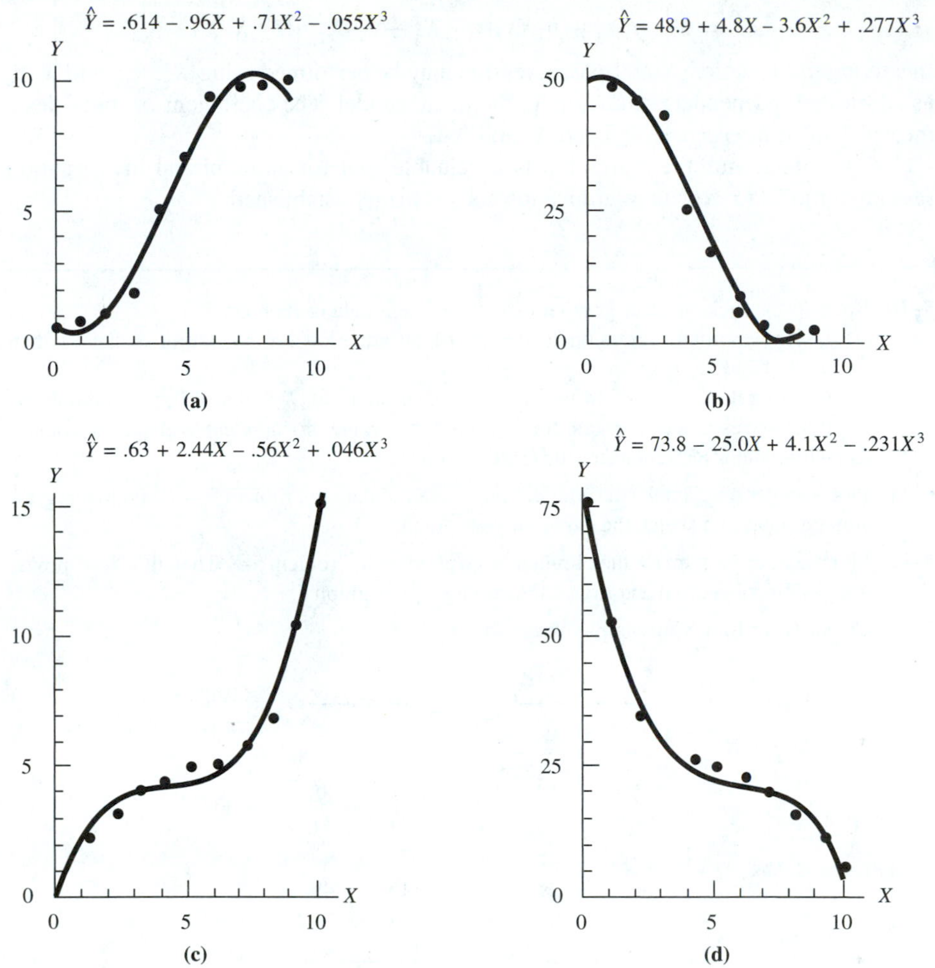

Figure 5–9
Scatter diagrams and regression curves for third-power polynomials.

explains the regression coefficients. For this reason polynomials involving powers higher than X^3 are rarely employed in regression analysis. Figure 5–9 shows a variety of curve shapes that might be encountered in third-power polynomial regression.

The least-squares procedure for higher-degree polynomial regression is really a simple extension of multiple regression, using X^2, X^3, and so on as additional independent variables. As we have seen, such regressions will usually be accomplished with the assistance of a digital computer.

Polynomial Multiple Regression

The linear multiple regression procedures described so far are based on a linear relationship between Y and two or more independent variables. In some applications the relationship between the several variables may be a nonlinear one. In such cases a model expressing Y as a higher-degree polynomial in the several independent variables might be employed. When there are two independent variables, the following second-degree polynomial might be fitted to the sample data:

$$\hat{Y} = a + b_1 X_1 + b_2 X_2 + b_{11} X_1^2 + b_{22} X_2^2 + b_{12} X_1 X_2$$

In solving this equation, a multiple regression may be performed using X_1^2, X_2^2, and $X_1 X_2$ as additional independent predictors in the linear model. The coefficient b_{12} provides a measure of the *interaction* between X_1 and X_2.

Polynomial multiple regression is a valuable tool for an empirical investigation seeking a model to explain relationships not yet firmly established.

Problems

5–10 Refer to the Rickenbacher gravitational experiment data in Problem 5–1.

(a) Use polynomial regression to determine an equation for predicting distance when $X_1 = t$ and $X_2 = t^2$.

(b) Ignoring the X_1 term and intercept, the coefficient of X_2—that is, of t^2—should be twice the estimated constant for g. What is the value of g applicable inside the cockpit while Major Rickenbacher flies inside loops?

5–11 Plot a scatter diagram for the aquifer contamination data in Problem 5–4. Then fit a regression parabola and sketch the curve on your graph.

5–12 Plot a scatter diagram for the bitumen viscosity data in Problem 5–8. Then fit a third-power regression polynomial and sketch the curve on your graph.

5–13 Consider the following sample observations:

Y	*X*	*Y*	*X*
50	10	55	38
61	18	41	41
72	25	21	43
69	29	15	49
70	37	18	55

(a) Plot these data on a scatter diagram.

(b) Fit a parabola to the data and sketch the resulting regression curve on your graph.

(c) Fit a third-power polynomial to the data and sketch the resulting regression curve on your graph.

(d) Which curve do you think provides the better fit?

5–14 The following stress-strain data were collected from testing a particular material:

Conventional Stress (ksi)	Strain (in./in.)	Conventional Stress (ksi)	Strain (in./in.)
18	.0015	25	.0091
23	.0022	22	.0109
29	.0029	18	.0118
28	.0054	19	.0119
33	.0069	15	.0125

Plot these data on a scatter diagram using conventional stress as the dependent variable. Then fit a regression parabola and sketch the resulting curve on your graph.

5–15 The following data apply to the material in Problem 5–14, where, after the onset of plastic instability, the smallest cross-sectional area of the metal bar is used in computing true stress.

True Stress (ksi)	Strain (in./in.)	True Stress (ksi)	Strain (in./in.)
18	.0015	39	.0091
23	.0022	40	.0109
30	.0029	49	.0118
32	.0054	58	.0119
35	.0069	69	.0125

Plot these data on a scatter diagram using true stress as the dependent variable. Then fit a third-power regression polynomial and sketch the resulting curve on your graph.

5–16 Consider the following data:

Y	X_1	X_2	Y	X_1	X_2
103	5	10	251	50	50
105	10	5	267	50	60
131	20	30	308	60	50
142	25	40	415	70	70
175	40	20	474	80	70

Extend the linear multiple regression procedure to fit a second-degree polynomial (including X_1^2, X_2^2, and $X_1 X_2$ terms) to these data.

5–4 Multiple Regression with Indicator Variables

Our discussions of multivariate relationships have so far been limited to quantitative variables such as time, weight, and volume. Regression analysis is concerned mainly with finding an appropriate functional relationship between such values. An interesting extension of multiple regression analysis considers a wider class of applications involving *qualitative* variables as well as quantitative ones.

A variety of qualitative variables might be important in engineering evaluations. A chemical engineer might use regression to predict chemical yield in response to different temperature or pressure settings. But slightly different raw feedstocks from two suppliers might provide a different response, so it may be helpful to further categorize each production run by *type* of feedstock. Or consider a computer designer simulating a proposed microcircuit to predict how processing time will respond to the number of inputs and volume of computations. Different results may be achieved for test runs involving main memory only as opposed to those needing peripheral access as well; thus, *memory classification* may be a crucial qualitative variable. As a further illustration, a tire-tread evaluation might be based on test runs involving cars from one or more manufacturers, so an important variable might be *car model*.

Qualitative factors may be incorporated into regression evaluations by using **indicator variables**. Such a quantity assumes a value of 0 or 1, depending upon which category applies in a two-way classification.

The Basic Multiple Regression Model

A typical application involves estimating the relationship between Y and a single quantitative independent variable X_1. But the sample observations fall into two categories, so an indicator variable X_2 is incorporated as the second independent variable. The estimated regression equation takes the form

$$\hat{Y} = a + b_1X_1 + b_2X_2$$

where $X_2 = 0$ if one category applies and $X_2 = 1$ if the second is indicated. Two estimated regression lines will apply:

$$\hat{Y} = a + b_1X_1 + b_2(0) = a + b_1X_1 \qquad \text{(when } X_2 = 0\text{)}$$

and

$$\hat{Y} = a + b_1X_1 + b_2(1) = (a + b_2) + b_1X_1 \qquad \text{(when } X_2 = 1\text{)}$$

Illustration: *Predicting Chemical Yields*

Consider the chemical processing data in Table 5–4, where the dependent variable Y is the final yield and the independent variable is the amount X_1 of active ingredient. An expensive catalyst is used on some of the sample runs. The following indicator variable is defined:

$$X_2 = \begin{cases} 0 & \text{if no catalyst is present} \\ 1 & \text{if catalyst is present} \end{cases}$$

The following estimated multiple regression equation applies:

$$\hat{Y} = 9.000 + .225X_1 + 9.875X_2$$

The partial regression coefficient $b_1 = .225$ indicates that .225 kg of yield is estimated for every 1 kg of active ingredient used in processing a batch of chemical. The second coefficient, $b_2 = 9.875$, signifies that batches run with the catalyst present will provide yields that are estimated to be 9.875 kg greater than those run with an identical quantity of active ingredient, but with the catalyst absent. This is presumed true regardless of the amount of active ingredient used.

Figure 5–10 should help demonstrate the advantages of performing the multiple regression with the indicator variable. The crosses represent those sample observations resulting from test runs where the catalyst was present. The circles apply to the outcomes where the catalyst is absent.

Table 5–4
Chemical Yield and Ingredient Data from Processing Test Batches Where Catalyst Is Present or Absent

Yield (kg) Y	Amount of Active Ingredient (kg) X_1	Catalyst	Indicator Variable X_2
21	50	Absent	0
33	100	Absent	0
41	90	Present	1
35	80	Present	1
24	60	Absent	0
20	60	Absent	0
27	70	Absent	0
23	50	Absent	0
24	80	Absent	0
26	90	Absent	0
39	90	Present	1
43	110	Present	1
47	130	Present	1
45	130	Present	1
53	140	Present	1
55	150	Present	1

The bottom yield regression line

$$\hat{Y} = 9.000 + .225X_1 \qquad \text{(catalyst absent)}$$

applies when the catalyst is absent. The higher parallel line

$$\begin{aligned}\hat{Y} &= (9.000 + 9.875) + .225X_1 \\ &= 18.875 + .225X_1 \qquad \text{(catalyst present)}\end{aligned}$$

is applicable when the catalyst is used. Notice that each line falls near the middle of the scatter for the respective data points for the category it represents. They were not, however, chosen so that the respective sums of squared deviations about them were minimized. These lines are instead the $X_1 - Y$ *projections* of perpendicular slices made through the estimated regression plane at $X_2 = 0$ and $X_2 = 1$, respectively. The least-squares criterion applies to that plane, not to the linear projections.

Advantages of Using Indicator Variables

The multiple regression using X_2 as a catalyst indicator variable is superior to the result that would be achieved with a single variable regression evaluation employing amount of active ingredient as the only predictor. Ignoring the catalyst effect, the dashed line in Figure 5–10 shows that a single line fits the sample data poorly.

But there is no reason why the effect of the catalyst should be ignored. Why not perform separate simple regression evaluations, the first based only on those points where no catalyst was used and the second for those involving the catalyst?

Upon reflection, the answer should be obvious. In any scientific sampling investigation, greater precision and reliability accrue as the number of observations increases.

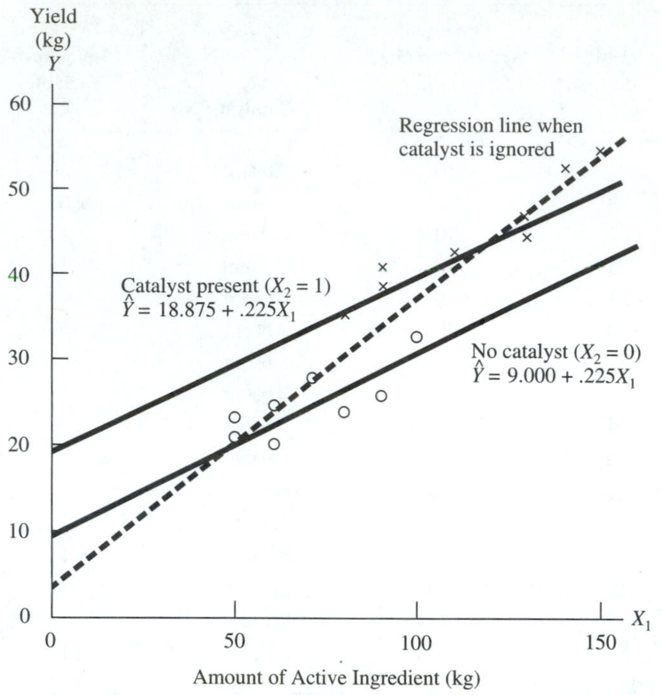

Figure 5–10 *Scatter diagram for chemical processing data showing linear projections from multiple regression with indicator variable for catalyst (no interaction assumed).*

The standard error obtained from the multiple regression will be very close in value to what would be obtained by averaging the pooled squared deviations about the two separate regression lines. But the multiple regression represents a greater amount of information per observation. It should therefore provide more precise confidence intervals for the conditional mean Y than what would be possible from either simple regression analysis alone. In effect, the multiple regression takes greater advantage of existing information.

Of course, the underlying model should have a meaningful rationale. It seems perfectly logical that the Y-intercept should be greater for the catalyst case than it is for no catalyst since the greater yields are expected when the catalyst is included. But the same slope applies to both linear projections. This can be true only if it is assumed that yield responds to the active-ingredient levels at a constant rate — regardless of the presence or absence of the catalyst. That assumption may be inappropriate, and a greater slope might apply when the catalyst is present.

We next consider a procedure even better suited to the chemical processing illustration because it allows both for different *slopes* and for different Y-intercepts.

Interactive Multiple Regression with Indicator Variables

As we have seen, a linear model may not be appropriate for characterizing the relationship between two or more variables. In the previous section we encountered a second-degree polynomial in two independent variables involving X_1^2, X_2^2, and X_1X_2 terms. As a special case, we may drop the squared terms, leaving X_1X_2 as the only additional term. This product characterizes the *interaction* between X_1 and X_2. When X_2 is a zero-one indicator variable, X_1X_2 becomes 0 when $X_2 = 0$; X_1X_2 becomes X_1 when $X_2 = 1$.

If we apply a linear multiple regression using X_1X_2 as the third independent variable, the following estimated multiple regression equation applies:

$$\hat{Y} = a + b_1X_1 + b_2X_2 + b_{12}X_1X_2$$

This regression surface gives two linear projections:

$$\hat{Y} = a + b_1X_1 \qquad \text{(when } X_2 = 0\text{)}$$

and

$$\hat{Y} = (a + b_2) + (b_1 + b_{12})X_1 \qquad \text{(when } X_2 = 1\text{)}$$

A computer run with the data in Table 5–4 gives the following estimated multiple regression equation:

$$\hat{Y} = 12.208 + .179X_1 + 4.031X_2 + .069X_1X_2$$

Setting X_2 at the respective indicator values, we find that this equation gives

$$\hat{Y} = 12.208 + .179X_1 \qquad \text{(catalyst absent)}$$

and

$$\hat{Y} = (12.208 + 4.031) + (.179 + .069)X_1$$
$$= 16.239 + .248X_1 \qquad \text{(catalyst present)}$$

Figure 5–11 shows these linear projections.

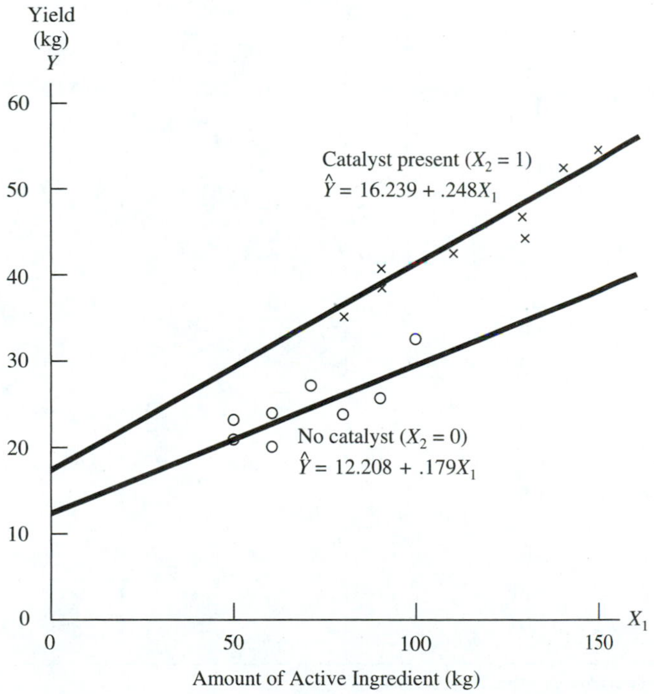

Figure 5–11
Scatter diagram for chemical processing data showing linear projections from multiple regression with catalyst indicator variable. Interaction between independent variables is assumed.

Illustration:* *Polymer Production of Oil Recovery Enhancement

Researchers conducted an experiment with xanthan biopolymer broth, a chemical used to thicken water for oil-recovery enhancement. The final chemical is produced from fermentation using a beginning cell concentration and feed sugar. Table 5–5 shows partial results from 22 test runs. The main variable of interest is the viscosity Y of the final product. The level is affected by four variables: percentage weight of feed sugar X_1, starting cell concentration X_2, presence or absence of logarithmic growth inoculum X_3 (indicator variable), and whether or not sugar is sterilized separately X_4 (indicator variable).

Figure 5–12 shows the results of a multiple regression analysis run with SAS. Included is a fifth indicator interaction variable X_3X_4. The following estimated multiple regression equation was obtained:

$$\hat{Y} = -4{,}232.4 + 5{,}043.9X_1 - 20.29X_2 + 408.6X_3 + 2{,}500.7X_4 - 935.0X_3X_4$$

This equation provides an estimated increase in viscosity of 408.6 when $X_3 = 1$ (logarithmic growth inoculum is present) or an increase of 2,500.7 when $X_4 = 1$ (sugar is

Table 5–5 *Results of Fermentation to Produce Polymer Used in Oil-Recovery Enhancement*

Test Run	Yield Viscosity (cp) Y	Feed Sugar (%) X_1	Beg. Cell Concent. $\times 10^{-8}$/ml X_2	Growth Inoculum (no = 0, yes = 1) X_3	Separate Sugar Steril. (no = 0, yes = 1) X_4	Interaction X_3X_4
1	6,300	2.00	5.2	0	0	0
2	3,900	1.50	2.2	0	0	0
3	4,900	2.00	2.7	0	0	0
4	3,000	1.50	3.6	0	0	0
5	8,000	2.00	4.6	1	1	1
6	5,200	1.50	0.9	1	1	1
7	3,700	1.25	3.6	1	1	1
8	2,400	1.25	4.0	1	0	0
9	3,900	1.25	2.8	1	1	1
10	4,200	1.25	1.9	1	1	1
11	4,400	1.25	1.5	1	1	1
12	4,000	1.25	1.5	1	1	1
13	2,700	1.25	2.0	1	1	1
14	4,100	1.25	1.9	1	1	1
15	4,300	1.25	12.0	0	1	0
16	4,400	1.25	10.0	0	1	0
17	4,200	1.25	1.2	1	1	1
18	4,600	1.25	1.8	1	1	1
19	4,500	1.25	0.2	1	1	1
20	3,500	1.25	0.2	1	1	1
21	3,300	1.25	1.8	1	1	1
22	4,900	1.25	1.0	1	1	1

Source: Norton et al., *Society of Petroleum Engineers Journal*, April 1981, pp. 205–217.

```
                        OIL-RECOVERY ENHANCEMENT

Model: MODEL1
Dep Variable: VISCOS

                          Analysis of Variance

                           Sum of          Mean
      Source      DF      Squares         Square       F Value        Prob>F

      Model        5 23931512.261 4786302.4523          13.280        0.0001
      Error       16 5766669.5569 360416.84730
      C Total     21 29698181.818

          Root MSE       600.34727      R-Square        0.8058
          Dep Mean      4290.90909      Adj R-Sq        0.7451
          C.V.            13.99114

                          Parameter Estimates

                       Parameter        Standard        T for H0:
Variable     DF         Estimate           Error      Parameter=0      Prob > |T|

INTERCEP      1     -4232.362222    1230.9139387           -3.438          0.0034
SUGAR         1      5043.922101    781.97732144            6.450          0.0001
CELL          1       -20.292395    136.68487861           -0.148          0.8838
GROWTH        1       408.629178    800.97083972            0.510          0.6169
SEPARATE      1      2500.675946    1384.4019803            1.806          0.0897
INTERACT      1      -935.049305    1804.2575445           -0.518          0.6114
```

Figure 5–12 *SAS multiple regression report for oil-recovery enhancement data.*

sterilized separately). The coefficient of the indicator-variable product is –935.0, which requires that the estimated viscosity be reduced by that amount when X_3 and X_4 each equal 1 (yes for both factors).

Problems

5–17 The following data have been obtained for the return signal achieved from radar bursts aimed at satellite targets:

Mean Signal-to-Noise Ratio	Target Altitude (miles)	Target in Earth's Shadow	Mean Signal-to-Noise Ratio	Target Altitude (miles)	Target in Earth's Shadow
43	90	Yes	27	212	Yes
42	45	Yes	39	78	Yes
17	250	No	34	85	No
11	285	No	24	135	No
25	155	No	30	215	Yes
26	258	Yes			

(a) Using an indicator variable (with value 0 if the target is not in Earth's shadow), determine the estimated multiple regression equation expressing mean signal-to-noise ratio as the dependent variable. Assume no interaction between the indicator and target altitude.

(b) Give the linear equation for predicting the mean signal-to-noise ratio when satellite targets are in the shadow.

(c) Give the linear equation for predicting the mean signal-to-noise ratio when satellite targets are not shadowed by Earth.

5–18 Refer to the reaction time and temperature data in Problem 4–24 for refining a chemical feedstock. A stabilizer was present for all test batches run below $250°$, but because that additive breaks down at the higher temperatures, it could not be used for the 5 test batches having the hottest retort temperatures.

(a) Using reaction time as the dependent variable, find the estimated multiple regression equation when an indicator variable X_2 is used for the stabilizer (having value 1 when the stabilizer is present). Ignore any interaction between the indicator and temperature.

(b) Determine the linear equations for predicting reaction time from retort temperature when (1) the stabilizer is not present and (2) the stabilizer is present.

5–19 Refer to the mileage and speed data in Problem 4–25 for automobile test-track runs. The first four runs were made in the rain.

(a) Using mileage as the dependent variable, find the estimated multiple regression equation when an indicator variable X_2 is used for rain (having value 1 for test runs made in the rain). Ignore any interaction between the indicator and speed.

(b) Determine the linear equations for predicting mileage from speed when it is (1) not raining and (2) raining.

5–20 The following data apply to the total processing time of various sizes of computer runs with a magneto-hydrodynamics software package. Only some runs required peripheral memory:

Processing Time (seconds)	Amount of Input (k bytes)	Peripheral Memory Needed	Processing Time (seconds)	Amount of Input (k bytes)	Peripheral Memory Needed
550	3.5	No	648	8.4	No
915	6.1	Yes	1,025	5.9	Yes
748	4.4	Yes	1,284	11.4	Yes
910	5.6	Yes	627	7.7	No
707	7.2	No	849	6.1	No
589	4.5	Yes	753	8.3	No
786	10.2	No	1,806	12.5	Yes
1,004	12.6	No	2,253	14.2	Yes

(a) Treating processing time as the dependent variable Y and amount of input X_1 as the independent variable, plot these data on a scatter diagram. Use crosses for those points involving no peripheral memory and circles for the rest.

(b) Determine the estimated regression equation using X_1 as the only independent variable. Plot this line on your scatter diagram.

(c) Let X_2 be an indicator variable having value $X_2 = 0$ if peripheral memory is not needed and $X_2 = 1$ if it is needed. Determine the estimated multiple regression equation when no interaction is assumed between X_1 and X_2.

(d) Determine the estimated multiple regression equation assuming interaction between amount of input and need for peripheral memory. Then determine the expression for the projected lines when $X_2 = 0$ and $X_2 = 1$. Plot these on your scatter diagram.

5–21 Refer to Problem 5–20. Separate the data into two sets: (a) the case of no peripheral memory and (b) computer runs made when peripheral memory is needed. Perform a simple regression analysis on the two data sets and determine the expressions for each estimated regression line. Determine for each the standard error of the estimate.

5–22 Refer to the polymer production illustration on page 148. The researchers were interested in a second dependent variable, fermentation time (hours). The following data apply:

Run	Time	Run	Time	Run	Time
1	42	9	32	16	33
2	32	10	34	17	40
3	33	11	30	18	36
4	26	12	31	19	36
5	48	13	47	20	34
6	35	14	31	21	27
7	31	15	33	22	42
8	23				

Find the estimated multiple regression equation using fermentation time, instead of viscosity, as Y.

Review Problems

5–23 Refer to the artificial intelligence experiment data in Table 5–1.

(a) Using the original X-Y units, determine the sample correlation coefficient.

(b) Duplicate (a) using instead X with $\ln Y$.

5–24 Repeat Problem 5–9 using instead the three-hour results and using logarithmic transformations to the base 10.

5–25 Consider the chemical feedstock refining data in Problem 4–24.

(a) Transforming reaction time, fit a logarithmic regression line to the data and determine the estimated regression curve.

(b) Plot the above curve on a scatter diagram representing the original data. Over the ranges of observation, does the curvilinear model seem to provide a good fit to the data?

5–26 Consider the following data:

Y	X_1	X_2	X_3	Y	X_1	X_2	X_3
40	5	2	8	82	5	10	15
39	6	5	5	73	7	8	12
42	10	10	2	52	10	5	5
55	4	6	8	50	5	5	10
42	5	2	10	69	4	8	12
41	1	9	3	112	12	16	20
36	3	1	9	61	15	10	5
29	4	4	4				

Determine the estimated multiple regression equations for each of the following cases:

Independent Variables
(a) X_1 and X_2
(b) X_1 and X_3
(c) X_2 and X_3
(d) X_1, X_2, and X_3

5–27 The following data apply to a random sample of warehouses constructed under the supervision of a particular architectural and engineering firm:

Heating and Cooling Cost Y	Floor Space (1,000 sq. ft.) X_1	Winter/Spring Mean Temp. (°F) X_2	Summer/Fall Mean Temp. (°F) X_3	Energy Cost Index X_4
$8,500	45	50	70	82
11,120	40	38	65	95
11,400	45	42	72	87
8,010	25	45	68	100
8,760	30	25	65	105
15,500	60	34	71	111
17,950	75	37	72	110
18,700	80	48	75	89
13,950	55	51	81	97
8,540	25	52	69	124
10,410	35	45	68	110
12,260	45	22	71	98
13,200	50	24	70	112
22,730	75	25	65	129
12,090	45	35	72	105

(a) Perform a simple regression analysis and find the estimated regression equation for the regression line expressing heating and cooling cost in terms of floor space. Then calculate the standard error of the estimate using that regression line.

(b) Using your regression equation, predict the mean heating and cooling cost for the following floor space levels (1,000 sq. ft.): (1) 25 and (2) 30.

5–28 Refer to the data in Problem 5–27.

(a) Perform a multiple regression analysis and determine the equation for the estimated regression plane using both floor space and winter/spring mean temperature as predictors of heating and cooling cost. Then calculate the standard error of the estimate.

(b) Using your regression equation, predict the mean heating and cooling cost for the following circumstances.

	Floor Space (1,000 sq. ft.)	Winter/Spring Mean Temp. (°F)
(1)	25	40
(2)	30	45
(3)	30	50
(4)	30	55

5–29 Refer to the data in Problem 5–27 and your answers to Problem 5–28.

(a) Perform a multiple regression analysis and determine the equation for the estimated regression hyperplane incorporating summer/fall mean temperature as the third predictor of heating and cooling cost. Then calculate the standard error of the estimate.

(b) Using your regression equation, predict the mean heating and cooling cost for the following circumstances.

	Floor Space (1,000. sq. ft.)	Winter/Spring Mean Temp. (°F)	Summer/Fall Mean Temp. (°F)
(1)	25	40	60
(2)	30	45	60
(3)	30	50	60
(4)	30	55	70

5–30 Refer to the data in Problem 5–27 and your answers to Problem 5–29.

(a) Perform a multiple regression analysis and determine the equation for the estimated regression hyperplane incorporating energy cost index as the fourth predictor of heating and cooling cost. Then calculate the standard error of the estimate.

(b) Using your regression equation, predict the mean heating and cooling cost for the following circumstances.

	Floor Space (1,000 sq. ft.)	Winter/Spring Mean Temp. (°F)	Summer/Fall Mean Temp. (°F)	Energy Cost Index
(1)	25	40	60	90
(2)	30	45	60	100
(3)	30	50	60	100
(4)	30	55	70	110

5–31 The following data were obtained from a strength-of-materials test, where the neighborhood of 50°C is considered a critical threshold temperature:

Maximum Elastic Stress (ksi)	Temperature (°C)	Maximum Elastic Stress (ksi)	Temperature (°C)
43	35	32	52
40	38	26	58
40	42	20	63
38	45	16	65
37	48	10	72

Using temperature as the independent variable and maximum elastic stress as the dependent variable, complete the following:

(a) Plot the raw data on a scatter diagram.

(b) Fit a regression parabola to the data. Then sketch the curve on your graph.

(c) Each sample observation falls into one category: below critical temperature or above critical temperature. For the data points in each range perform a separate, simple linear least-squares regression. Then plot each line on your graph.

(d) Which relationship do you think better fits the observed data, the parabola from (b) or the kinked linear function found in (c)?

5–32 An industrial engineering student has collected the following sample data for 10 skilled workers. Productivity is a dependent variable to be predicted from years of experience.

Productivity (units/hr)	Years of Experience	Sex	Productivity (units/hr)	Years of Experience	Sex
60	2	Male	74	6	Male
51	1	Female	73	6	Male
53	2	Female	78	4	Female
61	3	Female	75	7	Male
72	5	Male	91	7	Female

(a) Using different symbols for men and women, plot these data on a scatter diagram. Do your data suggest that sex might be a helpful predictor?
(b) Using all 10 data points and ignoring sex of worker, find a simple regression line and plot it on your graph.
(c) Consider using an indicator variable for sex. Assuming interaction between that variable and experience, find the expression for the estimated regression equation. Then determine separate equations by worker sex for the linear projections. Plot each of these on your graph.

5–33 Refer to the strength of materials data in Problem 5–31. Include an indicator variable for whether the observed temperature falls below or above the critical level of 50°C. Assuming interaction between the independent variables, determine the estimated multiple regression equation. Then express equations for the respective linear projections.

5–34 Consider the following data:

X	Y	X	Y
1.0	10	3.6	21
1.3	16	4.1	23
2.1	11	4.5	35
2.4	13	4.7	44
2.9	16	5.0	59

(a) Calculate the sample correlation coefficient.
(b) Consider the transformation $Y' = \ln Y$. Calculate the sample correlation coefficient under the transformation.
(c) Which linear regression would explain the greatest percentage of variation in the dependent variable, the one with the original units or the one with transformed units? Explain.

CHAPTER SIX

Probability

PROBABILITY PLAYS A key role in statistics, serving as the building block for much of the underlying theory. Much statistical methodology involves the deductive statistics we first encountered in Chapter 1, which begins with assumed population characteristics and establishes the likelihood for various possible sample results. For example, probability will answer the question: "If a robot welder misses .1% of its targets, what is the probability it will still successfully complete a series of 20 spot welds?" Different approaches to deductive statistics are taken, depending on the type of population assumed.

Probability theory in turn forms the basis for inductive or inferential statistics, such as estimating strain, predicting curing time, determining whether to accept or reject a batch of components, or answering such design questions as "How much redundancy is needed in system communication?"

6–1 Fundamental Concepts of Probability

Probability theory has two basic building blocks. The more fundamental is the generating mechanism, called the **random experiment**, that gives rise to uncertain outcomes. Selecting one part for inspection from an incoming shipment is a random experiment having two elements of uncertainty: which particular item gets picked and the quality of that part. The second structural element of probability theory consists of the random experiment's outcomes, referred to as **events**. During a quality-control inspection, the usual events of interest for tested items are "good" and "defective."

Elementary Events and the Sample Space

A preliminary step in a probability evaluation is to catalog the events that might arise from the random experiment. Such a listing is made up of **elementary events**, which are the most detailed events of interest. A complete listing provides us with the random experiment's **sample space**.

For example, if we are interested only in the particular *upside showing face* from a coin toss, then we would have

Coin toss sample space = {head, tail} (showing face)

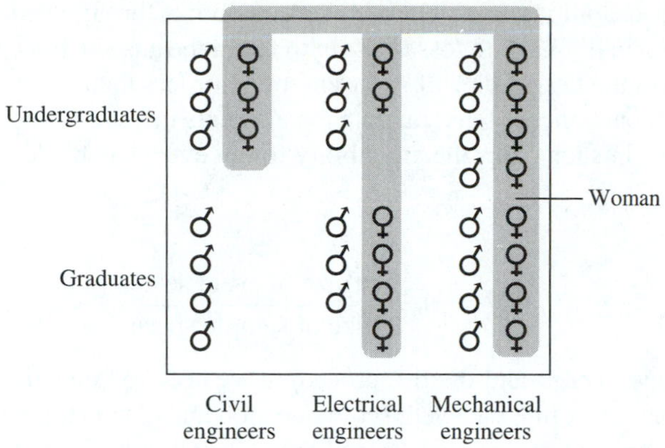

Figure 6–1
Sample space for random selection of a Tau Beta Pi member.

Alternatively, we might care only about the number of complete *revolutions*, and the same random experiment would yield an entirely different sample space:

$$\text{Coin toss sample space} = \{0, 1, 2, 3, \ldots\} \qquad \text{(revolutions)}$$

A sample space might be portrayed as a list, as above, or in some other convenient form. Sometimes it is conceptually helpful to use a picture, as in Figure 6–1, in which biological symbols are used to represent individual members of a Tau Beta Pi chapter. The random experiment is the selection of one member by lottery. Each person is an elementary event.

Event Sets

The immediate concern when finding a probability is properly identifying the event. This may be achieved by listing all the elementary events that give rise to the desired event. Such a collection is referred to as an **event set**.

Consider again the Tau Beta Pi lottery. A list of the names of female members represents the event set for the event "woman."

$$\text{Women} = \{\text{Ann, Barbara, Betty, Cheryl, Diane}, \ldots\}$$

This same event set might be pictured. Consider the tinted area in Figure 6–1.

Basic Definitions of Probability

The most common probability measure expresses the *long-run frequency* for an event occurring in many repeated random experiments. If a perfectly balanced coin is tossed *many times* without bias toward either side, we should obtain a head in about half the tosses. Thus, the long-run frequency of "head" is .50. We may then assume that .50 is the probability of "head," expressed symbolically as

$$\Pr[\text{head}] = .50$$

This value is an **objective probability**, and there should be no disagreement about how to find its value.

Objective probabilities can often be found, as above, through deductive reasoning alone. We don't actually have to toss the coin to reach the answer since we have no reason to believe that the head side will show any more or less than half the time.

Many commonly encountered random experiments involve elementary events that are *equally likely*. In such cases the probability for an event can be deduced directly.

PROBABILITY WHEN ELEMENTARY EVENTS ARE EQUALLY LIKELY

$$\Pr[\text{event}] = \frac{\text{Size of event set}}{\text{Size of sample space}}$$

To apply the above "count-and-divide" procedure we need to know only (1) how many elementary events constitute the event set in question and (2) the total number of possibilities. For example, we know that "head" is the 1 and only elementary event in the event set for that face, that it is equally likely to occur versus "tail," and that these 2 elementary events constitute the sample space for the showing side from a coin toss. Thus,

$$\Pr[\text{head}] = \frac{1}{2}$$

Consider again the Tau Beta Pi lottery summarized in Figure 6–1. We can apply this procedure to determine the probability that a woman will be selected. There are 17 women (that is, "woman" has an event set containing 17 elementary events, so that its size is 17). The sample space is made up of 38 equally likely elementary events, and

$$\Pr[\text{woman}] = \frac{17}{38} = .447$$

Certain and Impossible Events

The basic definitions of probability express the value as a fraction because an objective probability is a long-run frequency. As such, it must be a value between 0 and 1, inclusive. Events having the extreme probabilities provide two interesting cases.

An outcome having zero probability is called an **impossible event**, since it cannot occur. Returning to the lottery, we find that the empty set

$$\text{Female graduate civil engineer} = \{\quad\} = \emptyset$$

applies because there was no student (elementary event) in the above category. The event "female graduate civil engineer" is an impossible lottery event. Applying the basic definition for equally likely elementary events, we have

$$\Pr\ [\text{female graduate civil engineer}] = \frac{0}{38} = 0$$

All impossible events have zero probability.

As further illustration, we consider *mutually exclusive* events, only one of which may occur. That makes it impossible for them to occur jointly. For example, consider a part which may be classified as satisfactory or unsatisfactory, two mutually exclusive possibilities. Then for a single part, it follows

$$\Pr[\text{satisfactory } \textit{and} \text{ unsatisfactory}] = 0$$

Likewise a new satellite will either work as intended or fail. These outcomes are mutually exclusive, and their simultaneous joint occurrence is an impossible event, so that

$$\Pr[\text{work } \textit{and} \text{ fail}] = 0$$

An outcome bound to occur is a **certain event** and has a probability of 1. Returning to the lottery, all possible names to be drawn are those of students, who constitute the entire sample space. Hence,

$$\Pr[\text{student is selected}] = \frac{38}{38} = 1$$

In tossing a coin, just two faces, "head" and "tail," are possible. (Should the coin land on its edge or be lost, the random experiment would be incomplete and another toss would have to be made.) Thus,

$$\Pr[\text{head } \textit{or} \text{ tail}] = \frac{2}{2} = 1$$

Logically Deduced Probabilities

Example: *Quantum Mechanical Explanation of Photon Filtering*

One of the findings of quantum mechanics is that a photon can exist in an ambiguous state until it is measured. Thus, the measured state might be an uncertain event for which a probability might be found. Consider an ideally polarized film and the way in which it filters light that is incident on it at a right angle. Light that is polarized in parallel to the transmission axis is certain to pass through the film, while light that is polarized perpendicular to that axis is certain to be blocked.

Suppose that the light is polarized at some angle θ to the film's transmission axis. The probability that any particular photon will be transmitted is computed from

$$\Pr[\text{transmission}] = \cos\theta$$

Thus, since $\cos(0°) = 1$, parallel light is *certain* to be transmitted, and since $\cos(90°) = 0$, it is *impossible* for perpendicular light to be transmitted. For light polarized at $\theta = 45°$,

$$\Pr[\text{transmission with } \theta = 45°] = \cos(45°) = .70711$$

Source: Abner Shimony, "The Reality of the Quantum World," *Scientific American*, January 1988, pp. 46–53.

Finding Probabilities from Experimentation

It may be difficult or impossible to logically deduce probabilities for some events. Returning once more to the coin toss, suppose that we take a fair coin and make it lopsided by dropping a few blobs of solder on one face. We would no longer be able to establish the probability value by reasoning alone, and Pr[head] might be a number such as .37 or .59. Only by actually tossing the coin many times, and keeping records of what sides show, could we finally arrive at an estimate of the proper probability value.

Problems

6–1 Consider the Tau Beta Pi lottery with the sample space shown in Figure 6–1. Find probabilities for the following events:

(a) man (c) civil engineer

(b) graduate (d) mechanical engineer

6–2 Explain how you would go about finding the probabilities for the following events:

(a) A vessel bursts before a pressure of 5,000 psi is applied. It is one of many production items.

(b) One of the defective items from a batch of known size having a known percentage of defectives is selected at random.

(c) A yet-to-be-fabricated prototype will fail.

(d) Some member of your graduating class receives a job offer paying in excess of $50,000 per year.

6–3 Consider the population of 50 floppy disks in Figure 6–2, with the quality state given for each disk. One disk is selected at random.

(a) Find the following quality state probabilities.

(1) Pr[S] (2) Pr[U] (3) Pr[O] (4) Pr[I] (5) Pr[F] (6) Pr[B]

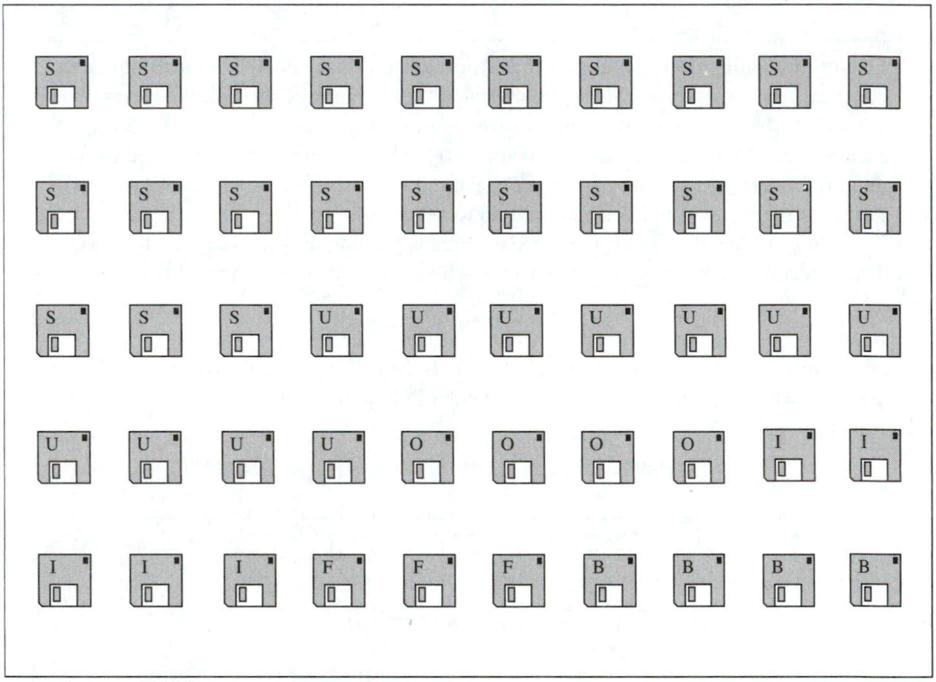

Figure 6–2
Population of disks as defined by their quality states.

Quality codes: S–Satisfactory for receiving any data
U–Unformatted but good
I–Incomplete with some bad tracks
O–Occupied with some data
F–Damaged wafer
B–Broken housing

(b) One disk is selected at random. Find the following probabilities.
  (1) workable (S, U, O, or I)
  (2) unworkable (F or B)
  (3) ready to receive small data set (S, O, or I)
  (4) reformatting required to receive maximum capacity (U or O)

6–4 A pilot plan has yielded chemical batches categorized as follows:

	Unusable	**Usable**
Low in impurities	3	24
High in impurities	21	7

Assuming this experience is representative of full-scale operation, determine estimated probabilities that a production batch will be

(a) low in impurities
(b) high in impurities
(c) unusable
(d) usable
(e) both low in impurities and unusable
(f) both high in impurities and unusable
(g) both low in impurities and usable
(h) both high in impurities and usable

6–5 Four students toss a lopsided coin. The following results are obtained:

	Number of Occurrences				**Total**
	a	*b*	*c*	*d*	*e*
Heads	6	7	24	49	86
Tails	4	8	18	37	67
Total	10	15	42	86	153

Determine each student's estimated probability for "head." Then use their combined experience to estimate the probability of "head."

6–6 Consider a deck of 52 playing cards, from which a card is randomly selected. Sketch the sample space, letting each card be represented by a dot on a grid that has one row for each suit and one column for each denomination. Identify the event sets for the following as areas on your sketch, and calculate the respective probabilities.

(a) ten (b) heart (c) black (d) face card (e) two or three

6–7 Consider the random experiment of tossing a pair of six-sided dice (a red and a white one). The elementary events are the number of dots (1 through 6) appearing on the up sides of each die. Each outcome may be represented as an ordered pair. For example, a 3 for the red die and a 5 for the white die is (3, 5). Give a complete listing of the sample space. Then establish probabilities for each possible *sum* of the dots on the showing faces.

6–8 Consider a photon arriving at a right angle to a filter film. Letting θ represent the polarization angle relative to the transmission axis, find the probability ($\cos\,\theta$) that the photon will be transmitted when

(a) $\theta = 30^\circ$ (b) $\theta = 60^\circ$ (c) $\theta = 1^\circ$ (d) $\theta = 89^\circ$

6–2 Probabilities for Compound Events

Much like electric circuits formed by combining components, probabilities may be found for complex, *compound events* by combining the probabilities for other events, called *component events*. There are two basic kinds of compound events in which the components are viewed in terms of their union (*or*) or their intersection (*and*). Probability theory allows us to compute probabilities for compound events by arithmetically combining the probabilities of individual components. These procedures are based on either the addition law or the multiplication law.

Applying the Basic Definition

Before we describe these laws, we must emphasize that every compound event is itself an outcome for which the basic definitions of probability apply. When elementary events are equally likely, it is usually possible to compute Pr[*A or B*] or Pr[*C and D*] by simply counting possibilities and dividing by the total number of elementary events.

For example, consider the student lottery discussed earlier and shown again in Figure 6–3. We can use the count-and-divide procedure to find the probability that a randomly selected student is either an undergraduate or an electrical engineering major:

$$\Pr[\text{UG } or \text{ EE}] = \frac{26}{38}$$

(We make sure to count each undergraduate electrical engineer just once.) Likewise, we may apply the basic definitions to determine the probability for the joint occurrence of the same events:

$$\Pr[\text{UG } and \text{ EE}] = \frac{5}{38}$$

Probabilities for joint events are usually referred to as **joint probabilities**.

But in some applications we may not be able to perform these types of calculations—because the elementary events are not equally likely (as with a lopsided coin or

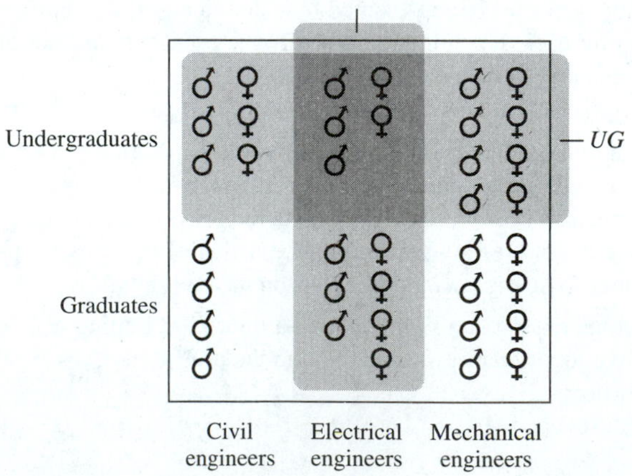

Figure 6–3 *Sample space for random selection of a Tau Beta Pi member.*

a crooked die cube), because we just don't have enough information, or because we cannot catalog the entire sample space. In such cases we might "synthesize" a compound event's probability by arithmetically manipulating the probabilities for its components (when these values are known).

The Addition Law

When a compound event is the union (*or*) of several components, the **addition law** allows us to find its probability by adding together the component event probabilities. There are two versions of this law. We begin with the simpler form.

ADDITION LAW FOR
MUTUALLY EXCLUSIVE EVENTS

$$\Pr[A \text{ or } B] = \Pr[A] + \Pr[B]$$

Illustration: *Power Cell Failure Probabilities*

As an illustration, suppose we know that the primary causes for failure of a space power cell are (1) solar radiation, (2) launch vibration, (3) material weakness, or (4) collision. For any particular failure, these causes may be considered as mutually exclusive events, since there can be only one primary reason for failing. Previous testing indicates that

$$\begin{aligned} \Pr[\text{solar radiation}] &= .10 \\ \Pr[\text{launch vibration}] &= .35 \\ \Pr[\text{material weakness}] &= .19 \\ \Pr[\text{collision}] &= .36 \end{aligned}$$

Causes (2) and (4) are mechanical accidents. To find the probability of this compound event, we apply the addition law:

$$\begin{aligned} \Pr[\text{mechanical accident}] &= \Pr[\text{launch vibration or collision}] \\ &= \Pr[\text{launch vibration}] + \Pr[\text{collision}] \\ &= .35 + .36 = .71 \end{aligned}$$

The addition law for mutually exclusive events extends to any number of components. For instance, consider the probabilities in Table 6–1 for the number of power cell failures in a fixed span of time. The addition law may be used to determine the probability for any particular compound event. For example, "at most 3" is an event occurring whether the number of failures is "exactly 0," "exactly 1," "exactly 2," or "exactly 3." Thus,

$$\begin{aligned} \Pr[\text{at most } 3] &= \Pr[0 \text{ or } 1 \text{ or } 2 \text{ or } 3] \\ &= \Pr[0] + \Pr[1] + \Pr[2] + \Pr[3] \\ &= .3679 + .3679 + .1839 + .0613 \\ &= .9810 \end{aligned}$$

Notice that the possible failure events given in Table 6–1 constitute a collectively exhaustive collection (one of these outcomes is bound to occur). The union of all the events is therefore certain and has probability 1. Notice also that the individual event probabilities in Table 6–1 sum to 1. For any collection of mutually exclusive events that are also collectively exhaustive, this must hold.

Table 6–1

Probabilities for the Number of Power Cell Failures

Number of Failures	Probability
0	.3679
1	.3679
2	.1839
3	.0613
4	.0153
5	.0031
6	.0005
7 or more	.0001
	1.0000

Application to Complementary Events

We are now ready again to consider complementary events (opposites). The union of any event A and its complement not A is certain, and by the addition law

$$\Pr[A \text{ or not } A] = \Pr[A] + \Pr[\text{not } A] = 1$$

From this we get the following:

$$\Pr[A] = 1 - \Pr[\text{not } A]$$

Thus, if we know the probability of an event's complement, we can subtract that value from 1 to get the probability value for the desired event.

Consider again the power cell failure events in Table 6–1. Suppose we want to find the probability that the number of failures is "at least 1." "At least 1" is the same as "some," which is the opposite of "none." Thus,

$$\begin{aligned}\Pr[\text{at least } 1] &= 1 - \Pr[0] \\ &= 1 - .3679 = .6321\end{aligned}$$

Here it is simpler to subtract a single quantity from 1 than to add together the probabilities of the seven component events for "at least 1." In fact, it is often easier and faster to reach an answer indirectly by evaluating the complementary event.

General Addition Law

The preceding version of the addition law works only for *mutually exclusive* component events. We should keep in mind that mutual exclusivity is the *exception*, so events typically encountered during probability evaluations do not share this property. In the more general case, we use the following.

GENERAL ADDITION LAW

$$\Pr[A \text{ or } B] = \Pr[A] + \Pr[B] - \Pr[A \text{ and } B]$$

Notice that the joint probability must be subtracted from the sum of the component event probabilities. This is to avoid double accounting of events.

***Example:* Metal Fabrication Faults**

Suppose we know that a metal fabrication process yields output with faulty bondings (FB) 10% of the time and excessive oxidation (EO) in 25% of all segments; 5% of the output has both faults. The following probabilities apply to a sample segment:

$$\Pr[FB] = .10$$

$$\Pr[EO] = .25$$

$$\Pr[FB \text{ and } EO] = .05$$

The general addition law can be used to find the probability that a sample segment has faulty bonding, excessive oxidation, or possibly both:

$$\Pr[FB \text{ or } EO] = \Pr[FB] + \Pr[EO] - \Pr[FB \text{ and } EO]$$

$$= .10 + .25 - .05 = .30$$

Thus, 30% of the output has at least one of the faults.

Statistical Independence

An important relationship between two events is statistical independence:

Definition

> Two events are **statistically independent** if the probability of any one of them is unaffected by the occurrence of the other.

Independence can ordinarily be assumed from the nature of the random experiment. For instance, consider Experiment 1 involving separate tosses of two fair coins, a dime and a quarter. The events "head for the dime" H_D and "head for the quarter" H_Q may be assumed to be statistically independent.

In Experiment 2 the dime is tossed first. If a head is obtained(H_D), a fair quarter is tossed. But if a tail is obtained (T_D), a crooked two-headed quarter is tossed instead. Thus,

$$Pr[H_Q] = \frac{1}{2} \quad \text{when } H_D \text{ occurs}$$

$$\Pr[H_Q] = 1 \quad \text{when } T_D \text{ occurs}$$

and

$$Pr[H_Q] = \frac{3}{4} \quad \text{in any repetition of the experiment}$$

(You may verify this last result using the procedures of Section 6–4.) Thus, the events H_D and H_Q for Experiment 2 are *not* statistically independent. We say that they are *dependent* events.

Statistical independence between events may be assumed in many sampling experiments, such as those sometimes encountered in quality assessment. When items are removed from a continuous production process, the probability for getting a satisfactory unit is generally assumed to be unaffected by the quality of the previous item. However, as we shall see, independence between quality events will not be the case when sampling instead from an existing production lot of fixed size.

When events are independent, a useful probability law applies.

The Multiplication Law for Independent Events

The second basic law of probability is intended for computing joint probabilities. This is achieved by multiplying together the probabilities of the component events.

MULTIPLICATION LAW FOR INDEPENDENT EVENTS

$$\Pr[A \textit{ and } B] = \Pr[A] \times \Pr[B]$$

To illustrate, suppose a perfectly balanced coin is fairly tossed twice in succession. We denote the showing faces from the first toss by H_1 and T_1 and those from the second by H_2 and T_2. The probability of two heads is the probability of the joint event H_1 *and* H_2. Since the events may be assumed to be independent, the multiplication law provides

$$\begin{aligned}\Pr[H_1 \textit{ and } H_2] &= \Pr[H_1] \times \Pr[H_2]\\ &= \frac{1}{2} \times \frac{1}{2} = \frac{1}{4}\end{aligned}$$

It is easy to verify this since the random experiment involves exactly four equally likely outcomes:

$$H_1 \textit{ and } H_2 \qquad H_1 \textit{ and } T_2 \qquad T_1 \textit{ and } H_2 \qquad T_1 \textit{ and } T_2$$

and $\Pr[H_1 \textit{ and } H_2]$ must be equal to 1 divided by 4.

Of course, if we could always count and divide we wouldn't need the multiplication law. Suppose instead that the coin is lopsided, having a 25% chance of "head." Assuming that independence still applies between successive toss outcomes, the multiplication law gives

$$\begin{aligned}\Pr[H_1 \textit{ and } H_2] &= \Pr[H_1] \times \Pr[H_2]\\ &= .25 \times .25 = .0625\end{aligned}$$

which is the simplest way to logically deduce the joint probability.

Example: *Noise-Induced Data-Transmission Errors*

Microwave telecommunication systems for data transmission are often subject to short bursts of spurious noise, which can render entire message segments useless. When identified as anomalies by parity-check codes, erroneous segments can sometimes be automatically retransmitted without human intervention. One such system stores the bits from millisecond-long segments in a buffer memory for retransmission should there be a noise interruption.

Suppose that 1% of such data segments suffer noise distortions (N), whereas the remainder are clear (C). Bursts of noise are so short that we may assume independence between C and N events for successive segments. For segment i, we have probability

.01 that there is noise (N_i) and .99 that the segment is clear (C_i). Consider just three segments, identified by 1, 2, and 3. The multiplication law for independent events allows us to evaluate the joint probability that all three segments are transmitted clearly:

$$\Pr[C_1 \textit{ and } C_2 \textit{ and } C_3] = \Pr[C_1] \times \Pr[C_2] \times \Pr[C_3]$$
$$= .99 \times .99 \times .99 = .9703$$

The probability that the second segment is the only one subject to noise is found by applying the multiplication law in evaluating the joint event C_1 *and* N_2 *and* C_3:

$$\Pr[C_1 \textit{ and } N_2 \textit{ and } C_3] = \Pr[C_1] \times \Pr[N_2] \times \Pr[C_3]$$
$$= .99 \times .01 \times .99 = .0098$$

The above joint events are also two of the elementary events in the sample space enumerating all three-segment transmission possibilities. Two more are (N_1 *and* C_2 *and* C_3) and (C_1 *and* C_2 *and* N_3), each having probability .0098. (As an exercise, you may find the remaining four elementary events and their probabilities.) To find the probability of exactly 1 noise interruption, we may apply the addition law to the respective mutually exclusive component events (the three elementary events involving exactly one N_i):

$$\Pr[\text{exactly 1 noise interruption}]$$
$$= \Pr[C_1 \textit{ and } N_2 \textit{ and } C_3] + \Pr[N_1 \textit{ and } C_2 \textit{ and } C_3] + \Pr[C_1 \textit{ and } C_2 \textit{ and } N_3]$$
$$= .0098 + .0098 + .0098 = .0294$$

Later in the chapter we will encounter a more general multiplication law that works even when the component events are dependent.

The above example shows that the multiplication law extends to any number of independent components. It can also be used to evaluate the probability for a joint event that may itself be the elementary event of a complex random experiment.

After using the multiplication law to establish elementary event probabilities, the *addition law* might then be applied to find the probability for a compound event. Consider the following example.

Example: *Catastrophe and the Space Shuttle*

U.S. manned space shuttle flights have proven to be extremely hazardous, with NASA officials seeing the "risk [probability] of catastrophe as roughly 1 in 145 missions for each ship."* It will take years to make major modifications that can reduce this probability, and those will not take place until the year 2005.

Assuming that one mission's success or failure is independent of the status of the others, we may use the multiplication law to find the probability that there will be one or more catastrophes in a given number of flights. This is done by first computing the probability that all flights are successful and then subtracting that probability from 1:

(1) *For two missions:*

$$\Pr[\text{at least 1 catastrophe}] = 1 - \Pr[\text{all missions succeed}]$$
$$= 1 - (144/145)(144/145)$$
$$= 1 - .986 = .014$$

(2) *For five missions:*

$$\begin{aligned}\Pr[\text{at least 1 catastrophe}] &= 1 - (144/145)^5 \\ &= 1 - .966 = .034\end{aligned}$$

(3) *For ten missions:*

$$\begin{aligned}\Pr[\text{at least 1 catastrophe}] &= 1 - (144/145)^{10} \\ &= 1 - .933 = .067\end{aligned}$$

(4) *For fifty missions:*

$$\begin{aligned}\Pr[\text{at least 1 catastrophe}] &= 1 - (144/145)^{50} \\ &= 1 - .707 = .293\end{aligned}$$

The above result is consistent with past history. There has been one catastrophe in the 49 flights through 1995.

---

*William J. Broad, "Risks Remain Despite NASA's Rebuilding," *New York Times*, January 28, 1996.

Problems

6–9 One Tau Beta Pi member is selected at random, and the sample space in Figure 6–3 applies. Find the following probabilities:
(a) Pr[graduate *or* CE]
(b) Pr[undergraduate *or* woman]
(c) Pr[ME *or* CE]
(d) Pr[CE *or* EE]

6–10 One Tau Beta Pi member is selected at random, and the sample space in Figure 6–3 applies. Use the count-and-divide method to find the following probabilities:
(a) Pr[graduate *and* woman]
(b) Pr[undergraduate *and* man]
(c) Pr[EE *and* man]
(d) Pr[CE *and* woman]

6–11 One Tau Beta Pi member is selected at random, and the sample space in Figure 6–3 applies.
(a) Find the probability that the selection will be an *undergraduate* assuming that (1) no other attributes are known and (2) the person's major is known to be mechanical engineering. What relationship applies between the events "undergraduate" and "mechanical engineer"?
(b) Using the count-and-divide method only, find the following:
(1) Pr[undergraduate]
(2) Pr[ME]
(3) Pr[CE]
(4) Pr[ME *and* undergraduate]
(5) Pr[CE *and* undergraduate]
(c) Use your answers to (b) above. Multiply (1) and (2). Comparing that product to (4), what do you notice? Explain.
(d) Use your answers to (b) above. Multiply (1) and (3). Comparing that product to (5), what do you notice? Explain.

6–12 Consider the data-transmission example on page 166. List all the elementary events for the transmission of three segments. Then use the multiplication law to determine the probability for each elementary event.

6–13 Using your answer to Problem 6–12 complete the following:
(a) Determine the following for the number of noise-interference events:
(1) Pr[exactly 0]
(2) Pr[exactly 1]
(3) Pr[exactly 2]
(4) Pr[exactly 3]

(b) Apply the addition law for mutually exclusive events to determine the following for the number of noise-interference events:
(1) Pr[exactly 1 *or* exactly 2] (2) Pr[at least 2] (3) Pr[at most 2]

6–14 Indicate for each of the following automobile event pairs, whether independence (1) can be confidently assumed, (2) might be true, or (3) is not expected.
(a) failure of car radio, failure of car battery
(b) transmission squeal, leaking transmission seal
(c) strong wheel vibration, high speed
(d) uneven tread wear, wheels misaligned
(e) transmission squeal, leaking fuel line

6–15 Items produced by a parts supplier have the following percentages of items satisfactory in the respective categories:

Dimension	80%
Weight	90
Texture	95

Assume an item's status is independent from category to category. Let D_S and D_U denote the events that the next item produced will be satisfactory or unsatisfactory in dimension, with W_S and W_U serving analogously to denote the events for weight and T_S and T_U for texture.
(a) Consider the sample space for the characteristics of any part. One elementary event is $D_S W_S T_S$ for getting a part satisfactory in all categories. List the sample space events.
(b) The probability of the elementary event $D_S W_S T_S$ equals Pr[D_S *and* W_S *and* T_S]. Use the multiplication law to find the probability for each elementary event listed in (a).

6–16 Problem 6–15 *continued.* A part may be classified in a variety of ways. In each case list the elementary events and then determine the probability that a part will fall in that category, using as needed the addition law.
(a) an *acceptable* part that is unsatisfactory in at most one category
(b) an *unacceptable* part that is not acceptable
(c) a *reworkable* part that is an unacceptable part that is satisfactory in dimension or satisfactory in weight
(d) a *salvageable* part that is satisfactory only in texture
(e) a part that is satisfactory at least in dimension
(f) a part that is acceptable or reworkable
(g) a part that is satisfactory at least in dimension or that is salvageable
(h) a part that is salvageable or reworkable

6–17 A production process yields 90% satisfactory items. The quality events pertaining to successive items are assumed to be independent. Find the probability of finding (1) all satisfactory, (2) none satisfactory, and (3) at least one satisfactory, when the number of selected items is (a) 2, (b) 3, (c) 4, and (d) 5.

6–18 Show that if A and B are mutually exclusive events and that neither is impossible, then they cannot be statistically independent.

6–19 For about $1 billion in new space shuttle expenditures, NASA has proposed to install new heat pumps, power heads, heat exchangers, and combustion chambers. These will lower the mission catastrophe probability to 1 in 200.*
(a) Assume that these changes are made. Then for (1) two, (2) five, (3) ten, and (4) fifty missions, calculate the probability of one or more catastrophes.
(b) Comparing your answer from (a) to the analogous probabilities on pages 167–68, do NASA's proposed changes substantially reduce the probabilities for catastrophe?

*William J. Broad, "Risks Remain Despite NASA's Rebuilding," *New York Times*, January 28, 1996.

6–3 Conditional Probability

We have now seen the important relationships between events, the main forms of compound events, and several ways to compute probabilities for a variety of outcome types. This section consolidates that knowledge and establishes a conceptual framework for further extensions of probability theory.

Consider how the probability value of one event is affected by the occurrence of another. It is more likely to rain tomorrow if there is a solid cloud cover than if there are patches of blue. Electronic equipment is more likely to fail in high temperatures than in moderate ones—and we might be uncertain which operating environment will apply. Items produced in antiquated plants will generally involve more rejects than those created in modern facilities (although we might not know where a particular item is made).

A probability value computed under the assumption that another event is going to occur is a **conditional probability**. For the events "rain" and "solid overcast," the following value might apply:

$$\Pr[\text{rain} \mid \text{solid overcast}] = .90$$

The event "rain" is listed first, and .90 is the conditional probability of that event; the condition—given event—"solid overcast," is listed second, and the vertical bar is a shorthand representation for "given that." A different rain probability applies given some other event:

$$\Pr[\text{rain} \mid \text{blue patches}] = .30$$

Should no stipulations be made regarding sky conditions, we might have

$$\Pr[\text{rain}] = .20$$

which is an **unconditional probability**.

It may instead be possible to compute a conditional probability directly from known probability values using the following:

CONDITIONAL PROBABILITY IDENTITY

$$\Pr[A|B] = \frac{\Pr[A \text{ and } B]}{\Pr[B]}$$

Consider again the events "undergraduate" and "electrical engineer" for the student lottery in Figure 6–3. We have

$$\Pr[UG \mid EE] = \frac{\Pr[UG \text{ and } EE]}{\Pr[EE]} = \frac{\frac{5}{38}}{\frac{12}{38}} = \frac{5}{12}$$

which expresses the probability that the selected student is an undergraduate given that he or she majors in electrical engineering. Notice that the numerator is the joint probability and the denominator is the given event's probability. A total reversal of the given and uncertain events is achieved by switching divisors:

$$\Pr[EE \mid UG] = \frac{\Pr[UG \text{ and } EE]}{\Pr[UG]} = \frac{\frac{5}{38}}{\frac{19}{38}} = \frac{5}{19}$$

Of course, the same results could have been achieved using the basic definition: With 5 out of 12 electrical engineering majors being undergraduates, we have

$\Pr[UG \mid EE] = \frac{5}{12}$. Likewise, $\Pr[EE \mid UG] = \frac{5}{19}$, since there are 5 electrical engineers out of 19 undergraduate students.

Although the foregoing identity is always true, it may not always be suitable for computing a conditional probability. We must know the values Pr[*A and B*] and Pr[*B*] in order to use it to evaluate $\Pr[A \mid B]$.

Example:* *Crooked and Fair Dice

A box contains four crooked (*C*) pairs of dice. One member of each pair has 3 dots on every face and the other has 4 dots on all sides. The box also contains six fair (*F*) pairs of dice. The faces of each of these dice have a different number of dots from 1 through 6.

One pair is selected at random and rolled. Given that the pair is fair, what is the conditional probability that a 7-sum results when the number of dots on the two up sides are added together?

The answer is

$$\Pr[\text{7-sum} \mid F] = \frac{6}{36} = \frac{1}{6}$$

This is found by *counting and dividing* the respective numbers of possibilities. (You may verify that there are 36 equally likely possibilities, six of which result in a sum of 7.)

The conditional probability identity cannot be used since it requires

$$\Pr[\text{7-sum } \textit{and } F]$$

in the numerator, and the value of this is not presently known.

Establishing Independence by Comparing Probabilities

Recall that two events are independent if the probability of one is unaffected by the occurrence of the other. We may apply conditional probability concepts to more precisely define independence.

Definition

Two events *A* and *B* are **statistically independent** whenever $\Pr[A|B] = \Pr[A]$ or $\Pr[B|A] = \Pr[B]$.

Stated less formally, *A* and *B* are independent if one event has unconditional probability equal to its own conditional probability given the other event.

We may use this fact to establish statistical independence or dependence by comparing probability values. (But keep in mind that the independence may be simply assumed to hold between some outcomes—such as the showing faces from two coin tosses—wherein the nature of the random experiments makes the relationship obvious.)

Returning to the student lottery in Figure 6–1, we can see that *W* (woman) and *G* (graduate student) are *dependent* because

$$\Pr[W \mid G] = \frac{8}{19} \quad \text{does not equal} \quad \Pr[W] = \frac{17}{38}$$

This is because the conditional probability of *W* given *G* differs from the unconditional probability of *W*. (You may also establish that $\Pr[G \mid W] \neq \Pr[G]$). The same relationship applies here for all sex and student-level event pairs.

Problems

6–20 Consider Experiment 2 on page 165. Find the following.
(a)(1)$\Pr[H_Q \mid H_D]$ (2)$\Pr[T_Q \mid H_D]$
(b)(1)$\Pr[H_Q \mid T_D]$ (2)$\Pr[T_Q \mid T_D]$

6–21 A floppy disk is randomly selected from those listed in Figure 6–2. Find the following for its quality state.
(a)(1)Pr[S] (2)Pr[S | workable (S, U, I, or O)]
(b)(1)Pr[B] (2)Pr[B | unworkable (F or B)]
(c)(1)Pr[ready to use (S, I, or O)] (2)Pr[ready to use | workable (S, U, I, or O)]
(d)(1)Pr[needs formatting (U or O)] (2)Pr[needs formatting | workable (S, U, I, or O)]

6–22 Refer to Problem 6–21 and your answers to that exercise. For each of the following event pairs indicate whether the events are (1) independent or (2) dependent.
(a) S workable
(b) B unworkable
(c) ready to use workable
(d) needs formatting workable

6–23 Consider a random experiment involving three boxes, each containing a mixture of red (R) and green (G) balls, with the following quantities:

A	B	C
5 R	14 R	8 R
5 G	6 G	12 G

The first ball will be selected at random from box A. If that ball is red, the second ball will be drawn from box B; otherwise, the second ball will be taken from box C.

Let R_1 and G_1 represent the color of the first ball, R_2 and G_2 the color of the second. Determine the following probabilities. (*Hint:* The conditional probability identity will not work.)

(a) $\Pr[R_1]$ (c) $\Pr[R_2 \mid R_1]$ (e) $\Pr[G_2 \mid G_1]$
(b) $\Pr[G_1]$ (d) $\Pr[R_2 \mid G_1]$ (f) $\Pr[G_2 \mid R_1]$

6–24 Consider the student lottery summarized in Figure 6–3. Find the following conditional probabilities:
(a) $\Pr[EE \mid G]$ (b) $\Pr[ME \mid G]$ (c) $\Pr[G \mid EE]$ (d) $\Pr[G \mid ME]$

6–25 Refer to Problem 6–24. Find the following conditional probabilities:
(a) $\Pr[ME \mid U]$ (b) $\Pr[CE \mid U]$ (c) $\Pr[U \mid ME]$ (d) $\Pr[U \mid CE]$

6–4 The Multiplication Law, Probability Trees, and Sampling

With basic probability concepts and tools firmly in hand, we set the stage for deductive statistics by considering the sampling process. We can now find probabilities for particular sample outcomes. Later in the text this background will prove useful in developing the concepts of inductive or inferential statistics.

In this section we introduce two new tools while laying the above groundwork. A *general* multiplication law is given that works whether or not component events are independent. To help explain this procedure, we introduce a useful display, the probability tree diagram, which conveniently organizes information about random experiments that involve a series of uncertainties.

The General Multiplication Law

The multiplication law introduced earlier requires that the component events be independent. But independence is often the exception. This is especially true in testing and evaluation—in which the desired event and the experimental result should be dependent. For example, a seismic survey result (an uncertainty) should be statistically dependent on the presence of oil. (Otherwise seismic predictions would be worthless in finding drilling locations.)

We have a general version of the multiplication law.

GENERAL MULTIPLICATION LAW

$$\Pr[A \text{ and } B] = \Pr[A] \times \Pr[B \mid A]$$

and

$$\Pr[A \text{ and } B] = \Pr[B] \times \Pr[A \mid B]$$

Notice that the probability for the first event in the product is an unconditional one, whereas the second is a conditional probability for the other event, given the first. The multiplication law given earlier is a special case of this because when A and B are independent, $\Pr[A \mid B] = \Pr[A]$ and $\Pr[B \mid A] = \Pr[B]$.

Illustration: *Quality-Control Inspector's Probabilities*

To illustrate, suppose that a quality-control inspector has established that 10% of all electric assemblies fail (F) the circuit test. Twenty percent of the failing items have poor (P) solder joints. This experience provides the following probabilities:

$$\Pr[F] = .10$$
$$\Pr[P \mid F] = .20$$

The multiplication law gives the joint probability that a particular assembly will fail and have poor solder joints:

$$\begin{aligned}\Pr[F \text{ and } P] &= \Pr[F] \times \Pr[P \mid F] \\ &= .10 \times .20 \\ &= .02\end{aligned}$$

Altogether, 2% of all assemblies will have both deficiencies.

Illustration: *Oil Wildcatter's Probabilities*

A second illustration is provided by the experience of an oil wildcatter. He has established that there is a 40% chance for oil (O) beneath a particular site. Furthermore, from past experience with seismic testing, the procedure is known to be 80% reliable. That is, for known oil-bearing sites there is an 80% probability of getting favorable (F) seismic results, and when the site is known to be dry, there is a 90% probability of getting an unfavorable (U) prediction. Thus,

$$\Pr[O] = .4$$
$$\Pr[\text{not } O] = 1 - .4 = .6$$
$$\Pr[F \mid O] = .8$$
$$\Pr[U \mid \text{not } O] = .9$$

The general multiplication law may be used to compute joint probabilities for geological and seismic events:

$$\Pr[O \textit{ and } F] = \Pr[O] \times \Pr[F \mid O] = .4 \times .8 = .32$$

$$\Pr[\text{not } O \textit{ and } U] = \Pr[\text{not } O] \times \Pr[U \mid \text{not } O] = .6 \times .9 = .54$$

Example:* *Using Probability to Help Identify Space Shuttle Improvements

One adviser has offered NASA some assistance in helping to improve the safety of the U.S. manned space shuttle flights, for which she has read an article reporting the probability of any single mission meeting with catastrophe as roughly 1 in 145. Given that a mission catastrophe occurs, the following conditional probabilities apply for its major cause.

Major Cause	Conditional Probability
(1) E_1: Faulty ignition or booster separation	.60
(2) E_2: Main engine cutoff or external tank separation	.23
(3) E_3: Auxiliary power shutdown	.01
(4) E_4: De-orbit accident	.02
(5) E_5: Reentry failure	.08
(6) E_6: Final flight accident	.02
(7) E_7: Landing accident	.04
	1.00

Source: William J. Broad, "Risks Remain Despite NASA's Rebuilding," *New York Times*, January 28, 1996.

Applying the multiplication law, the following joint probability applies:

$$\Pr[E_1 \textit{ and } \text{Catastrophe}] = \Pr[E_1 \mid \text{Catastrophe}] \times \Pr[\text{Catastrophe}] = .60(1/145) = .00414$$

Thus, there is about a .4% chance that any particular mission will fail because of faulty ignition or booster separation.

The adviser suggests that NASA make a list of possible shuttle system improvements and the corresponding reductions in the conditional probabilities listed in the above table. That might help establish budget priorities.

Multiplication Law for Several Events

The general multiplication law can be extended to several components. Consider events $A_1, A_2, \ldots, A_n$. We have

$$\Pr[A_1 \textit{ and } A_2 \textit{ and } A_3 \ldots \textit{ and } A_n] = \Pr[A_1] \times \Pr[A_2 \mid A_1] \times \Pr[A_3 \mid A_1 \textit{ and } A_2] \times \ldots \times \Pr[A_n \mid A_1 \textit{ and } A_2 \textit{ and } A_3 \ldots \textit{ and } A_{n-1}]$$

The Probability Tree Diagram

A convenient summary display is the **probability tree**. Figure 6–4 shows the probability tree diagram for the oil wildcatting illustration. Each event is represented as a **branch** in one or more **event forks**. A probability tree can be especially convenient for random experiments having events that occur at different times or stages.

In probability trees, time ordinarily moves from left to right. Since the geology events precede the seismic ones, the branches for oil (O) and not oil (not O) appear in the leftmost event fork. Each of these is followed by a separate fork representing the seismic events, with one branch for favorable (F) and one for unfavorable (U). The complete tree then exhibits each final outcome as a single *path* from beginning to end. The oil wildcatter's tree has four paths: oil-favorable, oil-unfavorable, not-oil-favorable, and not-oil-unfavorable. Each of these corresponds to a distinct joint event.

The probabilities for each event are placed alongside its branch. The values listed in the left fork—the geology events fork—are $\Pr[O] = .4$ and $\Pr[\text{not } O] = .6$. Since that event fork has no predecessor, these probabilities are unconditional ones.

Probabilities for events at later stages will all be conditional probabilities, with the branch or subpath leading to the branching point signifying the given events. The top

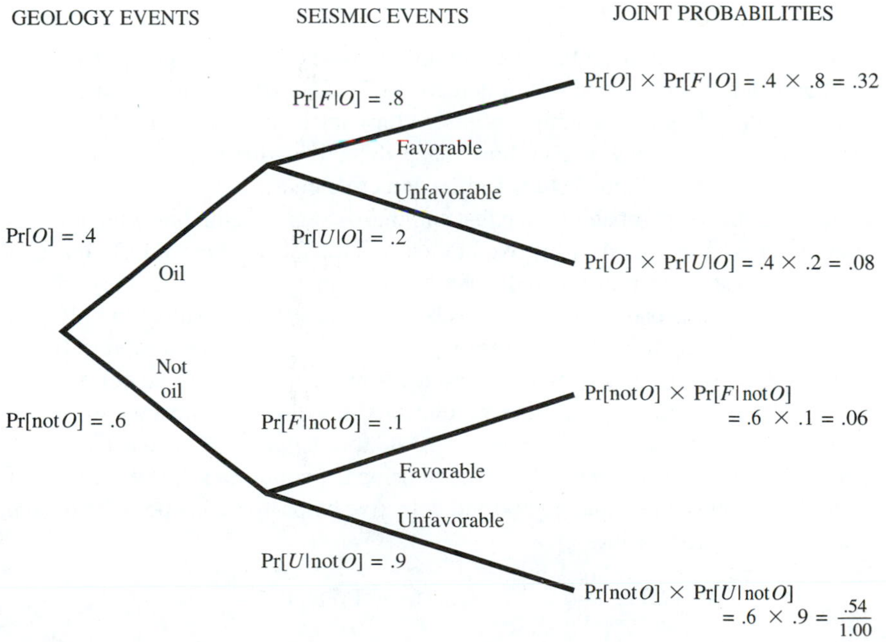

Figure 6–4 Probability tree diagram for oil wildcatter.

seismic events fork lists the probabilities $\Pr[F \mid O] = .8$ for favorable and $\Pr[U \mid O] = .2$ for unfavorable; since that fork is preceded by the oil (O) branch, both conditional probabilities involve O as the given event. A completely different set of conditional probabilities, $\Pr[F \mid \text{not } O] = .1$ and $\Pr[U \mid \text{not } O] = .9$, apply in the bottom event fork; that branching point is preceded by the not oil (not O) branch, so that not O is the given event.

The events emanating from a single branching point are mutually exclusive and collectively exhaustive, so that exactly one of them must occur. All the probabilities on branches within the same fork must therefore sum to 1.

The probability tree is very convenient for computing joint probabilities. Since each joint event is represented by a path through the tree, its probability is found by multiplying together all the individual branch probabilities along its path. For instance, the topmost path represents the outcome sequence oil-favorable. The corresponding joint probability is

$$\begin{aligned}\Pr[O \textit{ and } F] &= \Pr[O] \times \Pr[F \mid O] \\ &= .4(.8) \\ &= .32\end{aligned}$$

All of the oil wildcatter's joint probabilities are listed in Figure 6–4 at the terminus of the respective event path. Notice again that, because the joint events themselves are mutually exclusive and collectively exhaustive, they sum to 1.

Probability and Sampling

Probability is especially important in summarizing potential sampling results. Because multiple observations are usually involved in statistical evaluations, the probability calculations can be complicated. Probability trees can be helpful in organizing these computations.

Illustration: *Quality-Control Sampling with Memory Chips*

To illustrate, consider a shipment of 100 memory chips received by a computer manufacturer. Each chip is either good (G) or defective (D). The decision to accept or reject the shipment will be based on a sample of three chips selected at random. Although the inspector cannot know ahead of time how many defectives there are, let's consider a shipment having exactly 90 good chips and 10 defective ones.

Figure 6–5 shows a probability tree that summarizes the essential information. The first observation is represented by the two branches in the leftmost event fork. It is convenient to use the abbreviation G_1 to denote the event that "the first chip is good" and D_1 that "the first chip is bad." The subscripts help distinguish the results from different observations. Each branch for the first observation leads to a separate event fork for the second item, with both forks having a G_2 and a D_2 branch. Together, the initial observations create four distinct circumstances under which the third observation might occur, and each of these is represented by a separate fork having a G_3 and a D_3 branch.

The probabilities for each branch are found using the count-and-divide method, according to the number of remaining good and defective chips up to that point in the tree. The initial fork has probabilities

$$\Pr[G_1] = \frac{90}{100} \qquad \text{and} \qquad \Pr[D_1] = \frac{10}{100}$$

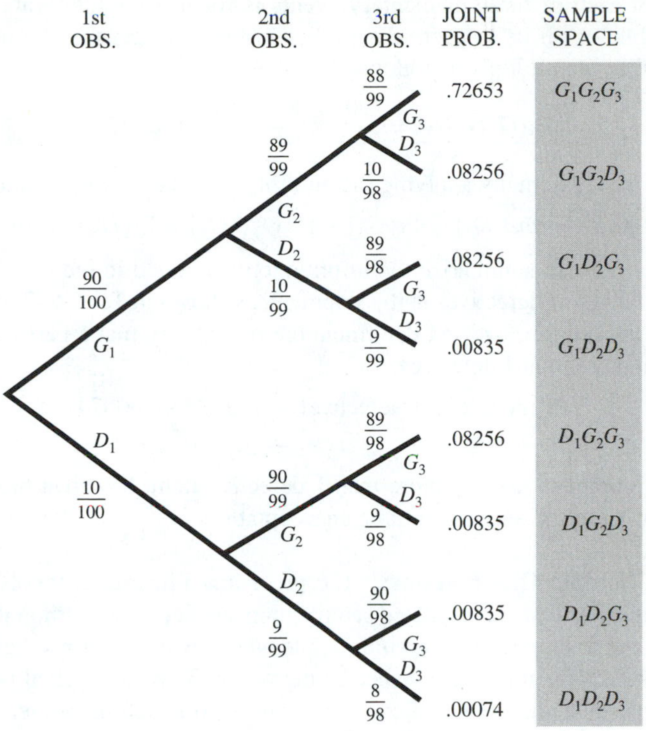

Figure 6–5 *Probability tree diagram for a shipment of items sampled without replacement.*

Because the quality mix of the remaining items varies at the second stage, the probabilities for G_2 and D_2 differ in the two event forks for that observation. The shown values are *conditional probabilities*. The top set applies when G_1 has occurred, when there are only 89 remaining good chips and 10 defective. With 99 chips then available for testing, the count-and-divide procedure provides

$$\Pr[G_2 \mid G_1] = \frac{89}{99} \qquad \text{and} \qquad \Pr[D_2 \mid G_1] = \frac{10}{99}$$

Likewise, the lower second-stage fork involves

$$\Pr[G_2 \mid D_1] = \frac{90}{99} \qquad \text{and} \qquad \Pr[D_2 \mid D_1] = \frac{9}{99}$$

since at that point—with one defective chip removed—the tree shows that 90 good chips and 9 defectives remain.

The third-stage forks involve probabilities reflecting the respective histories of good and defective items. The fork at the top right is preceded by G_1 and G_2 branches, so the conditional probabilities are

$$\Pr[G_3 \mid G_1 \textit{ and } G_2] = \frac{88}{98} \qquad \text{and} \qquad \Pr[D_3 \mid G_1 \textit{ and } G_2] = \frac{10}{98}$$

It is easy to verify the remaining third-stage probabilities.

Every path through the tree in Figure 6–5 corresponds to a distinct final outcome, each of which is summarized in the sample space listed in the box beside the tree. The probability for each of these elementary events is found by multiplying together the branch probabilities on its respective path. For instance, to get the second elementary event probability, we multiply as follows:

$$\Pr[G_1 G_2 D_3] = \frac{90}{100} \times \frac{89}{99} \times \frac{10}{98} = .08256$$

In doing this, we are actually applying the multiplication law for several joint events:

$$\Pr[G_1 \textit{ and } G_2 \textit{ and } D_3] = \Pr[G_1] \times \Pr[G_2 \mid G_1] \times \Pr[D_3 \mid G_1 \textit{ and } G_2]$$

The inspector can summarize the information contained in the probability tree in terms of the number of defectives in the shipment, as shown in Table 6–2. Using the values from the last column, we can determine the probability that a particular shipment would contain more than 1 defective:

$$\begin{aligned}\Pr[\text{more than 1 defective}] &= .02505 + .00074 \\ &= .02579\end{aligned}$$

If she rejects shipments having more than 1 defective item, less than three percent of shipments similar to the above would be unacceptable.

Independent Sample Observations: The preceding illustration involves **sampling without replacement**, since inspected sample units are set aside. Although it seems inherently wasteful to do so (and even impossible when testing destroys the items), some inspection schemes would put the inspected items back into the original population, allowing them an equal chance with the rest to be chosen for each subsequent observation. Those plans involve **sampling with replacement**. As we shall see, sampling with replacement simplifies the probability calculations.

Figure 6–6 shows the probability tree diagram that would apply if the inspector sampled with replacement. Notice that all the G branches have identical probabilities of

Table 6–2
Probabilities for the Number of Defective Memory Chips When Three Items Are Inspected in a Random Sample without Replacement

Number of Defectives	Corresponding Elementary Events	Elementary Event Probability	Defectives Probability
0	$G_1G_2G_3$	.72653	.72653
1	$D_1G_2G_3$	.08256	.24768
	$G_1D_2G_3$	.08256	
	$G_1G_2D_3$	.08256	
		.24768	
2	$D_1D_2G_3$	.00835	.02505
	$D_1G_2D_3$	.00835	
	$G_1D_2D_3$	.00835	
		.02505	
3	$D_1D_2D_3$	.00074	.00074
			1.00000

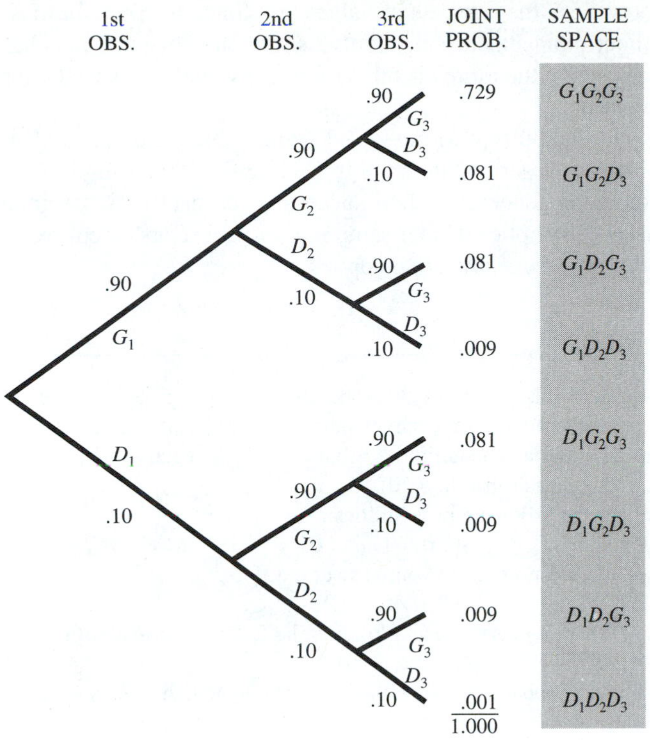

Figure 6–6
Probability tree diagram for items from a production process sampled with replacement.

.90, and similarly the D probabilities are all .10. Since sampled items are put back, the probabilities for later quality events are unaffected by what happens earlier. In effect, successive quality events are *statistically independent*. Independence between successive sample observations arises naturally when sampling from a continuing production line, when replacement or nonreplacement would yield identical probabilities due to the theoretically infinite population size.

The probabilities for the number of defectives found under sampling with replacement are given in Table 6–3. The probabilities are obtained in the same manner as before. To help in comparing the two procedures, the two sets of probabilities are given side by side.

Table 6–3
Probabilities for the Number of Defective Memory Chips When Three Sample Items Are Inspected

	Probabilities for Sampling	
Number of Defectives	**Without Replacement**	**With Replacement**
0	.72653	.729
1	.24768	.243
2	.02505	.027
3	.00074	.001
	1.00000	1.000

Notice how close the two sets of values are. Since the probabilities computed for sampling without replacement can involve some cumbersome calculations, they are often *approximated* by the more cleanly computed values that strictly apply only when there is replacement.

The sets of probability information in Table 6–3 are examples of *probability distributions*, which are described in detail in Chapter 7. When sampling is done with replacement—or more generally, when successive observations are independent—the *binomial distribution* applies. Under sampling without replacement from finite populations the *hypergeometric distribution* applies.

Problems

6–26 An oil wildcatter has assigned a .30 probability for striking gas (G) under a particular leasehold. He has ordered a seismic survey that has a 90% positive reliability (given gas it confirms gas [C] with a probability .90), but only 70% negative reliability (given no gas it denies gas [D] with probability .70).

(a) Establish the following probabilities:
(1) Pr[G *and* C] (2) Pr[no G *and* C] (3) Pr[no G *and* D] (4) Pr[G *and* D]

(b) Apply the addition law to your answer to (a) to find
(1) Pr[C] (2) Pr[D]

(c) Using your answers to (a) and (b), find the following probabilities:
(1) Pr[$G|C$] (2) Pr[no $G|D$]

(d) Construct a probability tree with gas events in the first stage and seismic events in the second.

6–27 A project manager is creating the design for a new engine. He judges that there will be a 50–50 chance that it will have high-energy (H) consumption instead of low (L). Historically, 30% of all high-energy engines have been approved (A) with the rest disapproved (D), while 60% of all low-energy engines have been approved.

(a) Construct a probability tree for the project manager's engine events.

(b) What is the probability that his design will result in an approved engine?

6–28 Repeat Problem 6–26 (a)-(c) for a new site where gas has a .20 probability, the positive reliability is 80%, and the negative reliability is 90%.

6–29 A quality-control inspector is testing sample output from a production process for widgets wherein 95% of the items are satisfactory (S) and 5% are unsatisfactory (U). Three widgets are chosen randomly for inspection. The successive quality events may be assumed independent.

(a) Construct a probability tree diagram for this experiment, identify all the sample-outcome elementary events, and compute the respective joint probabilities.

(b) Find the probabilities for the following numbers of unsatisfactory items:
(1) none (3) exactly 2 (5) at least 1
(2) exactly 1 (4) exactly 3 (6) at most 2

6–30 A user of the widgets receives a shipment of 5 dozen containing 95% satisfactory. A sample of 3 widgets is selected *without* replacement. Repeat (a) and (b) from Problem 6–29.

6–31 Two chapters of a civil engineering society each have 60% licensed members (the rest are unlicensed), with some on public payrolls (and the rest privately employed). A questionnaire is being sent to a random sample of the members. Consider the characteristics of any one of the members surveyed.

(a) Chapter A has just as many licensed engineers on public payrolls as on private ones. This makes any professional status event *independent* of any employment sector event. Construct a joint probability table having one row for each license category and one column for each sector by first finding the marginal probabilities and then using the multiplication law to obtain the joint probabilities, which in this case may be expressed as the products of the two respective marginal probabilities.

(b) In Chapter B only 40% of the licensed engineers are public employees. It follows that any professional status event is *dependent* on any employment sector event. This means that no joint probability is equal to the product of the respective marginal probabilities. Construct the joint probability table.

6–32 Suppose that major system changes may be made to reduce the single-mission cause-of-failure probabilities for the space shuttle. The respective percentage reductions in the joint probability for mission failure and the costs of the changes are listed below.

Major Cause i	Percentage Reduction in Pr[E_i *and* Catastrophe]	Cost of Change ($ billion)
(1) E_1: Faulty ignition or booster separation	50%	2.5
(2) E_2: Main engine cutoff or external tank separation	40	.5
(3) E_3: Auxiliary power shutdown	50	.2
(4) E_4: De-orbit accident	50	.5
(5) E_5: Reentry failure	30	.7
(6) E_6: Final flight accident	50	.2
(7) E_7: Landing accident	25	.1

(a) Using the data on page 174, compute the original joint probabilities for the major cause and catastrophe events.

(b) Assume that only one of the changes listed above is made. Reduce, one change at a time, the respective joint probabilities computed in (a). Then compute the new overall mission catastrophe probability by summing the original joint probabilities with the modified one.

(c) Which change is most cost effective? Which change is least cost effective?

6–5 Predicting the Reliability of Systems

An important application of probability is reliability analysis used to predict an overall system's reliability, using as building blocks the reliabilities of individual components. It is convenient to define reliability in terms of a survival probability, using the notation

$$R_i(t) = \Pr[\text{component } i \text{ survives beyond time } t]$$

The reliability of the system is then a function of the reliabilities of its components:

$$R_s(t) = f\{R_1(t), R_2(t), \ldots\}$$

When t is fixed at a specified level, a closed form may be obtained for $R_s(t)$ that coincides with the system logic. There are two primary system forms: one applies when the components are arranged in *parallel* and another when they are joined in a *series*. More complex cases involve modular subsystems as building blocks. We can portray the logic of systems schematically in similar fashion to electrical circuits.

Figure 6–7
Logic for system with series components.

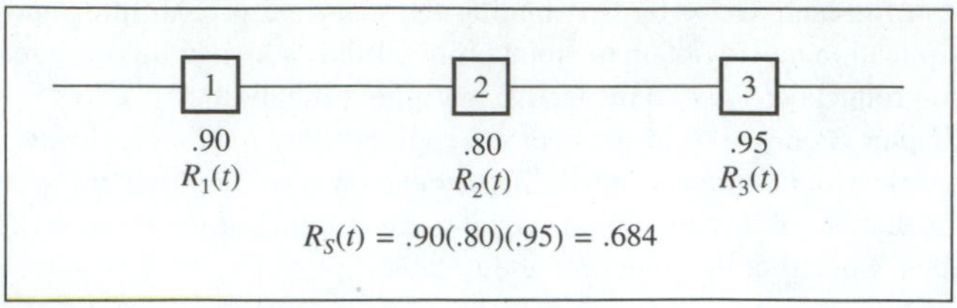

Systems with Series Components

A **series system** is one that performs satisfactorily as long as *all* components are fully functional. Figure 6–7 shows the logic of a small system. The system is analogous to an electrical circuit that works as long as current flows through all components. Failure of any component will cause the entire system to fail. Propulsion on the NASA space shuttle is an example of a series system since failure on any of the booster rockets will result in an aborted mission.

Under the series logic, the system will survive past time t only if its components do so. System reliability, therefore, is equal to the product of the component reliabilities:

$$R_S(t) = R_1(t)R_2(t)\dots R_n(t)$$

The above follows directly from the multiplication law for independent events.

The system in Figure 6–7 involves three components with reliabilities $R_1(t) = .90, R_2(t) = .80$, and $R_3(t) = .95$. The system reliability is

$$R_S(t) = .90(.80)(.95) = .684$$

Systems with Parallel Components

A **parallel system** is one that performs as long as *any one* of its components remains operational. Figure 6–8 is a schematic for a simple parallel system. The electrical analogy is a circuit through which current flows as long as any component works, providing an open path. All components must fail before the parallel system does. For example, batteries for a handheld calculator might be arranged in parallel, so that the calculator may function as long as at least one battery provides sufficient power.

A parallel system will survive past time t if at least one component survives. Failure of the system occurs only if all components fail. The probability of system failure is therefore the product of those failure probabilities (each being 1 minus the respective survival probability):

$$\Pr[\text{system failure at or before } t] = [1 - R_1(t)][1 - R_2(t)]\dots[1 - R_n(t)]$$

Thus, system reliability must be the complement to this, as follows:

$$R_S(t) = 1 - [1 - R_1(t)][1 - R_2(t)]\dots[1 - R_n(t)]$$

The above follows from the multiplication law for independent events and the property for computing complementary event probabilities.

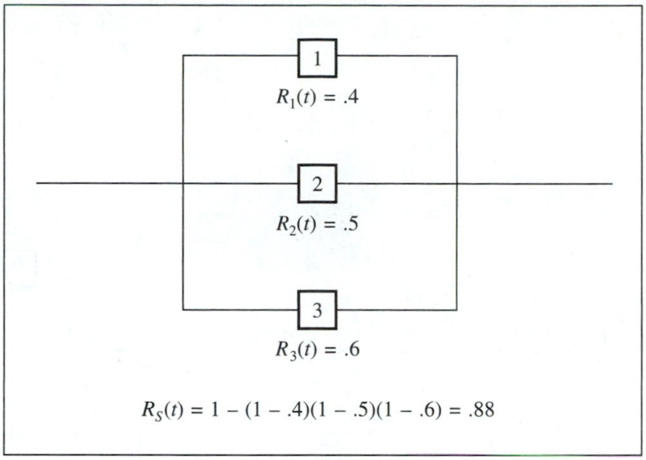

Figure 6–8
Logic for system with parallel components.

The parallel system shown in Figure 6–8 has component reliabilities of $R_1(t) = .4$, $R_2(t) = .5$, and $R_3(t) = .6$. That system's reliability is

$$R_S(t) = 1 - (1 - .4)(1 - .5)(1 - .6) = 1 - .12 = .88$$

Increasing System Reliability

A design issue in systems is how to increase system reliability. One way this might be accomplished is by raising the reliability of individual components (perhaps by substituting better materials, such as gold wiring instead of copper). To illustrate, consider the system in Figure 6–9(a). Suppose that the reliability of component 2 can be raised to .95, as in Figure 6–9(b), by substituting a more expensive item.

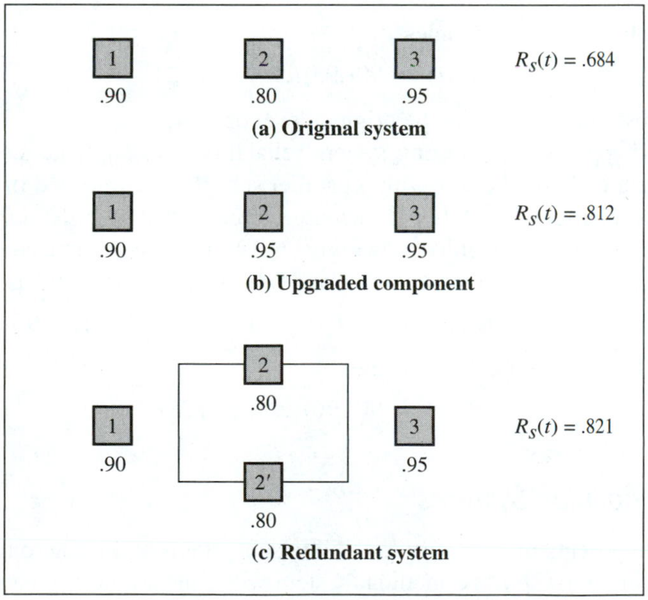

Figure 6–9
Two approaches for increasing system reliability.

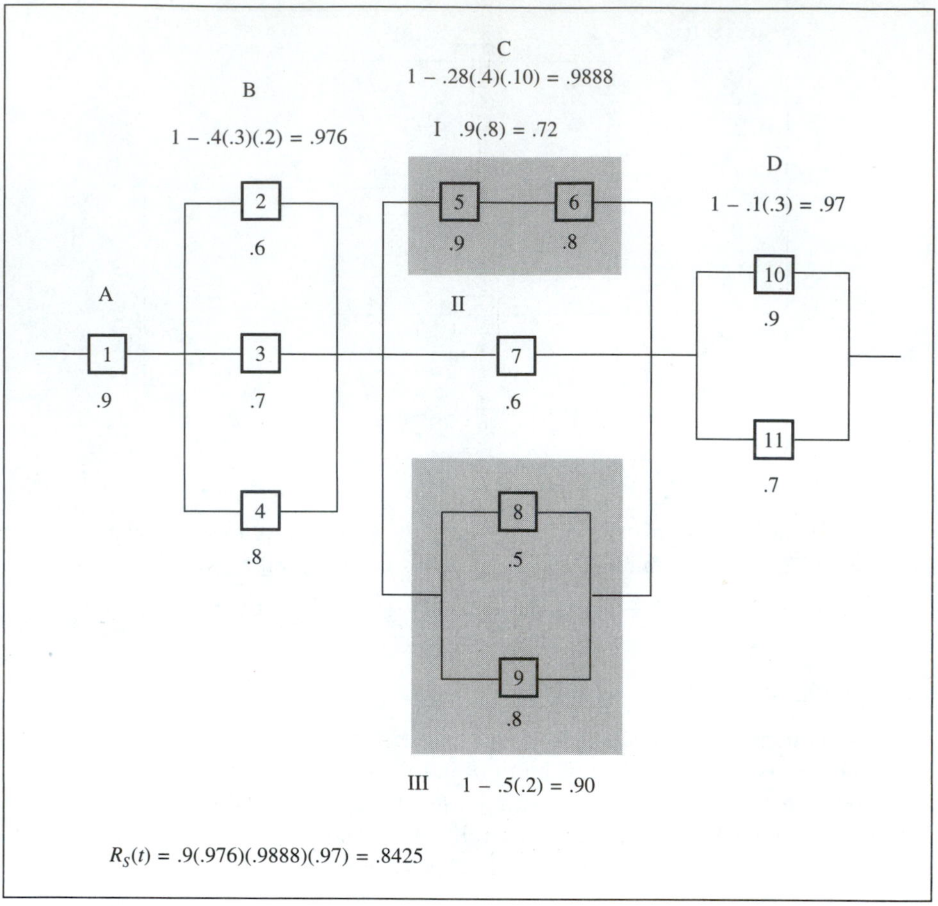

Figure 6–10 *Modular system with series systems.*

The system reliability then becomes

$$R_S(t) = .90(.95)(.95) = .812$$

which is a substantial boost over the original level of .684.

A second approach to boosting system reliability is to duplicate certain components, replacing individual items with a parallel subsystem composed of two or more identical units. In effect, reliability is increased through system *redundancy*. Consider the system again, but now modified as in Figure 6–12(c). In this system, component 2 has an identical partner 2′, joined in parallel. The new modular subsystem has reliability

$$1 - [1 - R_2(t)][1 - R_{2'}(t)] = 1 - (1 - .8)(1 - .8) = .96$$

and the overall system reliability becomes

$$R_S(t) = .90(.96)(.95) = .821$$

Complex Modular Systems

Individual components may be aggregated into subsystems according to their interrelationships. Figure 6–10 shows a modular system with four primary subsystems—A, B,

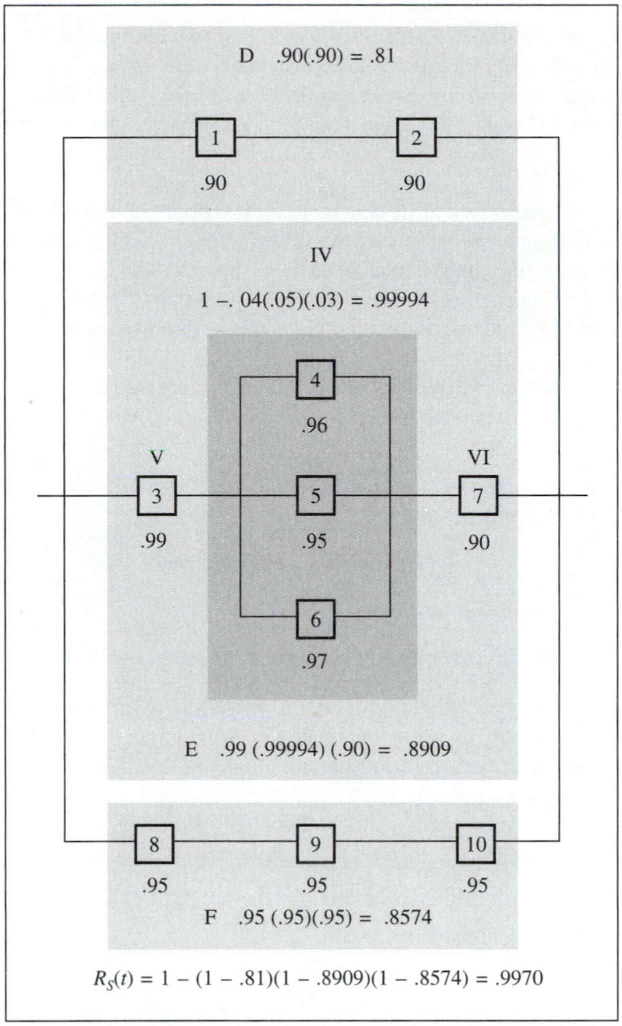

Figure 6–11 *Modular system with parallel subsystems.*

C, and D—arranged in series. To find the system reliability, we must first establish separate reliabilities for each subsystem. The most complex module is C, with three parallel components or subsystems of its own—I, II, and III. Before we can get the reliability of C, these must each be evaluated.

Figure 6–11 shows another module with three subsystems—D, E, and F—arranged in parallel. Each primary subsystem involves a series structure, and their reliabilities must be established before we can compute overall system reliability. Subsystem E has three components or subsystems, including IV, itself a parallel subsystem.

Example: *Reliability of a Personal Computer System*

A software engineer's personal computer system is shown in Figure 6–12. The reliabilities given there represent the probability that the component performs satisfactorily throughout a specific workday. This setup involves 5 components: the power supply (reliability .995), PC unit (.999), keyboard (.9999), peripheral memory subsystem, and printer subsystem. These are shown in *series* since all must function in order for the system to be usable.

The peripheral memory subsystem includes two floppy disk drives (A and B, both with .99 reliability) and one hard disk drive C (.95). The system logic shows those components in *parallel*, since the personal computer system will be functional as long as at least one of them is operational. The printer subsystem also has two parallel components, one laser (.99) and one dot-matrix (.999) printer; the system is usable as long as one of these works.

The two subsystem reliabilities must be found before the overall system reliability can be computed. For the peripheral memory subsystem, we have the reliability

$$1 - (1 - .99)(1 - .99)(1 - .95) = .999995$$

The reliability of the printer subsystem is

$$1 - (1 - .99)(1 - .999) = .99999$$

The overall system reliability is

$$R_S = .995(.999)(.9999)(.999995)(.99999) = .994$$

Notice that a system reliability of .994 is only slightly lower than the .995 reliability of the power supply by itself. Although minuscule improvements in system reliability might be achieved by substituting hardier components, no significant improvement in survival probability can be achieved without greater power reliability. Since power is supplied by a public utility, that might be achieved only by installing an auxiliary power system.

Figure 6–12 *Logic for personal computer system.*

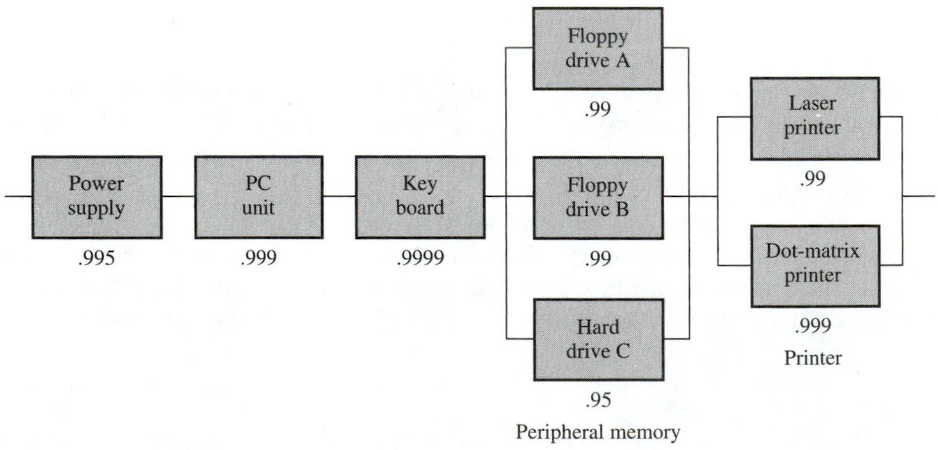

Problems

6–33 Consider a system having four components with reliabilities through time t of (1) .9, (2) .7, (3) .8, and (4) .6. Find the system reliability assuming (a) series logic and (b) parallel logic.

6–34 Which has a greater reliability: (a) a series system with 4 components, each having .99 reliability, or (b) a parallel system with 4 components, each having .90 reliability?

6–35 Consider a sound system with logic and single-hour listening reliability as given in Figure 6–13. Find the system reliability.

6–36 Consider the system in Figure 6–14. The reliabilities apply to the first 100 hours of operation.

(a) Determine the system reliability.

(b) Upgrade component 4 to .9 reliability. Recompute the system reliability.

(c) A redundant system may be created instead by substituting two type-4 items, each at the original reliability. Recompute the reliability for that system.

6–37 Refer to the sound system in Figure 6–13. Suppose that all components have independent exponential failure-time distributions, with the following mean failure rates:

(1) Power: .0001/hour
(2) Amplifier: .0002
(3) CD player: .0050
(4) Tuner: .0010
(5) Tape deck: .0020
(6) Turntable: .0090
(7,8) Each speaker: .0010

Find the system reliability for the following number of hours of play: (a) 5, (b) 10, and (c) 100. [Hint: Use $R_i(t) = e^{-\lambda t}$, where λ is the mean failure rate.]

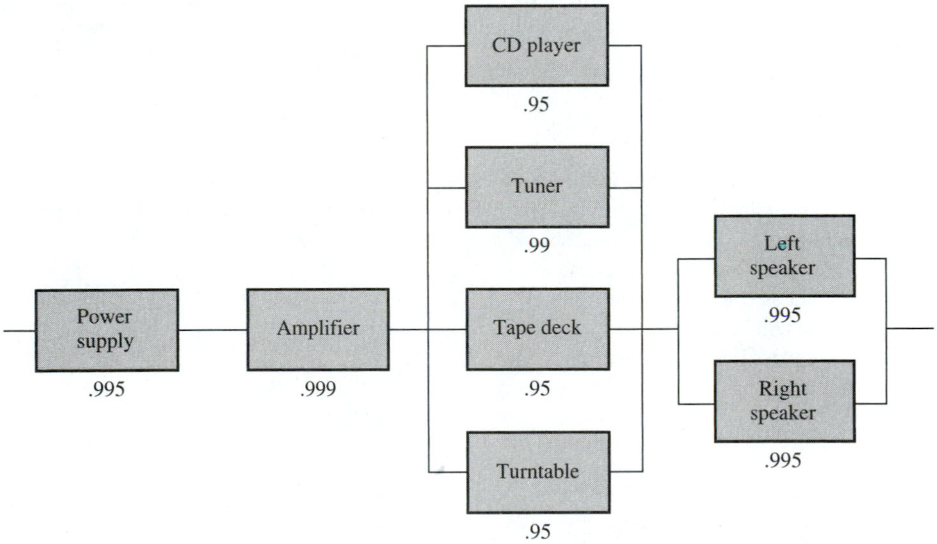

Figure 6–13 *Logic for sound system.*

Figure 6–14

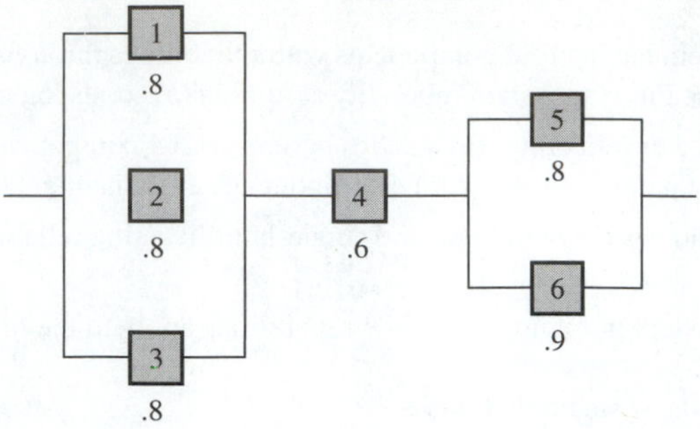

6–38 Consider the system in Figure 6–14. For each percentage point increase in system reliability there will be a $10 payoff. The following possibilities apply:

(a) Component 4 can be upgraded for $100 to reliability .9.

(b) Component 4 can be duplicated any number of times in a redundant system at a cost of $50 per extra item added.

What is the optimal solution?

6–39 Show that, for a system of n components in series, the system reliability is the product of the component reliabilities.

6–40 Show that, for a system of n components in parallel, the system reliability is one minus the product of the component failure probabilities.

Review Problems

6–41 One faculty member is chosen from those summarized in Figure 6–15. All persons have the same chance of being selected. Determine the following conditional probabilities:

(a) Pr[*CE* | woman]

(b) Pr[*EE* | man]

(c) Pr [*ME* | man *and FT*]

(d) Pr[*EE* | woman *and PT*]

(e) Pr[*FT* | woman]

(f) Pr[*FT and EE* | man]

(g) Pr[man | *CE or EE*]

(h) Pr[*PT and CE* | man]

6–42 Consider the faculty data in Figure 6–15. A lottery is held to select school membership on the university social committee. Three persons will be selected. Find the following probability that

(a) exactly 2 men will be chosen

(b) at least 1 woman will be chosen

(c) all will be from the electrical engineering department

(d) at least 1 part-timer will be chosen

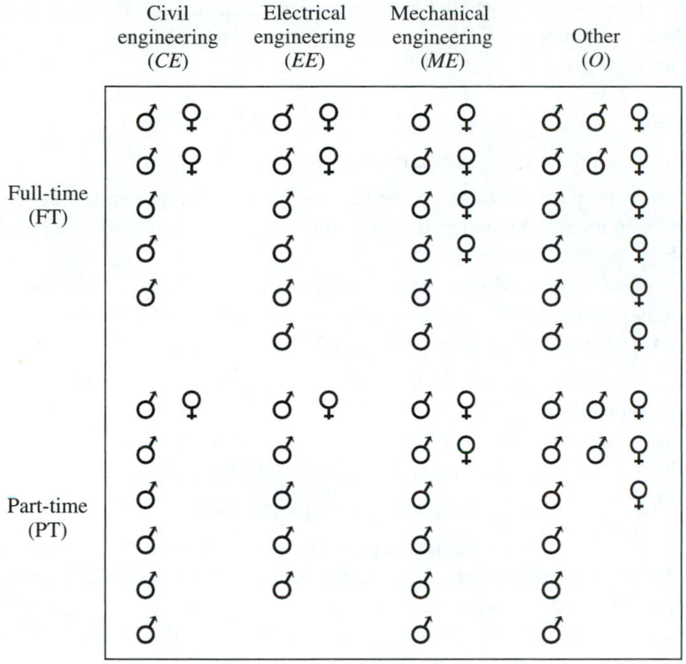

Figure 6–15 *Sample space for the characteristics of a randomly selected faculty member.*

6–43 The following facts are known regarding three events A, B, and C:

$$\Pr[A\ or\ B] = \frac{3}{4} \qquad \Pr[A\ and\ B] = \frac{1}{2} \qquad \Pr[B \mid C] = \frac{1}{2}$$

$$\Pr[B\ or\ C] = \frac{3}{4} \qquad \Pr[A\ and\ C] = \frac{1}{4} \qquad \Pr[A \mid B] = \frac{3}{4}$$

Find the following:

(a) $\Pr[B]$ (c) $\Pr[B \mid A]$ (e) $\Pr[B\ and\ C]$

(b) $\Pr[A]$ (d) $\Pr[C]$ (f) $\Pr[C \mid B]$

6–44 The probability is .95 that a driver in a particular community will survive the year without an automobile accident. Assuming that accident experiences in successive years are independent events, find the probability that a particular driver

(a) goes 5 straight years with no accident

(b) has at least 1 accident in 5 years

6–45 Consider the elementary events for the sampling experiment in Figure 6–6. Apply the addition law to compute the following compound event probabilities:
(a) exactly 1 defective
(b) exactly 2 defectives
(c) at least 1 defective

6–46 Repeat Problem 6–45 using the experiment in Figure 6–5.

6–47 The probability of one or more power failures on an automobile assembly line is .10 during any given month. Assuming power events in successive months are independent, find the probability that there will be
(a) no power failures during a 3-month span of time
(b) exactly 1 month involving a power failure during the next 4 months
(c) at least 1 power failure during the next 5 months

6–48 Consider the elementary events for the sampling experiment in Figure 6–6. Determine the following probabilities:

(a) $\Pr[G_1 \textit{ and } G_2]$	(c) $\Pr[G_1 \mid G_2]$	(e) $\Pr[G_3]$
(b) $\Pr[G_2]$	(d) $\Pr[G_2 \textit{ and } G_3]$	(f) $\Pr[G_2 \mid G_3]$

6–49 Repeat Problem 6–48 using the experiment in Figure 6–5.

6–50 A sample of 4 parts is selected from a production line where 20% are overweight.
(a) Construct a probability tree diagram for this experiment in which each sample observation is treated as a separate stage.
(b) Find the probability that the number of overweights in the sample is

(1) zero	(3) exactly 2	(5) exactly 4
(2) exactly 1	(4) exactly 3	(6) at least 2

6–51 Suppose that a customer inspects a sample of 4 parts taken without replacement from a shipment of 20 parts from the above plant. Repeat (a) and (b) of Problem 6–50.

CHAPTER SEVEN

Random Variables and Probability Distributions

This chapter focuses on the uncertain quantities that may arise from a random experiment. Because the particular level of a variable is uncertain and subject to chance, these quantities are referred to as **random variables**. Our present concern is the *number of events* of a particular type. For instance, we might believe that 99% of all circuit relays are acceptable. If so, how many acceptables will be identified in 100 relays inspected? Or, we might know that once installed, the acceptable relays will fail at an average rate that we might guess to be one per month. How many will fail and need to be replaced in 6 months' time? As we shall see, there is no single answer to these questions. Some quantities are more likely to occur than others, but many results are possible. We will see how to compute probabilities for each outcome. Collectively, the possible values for a random variable and their associated probability values establish a **probability distribution**.

Probability distributions fall into various *classes* or *families*. We will investigate those that have important engineering applications or play an important role in statistical methodology. This chapter introduces two probability distributions that are basic to statistics. The **binomial distribution** is helpful in a variety of statistical applications, especially quality control evaluations. Because of its connection with the sample mean, the **normal distribution** plays a unique role in statistical inference.

Probability distributions may involve random variables that assume *discrete* values. They may instead involve values that can fall anywhere within a specified range of real numbers. Continuous random variables are typical of physical measurements or time. It is desirable to have a systematic way to summarize and compare random variables. This is achieved by finding the average **outcome** or **expected value**. The concept of expected value provides a necessary framework for interpreting and evaluating statistical procedures.

7–1 Random Variables and Probability Distributions

We have seen that all random experiments culminate in an elementary event and that the sample space is the entire collection of these possibilities. Often the elementary events themselves are qualitative outcomes, such as a particular sequence of satisfactories and defectives for n sampled items or a series of successes and failures recorded at a test stand. We usually want to quantify these outcomes in terms of the number of results achieved in a particular category. Before the random experiment, the actual numerical result to be achieved is an unknown and must be treated as a *variable*. This quantity is uncertain, so before the experiment the particular outcome is subject to *random* chance. We use the term **random variable** for such uncertain quantities.

The Random Variable as a Function

The number of defective castings in a sample and the number of test failures of electronic components are examples of random variables. Mathematically, a random variable is a function that maps every elementary event in the sample space onto the real line. Figure 7–1 illustrates this concept. There X denotes the random variable, the number of defective castings in a sample of $n = 5$. Each observed unit is classified as good (G) or defective (D). Each of the 32 elementary events is represented by a string of Gs and Ds, with subscripts 1 through 5 designating the sequence of the item. Although there are 32 distinguishable outcomes, there are only 6 levels for X, ranging from 0 (no defectives) to 5 (all defectives).

As a practical matter, statistical applications don't require that we perform any mathematical operations to establish probabilities for X. We can usually match elementary events to a level for X and apply special counting methods to help find how many outcomes apply to any possible value. This requires a meaningful arrangement of the probability information so that a probability value can be established at each level of X.

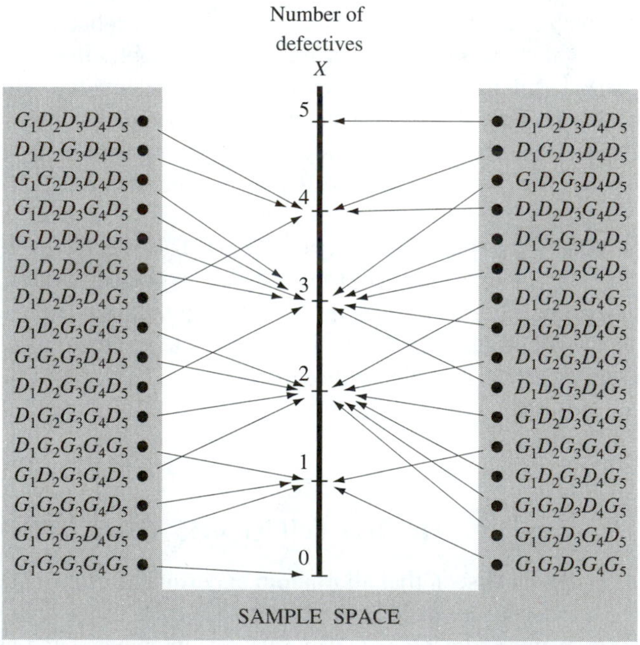

Figure 7–1
Random variable X (number of defective castings in a sample of size 5) is a function mapping the elements of the sample space onto the real line.

The Discrete Probability Distribution

The levels for a random variable together with their corresponding probabilities constitute a **probability distribution**. Table 7–1 provides the probability distribution for the number of equipment failures Y that might be encountered over a one-month span of time. There we see that exactly 1 failure has probability $\Pr[Y = 1] = .2707$.

It is standard notation to distinguish the random variable itself by an uppercase letter. Thus, Y is the actual number of failures (whose value is presently unknown). The events from the random experiment are the mutually exclusive and collectively exhaustive outcomes "$Y = 0$," "$Y = 1$," "$Y = 2$,". . . . To avoid repetition, they may all be summarized as "$Y = y$" with some specification as to what levels apply for the dummy variable y, which represents any of the possibilities, 0, 1, 2, ...

The probabilities for the outcomes constitute a **probability mass function**, which assigns a number in [0, 1] to each level for the random variable. Denoting this function by p, we have

$$p(y) = \Pr[Y = y] \quad \text{where} \quad y = 0, 1, 2 \ldots$$

For any level of y not listed the probability is $p(y) = 0$, reflecting that those values are impossible. (In the equipment example, only nonnegative integers are listed for y; outcomes such as 1.5 or –5 failures are impossible.) Because $p(y)$ is a probability, for any possible level y

$$0 \leq p(y) \leq 1$$

and because the possible y's correspond to mutually exclusive and collectively exhaustive events, it follows that

$$\sum_{y=0}^{\infty} p(y) = 1$$

A probability distribution may also be summarized by a graph, as in Figure 7–2, which shows the probability mass function for the number of equipment failures Y.

Table 7–1 *Probability Distribution for the Number of Equipment Failures over a One-Month Time Span*

Possible Number of Failures y	Probability $p(y)$
0	.1353
1	.2707
2	.2707
3	.1804
4	.0902
5	.0361
6	.0121
7	.0034
8	.0009
9	.0002
	1.0000

Figure 7–2
Probability mass function for the number of equipment failures.

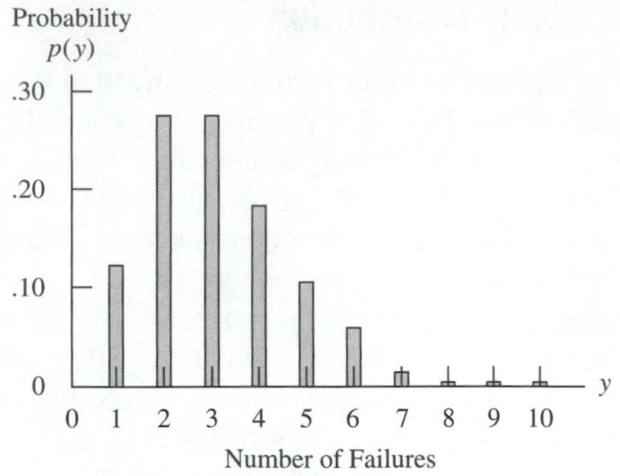

There, each possible level y is represented by a spike. The ordinate provides the corresponding probability for each possible number of failures. We see that $p(y)$ is a discrete function that "concentrates its mass" only at those y's that are possible, with the more likely levels being assigned greater probability mass and having taller spikes.

For many classes of probability distributions, the mass function may be characterized by a mathematical expression. Each specific distribution is distinguished by a particular level for one or more parameters. The failure probabilities in Table 7–1 were calculated from

$$p(y) = \frac{e^{-2}(2)^y}{y!} \qquad y = 0, 1, 2, \ldots$$

where the parameter is the mean rate of failures (here set at a value of 2) per month and $e = 2.7182$ is the base of the natural logarithms. A different set of probabilities will apply for some other rate. This probability mass function is for a Poisson distribution (to be discussed in Chapter 8).

The preceding probability distributions apply to **discrete random variables** that assume one value on a specific list of possibilities. A second class of random variables may assume any value within a specified *range*. These are **continuous random variables**.

Continuous Random Variables

To illustrate the concept of continuous random variables, we consider an application from chemical engineering. Chemical engineers often keep detailed records of the inputs and outputs of various monitored processes. Figure 7–3 shows the frequency curve representing several hundred batches of a technical chemical run at a particular plant. Because of varying characteristics of raw materials and fluctuating operating conditions there was considerable variation in the final product. Each batch resulted in a particular level of active ingredient, expressed in grams per liter.

The detailed data from a randomly chosen batch were to be evaluated with a computer model to determine how closely the model predicts actual yield. Our present concern is the yield X from that selected batch.

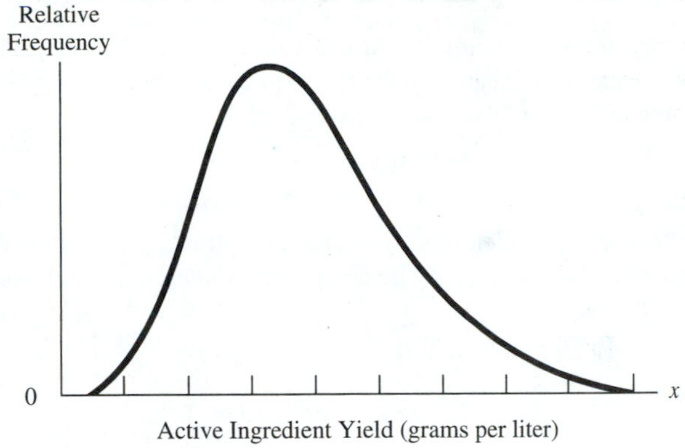

Figure 7–3
Frequency curve for population of active-ingredient yield from batches of raw chemicals processed.

The Probability Density Function

Every continuous random variable has a density function that completely specifies its probability characteristics. Figure 7–4 shows the probability density function for X, which we denote by $f(x)$. Notice that $f(x) \geq 0$ for all levels x; a probability density function never assumes negative values.

The probability that a continuous random variable will fall within an interval is equal to the *area* under the density curve over that range:

$$\Pr[a \leq X \leq b] = \int_a^b f(x)dx$$

The desired probability is found by integrating $f(x)$ over the interval of interest.

Because there is no area over a single point (which has no width), the probability is theoretically 0 that a continuous random variable will be precisely equal to some value. For example, the probability that the chosen batch has a yield of exactly 30 grams per liter may be assumed to be 0. This implies that every batch yield will be above or below 30 grams. Although some batches may have yields very close to 30, under extremely precise measurement there is only an infinitesimal chance that any one yield will be found exact. And this likelihood will more closely approach 0 when under more precise measurement.

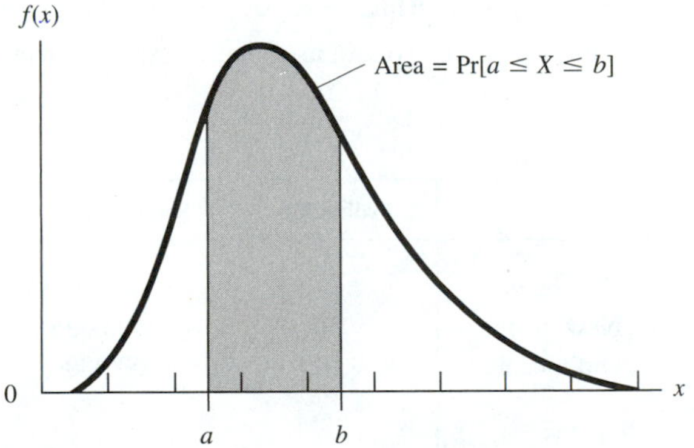

Figure 7–4
Graph of the probability density function for the active-ingredient yield of a randomly selected batch of a chemical.

Example:* *Distance between Flaws on Magnetic Tape

A radar telemetry tracking station requires a vast quantity of high-quality magnetic tape. It has been established that the distance X (in inches) between tape-surface flaws has the following probability density function:

$$f(x) = \begin{cases} .01e^{-.01x} & x \geq 0 \\ 0 & \text{otherwise} \end{cases}$$

Suppose one flaw has been identified. The probability that an additional flaw is found within the next 50 inches of tape may be computed by integrating $f(x)$ from 0 to 50:

$$\Pr[0 \leq X \leq 50] = \int_0^{50} .01e^{-.01x}dx = -e^{-.01x}\Big]_0^{50}$$
$$= e^0 - e^{-.50} = 1 - e^{-.50}$$
$$= 1 - .6065 = .3935$$

The probability that the next flaw will occur after at least 100 inches involves evaluating the integral at limits of 100 and ∞:

$$\Pr[100 \leq X \leq \infty] = \int_{100}^{\infty} .01e^{-.01x}dx = -e^{-.01X}\Big]_{100}^{\infty}$$
$$= e^{-1} - e^{-\infty} = e^{-1} - 0$$
$$= .3679$$

When the integration is taken over the entire real line, all possible levels of the random variable are included. Thus,

$$\Pr[-\infty < X < \infty] = \int_{-\infty}^{\infty} f(x)dx = 1$$

This reflects the fact that X is certain to be equal to some value, so the area under the entire density curve always equals 1.

Problems

7–1 Consider the quality-control illustration in Figure 6–6. Determine which elementary events apply to the random variable X, the number of defectives. Then construct a table for the probability distribution for X using the addition law to find the respective probabilities. Sketch this on a graph.

7–2 A computer manufacturer is expecting one of its clients to order a system. The following data apply:

	Probability	**Selling Price**
CPU (one)		
New model	.50	$250,000
Old model	.50	200,000
RAM (one)		
Small	.10	50,000
Medium	.50	100,000
Large	.40	150,000

Assuming that each of the choices is an independent event, construct a table showing the probability distribution for the total selling price of the ordered system.

7–3 Consider an experiment wherein a fair coin is tossed 3 times. Construct a table for the probability distribution for the number of heads obtained.

7–4 An engineering student has been assigned the following probabilities for final grades:

	A	B	C
Thermodynamics	.60	.40	0
Statistics	.50	.50	0
Circuit design	.90	.10	0
Economics	0	.50	.50

(a) Ignoring impossible outcomes, list the sample space for the student's final grades. Assuming independence between course grades, find the probability for each elementary event.

(b) For each elementary event in (a) determine the grade-point average, assuming equal weight for each course. Use A = 4 points, B = 3 points, and C = 2 points. Then construct a table summarizing the probability distribution for the student's GPA X.

7–5 Refer to the dice-tossing random experiment in Problem 6–7. Consider as a random variable the *range* in showing dots (largest value–smallest value).

(a) List the elementary events that correspond to each level of the random variable and determine the corresponding probability value.

(b) Summarize your findings in a table that shows the complete probability distribution.

7–6 Consider the random variable X having probability density function

$$f(x) = \begin{cases} \dfrac{x}{2} & 0 \le x \le 2 \\ 0 & \text{otherwise} \end{cases}$$

Determine the following:

(a) $\Pr[1 \le X \le 1.5]$ (c) $\Pr[X \le .75]$

(b) $\Pr[X > .25]$ (d) $\Pr[X < 4]$

7–7 The following cannot be probability density functions. Explain for each why this is so.

(a) $f(x) = \begin{cases} x & 1 \le x \le 4 \\ 0 & \text{otherwise} \end{cases}$ (c) $f(x) = \begin{cases} -1 & 1 \le x \le 2 \\ 0 & \text{otherwise} \end{cases}$

(b) $f(x) = \begin{cases} 1 & x = 10 \\ 0 & \text{otherwise} \end{cases}$ (d) $f(x) = \begin{cases} \dfrac{x^2 - 2}{3} & 0 \le x \le 3 \\ 0 & \text{otherwise} \end{cases}$

7–8 Scrap tubing is left over from fabricating a compressor. Each piece has a length (in inches) represented by the density function

$$f(x) = \begin{cases} \dfrac{1}{10} & 0 \le x \le 10 \\ 0 & \text{otherwise} \end{cases}$$

Determine the probability that a randomly selected tube will be
(a) less than 4″
(b) greater than 6″
(c) between 5.5″ and 7.5″

7–9 Consider the level of a random decimal, which is a quantity equally likely to fall anywhere between 0 and 1, inclusive. Give the expression for the probability density function.

7–10 For each random variable having the following probability density functions, determine $\Pr[.5 \leq X \leq 1.5]$.

(a) $f(x) = \begin{cases} \frac{1}{12} & 0 \leq x \leq 12 \\ 0 & \text{otherwise} \end{cases}$ (c) $f(x) = \begin{cases} \frac{3x^2}{7} & 1 \leq x \leq 2 \\ 0 & \text{otherwise} \end{cases}$

(b) $f(x) = \begin{cases} 5e^{-5x} & 0 \leq x \leq \infty \\ 0 & \text{otherwise} \end{cases}$ (d) $f(x) = \begin{cases} x & 0 \leq x \leq 1 \\ 2-x & 1 \leq x \leq 2 \\ 0 & \text{otherwise} \end{cases}$

7–11 Determine levels for the constants that will make $f(x)$ a probability density function.

(a) $f(x) = \begin{cases} \frac{x}{a} & 0 \leq x \leq 4 \\ 0 & \text{otherwise} \end{cases}$ (c) $f(x) = \begin{cases} e^{-cx} & 0 \leq x \leq \infty \\ 0 & \text{otherwise} \end{cases}$

(b) (b) $f(x) = \begin{cases} bx^2 & 0 \leq x \leq 1 \\ 0 & \text{otherwise} \end{cases}$ (d) $f(x) = \begin{cases} d(x-.5)^2 & 0 \leq x \leq 1 \\ 0 & \text{otherwise} \end{cases}$

7–12 Refer to Problem 7–11 and your answers to that problem. Determine for each random variable $\Pr[.25 \leq X \leq .75]$.

7–2 Expected Value and Variance

Although a detailed description of a random variable is provided by its probability distribution, it is often desirable to summarize that information in a concise form. Random variables commonly encountered in engineering may be summarized by a measure of central tendency or mean, referred to as their **expected values**. Viewing probabilities themselves as long-run frequencies, this number is the long-run average level of the random variable. For example, in a series of 10 coin tosses the average number of heads will be 5. Although many may reach this conclusion by intuition, there is a general procedure for computing expected values directly from the probability distribution.

Another useful summary characterizes how the random variable values tend to differ from each other. A measure of dispersion is provided by the **variance**. Like the expected value, this quantity may be found by an arithmetical operation with the probability distribution.

Table 7–2
Expected Value Calculation for the Number of Equipment Failures in One Month

Possible Number of Failures y	Probability $p(y)$	Weighted Value $yp(y)$
0	.1353	0
1	.2707	.2707
2	.2707	.5414
3	.1804	.5412
4	.0902	.3608
5	.0361	.1805
6	.0121	.0726
7	.0034	.0238
8	.0009	.0072
9	.0002	.0018
		$E(Y) = 2.0000$

Expected Value

The expected value of a random variable is simply a weighted average of the possible values, using the respective probabilities as weights. The following expression is used to compute the expected value of a discrete random variable.

EXPECTED VALUE OF A DISCRETE RANDOM VARIABLE

$$E(X) = \sum xp(x)$$

In the expression the summation is taken over all possible levels for the random variable. The expected value measures the random variable's **central tendency**, and $E(X)$ is often referred to as the **mean** of X. The foregoing shows that the expected value is analogous to the *first moment* of physics, and this quantity is sometimes identified by that term.

As an illustration of the expected value calculation, consider again the probability distribution for the number of equipment failures, here shown in Table 7–2. The expected value is $E(Y) = 2.0$, indicating that the mean number of failures is 2 per month.

The usual interpretation of an expected value is that it represents the long-run average result from a series of repeated random experiments. Thus, over several years of similar operation the actual tally of month-by-month failures should average to nearly 2.

Variance and Standard Deviation of a Random Variable

The common measure of dispersion for a random variable is its variance. This measure is also a weighted average, wherein the quantities involved indicate how much individual values differ from the center of the distribution. These quantities are the squared deviations of each possible level from the expected value.

The variance is an important statistical concept because it provides a systematic summary of individual differences and because of its convenient mathematical properties. In general, we will use the following expression.

Table 7–3
Variance Computation for the Number of Equipment Failures in One Month

Possible Number of Failures y	Probability $p(y)$	Squared Deviation $[y - E(Y)]^2$	Weighted Value $[y - E(Y)]^2 p(y)$
0	.1353	$(0 - 2.0)^2 = 4.0$	.5412
1	.2707	$(1 - 2.0)^2 = 1.0$	.2707
2	.2707	$(2 - 2.0)^2 = 0.0$	.0000
3	.1804	$(3 - 2.0)^2 = 1.0$	.1804
4	.0902	$(4 - 2.0)^2 = 4.0$	.3608
5	.0361	$(5 - 2.0)^2 = 9.0$	.3249
6	.0121	$(6 - 2.0)^2 = 16.0$	.1936
7	.0034	$(7 - 2.0)^2 = 25.0$	.0850
8	.0009	$(8 - 2.0)^2 = 36.0$	.0324
9	.0002	$(9 - 2.0)^2 = 49.0$	.0098
			$\text{Var}(Y) = 1.9988$

VARIANCE OF A DISCRETE RANDOM VARIABLE

$$\text{Var}(X) = \sum [x - E(X)]^2 \, p(x)$$

Here, as before, the summation is taken over all possible levels for the random variable. We see that Var(X) is a measure similar to the *second moment* of physics.

Continuing with the example of number of equipment failures Y, Table 7–3 shows the variance computation. The rounded level is Var(Y) = 2.0. In Chapter 2 we noted that the variance will be in *squared* units. Thus, we have Var(Y) = 2.0 square failures. As a practical matter, the square root of the variance, or the **standard deviation**, is sometimes used to express dispersion:

$$SD(X) = \sqrt{\text{Var}(X)}$$

We have $SD(Y) = \sqrt{2.0} = 1.41$ failures; this equation conveys the same information as the variance, but in more convenient units.

Some Important Properties of Expected Value and Variance

Var(X) is the expected value of a function of X:

$$[X - E(X)]^2$$

In general,

$$E[g(X)] = \sum g(x) p(x)$$

We can apply the above for variance. Note that the variance is itself the expected value of a function of X.

$$\text{Var}(X) = \sum [X - E(X)]^2 p(x) = E([X - E(X)]^2)$$

Three of the key properties of expected value are listed.

1. $E(c) = c$
2. $E(cX) = cE(X)$
3. $E(a + bX) = a + bE(X)$

Caution: Although true for the above, generally

$$E[g(X)] \neq g[E(X)]$$

when $g(X)$ is a nonlinear function.

It is easy to show that

$$\begin{aligned} E([X - E(X)]^2) &= E[X^2 - 2XE(X) + E(X)^2] \\ &= E(X^2) - E(X)^2 \end{aligned}$$

which provides the

EQUIVALENT EXPRESSION FOR THE VARIANCE

$$\text{Var}(X) = E(X^2) - E(X)^2$$

Two important properties of the variance are listed.

1. $\text{Var}(c) = 0$
2. $\text{Var}(a + bX) = b^2\text{Var}(X)$

Expected Value and Variance of a Continuous Random Variable

Recall that the expected value of a discrete random variable expresses its central tendency. It is a weighted average computed by summing the products of each possible level times the respective probability. The expected value of a continuous random variable is found analogously by integrating the product of the dummy variable and the density over the entire span of possibilities.

EXPECTED VALUE OF A CONTINUOUS RANDOM VARIABLE

$$E(X) = \int_{-\infty}^{\infty} xf(x)dx$$

As an illustration, consider the distance X between magnetic tape flaws in the preceding example. Because $f(x) = 0$ for all $x > 0$ the lower limit of integration will be taken as 0. We have

$$E(X) = \int_0^{\infty} x(.01e^{-.01x})dx = (-100 - x)e^{-.01x}\Big]_0^{\infty} = 100$$

This means that on the average there will be $E(X) = 100$ inches of tape between each flaw encountered.

The method of finding the variance of a continuous random variable is also analogous to that of finding its discrete counterpart. Recall that Var(X) was defined in terms of the weighted average of squared deviations, $[x - E(X)]^2$. For continuous random variables we find the variance by evaluating the integral

$$\text{Var}(X) = \int_{-\infty}^{\infty} [x - E(X)]^2 f(x)dx$$

In practice, the foregoing is cumbersome to work with, so we employ the fact that

$$\text{Var}(X) = E(X^2) - E(X)^2$$

and use instead the following expression.

VARIANCE OF A CONTINUOUS RANDOM VARIABLE

$$\text{Var}(X) = \int_{-\infty}^{\infty} x^2 f(x)dx - E(X)^2$$

Continuing with the magnetic tape illustration, the variance in distance between flaws is

$$\begin{aligned}\text{Var}(X) &= \int_{0}^{\infty} x^2(.01)e^{-.01x}dx - E(X)^2 \\ &= (.01)\int_{0}^{\infty} x^2 e^{-.01x}dx - (100)^2 \\ &= (.01)\left(\frac{x^2 e^{-.01x}}{-.01}\bigg]_0^{\infty} - \frac{2}{(-.01)}\int_{0}^{\infty} xe^{-.01x}dx\right) - 10{,}000 \\ &= 0 + 2\left[\frac{e^{-.01x}}{(-.01)^2}([-.01]x - 1)\right]_0^{\infty} - 10{,}000 \\ &= 20{,}000 - 10{,}000 = 10{,}000\end{aligned}$$

This variance is in *square* inches, not a very helpful measure of distance. By taking the square root, we obtain the standard deviation

$$SD(X) = \sqrt{\text{Var}(X)} = \sqrt{10{,}000} = 100 \text{ inches}$$

Problems

7–13 The probability distribution for the number of cars X arriving at a toll booth during any minute is as follows:

x	$p(x)$
0	.37
1	.37
2	.18
3	.06
4	.02

Calculate the expected value, variance, and standard deviation for X.

7–14 The probability distribution for the number of defective items in a random sample is as follows:

x	$p(x)$
0	.35
1	.39
2	.19
3	.06
4	.01

Calculate the expected value, variance, and standard deviation for X.

7–15 The following probability distribution applies to the high temperature X to be experienced in a city when the morning reading is 24 °C.

x (°C)	$p(x)$
30	.05
31	.13
32	.27
33	.36
34	.14
35	.05

Calculate the expected value, variance, and standard deviation for the day's high temperature as expressed in

(a) Celsius degrees

(b) Fahrenheit degrees, using the transformation F° = 32 + 1.8 C°

7–16 An environmental test shows that transformer failure characteristics are different at various ambient temperature levels. The following probability distributions apply for the number of failures at low (X) and high (Y) temperature settings.

Number of Failures x, y	**Low Setting** $p(x)$	**High Setting** $p(y)$
0	.02	.00
1	.07	.02
2	.15	.04
3	.20	.09
4	.20	.13
5	.16	.16
6	.10	.16

Number of Failures x, y	Low Setting $p(x)$	High Setting $p(y)$
7	.06	.14
8	.03	.10
9	.01	.07
10	.00	.04
11	.00	.02
12	.00	.02
13	.00	.01

Determine the expected number of failures under each setting. On the average, which temperature setting will result in the greater number of failures?

7–17 Consider the sum of the showing dots X from rolling two die cubes. (You may refer to your answer to Problem 6–7.) Calculate the expected value, variance, and standard deviation.

7–18 Refer to the random variable in Problem 7–6. Determine the expected value and the variance.

7–19 Refer to the random variable in Problem 7–8. Determine the expected value and the variance.

7–20 Refer to the random variables in Problem 7–10. Determine for each (1) $E(X)$, (2) $\text{Var}(X)$, and (3) $SD(X)$.

7–21 Refer to Problem 7–12 and your answers to that problem. Determine for each random variable (1) $E(X)$, (2) $\text{Var}(X)$, and (3) $SD(X)$.

7–3 The Binomial Distribution

The binomial distribution has wide application in engineering and statistics because it provides a basis for evaluating samples from *qualitative* populations. In such investigations there are ordinarily two complementary categories of major interest—such as defective versus good, satisfactory versus unsatisfactory, operable versus inoperable, or success versus failure. The binomial distribution provides probabilities for the number of observations falling into a particular category.

The binomial distribution is concerned with the events that arise from a *series* of random experiments and is based on several assumptions. These experiments constitute a **Bernoulli process**, named in honor of the pioneer mathematician who first published on the subject.

The Bernoulli Process

A Bernoulli process is epitomized by coin tossing. It is useful to apply the coin analogy in describing the process. The following assumptions can be made:

1. Each random experiment is referred to as a **trial**. The Bernoulli process is a series of trials (such as coin tosses), each having two complementary outcomes.

Although designations are arbitrary, one outcome is called a **success** and the other a **failure**.

2. The trial success probability is some constant value for all trials. In coin tossing, Pr[head] is the same for all tosses (although the coin may be lopsided, so that the head probability need not be 1/2).
3. Successive trial outcomes are statistically *independent*. No matter how many successes or failures have been achieved, the probability of success on the next trial cannot vary. (This is an obvious property of coin tosses.)

The Bernoulli process is commonly encountered in engineering applications involving quality assurance. Each new item created in a production process may be considered as a trial resulting in a good or defective unit.

Binomial Probabilities

The binomial distribution is concerned with how many successes will be achieved in a specified number of trials R of a Bernoulli process. Two parameters establish a particular member of this distribution family. One is the trial success probability, denoted by π, and the other parameter is the number of trials (n).

Illustration: *Quality-Control Sampling with Car Parts*

A quality-control illustration will help us develop the procedure used in finding the probabilities. Consider a production line for car parts wherein 10% of the output is defective and must be scrapped, and the rest is good. Treating a defective as a "success," the probability that any particular part is defective is therefore $\pi = .10$. A random sample of $n = 5$ parts is taken.

Figure 7–5 is the probability tree diagram for the car parts investigation, wherein all 5 observation trials have been treated as a combined random experiment. There are $2^5 = 32$ elementary events, distinguishable in terms of good (G) and defective (D) for each successive part. For each part the probabilities are .10 for defective and .90 for good; these values are entered onto the respective branches. The quality events between sample parts are independent, so these same probability values apply for each observed part.

The product of the branch probabilities on each path leading to the respective end position gives the probability for the culminating elementary event. For example, we have for the fourth elementary event in the sample space

$$\begin{aligned}\Pr[G_1G_2G_3D_4D_5] &= (.90)(.90)(.90)(.10)(.10)\\ &= (.90)^3(.10)^2 = .00729\end{aligned}$$

The number of defectives for each of the compound results in Figure 7–5 has been determined and listed in a column next to the probabilities. The final grouping of columns gives a tally of the number of ways in which a particular number of defectives might occur.

The following list shows all the elementary events in which exactly 2 defectives occur.

$$\begin{array}{ll} G_1G_2G_3D_4D_5 & G_1D_2D_3G_4G_5 \\ G_1G_2D_3G_4D_5 & D_1G_2G_3G_4D_5 \\ G_1G_2D_3D_4G_5 & D_1G_2G_3D_4G_5 \\ G_1D_2G_3G_4D_5 & D_1G_2D_3G_4G_5 \\ G_1D_2G_3D_4G_5 & D_1D_2G_3G_4G_5 \end{array}$$

Figure 7–5
Probability tree diagram for a sample from a car parts production line.

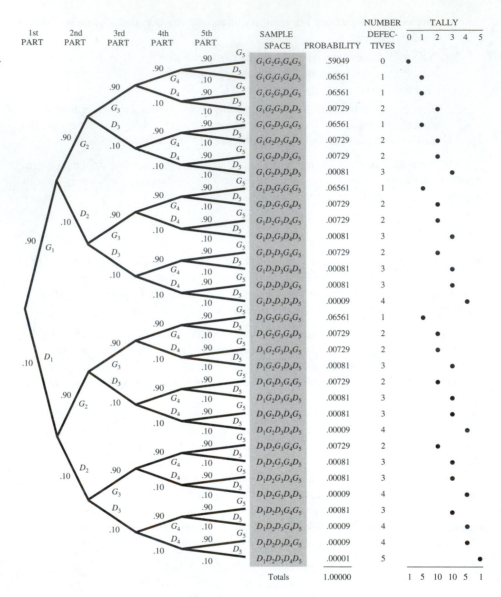

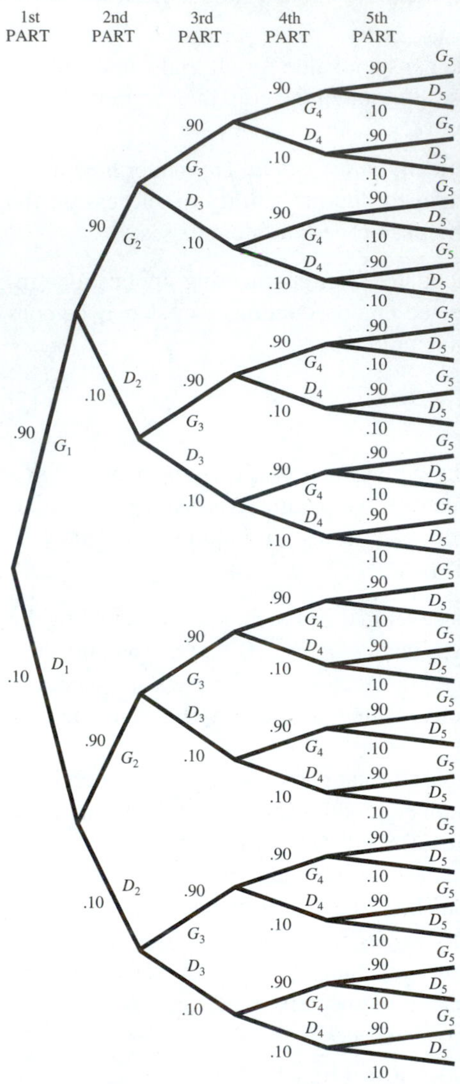

SAMPLE SPACE	PROBABILITY	NUMBER DEFECTIVES	TALLY 0	1	2	3	4	5
$G_1G_2G_3G_4G_5$	.59049	0	•					
$G_1G_2G_3G_4D_5$	.06561	1		•				
$G_1G_2G_3D_4G_5$	.06561	1		•				
$G_1G_2G_3D_4D_5$	.00729	2			•			
$G_1G_2D_3G_4G_5$	.06561	1		•				
$G_1G_2D_3G_4D_5$	.00729	2			•			
$G_1G_2D_3D_4G_5$	.00729	2			•			
$G_1G_2D_3D_4D_5$	.00081	3				•		
$G_1D_2G_3G_4G_5$	.06561	1		•				
$G_1D_2G_3G_4D_5$	.00729	2			•			
$G_1D_2G_3D_4G_5$	.00729	2			•			
$G_1D_2G_3D_4D_5$	.00081	3				•		
$G_1D_2D_3G_4G_5$	.00729	2			•			
$G_1D_2D_3G_4D_5$	.00081	3				•		
$G_1D_2D_3D_4G_5$	.00081	3				•		
$G_1D_2D_3D_4D_5$	.00009	4					•	
$D_1G_2G_3G_4G_5$	.06561	1		•				
$D_1G_2G_3G_4D_5$	.00729	2			•			
$D_1G_2G_3D_4G_5$	.00729	2			•			
$D_1G_2G_3D_4D_5$	.00081	3				•		
$D_1G_2D_3G_4G_5$	.00729	2			•			
$D_1G_2D_3G_4D_5$	.00081	3				•		
$D_1G_2D_3D_4G_5$	.00081	3				•		
$D_1G_2D_3D_4D_5$	.00009	4					•	
$D_1D_2G_3G_4G_5$	.00729	2			•			
$D_1D_2G_3G_4D_5$	.00081	3				•		
$D_1D_2G_3D_4G_5$	.00081	3				•		
$D_1D_2G_3D_4D_5$	.00009	4					•	
$D_1D_2D_3G_4G_5$	.00081	3				•		
$D_1D_2D_3G_4D_5$	.00009	4					•	
$D_1D_2D_3D_4G_5$	.00009	4					•	
$D_1D_2D_3D_4D_5$	.00001	5						•
Totals	1.00000		1	5	10	10	5	1

It is easy to get the probability of any particular path. It will always be a product involving r D's and $n - r$ G's. Rearranging terms in the probability product, we can get

$$\Pr[\text{particular path with } r \ Ds \text{ and } n - r \ \ Gs]$$

$$= \underbrace{\Pr[D_i] \times \ldots \times \Pr[D_i]}_{r \text{ terms}} \times \underbrace{\Pr[G_k] \times \ldots \times Pr[G_m]}_{n-r \text{ terms}}$$

$$= \Pr[D]^r \Pr[G]^{n-r}$$

$$= \Pr[D]^r (1 - \Pr[D])^{n-r}$$

For any level of r and n we can always find an individual path probability (even when the tree is too big to draw and is only imagined).

But how many paths are there having exactly r Ds and $n - r$ Gs? To answer this question, we need to look at *combinations* of sequence positions involving such outcomes and find the number of combinations that applies.

Counting Paths of Like Type: Factorials and Combinations

We should be able to count the number of two-defective outcomes without even looking at the entire tree. Let's consider a succession of forks, one extracted from each stage in the tree:

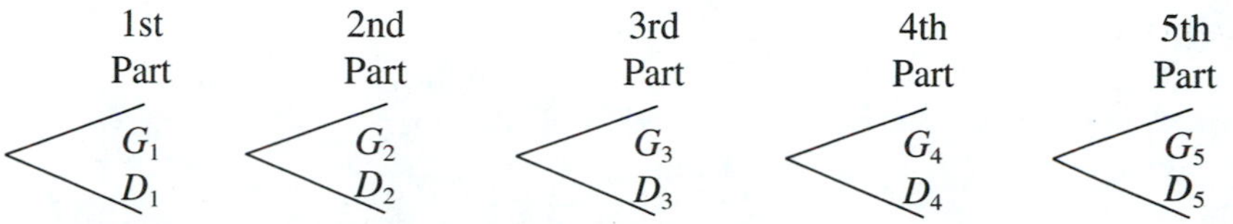

We want to select one branch from each fork. Two of the selected branches will be Ds and three will be Gs. Our problem is to determine how many ways there are to identify two D branches from the five.

If we pick the D branches one at a time, we have 5 possibilities for the first choice. No matter which particular D branch is selected first, there remain 4 D-branch possibilities for the second choice. Multiplying these possibilities, we have

$$5 \times 4 = 20$$

Of course, the above counts each possibility twice. For example,

$$D_1 D_5 \qquad \text{and} \qquad D_5 D_1$$

are really the same outcome or combination. Dividing by 2, we achieve the proper count:

$$\frac{5 \times 4}{2} = 10$$

This type of calculation finds the **number of combinations** for a variety of situations. To illustrate the general expression, we re-express the above fraction in the equivalent form

$$\frac{5 \times 4}{2} = \frac{5 \times 4}{2 \times 1} = \frac{5 \times 4 \times 3 \times 2 \times 1}{2 \times 1 \times 3 \times 2 \times 1} = 10$$

Multiplying both the numerator and the denominator of the second fraction by 3 and then by 2 and finally by 1, we get an equivalent result.

The final fraction contains factorial terms. A **factorial** is the product of successive integer values ending with 1. Such a product is denoted by placing an exclamation point after the highest number.

$$2! = 2 \times 1 \qquad (= 2)$$

$$3! = 3 \times 2 \times 1 \qquad (= 6)$$

$$5! = 5 \times 4 \times 3 \times 2 \times 1 \qquad (= 120)$$

We define

$$0! = 1$$

$$1! = 1$$

In factorial notation the number of two-defective sequences in 5 successive parts is

$$\frac{5 \times 4 \times 3 \times 2 \times 1}{2 \times 1 \times 3 \times 2 \times 1} = \frac{5!}{2!3!} = 10$$

This result suggests the general procedure for finding the number of combinations.

Number of Combinations

The following expression is used to compute the number of combinations of r objects taken from n objects.

NUMBER OF COMBINATIONS

$$C_r^n = \frac{n!}{r!(n-r)!}$$

In our illustration there are $n = 5$ parts, and we are considering exactly $r = 2$ defectives occurring in those parts. Thus, the number of two-defective combinations is

$$C_2^5 = \frac{5!}{2!(5-2)!} = \frac{5!}{2!3!} = 10$$

We see that exactly 2 defectives can happen in 10 different ways. Since the elementary event in each of these cases has probability .00729, it follows that

$$\begin{aligned} \text{Pr[exactly 2 defectives]} &= \Pr[R = 2] \\ &= 10(.00729) = .07290 \end{aligned}$$

Of course, it would be impractical to construct a probability tree diagram every time we want to find the probability of a particular level for R. (A tree drawn on the same scale as Figure 7–5 but involving twice the number of trials would be over 20 feet tall.)

The Probability Mass Function

Instead, we can use the following expression for the probability mass function of the binomial distribution.

BINOMIAL DISTRIBUTION

$$b(r; n,\pi) = C_r^n \pi^r (1-\pi)^{n-r} \qquad r = 0,1,2,\ldots,n$$

Because we use it so often, we reserve the letter b to represent the binomial probabilities. To help distinguish particular binomial distributions, the parameters n and π are often listed after the dummy variable r. We may confirm the foregoing expression by evaluating our earlier result with $\pi = .10$ and $n = 5$:

$$\begin{aligned} \Pr[R = 2] = b(2; 5,.10) &= C_2^5(.10)^2(1-.10)^{5-2} \\ &= 10(.10)^2(.90)^3 \\ &= .07290 \end{aligned}$$

Table 7–4
Binomial Distribution for the Number of Defective Parts When n = 5 and $\pi = .10$

Number of Defectives r	Probability $b(r; n, \pi)$
0	$C_0^5(.10)^0(.90)^5 = .59049$
1	$C_1^5(.10)^1(.90)^4 = .32805$
2	$C_2^5(.10)^2(.90)^3 = .07290$
3	$C_3^5(.10)^3(.90)^2 = .00810$
4	$C_4^5(.10)^4(.90)^1 = .00045$
5	$C_5^5(.10)^5(.90)^0 = .00001$
	1.00000

The product $(.10)^2(.90)^3$ is the same result achieved by multiplying together all the branch probabilities on any one of the paths having two defectives. Each of those outcomes involves 2 D and 3 G branches. The quantity $C_2^5 = 10$ is the number of such paths and represents the number of ways of getting 2 particular items (as the defective D parts) from 5. In general, there will be C_r^n outcomes involving exactly r successes, each of which has a common probability, $\pi^r(1-\pi)^{n-r}$.

Table 7–4 gives the complete binomial probability distribution for the number of defective parts. This distribution is graphed in Figure 7–6.

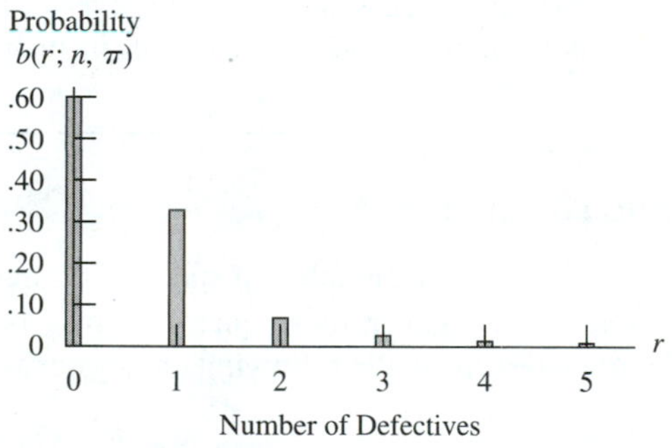

Figure 7–6
Binomial distribution for the number of defective car parts.

Example:
Opinions Regarding Computer Terminal Display

A computer manufacturer had to decide the final terminal display configuration for its consumer line of minicomputers. Two possibilities were (1) black-and-white monitor with a printer or (2) color monitor with no printer. The final choice depended on the proportion π of potential buyers who would have a definite need for hard copy. A sample of these would-be customers was used to help determine whether that proportion was high or low.

Since the population of potential customers was huge, the opinions of successive persons in the random sample could be assumed independent. This made the binomial distribution appropriate for establishing probabilities for the number of respondents R expressing a need for printing.

Using a sample size of $n = 5$, and assuming that

$$\pi = \Pr[\text{ a person requires a printer }] = .70$$

the probability that 4 persons would ask for a printer is

$$b(4; 5, .70) = C_4^5(.70)^4(.30)^1 = .3602$$

A larger sample size of $n = 20$ was selected in the actual investigation. For the same π, the probability that 15 persons would require the printer is

$$b(15; 20, .70) = C_{15}^{20}(.70)^{15}(.30)^5 = .1789$$

Of course, π itself was unknown. The actual sample outcome resulted in 8 out of 20 requesting the printer. Using the binomial distribution for various levels of π, the probabilities of getting as few or fewer favorable responses are as shown:

Possible Proportion π	**Probability** Pr[R ≤ 8]
.30	.8867
.40	.5956
.50	.2517
.60	.0565
.70	.0051
.80	.0001

For high levels of π it is quite improbable that so few favorable responses could have been achieved. The sample evidence tends to favor a lower level for π. Because of this, management decided to drop the printer and use the colored monitor.

Expected Value and Variance

The expected value and variance for the number of successes R may be computed using the procedures of Section 7–2. However, these quantities may be more directly computed from the stated parameters using the following expressions.

EXPECTED VALUE AND VARIANCE FOR A BINOMIAL DISTRIBUTION

$$E(R) = n\pi$$
$$\text{Var}(R) = n\pi(1 - \pi)$$

Thus, the expected number of heads in $n = 100$ tosses of a fair coin, where $\pi = \text{Pr[head]} = .5$, is

$$E(R) = n\pi = 100(.5) = 50$$

and the variance is

$$\text{Var}(R) = n\pi(1 - \pi) = 100(.50)(.50) = 25$$

It will be helpful to show the derivation for $E(R)$. By the definition of expected value,

$$E(R) = \sum_{r=0}^{n} rb(r; n, \pi) = \sum_{r=0}^{n} rC_r^n \pi^r (1 - \pi)^{n-r}$$
$$= \sum_{r=0}^{n} r\frac{n!}{r!(n - r)!}\pi^r(1 - \pi)^{n-r}$$

The term involving $r = 0$ is zero and may be dropped. Factoring $n\pi$ out of all terms and canceling r's, the preceding is equal to

$$n\pi \sum_{r=1}^{n} \frac{(n-1)!}{(r-1)!(n-r)!} \pi^{r-1}(1-\pi)^{n-r}$$

Representing $r - 1$ by s and $n - 1$ by m, so that $n - r = (n - 1) - (r - 1) = m - s$, the above is equivalent to

$$n\pi \sum_{s=0}^{m} \frac{m!}{s!\,(m-s)!} \pi^{s}(1-\pi)^{m-s}$$

The summation includes all binomial probabilities when there are m trials and is therefore equal to 1. Thus,

$$E(R) = n\pi(1) = n\pi$$

It will be left as an exercise for the reader to derive the expression for Var(R).

Notice that $E(R)$ is directly proportional to n and to π, so that greater levels in either parameter will yield an increased level for the expected number of successes. Var(R) also increases in direct proportion to n. However, for a fixed n, Var(R) is greatest when $\pi = .5$, assuming lower levels when π is closer to 0 or to 1.

Cumulative Probability and the Binomial Probability Table

When n is small, it is easy to compute binomial probabilities using a calculator. But when n is large, this becomes an onerous and error-prone chore. To ease the computational burden, the more common binomial probabilities have been tabled.

Such tables are usually constructed in terms of **cumulative probabilities**. A cumulative probability is found by summing the individual probability terms applicable to all the levels of the random variable that fall at or below a specified point. The resulting sums themselves form the **probability distribution function**, defined for discrete distributions in terms of the cumulative sum of the probability mass function. The following expression applies.

BINOMIAL PROBABILITY DISTRIBUTION FUNCTION

$$B(r; n,\pi) = \Pr[R \le r] = \sum_{x=0}^{r} b(x; n,\pi)$$

We use an uppercase $B(r)$ to distinguish the cumulative binomial distribution function from the binomial probability mass function $b(r)$, which provides individual probability terms.

Table 7-5 shows the distribution function constructed for the number of defective car parts considered earlier.

As an example, the cumulative probability for 2 defectives or fewer is found by adding together the individual probability terms for 0, 1, and 2 defectives:

$$\begin{aligned} \Pr[R \le 2] &= \Pr[R = 0] + \Pr[R = 1] + \Pr[R = 2] \\ &= .59049 + .32805 + .07290 \\ &= .99144 \end{aligned}$$

Figure 7–7 shows the probability mass function and the probability distribution function graphs of the binomial distribution for the number of successes in $n = 10$ tri-

Table 7–5 *Binomial Probability Distribution Function Constructed for the Number of Defective Car Parts When $n = 5$ and $\pi = .10$*

Number of Defectives r	Probability $\Pr[R = r] = b(r; n, \pi)$	Cumulative Probability $\Pr[R \leq r] = B(r; n, \pi)$
0	.59049	.59049
1	.32805	.91854
2	.07290	.99144
3	.00810	.99954
4	.00045	.99999
5	.00001	1.00000

als when $\pi = .30$. Notice that the probability distribution function plots as a stairway, wherein the cumulative probability at any r is the height of the stairway above ground level at that point. The size of each step equals the probability for getting exactly r successes and is the same size as the respective spike of the probability mass function. For example, the two cumulative probabilities for 2 and 3 successes are $B(2; 10, .30) = .3828$ and $B(3; 10, .30) = .6496$. The step size at 3 is $.6496 - .3828 = .2668$, and is equal to the probability of exactly 3 successes, $b(3; 10, .30)$.

Appendix Table A lists cumulative binomial probabilities for several levels of n and π. This table is a convenient source for binomial probabilities. Cumulative terms are tabled instead of individual terms because most applications consider R events spanning a range of successes.

Illustration: *Preferences for Dash Panel Displays*

As an illustration involving these tables, suppose that a random sample of $n = 100$ persons are asked to state their preferences for two alternative automobile dash panel displays: a sleek, modernistic one and a classical one. The population is sufficiently large for the binomial distribution to represent probabilities for the number of respondents R, favoring the sleek display over the classical one. The assumed proportion favoring the sleek display is 40%, so that $\pi = .40$ serves as the trial success probability.

The probability that 35 or fewer favor the sleek design is

$$\Pr[R \leq 35] = B(35; 100, .40) = .1795$$

whereas the probability that at least 50 like that display better is the complement of 49 or fewer:

$$\begin{aligned}\Pr[R \geq 50] &= 1 - \Pr[R \leq 49] \\ &= 1 - B(49) \\ &= 1 - .9729 = .0271\end{aligned}$$

The probability that between 40 and 50, inclusively, prefer the sleek version is found by subtracting from the cumulative probability for 50 the corresponding value for the unwanted levels 39 or below:

$$\begin{aligned}\Pr[40 \leq R \leq 50] &= \Pr[R \leq 50] - \Pr[R \leq 39] \\ &= B(50) - B(39) = .9832 - .4621 = .5211\end{aligned}$$

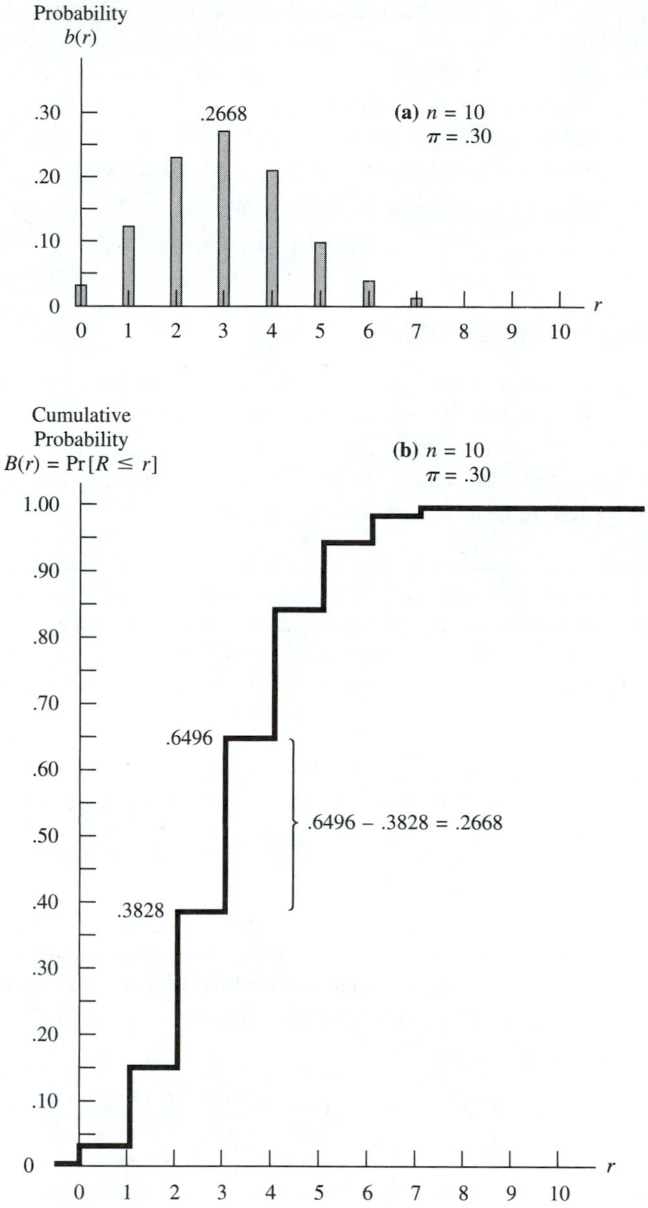

Figure 7–7 *Binomial probability mass function and corresponding probability distribution function.*

To find the probability that exactly 38 respondents favor that design, we take the difference in two successive cumulative probability entries:

$$\begin{aligned}\Pr[R = 38] &= \Pr[R \le 38] - \Pr[R \le 37] \\ &= B(38) - B(37) \\ &= .3822 - .3068 = .0754\end{aligned}$$

At these parameters, tabled values for $B(21)$ or lower are zero to four places, indicating just how rare that few successes would be for the given parameter settings. Likewise, any entry $B(60)$ or higher is a cumulative probability rounding at four places to 1.

Table A stops at $\pi = .50$. Should the trial success probability be greater than .50, then values may be read from the mirror-image distribution with the event reversed. For example, suppose that a lopsided coin having .60 as Pr[head] is tossed $n = 20$ times. What is the probability of getting at least 15 heads?

This may be evaluated by defining a *tail* as a success, so that the random variable is the number of tails and the trial success probability is $\pi = .40$. Thus,

$$\begin{aligned}\text{Pr[at least 15 heads]} &= \text{Pr}\,[R \leq 5 \text{ tails}] \\ &= B(5; 20, .40) = .1256\end{aligned}$$

Binomial Distribution and Sampling

The binomial distribution plays an important role in sampling from qualitative populations, where the number of selected attributes having a particular characteristic is a key random variable. An important engineering application is quality-control evaluations. There the proportion of defectives defines the overall level of quality, and binomial probabilities are used to determine what actions should be taken.

Example:
Air Force Acceptance Sampling

The U.S. Air Force contracts for a wide variety of services, such as vehicle maintenance and mess hall operations. In administering these contracts, inspections are required of units randomly selected from the defined *lot* (statistical population). One food service contract stipulates an *acceptable quality level* (AQL) for equipment cleanliness of 6.5% during meal preparations. The allowable proportion of unclean equipment is then $\pi = .065$.

Based on the above AQL, the Air Force requires that a sample size of $n = 13$ be used and that the lot be accepted or rejected according to the following *decision rule* for the observed number of defectives R (mealtimes during which some dirty equipment is found):*

$$\text{Accept lot if } R \leq 2$$

$$\text{Reject lot if } R > 2$$

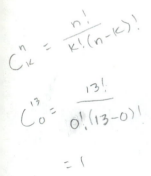

The value 2 is the *acceptance number,* the maximum number of defectives at which accepting the lot is allowed. If we apply the binomial distribution, we may compute several probabilities.** This decision rule was established so that there will be at least a 95% chance of accepting a lot with the stipulated AQL. We have

$$\begin{aligned}&\text{Pr[[Accept} \mid \text{AQL applies } (\pi = .065)] \\ &\quad = C_0^{13}(.065)^0(.935)^{13} + C_1^{13}(.065)^1(.935)^{12} + C_2^{13}(.065)^2(.935)^{11} \\ &\quad = .4174 + .3772 + .1573 = .9519\end{aligned}$$

The decision rule is not perfect, since two types of errors can occur: bad lots might be accepted or good ones rejected. For example, consider a poor-quality meal service where the equipment is actually dirty 20% of the time ($\pi = .20$). There is a substantial probability that poor service will be accepted:

$$\begin{aligned}&\text{Pr[Accept} \mid \text{bad lot } (\pi = .20)] \\ &\quad = C_0^{13}(.20)^0(.80)^{13} + C_1^{13}(.20)^1(.80)^{12} + C_2^{13}(.20)^2(.80)^{11} \\ &\quad = .0550 + .1787 + .2680 = .5017\end{aligned}$$

**Military Standard: Sampling Procedures and Tables for Inspection by Attributes* (MIL-STD–105D), 29 April, 1963.

**The Air Force tables are based on the binomial distribution and are only approximately correct. When sampling is done without replacement from a small population, the hypergeometric distribution described in Section 8–5 should be used in computing probabilities for R.

Also, consider a high-quality meal service where the equipment is dirty only 4% of the time. There is a small probability that the good service will be rejected:

$$\Pr[\text{Reject} \mid \text{high quality } (\pi = .04)] = C_3^{13}(.04)^3(.96)^{10} + \ldots + C_{13}^{13}(.04)^{13}(.96)^0$$
$$= .0135$$

These types of errors are unavoidable in sampling decisions. A good decision rule achieves an acceptable balance between them.

Problems

7–22 A food-packaging apparatus underfills 10% of the containers. Find the probability that for any particular 5 containers the number of underfilled will be
(a) exactly 3 (b) exactly 2 (c) zero (d) at least 1

7–23 Indicate for each of the following situations whether or not the assumptions of a Bernoulli process are met. In all cases, items are produced that are either within or not within tolerance.
(a) To avoid tedium, a lathe operator switches back and forth between "easy" and "hard" types of jobs.
(b) Another machinist works only with a single type of item but gets very sloppy just before breaks, after lunch, and around quitting time.
(c) Each machinist in the plant checks his own work for size. One worker becomes overly careful just after finding himself out of tolerance.
(d) Some equipment has automatic settings that gradually work away from the intended levels.

7–24 Suppose that 20% of the applicants to an engineering school are women. An admissions committee reviews applications in groups of 100. Find the probability that the number of women in the next group is
(a) less than or equal to 15
(b) greater than 25
(c) exactly 27
(d) between 18 and 26, inclusively
(e) less than 10
(f) greater than 40

7–25 Compute the expected number of defectives and the variance when random samples are taken from a production process and the following parameters apply:
(a) $n = 50$ $\pi = .10$
(b) $n = 40$ $\pi = .20$
(c) $n = 20$ $\pi = .50$
(d) $n = 100$ $\pi = .10$

7–26 Construct a table of probabilities for the binomial distribution with $n = 5$ and $\pi = .15$ (rounded to four places). Then compute the cumulative probability for each possible number of successes.

7–27 The proportion π of defective transformer coils wound on automatic spindles is unknown. A quality-control inspector requests tension readjustments whenever more than two coils are found to have broken wires in a sample of $n = 100$. Determine the probability of tension readjustment, assuming
(a) $\pi = .01$ (b) $\pi = .02$ (c) $\pi = .03$ (d) $\pi = .05$ (e) $\pi = .10$

7–28 A student selects his answers on a true/false examination by tossing a coin (so that any particular answer has a .50 probability of being correct). He must answer at least 70% correctly in order to pass. Find his probability of passing when the number of questions is (a) 10, (b) 20, (c) 50, and (d) 100.

7–29 A lopsided coin has a 70% chance of "head." It is tossed 20 times. Determine the following probabilities for the possible results:
(a) at least 10 heads
(b) exactly 12 heads
(c) fewer than 9 heads
(d) at most 13 heads
(e) between 8 and 14 heads

7–30 A quality-control inspector rejects any shipment of printed circuit boards whenever 3 or more defectives are found in a sample of 20 boards tested. Find the (1) expected number defective and (2) the probability of rejecting the shipment when the proportion of defectives in the entire shipment is

(a) $\pi = .01$ (b) $\pi = .05$ (c) $\pi = .10$ (d) $\pi = .20$

7–31 Consider again the coils in Problem 7–27. The quality-control department is also interested in the uniformity of the wire used in the winding. Only a coil having broken wires is unwound to determine whether it was nonuniform. Suppose that 10% of all coils have spindle-caused broken wires and that 20% of all coils have been wound with nonuniform wire. Altogether, 100 coils are selected at random. Determine the probability that the number of coils found to have nonuniform wire is

(a) 0 (b) 1 (c) 2 (d) 3

7–32 An engineer has designed a modified welding robot. The robot will be considered good enough to manufacture if it misses only 1% of its assigned welds. And it will be judged a poor performer if it misses 5% of its welds. (In-between possibilities are not considered.) A test is performed involving 100 welds. The new design will be accepted if the number of missed welds R is 2 or fewer and rejected otherwise.

(a) What is the probability that a good design will be rejected?

(b) What is the probability that a poor design will be accepted?

7–33 A sample of $n = 4$ items is selected with replacement from a lot having $\pi = .20$ defective. Construct a probability tree diagram for this experiment. Then, reading the joint probabilities from your tree, compute the values for $b(r; 4, .20)$ using the basic probability concepts of Chapter 6.

7–34 Consider the Air Force acceptance sampling example again. Compute the probability of (1) accepting a bad lot (with $\pi = \pi_b$) and (2) rejecting a good lot (with $\pi = \pi_g$) for each of the following cases:

	(a)	(b)	(c)	(d)
Sample size	$n = 20$	$n = 50$	$n = 125$	$n = 200$
Acceptable quality level	AQL = 4%	AQL = 6.5%	AQL = .4%	AQL = 1%
Acceptance number	2	7	1	3
π_b	.05	.10	.01	.05
π_g	.02	.05	.001	.005

Source: MIL-STD–105D.

7–4 The Normal Distribution

Most frequently encountered in statistics is the **normal distribution**, often referred to as the *Gaussian distribution.* The normal distribution is used so often partly because many physical measurements, such as human height or diameters of ball bearings, provide frequency distributions that closely approximate a **normal curve**.

But the high incidence of its natural occurrence is not the primary reason why the normal distribution is so important in statistics. As we shall see, it may be used to provide probabilities for levels of the sample mean $\bar{X}$.

The Normal Distribution and the Population Frequency Curve

Envision the measured diameters of a large quantity of precision ball bearings for which the population histogram was constructed and then approximated by a smoothed curve. Figure 7–8 shows the resulting frequency curve. This curve not only describes the underlying population frequency distribution but also takes the shape of a normal curve.

The height of a normal curve above any point x is obtained from the following.

FREQUENCY CURVE AND PROBABILITY DENSITY FUNCTION FOR A NORMAL DISTRIBUTION

$$f(x) = \frac{1}{\sqrt{2\pi}\,\sigma} e^{-\frac{(x-\mu)^2}{2\sigma^2}} \qquad -\infty < x < \infty$$

This expression defines the probability density function for the value X of a random observation from the population.

When plotted, the foregoing function takes the familiar "bell shape." The two parameters μ and σ entirely specify a particular normal curve. The normal curve for diameters of the precision ball bearings in our illustration has parameters $\mu = 1''$ and $\sigma = .001''$, the levels established for the population mean and standard deviation.

Every normal frequency curve is centered on the population mean μ and is symmetrical about this point. Also, this location is both the median and the mode. The tails taper off fairly rapidly for levels of x very far above or below the mean. But it is obvious that the normal curve will never touch the x-axis, since $f(x)$ will be nonzero over the entire real line, from $-\infty$ to $+\infty$.

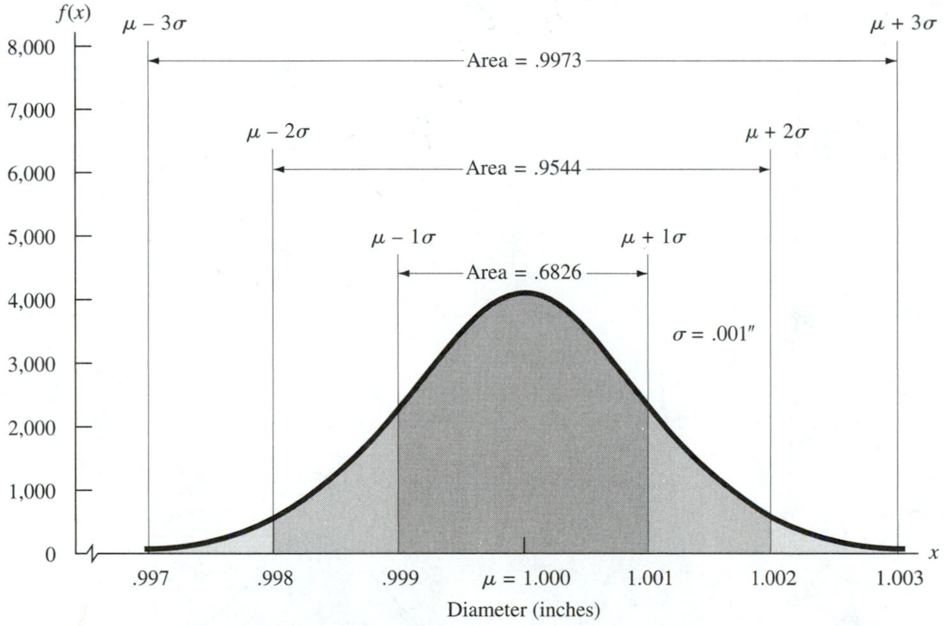

Figure 7–8
Normal curve for diameters of 1″ precision ball bearings.

As we have seen, it is areas under density functions that provide probabilities for continuous random variables. Normal curves have a very convenient property that facilitates finding these: The area under the normal curve between μ and any point depends only on the distance separating that point from the mean, as expressed in units of σ. For instance, the area within the interval $\mu \pm \sigma$ is .6826. This means that about 68% of the precision ball bearing diameters fall within the range $1'' \pm .001''$, or between $.999''$ and $1.001''$. Also, the area spanning $\mu \pm 2\sigma$ is .9544 and that covering $\mu \pm 3\sigma$ is .9973. Over 95% of the precision ball bearings will have diameters of $1'' \pm .002''$, with 5 out of 100 exceeding those limits. Only about 3 out of 1,000 will fall outside of $1'' \pm .003''$.

You may recognize the above from the empirical rule encountered in Chapter 2, which tells us that *approximately* 95% of data values will lie within the range from (the mean -2 standard deviations) and (the mean $+2$ standard deviations). That rule follows from the above property of the normal distribution. Of course, the empirical rule works well only with data whose frequency distribution takes an approximate bell shape and does not provide good limits when a normal distribution is not well approximated.

Here we are dealing with the case of the normal curve being an exact fit. The two parameters completely specify the location and scale of a normal curve. The mean μ locates the center, whereas the standard deviation σ provides the degree of dispersion. Figure 7–9 shows three normal curves, each representing the frequency distribution of a different population. Each normal curve is centered at the respective population mean. A low level for the standard deviation reflects low dispersion in the underlying population values. This is portrayed by a normal curve having a tight, peaked "bell." As σ gets

Figure 7–9
Normal curves for three different populations, each having different mean and standard deviation.

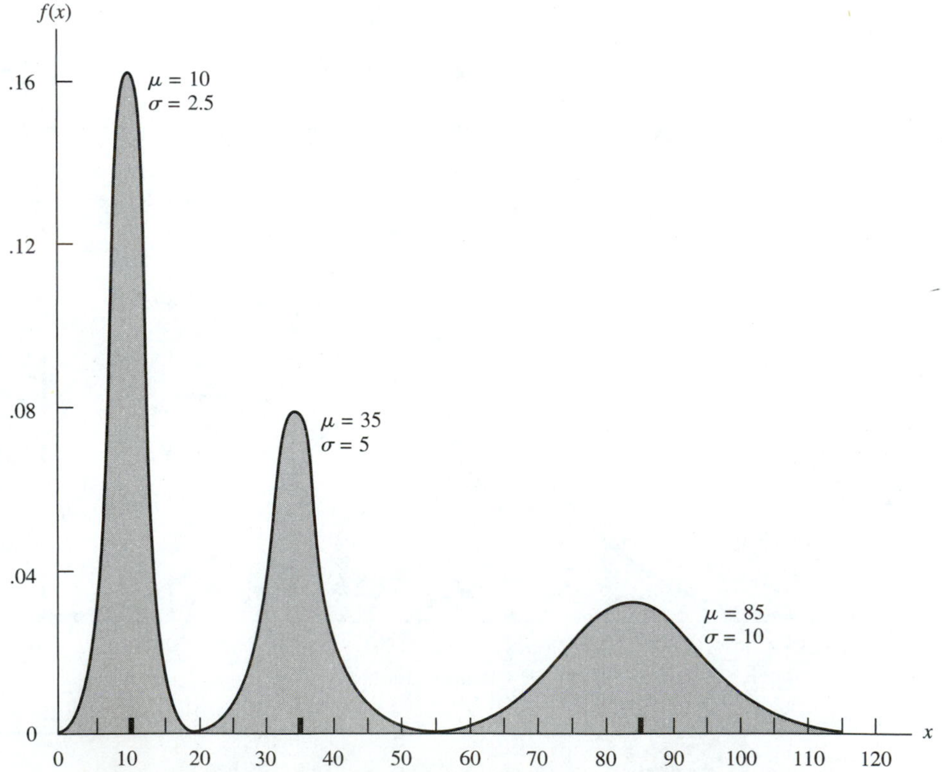

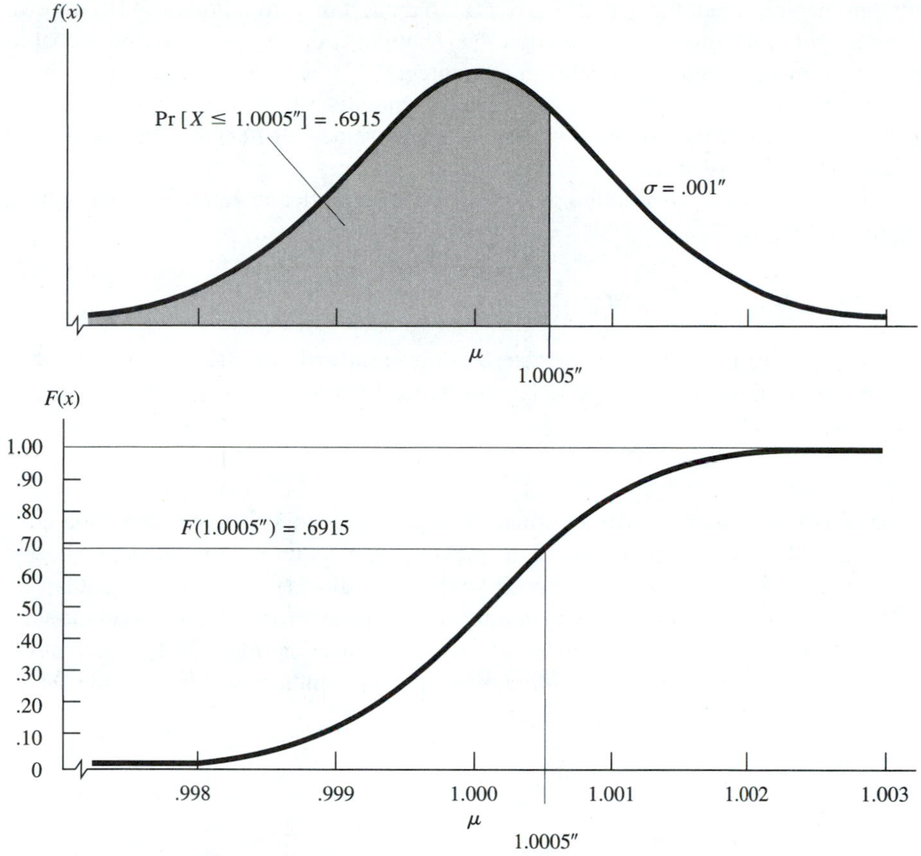

Figure 7–10 *Normal probability density function (top) and normal probability distribution function (bottom) for the diameter of a randomly selected precision ball bearing.*

larger, reflecting a population with greater dispersion, the corresponding normal curve becomes flatter and more spread out.

The Normally Distributed Random Variable

Consider a sample observation from a population whose frequency distribution may be represented by a normal curve. The value X obtained is referred to as a **normally distributed random variable**. Probabilities may be found for X by determining the respective areas under the normal curve. Since a normally distributed random variable is continuous, nonzero probabilities for X can be found only for possible levels falling within intervals (rather than at single points). It is therefore convenient to work with the following expression.

NORMAL PROBABILITY DISTRIBUTION FUNCTION

$$F(x) = \Pr[X \le x] = \int_{-\infty}^{x} \frac{1}{\sqrt{2\pi}\,\sigma} e^{-\frac{(y-\mu)^2}{2\sigma^2}}\,dy$$

Figure 7–10 shows the normal probability density and distribution functions for the diameter X of a randomly selected precision ball bearing. The figure shows that

$$F(1.0005'') = \Pr[X \le 1.0005''] = .6915$$

and that the area under the normal curve $f(x)$ to the left of x (top, Figure 7–10) is equal to the height of the distribution function $F(x)$ (bottom). As we have seen, the probability value may be obtained by evaluating the integral of $f(x)$ over the appropriate limits. But the normal density function has no closed-form integral, so $f(x)$ must be integrated numerically. That work has already been done, and values for $F(x)$ may be read from a table of normal curve areas.

Before describing that table, we will consider a particular random variable from the normal family.

The Standard Normal Distribution

A very important normal curve characterizes the **standard normal random variable**. Denoted by Z, this random variable is represented by a normal curve with parameters $\mu = 0$ and $\sigma = 1$.

Illustration: *Satellite Ranging Errors*

As an illustration, suppose that a normal curve applies for the ranging prediction error experienced by a ground station when its tracking radar acquires a satellite target during a particular orbit. The error could be positive (if the predicted range were higher than actual) or negative (prediction lower than actual). The mean error could be 0, and the standard deviation in the population of errors might be 1 nautical mile (NM). The ranging prediction error would then have a *standard* normal distribution. Appendix Table D provides values for this.

STANDARD NORMAL PROBABILITY DISTRIBUTION FUNCTION

$$\Phi(z) = \Pr[Z \le z] = \int_{-\infty}^{z} \frac{1}{\sqrt{2\pi}} e^{-\frac{y^2}{2}} dy$$

Figure 7–11 shows the areas under the standard normal curve corresponding to several possible cases with the satellite ranging error. The values were obtained from Table D by reading the respective cumulative probability at z, which is also the area to the left of z.

For instance, the probability that the ranging error is less than or equal to 1.50 NM is thus

$$\Pr[Z \le 1.50] = \Phi(1.50) = .9332$$

The same probability applies to Pr[Z < 1.50], because Z is continuous and a single point has no probability, so that Pr[Z = 1.50] = 0. (With continuous random variables, no distinctions are made in evaluating outcomes having "<" instead of "≤" or ">" instead of "≥". The probability that a random variable exceeds a particular amount, such as 2.00 NM, may be found by reading the cumulative probability for the complementary event from Table D and subtracting from 1:

$$\Pr[Z > 2.00] = 1 - \Phi(2.00) = 1 - .9772 = .0228$$

And to calculate the probability that the standard normal random variable falls within a range—say, between –.10 NM and .65 NM—the cumulative probability for the smaller z is subtracted from that for the larger z:

$$\begin{aligned}\Pr[-.10 \le Z \le .65] &= \Phi(.65) - \Phi(-.10)\\ &= .7422 - .4602 = .2820\end{aligned}$$

Figure 7–11 *Areas under the normal curve that represent probabilities for the standard normal random variable.*

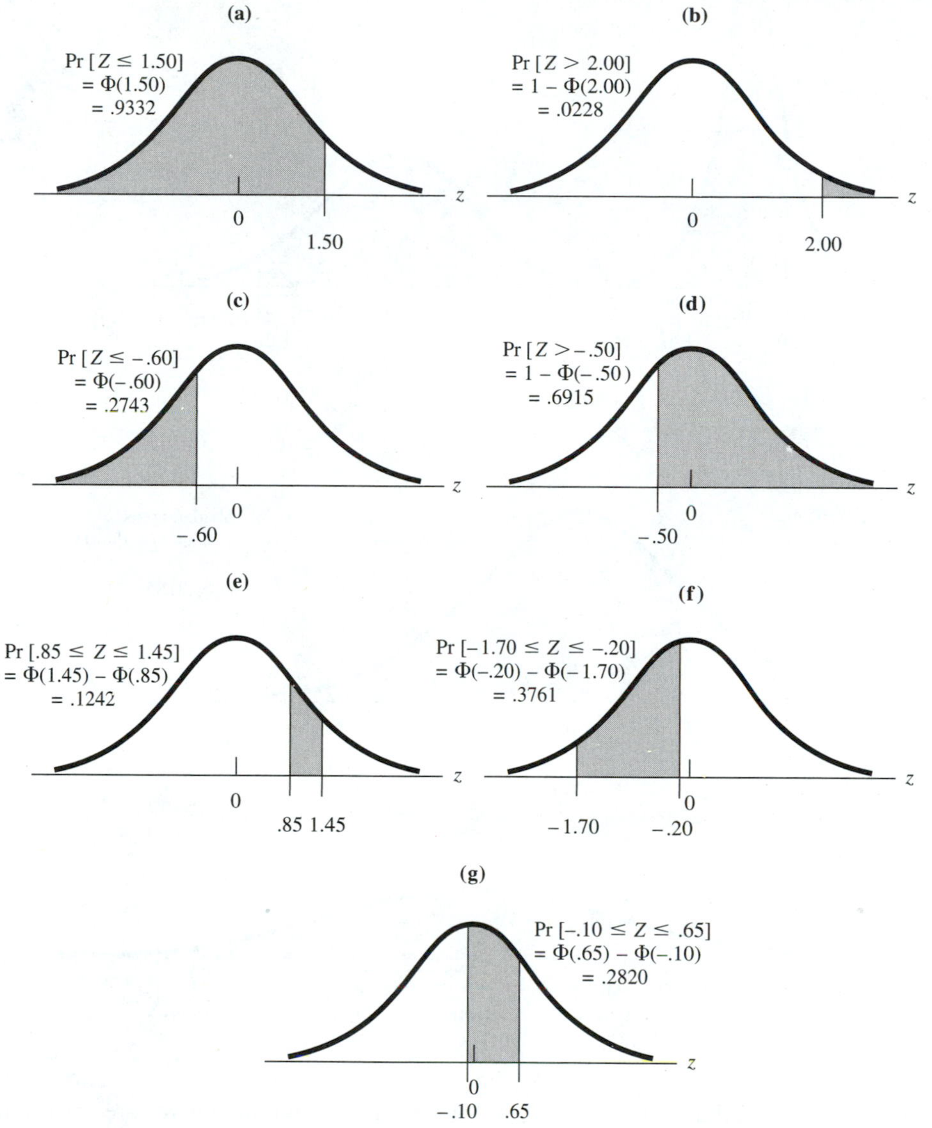

Probabilities for Any Normal Random Variable

The standard normal distribution may be used to obtain probabilities applicable to any normal distribution. This is because the area under any normal curve between two points depends only on how far those points are away from the mean, as measured in standard deviation units. And because the standard normal distribution has a unit parameter $\sigma = 1$, a point's standard deviation distance above or below the center of its normal curve may be directly related to a level of z.

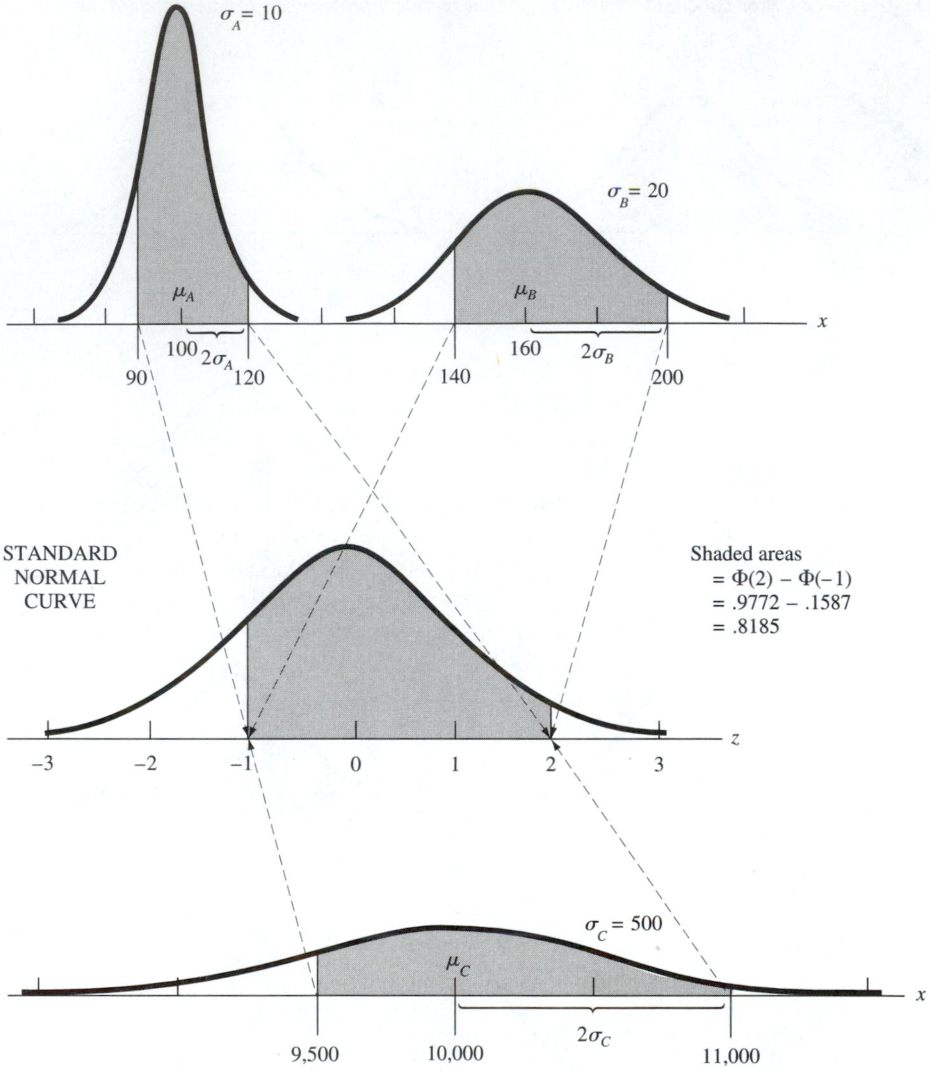

Figure 7–12
Illustration of how areas for various normal curves match areas under the standard normal curve. In all cases, the shaded areas span a range from 1 standard deviation below the mean to 2 standard deviations above the mean.

Figure 7–12 illustrates this concept. Three differing normal curves are portrayed there, with the shaded areas of each covering levels of x from a point 1 standard deviation below the mean to 2 standard deviations above it. All these areas are identical, and they match the area under the standard normal curve between $z = -1$ and $z = 2$.

Thus, to find probabilities, we need only identify the matching areas under the standard normal curve. This involves finding one or more levels for z, referred to as **normal deviates**.

The following transformation is used to identify each.

NORMAL DEVIATE

$$z = \frac{x - \mu}{\sigma}$$

In effect, the probability distribution function for any normal random variable X may be obtained by making the above transformation into the standard normal random variable:

$$F(x) = \Pr[X \leq x] = \Pr\left[\frac{X-\mu}{\sigma} \leq \frac{x-\mu}{\sigma}\right]$$
$$= \Pr\left[Z \leq \frac{x-\mu}{\sigma}\right] = \Phi\left(\frac{x-\mu}{\sigma}\right)$$

Continuing with the ball bearing illustration, we may find the probability that a random diameter X falls at or below 1.0015″:

$$\Pr[X \leq 1.0015''] = \Phi\left(\frac{1.0015'' - 1.000''}{.001''}\right)$$
$$= \Phi(1.5) = .9332$$

that it exceeds .9995″:

$$\Pr[X > .9995''] = 1 - \Phi\left(\frac{.9995'' - 1.000''}{.001''}\right)$$
$$= 1 - \Phi(-.50)$$
$$= 1 - .3085 = .6915$$

and that it lies between .9998″ and 1.0004″:

$$\Pr[.9998'' \leq X \leq 1.0004''] = \Phi\left(\frac{1.0004'' - 1.000''}{.001''}\right) - \Phi\left(\frac{.9998'' - 1.000''}{.001''}\right)$$
$$= \Phi(.40) - \Phi(-.20)$$
$$= .6554 - .4207 = .2347$$

Example: *Human Engineering and Nuclear Reactor Controls*

The Three-Mile Island nuclear reactor accident highlighted major deficiencies in the state of the art for configuring control panels. Accident investigators reported that operators had far too many dials, lights, and meters for a human to monitor adequately. Instrumentation displays are much more advanced in aircraft controls, in which for many years specialists called *human engineers* have segregated the mundane from the critical in designing cockpit layouts. It was recommended that future reactor control rooms be human-engineered.

One consulting engineer proposed setting up a control room simulator where various locations for key warning signals could be evaluated in detail. For instance, "hot lights" for reactor cooling-system components should be located where they can be easily seen and quickly reacted to. He devised a plan for simulating various hot-light locations by generating hypothetical danger events and timing various operator reactions. Locations could be adjusted until some optimal balance was achieved.

For each panel configuration, the simulation plan called for monitoring the average reaction time between activation of a warning light and the operator's sounding a buzzer alarm. Suppose this random variable X is normally distributed with a mean of $\mu = 45$ seconds and a standard deviation of $\sigma = 8$ seconds. The probability that the alarm is sounded within an average of 1 minute (60 seconds) after a hazardous event is

$$\Pr[X \leq 60] = \Phi\left(\frac{60-45}{8}\right) = \Phi(1.88) = .9699$$

whereas the probability that more than 2 minutes (120 seconds) are required is negligible and corresponds to a normal deviate off the table:

$$\Pr[X > 120] = 1 - \Phi\left(\frac{120 - 45}{8}\right) = 1 - \Phi(9.4) \approx 0$$

Expected Value, Variance, and Percentiles

The following facts may be established for a normally distributed random variable:

$$E(X) = \int_{-\infty}^{\infty} x f(x) dx = \mu$$

$$\text{Var}(X) = \int_{-\infty}^{\infty} x^2 f(x) dx - \mu^2 = \sigma^2$$

where $f(x)$ is the normal probability density function defined earlier. The mean level $E(X)$ equals the center μ of the normal curve, and Var(X) matches the square of the dispersion parameter, σ^2. In effect, specification of the parameters μ and σ completely specifies the particular normal distribution that applies.

The normal curve provides a close fit to the frequency curves for measurements of many dimensions and human characteristics. It may therefore answer important questions pertaining to **percentiles**. A population percentile is *the value at or below which the stated percentage of units lie.* When the population has a normal frequency curve, we may use the following expression.

PERCENTILE VALUE FOR A NORMAL DISTRIBUTION

$$x = \mu + z\sigma$$

Here the normal deviate z corresponds to the stated percentile.

For example, consider one final time the ball bearings, where $\mu = 1.000''$ and $\sigma = .001''$. The 95th percentile is the dimension at or below which 95% of the ball bearing diameters will lie. This is found by reading from Table D the z that corresponds to a cumulative probability of .95. We read the table in reverse, getting*

$$z = 1.64$$

The 95th percentile is thus

$$x = 1.000'' + 1.64(.001'') = 1.00164''$$

This value tells us that 95% of all ball bearings from the underlying population will have diameters $\leq 1.00164''$.

*The cumulative probability entry for the above is .9495. The tabled value for $z = 1.65$ is just as close, but in the case of a tie we will pick the lower z. Although an interpolation procedure will provide more accuracy, to ease our discussions in this book we will just find the closest tabled value.

Example: *Anthropometric Data and Equipment Design*

The man–machine interface is an important engineering consideration. Controls should be designed so that it doesn't take an "eight-armed superman" to operate them. (And today, most *women* should be able to operate equipment, too.) Anthropometric data are often used in establishing distances between controls where space is critical, as in aircraft cockpits.

One of the most important variables is *maximum reach.* Assuming that this quantity for adult males is normally distributed, with a mean $\mu = 32.33''$ and a standard deviation of $\sigma = 1.63''$, the following percentiles apply:+

	1st	**5th**	**50th**	**95th**	**99th**
z	−2.33	−1.64	0	1.64	2.33
$x = \mu + z\sigma$	28.5″	29.7″	32.3″	35.0″	36.1″

+*Source:* H. T. E. Hertzberg, G. S. Daniels, and E. Churchill, *Anthropometry of Flying Personnel,* WADC Technical Report 52–321, USAF, September 1954.

Thus, if design specifications require that a particular control be placed so that 95% of all seated male pilots can reach it without bending, the control must be located no farther than 29.71″ from the seat backrest.

Practical Limitations of the Normal Distribution

The normal distribution has wide application and is often used to represent physical measurements. As a practical matter, it is nearly always an *approximation* to reality rather than a totally realistic representation. For instance, the normal distribution is used to characterize human height data. The height of adult American males might be represented by a normal frequency curve with mean $\mu = 69''$ and standard deviation $\sigma = 2.5''$. Remember that the tails of the normal curve extend indefinitely, so that the height X of a randomly selected male has a finite probability under the normal distribution of exceeding any level—even 10 feet or one mile! Worse yet, there is a probability that X can be negative.

These absurd results apply to levels of the random variable beyond the region of reasonable approximation, ordinarily taken to be $\mu \pm 5\sigma$. Appendix Table D does not provide probabilities for $|z| < 5$ (A mile-tall man would correspond to a normal deviate $z > 25{,}000$.)

Problems

7-35 A satellite range prediction error has the standard normal distribution with mean 0 NM and standard deviation 1 NM. Find the following probabilities for the prediction error:

(a) less than 1.25 NM
(b) greater than −.75 NM
(c) less than or equal to −2.50 NM
(d) greater than or equal to 3.5 NM
(e) between 1.5 NM and 1.75 NM
(f) between −.50 NM and −.25 NM
(g) less than −2.80 NM or greater than .65 NM
(h) less than .65 NM or greater than .75 NM

$\mu = 0$, $\sigma = 1$, $z = \frac{(x-\mu)}{\sigma} = \frac{1.25-0}{1} = 1.25$ (for a)

Look up z on pg 536-537

7–36 Find the following percentiles for the satellite range prediction error described in Problem 7–35:
(a) 10th (b) 25th (c) 85th (d) 95th (e) 99th

7–37 The sitting height of adult males is normally distributed, with mean $\mu = 35.94''$ and standard deviation $\sigma = 1.29''$.
(a) Find the probability that a randomly chosen man's sitting height is (1) less than 33″, (2) between 34″ and 35.5″, (3) between 33″ and 37″, and (4) greater than 40″.
(b) Find the following percentiles: (1) 5th, (2) 10th, (3) 75th, and (4) 95th.

7–38 The average active-ingredient yield per liter of raw material for samples of vials may be approximated by a normal distribution with mean $\mu = 30$ grams and standards deviation $\sigma = .2$ gram.
(a) Find the probability that the average yield of a sample is (1) less than 29.55 grams, (2) between 29.5 and 30.25 grams, (3) greater than 30.45 grams, and (4) between 30.15 and 30.35 grams.
(b) Find the following percentiles: (1) 10th, (2) 20th, (3) 90th, and (4) 99th.

7–39 The forearm–hand length of adult males is normally distributed, with mean $\mu = 18.86''$ and standard deviation $\sigma = .81''$.
(a) Find the probability that a randomly chosen man's forearm–hand length is (1) less than 19″, (2) between 16.5″ and 17″, (3) between 18″ and 20″, and (4) greater than 21″.
(b) Find the following percentiles: (1) 1st, (2) 5th, (3) 50th, and (4) 60th.

7–40 The mean elongation of a steel bar under a particular tensile load has been established to be normally distributed, with parameters $\mu = .06''$ and $\sigma = .008''$. Assuming the same distribution applies to a new bar, find the probability that the mean elongation falls
(a) above .08″ (c) somewhere between .05″ and .07″
(b) below .055″ (d) either below .045″ or above .065″

7–41 A deflection test is performed on several cantilever beams fabricated from a new steel alloy.
(a) Assuming that the mean angle of rotation is normally distributed with parameters $\mu = .035$ and $\sigma = .007$ (both in radians), find the probability that this quantity falls
(1) below .050 (3) between .03 and .05
(2) above .045 (4) between .04 and .055
(b) The mean angle of deflection is also normally distributed with parameters $\mu = .08$ and $\sigma = .02$. Find the probability that this quantity falls
(1) above .11 (3) between .06 and .075
(2) below .12 (4) between .085 and .095

7–42 A ceramics engineer plans to conduct a test on sample scraps of two epoxy substances to determine the average setting time until bonding. Neither the mean nor the standard deviation for this random variable is known although a normal distribution may be assumed to apply. Find the probability that the average bonding time deviates from the unknown μ by no more than 5 hours, assuming that the standard deviation has value
(a) $\sigma = 2$ (b) $\sigma = 4$ (c) $\sigma = 8$ (d) $\sigma = 16$

7–43 A chemical plant superintendent orders a process shutdown and setting readjustment whenever the pH of the final product falls below 6.90 or above 7.10. The sample pH is normally distributed with unknown μ and standard deviation $\sigma = .05$. Determine the probability
(a) of readjusting when the process is operating as intended and $\mu = 7.0$
(b) of readjusting when the process is slightly off target and the mean pH is $\mu = 7.02$
(c) of failing to readjust when the process is too alkaline and the mean pH is $\mu = 7.15$
(d) of failing to readjust when the process is too acidic and the mean pH is $\mu = 6.80$

Review Problems

7–44 A new amplifier design is being tested for a radio telescope. Suppose that under a variety of test-signal patterns the average frequency until distortion is normally distributed with mean 35 kilohertz and a standard deviation of 2.5 kilohertz.

(a) Find the 95th percentile for the average distortion frequency.

(b) Determine the probability that the average distortion frequency level falls above that of the present amplifier, in which distortion takes place at an average frequency of 32 kilohertz.

7–45 Several sample observations will be obtained of paint-drying times. The average drying time may be assumed to be normally distributed with a mean of 2 hours and an unknown standard deviation.

(a) Assuming the standard deviation for average drying time is 15 minutes, determine the following percentiles: (1) 25th, (2) 90th, (3) 95th, and (4) 99th.

(b) Find the probability that the average drying time will lie between 1.85 and 2.25 hours assuming the standard deviation for that variable is (1) 15 minutes, (2) 10 minutes, and (3) 20 minutes.

7–46 The functional reach of adult men is a normally distributed random variable with mean 32.33″ and standard deviation 1.63″. Consider a randomly chosen man.

(a) Find the probability that his functional reach exceeds 35″.

(b) A second man is chosen. What is the probability that both have reach exceeding 35″?

7–47 Repeat Problem 7–46, substituting the event "shorter than 30 inches."

7–48 The *measurement signal-to-noise ratio* of the random variable X is defined by

$$\frac{|E(X)|}{SD(X)}$$

Determine this quantity for each of the following:

(a) a normally distributed random variable with parameters $\mu = 10$ and $\sigma = 5$

(b) a binomially distributed random variable with parameters $\pi = .5$ and $n = 5$

(c) a binomially distributed random variable with parameters $\pi = .5$ and $n = 100$

7-36 → 10th percentile means the point or value in a group of observed data where 10% of the data will have less than or equal to that point or value, or, in other words, the probability that all the data are less tan this is 10%

Look in back of book (pg 536) to find a) 0.10, b) 0.25, c) 0.85 } Find corresponding z value
a) −1.28
b) −0.675.

7-37 → a) $z = \frac{(33-35.94)}{1.29}$ ($\frac{x-\mu}{\sigma}$) $= -2.279$

$Pr[x \leq 33] = \phi(-2.279)$
$= 0.0113$

same as 7-35

7-37b) find z value like in 7-36. but then change it to a random var, "x"

$X = \mu + z\sigma$

CHAPTER EIGHT

Important Probability Distributions in Engineering

THE BINOMIAL AND normal distributions described in Chapter 7 are fundamental to all statistical applications. This chapter introduces five further probability distributions that have proven to be especially important in engineering applications.

Paramount in importance are the **Poisson distribution** and the **exponential distribution**. These provide alternative probabilistic interpretations of an underlying Poisson process, commonly associated with queuing and reliability applications. Those situations are sometimes evaluated with a third distribution, the **gamma distribution**.

The exponential and gamma distributions are helpful in reliability analysis, for which they may serve as **failure-time distributions**. Another common failure-time distribution is the **Weibull distribution**.

The chapter concludes with an introduction to the **hypergeometric distribution**, which serves as a more realistic alternative to the binomial because the latter is only approximately correct for many sampling situations.

8–1 The Poisson Distribution

The Poisson distribution family is common in waiting-line evaluations, in which it gives probabilities for the number of arrivals at a service facility. It is also frequently used in reliability analysis, in which it can give probabilities for the number of failures.

It will be useful to relate the Poisson distribution to the binomial, which is concerned with the number of successes in a fixed series of trials, each of which can culminate in one of the complementary outcomes. The Poisson distribution concerns only one event (instead of two), such as an arrival at a toll booth or a failure of a satellite power cell. It provides probabilities for *how many events* will occur. The number of Poisson events is uncertain rather than fixed, as with the binomial distribution, where each trial (specified to be n in number) results in one event (a success or a failure). The Poisson events will occur at some average rate over a set time span or within some prescribed space. Thus, the Poisson distribution might provide the probability that 3 accidents will occur during one hour along a stretch of highway, that a data-processing clerk will commit 3 errors while making 1,000 teleprocessing entries, or that a ship will encounter 3 icebergs while traversing the North Atlantic in July.

Like the binomial distribution, the Poisson distribution applies to an underlying stochastic process. Named for the eighteenth-century physicist and mathematician, Simeón Poisson, a **Poisson process** characterizes a random experiment having the special properties described next.

The Poisson Process

A Poisson process gives rise to a series of events occurring at a *mean rate* over time or space. These occur at random, so that no pattern is discernible. Figure 8–1 shows a time record of street lamp burnouts in a city, as recorded by special telemetry equipment. Each lamp failure is an event in an underlying Poisson process wherein lamps have burned out at a historical mean rate of $\lambda = 1$ per day. These occurrences are random over time and without pattern.

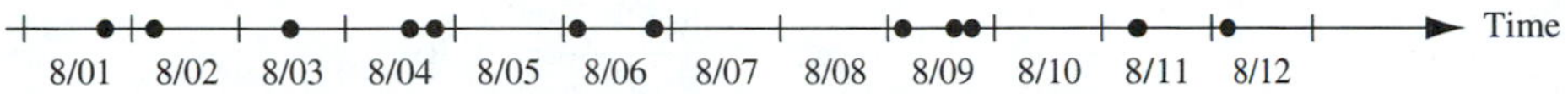

Figure 8–1 *Time record of street lamp burnouts.*

All Poisson processes share the following properties and differ from each other mathematically only by the level for λ.

1. The number of events occurring in one segment of time or space is independent of the number of events in any nonoverlapping segment. A Poisson process has *no memory*.
2. The mean process rate λ *must remain constant* for the entire time span or space considered.
3. The smaller the segment of time or space, the less likely it is for more than one event to occur in that segment. As the segment size tends to 0, the probability of 2 or more occurrences approaches 0.

The Poisson process gives rise to the Poisson distribution discussed here, wherein the concern is how many events will occur in a fixed period of time. It also generates the *continuous exponential distribution* (described in the next section), which has as its random variable the amount of time (or space) between events.

Poisson Probabilities

Poisson probabilities may be readily computed for any possible number of events X. These will depend on the specification of the mean process rate λ and the time span t (which may also represent the space expanse). We use the following expression.

PROBABILITY MASS FUNCTION FOR POISSON DISTRIBUTION

$$p(x; \lambda, t) = \Pr[X = x] = \frac{(\lambda t)^x e^{-\lambda t}}{x!} \qquad x = 0, 1, 2, \ldots$$

In keeping with our earlier notation, we use p to represent the probability mass function and list the parameters λ and t after the dummy variable x. The Poisson distribution places no upper limit on the number of events possible, so X may turn out to be any nonnegative integer value.

Illustration: *Arrivals at a Toll Booth*

As an illustration, consider the peak morning rush hour at the toll plaza for one of the bridges over San Francisco Bay. Suppose during this period cars arrive at a mean rate of $\lambda = 600$ per hour. We are interested in the number of arrivals X during a 12-second time span ($\frac{1}{5}$ minute), so that $t = (\frac{1}{5})(\frac{1}{60}) = \frac{1}{300}$ hour. (Parameters t and λ must be in compatible units.) We have $\lambda t = 600(\frac{1}{300}) = 2$. Values of $e^{-\lambda t}$ may be read from Appendix Table B at the back of the book. The following probability list is computed:

$$\Pr[X = 0] = p(0; 600, \tfrac{1}{300}) = \frac{(2)^0 e^{-2}}{0!} = .1353$$

$$p(1) = \frac{(2)^1 e^{-2}}{1!} = .2707$$

$$p(2) = \frac{(2)^2 e^{-2}}{2!} = .2707$$

$$\vdots$$

$$p(5) = \frac{(2)^5 e^{-2}}{5!} = .0361$$

$$\vdots$$

$$p(10) = \frac{(2)^{10} e^{-2}}{10!} = .0000382$$

$$\vdots$$

This list can never be completed, since X has no upper limit. But Poisson probabilities rapidly converge to 0 as X becomes large.

Parameter Levels and Poisson Probabilities

The probability mass function for the car arrivals is plotted in Figure 8–2. Notice the skew in the pattern of spikes, with probabilities gradually tapering off for larger X's. The degree of skew is most pronounced for the low levels of the parameter product λt. For very large levels of λt (such as 10) the Poisson distribution is nearly symmetrical.

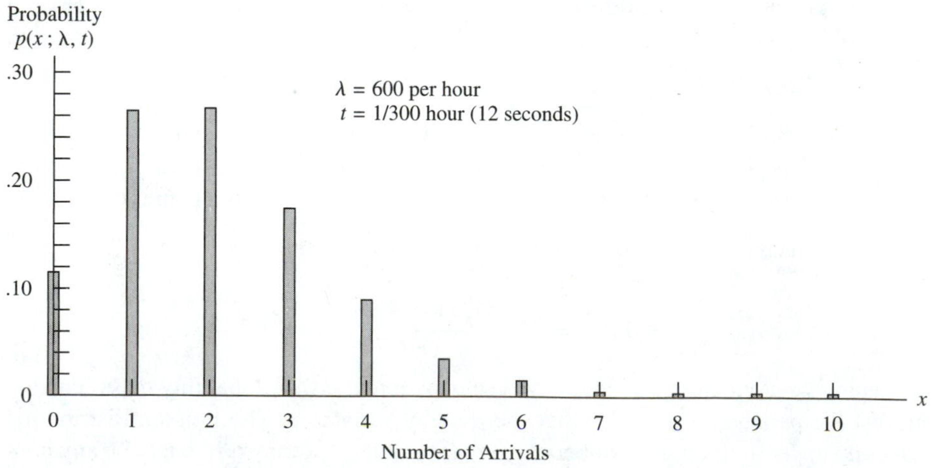

Figure 8–2
Poisson distribution for the number of cars arriving at a toll plaza.

Should the time span be increased, a different parameter applies for t and a new Poisson distribution will apply. For example, raising the time span to 30 seconds, with $t = (\frac{1}{2})(\frac{1}{60}) = \frac{1}{120}$ hour, we have $\lambda t = 600(\frac{1}{120}) = 5$. Thus,

$$\Pr[X = 2] = p(2; 600, \tfrac{1}{120}) = \frac{(5)^2 e^{-5}}{2!} = .0843$$

And reducing the time span to 6 seconds makes $t = (\frac{1}{10})(\frac{1}{60}) = \frac{1}{600}$ hour, so that $\lambda t = 600(\frac{1}{600}) = 1$. We have another new probability mass function, and

$$\Pr[X = 2] = p(2; 600, \tfrac{1}{600}) = \frac{(1)^2 e^{-1}}{2!} = .1839$$

In these three cases λ remained the same as we changed t, so that the underlying Poisson process was unchanged. A new setting for λ will affect both the probability mass function and the process itself. For instance, the intensity of bridge traffic will fluctuate throughout the day. Thus, some other mean arrival rate will apply during nonrush periods, such as 10 A.M. to 11 A.M. on a weekday. Suppose during this hour the rate is only $\lambda = 150$ cars per hour. Keeping t at $\frac{1}{300}$ hour, $\lambda t = 150(\frac{1}{300}) = .5$ and a completely different probability distribution applies. We have

$$\Pr[X = 2] = p(2; 150, \tfrac{1}{300}) = \frac{(.5)^2 e^{-.5}}{2!} = .0758$$

Importance of Poisson Assumptions

Care must be taken that λ not fluctuate over the time span considered, since that would violate one of the Poisson process assumptions. (We could not use a single Poisson distribution to find the probability that in a $t = 8$-hour time span there would be $X = 1{,}000$ arrivals based on an average rate of $\lambda = 200$ cars per hour.) The assumptions of the Poisson process also make the X level in one time period independent of a preceding X. Thus, given an abnormal "flood" of cars, the overall arrival pattern thereafter will be the same as if that event hadn't occurred (or if there had instead been a temporary lull in traffic). There is no "catch-up" effect. Also, although 2 or more cars may arrive close together, the probability that they will arrive at exactly the same time is assumed to be 0. (Several milliseconds will separate the arrivals.) Examples exist wherein these assumptions don't strictly apply—but keep in mind that the Poisson process is a theoretical model. Its practical value lies in how closely the theory approximates the reality.

Example: *Probabilities of Tornado Destruction*

The occurrences of tornadoes in the midwestern United States closely fit a Poisson process. In one 5-million-acre region a meteorologist has found that annual swaths traced by tornadoes have historically encompassed an area totaling 500 acres. The annual mean tornado intensity for the region is thus $\lambda = .0001$ tornado per acre.

Consider a city located in this tornado belt having an area of 8,000 acres. What is the probability that it will be hit exactly once during the year?

Setting $t = 8{,}000$ acres, we wish to find the probability that exactly 1 tornado will occur. We have $\lambda t = .0001(8{,}000) = .8$, and the answer is

$$\text{Pr[damage from exactly 1 tornado]} = p(1; .0001, 8{,}000)$$
$$= \frac{(.8)^1 e^{-.8}}{1!} = .3595$$

so there is almost a 36% chance that the city will suffer some damage from one tornado.

But the city could be hit by more than one tornado. Being hit at least once is complementary to no hits, so

$$\text{Pr [no tornado damage]} = \Pr[X = 0] = \frac{(.8)^0 e^{-.8}}{0!} = .4493$$

and

$$\text{Pr [some tornado damage]} = 1 - .4493 = .5507$$

The probability that no tornado damage occurs within 10 years is (by the multiplication law for independent events)

$$\text{Pr [no tornado damage for 10 years]} = (.4493)^{10} = .000335$$

so the probability of being hit by at least one tornado sometime within the next 10 years is $1 - .000335 = .999665$, an almost certainty.

Poisson Distribution Function and Probability Table

The following expression gives the Poisson probability distribution function.

POISSON PROBABILITY DISTRIBUTION FUNCTION

$$P(x; \lambda, t) = \Pr[X \le x] = \sum_{k=0}^{x} p(k; \lambda, t)$$

In keeping with our notation, we use an uppercase P to represent cumulative Poisson probabilities and a lowercase p for the individual terms obtained from the probability mass function.

Like the binomial distribution, Poisson probabilities can be a chore to compute by hand. Appendix Table C provides cumulative Poisson probability values for common levels of the product λt.

As an illustration, suppose that significant noisy spots occur on videotape at a mean rate of $\lambda = .1$ per foot. Consider a segment of tape $t = 200$ feet long. We have $\lambda t = .1(200) = 20$. Using Table C we find various probabilities for the number of noisy spots encountered.

The probability that tape noise is encountered less than 15 times is

$$\Pr[X < 15] = P(14; .1, 200) = .1049$$

Whereas the probability that more than 20 of these blemishes will be found is

$$\Pr[X > 20] = 1 - P(20) = 1 - .5591 = .4409$$

and the probability that tape noise will be encountered exactly 20 times is

$$\Pr[X = 20] = P(20) - P(19)$$
$$= .5591 - .4703 = .0888$$

Expected Value and Variance

As with the binomial distribution, the expected value and variance for the number of events may be computed by direct manipulation of the Poisson distribution parameters. We use the following expressions.

EXPECTED VALUE AND VARIANCE FOR A POISSON DISTRIBUTION

$$E(X) = \lambda t$$

$$\text{Var}(X) = \lambda t$$

The expected number of events is found by multiplying the specified span t times the mean rate λ at which these events occur.

The expected value identity is easily derived. We have

$$E(X) = \sum_{x=0}^{\infty} x p(x; \lambda, t) = \sum_{x=0}^{\infty} x \frac{(\lambda t)^x e^{-\lambda t}}{x!}$$

The first term in this summation is 0 when $x = 0$, so that dropping that term, we have

$$E(X) = \sum_{x=1}^{\infty} x \frac{(\lambda t)^x e^{-\lambda t}}{x!}$$

Factoring λt and canceling terms, the above is equal to

$$\lambda t \sum_{x=1}^{\infty} \frac{(\lambda t)^{x-1} e^{-\lambda t}}{(x-1)!}$$

Letting $y = x - 1$, the above equals

$$\lambda t \sum_{y=0}^{\infty} \frac{(\lambda t)^y e^{-\lambda t}}{y!}$$

The summation includes all Poisson probabilities for λ and t and is therefore equal to 1. This establishes that $E(X) = \lambda t$.

It will be left as an exercise for the reader to derive the identity for Var(X).

Considering again the cars arriving at a toll booth within a specific hour, so that $t = 1$, we see that this is plausible. Twice as many cars are expected during twice that duration, when $t = 2$ hours, and half as many will be expected when the duration is cut by half to $t = .5$ hour. And during commuting times when λ may be triple the off-peak rate, we would expect triple the number of arrivals within equal time spans.

That the variance also equals λt is less obvious, although it is easy to see that long t's will tend to result in greater variation in X than short ones. Likewise, X should vary more with large λs than with smaller ones.

Problems

8–1 Transmission line interruptions in a telecommunications network occur at an average rate of 1 per day. Find the probability that the line experiences

(a) no interruptions in 5 days
(b) exactly 2 interruptions in 3 days
(c) at least 1 interruption in 4 days
(d) at least 2 interruptions in 5 days

8–2 Cars arrive at a toll plaza according to a Poisson process.

(a) Find the probability of exactly 5 arrivals in 1 minute when the mean rate of arrivals per minute is

(1) $\lambda = 1$ (2) $\lambda = 2$ (3) $\lambda = 5$ (4) $\lambda = 10$

(b) During the rush period the mean rate is $\lambda = 10$ cars per minute. Find the probability of exactly 5 arrivals in an interval of

(1) $t = .3$ min (2) $t = .5$ min (3) $t = 1$ min (4) $t = 2$ min

8–3 Service calls received by a photocopier maintenance center constitute a Poisson process with a mean rate of 2.5 per hour.

(a) Construct a complete probability distribution for the number X of calls received during 10 to 11 A.M. on a particular Wednesday. Give answers accurate to four places, stopping at the first x whose probability rounds to zero.

(b) Construct the cumulative probability distribution function, using your results from (a).

(c) Compute (1) the expected number of calls and (2) the variance.

(d) Plot on graphs the distributions found in (a) and (b).

8–4 A computer time-sharing system receives teleport inquiries at an average rate of .1 per millisecond. Find the probabilities that the number of inquiries in a particular 50-millisecond stretch will be

(a) less than or equal to 8

(b) greater than 6

(c) between 5 and 12, inclusively

(d) equal to 7

(e) equal to 10

8–5 Use a Poisson distribution to approximate the following binomial probabilities:

(a) $b(5; 100, .05)$

(b) $b(2; 1{,}000, .01)$

(c) $b(9; 500, .04)$

(d) $b(6; 100, .01)$

8–6 An agronomist has found an average of 2 shrubs per acre of jojoba, which serves as the source for transmission oil. Find the following probabilities:

(a) that 5 plants will be found when the number of acres searched is (1) one; (2) two; or (3) three

(b) that no plants will be found in searching the next (1) .5 acre; (2) 1 acre; or (3) 1.5 acres

8–7 A traffic engineer believes that the accidents encountered along a particular stretch of road during rush hours is a Poisson process. Over a duration of 50 weeks, find (1) the expected number of accidents and (2) the probability there will be at least 5 accidents when

(a) there are no roadbed improvements and $\lambda = .1$ rush-hour accident per week

(b) the roadbed is improved, bringing λ down to .04 rush-hour accident per week

8–8 Consider two successive nonoverlapping process time segments t_1 and t_2. Show that $p(0; \lambda, t_1 + t_2) = p(0; \lambda, t_1)p(0; \lambda, t_2)$.

8–9 Government approval for a nuclear power plant on the California coast requires a hazard evaluation. Included is a probability analysis of various potentially damaging accidents or natural disasters. Compute the probability of at least one occurrence, (1) in a single year and (2) sometime in the next 100 years, from each of the following potentially damaging events, all of which arise from independent Poisson processes:

(a) impact from an airplane crash, presumed to occur in the vicinity of the generator site at a mean annual rate of 10^{-6}

(b) being hit by a large *tsunami* (tidal wave), known to occur once every 1,000 years with a further chance of $\frac{1}{500}$ of hitting a particular location the width of the generator site

(c) an earthquake causing rupture in the reactor cooling system. This could be only from a Richter-8 or greater shock whose epicenter falls near the generator site. This event is judged to have a mean rate of 10^{-5} per year.

8–10 Cars arriving at a particular highway patrol safety checkpoint form a Poisson process with mean rate $\lambda = 100$ per hour. Each car encountered during a 1-hour period is tested for mechanical defects, and the drivers are informed of needed repairs. Suppose that 10% of all cars in the state need such repairs.

(a) What is the probability that at least 10 cars will be found in need of repairs?

(b) To test the effectiveness of the warnings, the highway patrol will trace the defective cars to determine if the necessary repairs were eventually made. Supposing that only half of the originally warned operators ever fix their cars, what is the probability that at least 6 cars will have been repaired?

8–11 An electronic switching module is expected to malfunction some of the time, misdirecting messages. It must be replaced if the rate of such errors becomes too high. Letting λ represent the mean rate of misdirected messages, the switch is operating according to specifications when $\lambda = .10$ per hour. Should λ reach .50 error per hour, the module ought to be replaced.

There is no way to know λ precisely. A monitoring device may record the number of misdirections in a 10-hour test. Policy is to replace any module that causes more than 2 errors in the test, and otherwise to leave it in place.

(a) What is the probability that a module needing replacement is retained?

(b) What is the probability that a module is replaced even though it is operating according to specifications?

20 × 6

8–2 The Exponential Distribution

Having wide application in engineering evaluations is the **exponential distribution**. This distribution provides probabilities for the amount of time or space between successive events occurring in a Poisson process. The exponential distribution has been used for the times between arrivals at service facilities, making it a centerpiece in the theory of queues or waiting lines. It plays a similar role in reliability evaluations, in which the exponential distribution gives probabilities for the times to failure. Another major application is in setting production inventories, in which the time between requests for stock is often characterized by the exponential distribution.

All the assumptions of a Poisson process described in the preceding section are assumed to hold in those situations where the exponential distribution is applied. Recall that the Poisson process also gives rise to the *discrete* Poisson distribution, which provides probabilities for the number of events in a fixed span of time or space. The exponential distribution treats the span itself as the *continuous* random variable.

Finding Exponential Probabilities

Illustration: *Times Between Arrivals at a Toll Booth*

Figure 8–3 shows a graphical representation of the exponential distribution for the time T between successive arrivals of cars at a toll booth at one of the San Francisco Bay bridges. The random variable applies during a busy time of the day when arrivals occur on the average of 5 per minute. We see from the probability density curve in (a) that the modal value is 0 and that there is a single tail extending indefinitely to the right. As with all continuous probability distributions, probabilities for T are represented by areas under this curve. The area to the left of t provides the ordinate for the exponential prob-

Figure 8–3 *Curves for exponential distribution representing time T between successive arrivals of cars at a toll booth.*

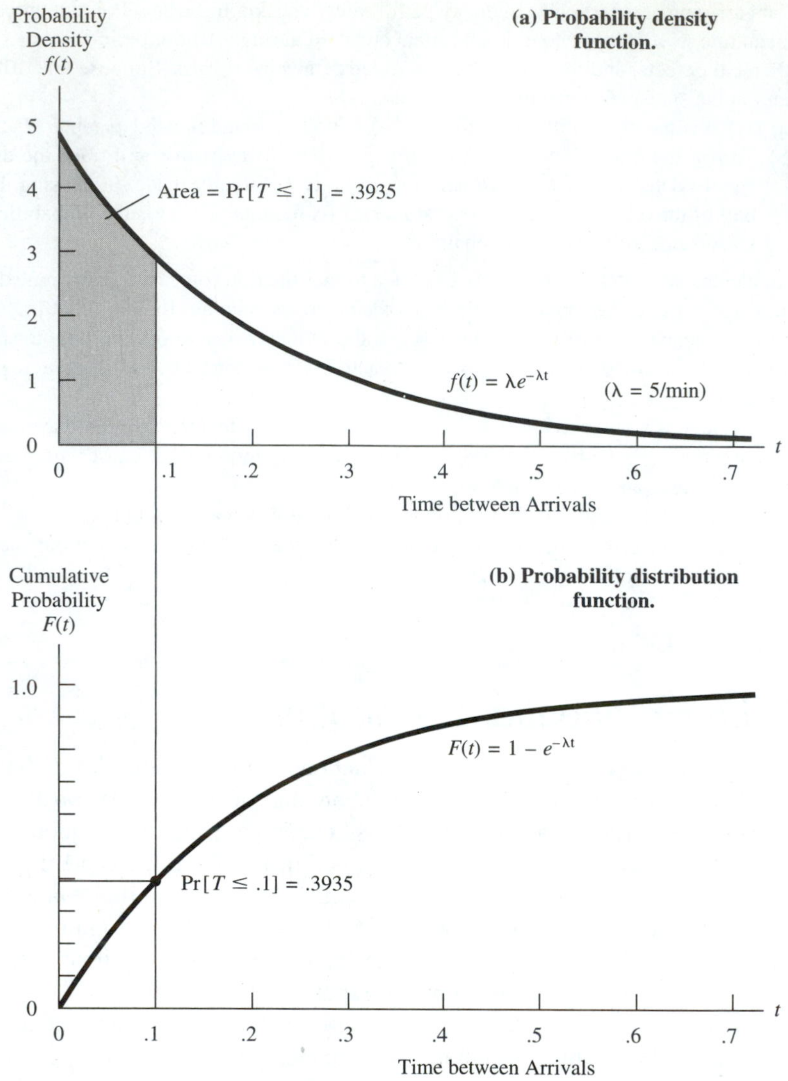

ability distribution curve at the bottom (b). The latter curve provides cumulative probabilities, which approach 1 as t becomes large.

An exponential distribution has a single parameter. This is the **mean process rate** λ, which expresses on the average how many events occur per unit of time. For the toll booth illustration this level is $\lambda = 5$ per minute. For a specific λ we use the following expression.

PROBABILITY DENSITY FUNCTION FOR THE EXPONENTIAL DISTRIBUTION

$$f(t) = \begin{cases} \lambda e^{-\lambda t} & t \geq 0 \\ 0 & \text{otherwise} \end{cases}$$

For the toll booth arrivals, $\lambda = 5$ and $f(t) = 5e^{-5t}$ when $t \geq 0$.

The probability that T will fall inside any interval may be found by integrating the function over that interval. We have

$$\Pr[a \leq T \leq b] = \int_a^b f(t)dt$$

When the lower limit is 0 and the upper limit is t, we have

$$\Pr[0 \leq T \leq t] = \int_0^t \lambda e^{-\lambda \tau} d\tau$$

$$= \int_{-\lambda t}^0 e^x dx = e^x \Big]_{-\lambda t}^0 = 1 - e^{-\lambda t}$$

This gives the following function.

EXPONENTIAL PROBABILITY DISTRIBUTION FUNCTION

$$F(t) = \Pr[T \leq t] = 1 - e^{-\lambda t}$$

which applies for all $t \geq 0$. Values for e^{-y} may be read from Appendix Table B.

We may use the foregoing to calculate the probability that the time T between successive car arrivals at the toll plaza is less than or equal to 6 seconds, or .1 minute:

$$\Pr[T \leq .1] = F(.1) = 1 - e^{-5(.1)} = 1 - e^{-.5}$$
$$= 1 - .6065 = .3935$$

With $\lambda = 5$ cars per minute, it is quite rare for more than a full minute to transpire between arrivals:

$$\Pr[T > 1] = 1 - F(1) = e^{-5(1)} = .0067$$

But the exponential density has no upper limit, so there is finite probability that several minutes could go by without any arrivals (although this quantity would be tiny).

Expected Value, Variance, and Percentiles

The average time or space between events is equal to the *reciprocal* of the mean rate at which they occur. And the variance is the square of this quantity. We use the following expression.

EXPECTED VALUE AND VARIANCE FOR AN EXPONENTIAL DISTRIBUTION

$$E(T) = \frac{1}{\lambda}$$

$$\text{Var}(T) = \frac{1}{\lambda^2}$$

For the car illustration, $\lambda = 5$ per minute. Thus, the expected time between arrivals is

$$E(T) = \frac{1}{5} = .2 \text{ minute}$$

and the variance is

$$\text{Var}(T) = \frac{1}{5^2} = .04$$

Taking the square root of the variance, we see that the standard deviation of T is equal to the expected value:

$$SD(T) = \sqrt{\frac{1}{\lambda^2}} = \frac{1}{\lambda} = E(T)$$

Percentiles for the exponential distribution may be found by reading Appendix Table B in reverse and using the fact that for any specified cumulative probability p, there is a corresponding negative power of e:

$$e^{-y} = 1 - p$$

so that

$$t = \frac{y}{\lambda}$$

The level for t that corresponds is the ***p*th percentile**. In our car arrival illustration we might set $p = .90$. Reading from Table B, the nearest negative power of e to $1 - .90 = .10$ is $y = 2.30$, so that

$$t = \frac{2.30}{\lambda} = \frac{2.30}{5} = .46 \text{ minute}$$

is the 90th percentile for the time between arrivals. This means that the probability is .90 that the time between any two successive arrivals will be .46 minute or less.

Example:* *Major Electrical Power Disruptions

One electrical utility has infrequently experienced major disruptions to its power grid that cause temporary shutdowns of the entire system. These disturbances are caused by random events such as lightning, transformer failures, forest fires, and airplane crashes. Historically, the grid shutdowns fit the Poisson process, occurring on the average once every 2.5 years. In reliability such a value is often referred to as the **mean time between failures** (MTBF). The mean rate of major power disruption is thus

$$\lambda = \frac{1}{\text{MTBF}} = \frac{1}{2.5} = .40 \text{ per year}$$

What is the probability of at least one disruption within the next year?

The desired event is equivalent to the time of the next disruption being 1 year or less. We have

$$\begin{aligned}\Pr[T \leq 1] = F(1) &= 1 - e^{-.4(1)} = 1 - e^{-.4} \\ &= 1 - .6703 = .3297\end{aligned}$$

or slightly less than a 33% chance.

About half the disruptions occur because of a cascading effect arising from disturbances originating outside the company network. Special equipment can eliminate these, but the cost would be \$1 million. The new investment would reduce λ to .20, so that the annual probability of disruption becomes

$$\begin{aligned}\Pr[T \leq 1] = F(1) &= 1 - e^{-.2(1)} = 1 - e^{-.2} \\ &= 1 - .8187 = .1813\end{aligned}$$

Company management decided, however, that the benefit in increased reliability did not justify the increased cost.

Applications of the Exponential Distribution

Recall that one property of a Poisson process is that occurrences in successive time periods are statistically independent events. In effect, *a Poisson process has no memory*. When applied to equipment failures, this requirement stipulates that, given an earlier failure, the chance of failure in the next 100 hours of operations must be the same as it would be if there had been no failure or if the operating history were unknown. It happens that many real-life reliability data demonstrate this characteristic, which is why the exponential distribution is so widely used in reliability evaluations.

The Poisson process not only provides a good representation for car arrivals at toll booths but characterizes an entire class of arrival patterns encountered in queuing analysis. In all cases the times between arrivals have an exponential distribution. An interesting special case arises when the time for service is also exponentially distributed.

Example: *Waiting Times in Telephone Switching*

As telephone calls are placed, they must be accepted by the telecommunications network and be routed to their destinations via clear channels. The switching circuitry encountered when a new call is placed may be busy processing earlier calls. While the circuits are busy, all incoming calls must be placed in a holding mode, in effect placed in a queue, until those earlier calls have been processed.

One central telephone exchange receives incoming calls at a rate of $\lambda = 2$ per second. An exponential distribution applies to the time between arriving calls. The equipment has the capacity of switching a call in an average time of .25 second, although with speed-of-light solid-state equipment the most frequent switching time is very near to 0. This feature, plus near adherence to a Poisson process, allows the switching time to be closely approximated by an exponential distribution with a mean rate of $\mu = 4$ calls per second.

Since the interarrival times and service times are both exponentially distributed, it can be established that the waiting time T_W of a particular call for switching to begin is also exponentially distributed, with a mean rate equal to the difference in service and arrival rates, $\mu - \lambda$. Thus, we may evaluate probabilities that an incoming call experiences various waiting times. In the present telephone system, we have $\mu - \lambda = 4 - 2 = 2$ so that the probability that less than 1 second of waiting time will be required is substantial,

$$\Pr[T_W \leq 1] = F(1) = 1 - e^{-2(1)} = 1 - e^{-2}$$
$$= 1 - .1353 = .8647$$

whereas the probability that there will be more than 3 seconds of waiting time is tiny:

$$\Pr[T_W > 3] = 1 - F(3) = e^{-2(3)} = e^{-6} = .0025$$

There is a danger in using the procedure in the foregoing example when evaluating more general queuing situations. Few service patterns are suitably represented by an exponential distribution. (Imagine a tool cage or barbershop where the most likely service time is zero!)

Problems

8–12 The time between accidents at a congested intersection has an exponential distribution with unknown parameter λ, which may assume values of (1) 1/hour, (2) 2/hour, or (3) 3/hour. Complete the following for each of these levels.
(a) Determine the probability that an accident will happen within 1.5 hours.
(b) Find the expected time between accidents.
(c) Find the variance in time between accidents.
(d) Find the 10th percentile for the time to an accident.
(e) Find the 90th percentile for the time to an accident.

8–13 Cars arrive at a toll plaza at a mean rate of 5 per minute between 7 A.M. and 8 A.M. Find the probability that starting at 7:30 A.M. the toll taker gets no cars until after (a) 7:31, (b) 7:32, (c) 7:30 plus 12 seconds, and (d) 7:30 plus 30 seconds.

8–14 Suppose the telephone switching circuit described in the example is improved, so that service is faster, with an average rate of $\mu = 5$ calls per second. Find the probability that a call's waiting time is
(a) less than 1 second
(b) greater than 2 seconds
(c) between .2 and .8 second
(d) greater than .5 second

8–15 Referring to the power example in the text, suppose that a disruption causes an average damage loss of $500,000.
(a) Find the expected loss per year under (1) the present generating system and (2) the proposed improved system.
(b) What are the expected annual savings in damage claims with the improved equipment?
(c) How much time is expected before the proposed equipment improvements are paid for?

8–16 A hospital presently has one backup emergency electric generator connected to auxiliary circuits that supply power to critical areas. During a blackout, such a unit allows continuation of surgical, emergency, and life-support systems. The mean time between failures ($1/\lambda$) of the generator is 100 hours.
(a) Find the probability that the emergency generator will fail during a 10-hour power blackout.
(b) Suppose a second identical emergency generator is used to provide service in parallel. They operate independently. Both must fail before critical service is halted. Find the probability that this event will occur during the next 10-hour blackout.

8–17 The central processor of a computer may be viewed as a service facility where successive microinstructions are arriving "customers." Suppose these form a Poisson process with mean arrival rate $\lambda = 10^6$ per second. While the processor is handling earlier arrivals, new instructions stack up in a buffer and must wait their turn. The time for completing an instruction (service) may be assumed to be exponentially distributed with a mean rate of 10^7 per second.

Find the probability that the time a particular microinstruction spends in the buffer is
(a) less than 100 nanoseconds
(b) more than 500 nanoseconds
(c) more than 1 microsecond
(d) between 300 and 400 nanoseconds

8–18 Find the exact probability that any exponentially distributed random variable T will fall within $E(T) \pm kSD(T)$ when (a) $k = 1$, (b) $k = 2$, and (c) $k = 3$.

8–19 Molecules of a toxic chemical eventually decompose into inert substances. Suppose the decomposition time is exponentially distributed with a mean of $1/\lambda$. The half-life of such a persistent poison is that time beyond which the probability is .50 that a particular molecule will remain toxic.

Find the half-life for chemicals whose molecules have an average decomposition time of (a) 1 year, (b) 5 years, (c) 25 years, and (d) 100 years.

8–20 The n power cells in a satellite will be arranged in parallel and will fail at a mean rate of .01 per day. They have independent lifetimes.

(a) Find the probability that any specific cell will fail on or before (1) 100 days, (2) 150 days, and (3) 500 days.

(b) Find the probability that there will be at least 1 cell still working after 600 days, assuming (1) $n = 2$, (2) $n = 5$, and (3) $n = 10$.

(c) What should n be to provide a 90% chance that the satellite's power source will survive at least 600 days?

8–21 Flaws in a reel of high-fidelity radar recording tape occur on the average of once every 10 feet.

(a) Determine the probability that the next recording will begin on a flawless stretch of tape over (1) 5 feet long, (2) 10 feet long, (3) 20 feet long, and (4) 50 feet long.

(b) Over what distance is there a 90% chance that an error will be encountered since the last one?

8–22 Show that for an exponentially distributed variable,

$$\Pr[T > t + h \mid T > t] = \Pr[T > h]$$

which illustrates that the Poisson process lacks a memory.

8–3 The Gamma Distribution

The **gamma distribution** is important because it includes a wide class of specific distributions, some of which underlie fundamental statistical procedures. In addition to serving as a utility distribution, the gamma provides probabilities for yet another random variable associated with Poisson processes. And, as we shall see, the exponential distribution itself is a member of the gamma distribution family.

The Gamma Function

In order to describe the gamma distribution in detail, we must first consider a useful mathematical function, the gamma function.

GAMMA FUNCTION

$$\Gamma(r) = \int_0^\infty x^{r-1}e^{-x}dx \qquad r > 0$$

This is used in defining several random variables. The symbol Γ (Greek uppercase *gamma*) is reserved for this function.

We may evaluate the gamma function for various levels of r. When $r = 1$, we have

$$\Gamma(1) = \int_0^\infty e^{-x}dx = -e^{-x}\Big]_0^\infty = 1$$

Next, consider $\Gamma(r + 1)$. Integrating by parts, it is easy to establish that

$$\Gamma(r + 1) = r\Gamma(r)$$

and similarly,

$$\Gamma(r) = (r-1)\Gamma(r-1)$$

Thus, for any nonnegative integer k, it follows that

$$\Gamma(k+1) = k!$$

Noninteger values may apply for r. An important class involves values with halves. We have

$$\Gamma(\tfrac{1}{2}) = \sqrt{\pi}$$

and for any positive integer k

$$\Gamma(k + \tfrac{1}{2}) = \frac{1 \cdot 3 \cdot 5 \cdot \cdots \cdot (2k-1)}{2^k}\sqrt{\pi}$$

The Probability Density Function

The following expression gives the probability density function for a gamma distribution.

PROBABILITY DENSITY FUNCTION FOR A GAMMA DISTRIBUTION

$$f(x) = \begin{cases} \dfrac{\lambda^r}{\Gamma(r)} x^{r-1} e^{-\lambda x} & x \geq 0 \\ 0 & \text{otherwise} \end{cases}$$

The two parameters λ and r may be any nonnegative values.

A special case of this function occurs when $r = 1$. We have

$$f(x) = \frac{\lambda^1}{\Gamma(1)} x^{1-1} e^{-\lambda x} = \lambda e^{-\lambda x}$$

which is the density function for the exponential distribution. The gamma gives rise to the chi-square distribution (described in Chapter 9) when $\lambda = 1/2$ and $r = d/2$, with d being a nonnegative integer parameter.

The expected value and variance may be computed from

$$E(X) = \frac{r}{\lambda} \qquad \text{Var}(X) = \frac{r}{\lambda^2}$$

Relation to Poisson Process

Recall that a Poisson process characterizes many circumstances wherein events occur over time or space. As we have seen, the time or space encountered between any two events has the exponential distribution. We might consider the segment of time or space occurring until some specified number of events has transpired. The size X of such a segment is a random variable having a gamma distribution with parameters λ representing the mean process rate and r denoting the specific number of events that must transpire as X is reached.

For instance, suppose $r = 2$. We have

$$\begin{aligned}\Pr[X \le x] &= \int_0^x \frac{\lambda^2}{1!} y^{2-1} e^{-\lambda y} dy \\ &= \lambda^2 \int_0^x y e^{-\lambda y} dy = \lambda^2 \left[\frac{e^{-\lambda y}}{\lambda^2}(-\lambda y - 1) \right]_0^x \\ &= [-\lambda y e^{-\lambda y} - e^{-\lambda y}]_0^x = -\lambda x e^{-\lambda x} - e^{-\lambda x} - 0 + 1 \\ &= 1 - e^{-\lambda x}(1 + \lambda)\end{aligned}$$

Suppose car arrivals at a toll booth are a Poisson process with mean $\lambda = 5$ cars per minute. The probability that up to 1 minute will elapse until two cars have arrived is

$$\begin{aligned}\Pr[X \le 1] &= 1 - e^{-5(1)}(1 + 5) \\ &= 1 - 0.006738(6) = .96\end{aligned}$$

Problems

8–23 Determine the following values:

(a) $\Gamma(5)$ (c) $\Gamma(2)$ (e) $\Gamma(1.5)$ (g) $\Gamma(7.5)$
(b) $\Gamma(6)$ (d) $\Gamma(8)$ (f) $\Gamma(3.5)$ (h) $\Gamma(10.5)$

8–24 Find (1) the expected value, (2) the variance, and (3) the standard deviation for the gamma distributions having the following parameters:

(a) $\lambda = 2, r = 2$ (c) $\lambda = 5, r = 5$
(b) $\lambda = 2, r = 4$ (d) $\lambda = 5, r = 25$

8–25 Find the probability that the time taken for the next two cars to arrive at a toll booth will be 1 minute or less when (a) $\lambda = 1$ per minute, (b) $\lambda = 2$ per minute, and (c) $\lambda = 4$ per minute.

8–26 Consider the gamma distribution with $\lambda = 1$ and $r = 5$. Find the probability that

(a) $X \le 5$ (b) $X \le 10$ (c) $5 \le X \le 10$

8–27 Consider the gamma distribution with $\lambda = 4$ and $r = 3$. Find the probability that

(a) $X \le 1$ (b) $X \le 2$ (c) $1 \le X \le 2$

8–4 Failure-Time Distributions: The Weibull

Time to failure is the key variable in reliability analysis. A variety of probability distributions characterize how long an entity will survive. One aspect of reliability analysis is finding a distribution model that provides an appropriate fit to the item or system under investigation. Among the important probability distributions are the exponential and gamma. Those distributions have attractive mathematical properties, but they are based on assumptions that may not apply to the lifetimes of certain items and systems. A more flexible distribution that is quite useful in reliability analysis is the Weibull distribution. Failure time T is a continuous random variable with probability density function $f(x)$. The failure-time cumulative probability distribution function

$$F(t) = \Pr[T \le t] = \int_0^t f(x)dx$$

establishes the reliability function.

RELIABILITY FUNCTION

$$\bar{F}(t) = \Pr[T > t] = 1 - F(t)$$

Since this gives the probability that the item or system survives longer than time t, $\bar{F}(t)$ is sometimes referred to as the *survival function*.

Exponential Failure-Time Distribution

The exponential distribution is one of the common failure-time distributions. This family of distributions has probability density function

$$f(x) = \lambda e^{-\lambda x}$$

The failure-time cumulative distribution function is

$$F(t) = 1 - e^{-\lambda t}$$

and the reliability function is therefore

$$\bar{F}(t) = e^{-\lambda t}$$

To illustrate, suppose that a particular type of printed circuit board has a mean lifetime of 10,000 hours, so that the mean failure rate is

$$\lambda = \frac{1}{10{,}000} = .0001 \text{ per hour}$$

The following reliabilities apply:

$$\begin{aligned}
\bar{F}(1{,}000) &= \exp[-.0001(1{,}000)] &&= .9048\\
\bar{F}(5{,}000) &= \exp[-.0001(5{,}000)] &&= .6065\\
\bar{F}(10{,}000) &= \exp[-.0001(10{,}000)] &&= .3679\\
\bar{F}(50{,}000) &= \exp[-.0001(50{,}000)] &&= .0067
\end{aligned}$$

The exponential distribution is appropriate to a large class of lifetimes. All of these involve failures that occur randomly over time. Recall that the exponential distribution applies to a Poisson process and that failures in one time period are statistically independent of those in any other nonoverlapping period. In effect, the process has no memory, so the remaining lifetime expected for an old printed circuit board of the above type is the same as that for a new one. If the exponential distribution literally applies to an item's lifetime, then the tendency for it to fail remains unchanged as it ages.

Human mortality provides an analogy for explaining this critical assumption. If the exponential distribution were a valid model for representing time until death, then people would die only from random causes—such as an exotic disease or an accident—and the probability of death would be the same for a baby as a 100-year-old or 500-year-old. (There would be no such thing as "old age" as the cause of death.) The death rate would be the same for all ages, and everybody would sip from a perpetual fountain of youth.

The Failure Rate Function

Of course, human mortality is not constant. As people become older, they wear out and eventually meet their end because of old age. Many products, machines, and complex systems exhibit lifetimes similar to those of people, so their failure rates change over time. More generally, the failure rate is a function of an entity's lifetime so far.

FAILURE RATE FUNCTION

$$h(t) = \frac{f(t)}{1 - F(t)} = \frac{f(t)}{\overline{F}(t)}$$

This function is found by dividing the density function at time t by the reliability for that duration. It is sometimes referred to as the *hazard-rate function.*

Figure 8–4 shows the representative shape for a failure-rate function. Early in a system's life the failure rate is high due to the presence of defective original components. As these are replaced, the failure rate drops steeply. For electronic systems this initial stage of life systems is commonly referred to as the "burn-in" period. The next phase is a long period of stable or nearly constant failure rate. Eventually, components start to deteriorate physically, and the failure rate rises rapidly. The representative failure rate function creates a plot called the *bathtub curve*, as suggested by its shape. Human lifetimes follow this pattern, with the highest mortality rates occurring during early infancy, falling throughout childhood, and stabilizing during early adulthood through middle age, when causes of death are largely random. In later years, people experience increasing mortality as physical deterioration progresses.

The exponential distribution provides a *constant* failure rate over time, so that $h(t) = \lambda$. It empirically fits the actual experience of many systems only over the flat portions of their bathtub curves—that is, after they have survived the initial burn-in period and before the onset of physical deterioration. The applicability of the exponential distribution is limited (although it is favored because of its nice mathematical properties). Since a reliability analysis may involve items or systems with decreasing or increasing failure rates, a more general distribution is important to reliability analysis. The *Weibull distribution* fills this bill.

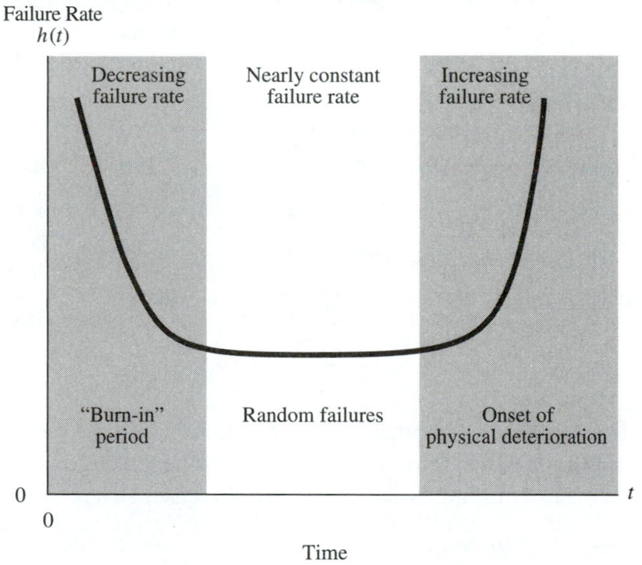

Figure 8–4 *Representative failure rate over time, or bathtub curve.*

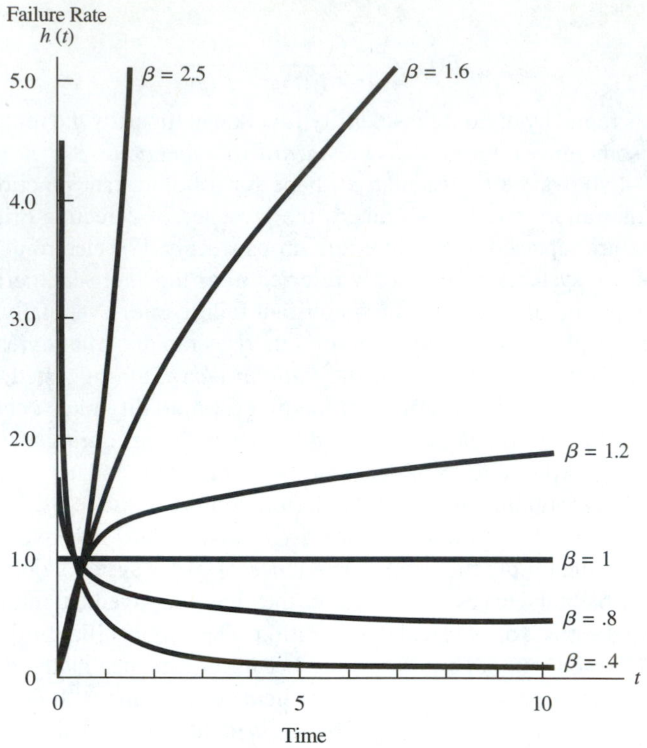

Figure 8–5
Failure rate curves for Weibull distribution with $\lambda = 1$.

The Weibull Distribution

Along with the exponential, the Weibull is one of the most useful failure-time distributions in reliability analysis. The following expressions are used.

WEIBULL PROBABILITY DISTRIBUTION

$$f(t) = \lambda\beta(\lambda t)^{\beta-1}e^{-(\lambda t)^\beta}$$
$$F(t) = 1 - e^{-(\lambda t)^\beta}$$

where the parameters λ and β are nonnegative constants. The reliability function is

$$\bar{F}(t) = e^{-(\lambda t)^\beta}$$

The Weibull distribution has the following failure-rate function:

$$h(t) = \frac{\lambda\beta(\lambda t)^{\beta-1}e^{-(\lambda t)^\beta}}{e^{-(\lambda t)^\beta}} = \lambda\beta(\lambda t)^{\beta-1}$$

Example: *Survival of Satellite Power System*

Figure 8–5 shows the failure rates for selected members of the Weibull family. Notice that depending on the level for the parameter β, as t becomes greater, $h(t)$ decreases (when $\beta < 1$), remains constant (when $\beta = 1$), or increases (when $\beta > 1$).

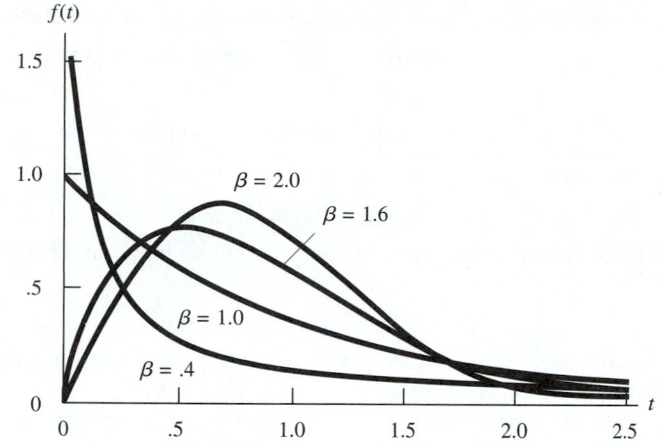

Figure 8–6
Weibull distributions with selected values for shape parameter ($\lambda = 1$).

The Weibull density function may take a variety of forms. The parameter β is the *shape parameter*. Figure 8–6 shows Weibull distributions for selected levels of β when $\lambda = 1$. Notice that the curves may be one-tailed or two-tailed with a positive skew. The second parameter λ is the *scale parameter*. Figure 8–7 shows Weibull distributions for several λs when $\beta = 2$.

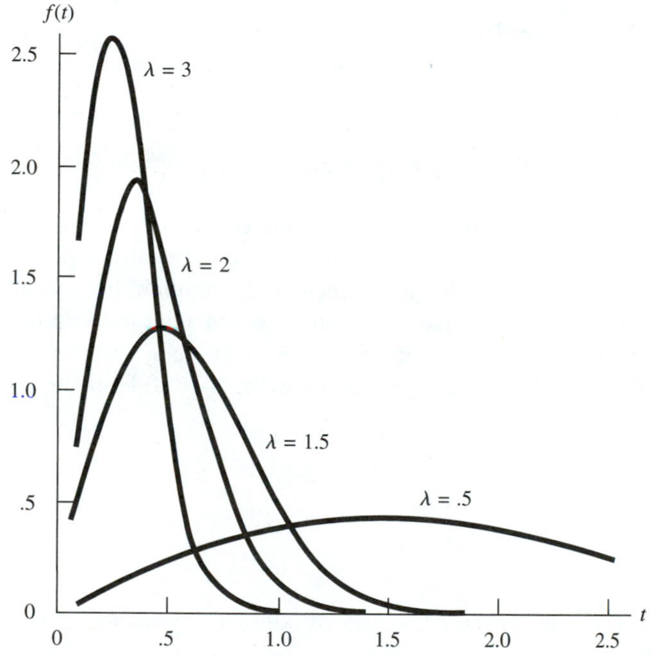

Figure 8–7
Weibull distributions with selected values for scale parameter ($\beta = 2$).

Example: *Fuse Lifetimes*

A reliability engineer assumes that the lifetimes of a particular fuse will follow the Weibull distribution, with parameters $\lambda = .005$ per year and $\beta = .10$. He determines the probability that a randomly chosen fuse will function for more than $t = 10$ years.

The reliability function provides the answer:

$$\bar{F}(10) = \exp\{-[.005(10)]^{.10}\}$$

He first evaluates $.005(10) = .05$ to the .10 power by taking the natural logarithm

$$\ln[(.05)^{.10}] = .10\ln(.05) = .10(-2.99573) = -.299573$$

so that

$$(.05)^{.10} = \exp(-.299573) = .74113$$

Using the above,

$$\bar{F}(10) = \exp(-.74113) = .4766$$

In a similar fashion the engineer obtained probabilities for survival beyond 5 years and past 15 years:

$$\bar{F}(5) = .5008 \qquad \bar{F}(15) = .4622$$

He was surprised to see that the probability that the fuse will survive beyond 5 years is so close to the probability for a longer-than-15-year life.

He found an explanation for this by first computing the failure rates when $t = 5$,

$$h(5) = .005(.10)[.005(5)]^{.10-1} = .0138$$

and when $t = 10$ and $t = 15$:

$$h(10) = .0074 \qquad h(15) = .0051$$

The failure rates are decreasing (always the case for $\beta < 1$) and are quite small, so that failure within a short time span is a rare event. Notice that from $t = 10$ years to $t = 15$ years, $h(t)$ changes very little.

An interesting property of the Weibull distribution is that it becomes the exponential distribution when $\beta = 1$. It also includes as a second special case—the *Rayleigh distribution*—when $\beta = 2$.

The Gamma as a Failure-Time Distribution

The gamma distribution can have decreasing, increasing, or constant failure rates, depending on the parameter settings. This distribution is especially important in reliability analysis because of its relation to the exponential distribution. In a system with several independent components, each having identical exponential failure-time distributions, the gamma distribution provides probabilities for that time by which a specified number of these will fail. The following expression may be used to compute gamma survival probabilities.

GAMMA RELIABILITY FUNCTION
(EXPONENTIAL COMPONENTS)

$$\bar{F}_r(t) = \sum_{k=0}^{r-1} \frac{(\lambda t)^k}{k!} e^{-\lambda t}$$

where $\lambda > 0$ and r is the specified number of failing components.

Example: *Survival of Satellite Power System*

A satellite is powered by 10 solar cells connected in a parallel system. There will be enough power for full operation if at least 3 cells are working. The cell lifetimes are independent random variables, each having exponential distribution, with failure rate $\lambda = 1$ per year. What is the probability that the useful satellite life is at least 2 years?

The system lifetime has a gamma distribution with parameters $\lambda = 1$ and $r = 7$ failures. Using $\lambda t = 1(2) = 2$, the probability is

$$\begin{aligned}\Pr[T > 2] &= \bar{F}_7(2)\\ &= \left(\frac{2^0}{0!} + \frac{2^1}{1!} + \frac{2^2}{2!} + \frac{2^3}{3!} + \frac{2^4}{4!} + \frac{2^5}{5!} + \frac{2^6}{6!}\right) e^{-2}\\ &= (1 + 2 + 2 + 1.3333 + .6667 + .2667 + .0889)(.1353)\\ &= .9952\end{aligned}$$

Exponential Series Systems

One reason why the exponential distribution is so common in reliability analysis is its nice mathematical properties. We see this in a *series system* having N types of components. We assume that each has an exponential distribution, with the failure rate for each type i component denoted as λ_i. There may be several items of each type, with n_i denoting the quantity of each. We further assume that the lifetimes of all are *statistically independent*. Then *the system failure-time distribution will itself be exponentially distributed*, with mean failure rate

$$\lambda_S = \sum_{i=1}^{N} n_i \lambda_i$$

To illustrate, suppose that a printed circuit board contains the following components:

Item	Quantity	Failure Rate
1. Chip	5	.00010/hour
2. Resistor	20	.00002
3. Capacitor	10	.00050
4. Switch	2	.00025
5. Relay	5	.00015

The system's mean failure rate is

$$\begin{aligned}\lambda_S &= 5(.00010) + 20(.00002) + 10(.00050) + 2(.00025) + 5(.00015)\\ &= .00715\end{aligned}$$

The mean time to failure (MTTF) is

$$\text{MTTF} = \frac{1}{\lambda_S} = \frac{1}{.00715} = 139.9 \text{ hours}$$

and the $t = 100$-hour reliability is

$$R_S(100) = \bar{F}(100) = \exp[-.00715(100)] = .489$$

Exponential Parallel Systems: Gamma Reliability

The gamma provides the reliability for a *parallel* system of components, each sharing a common exponential failure-time distribution with failure rate λ.

For a system that survives as long as at least one out of n components does, the gamma reliability function gives

$$R_S(t) = \bar{F}_n(t) = \sum_{k=0}^{n-1} \frac{(\lambda t)^k}{k!} e^{-\lambda t}$$

To illustrate, suppose that a redundant system has $n = 4$ components, each with mean failure rate $\lambda = .01$ per hour. The $t = 200$-hour system reliability is, using $\lambda t = 2$,

$$\begin{aligned} R_s(200) = \bar{F}_4(200) &= \left[\frac{2^0}{0!} + \frac{2^1}{1!} + \frac{2^2}{2!} + \frac{2^3}{3!}\right] e^{-2} \\ &= (1 + 2 + 2 + 1.3333)(.1353) \\ &= .857 \end{aligned}$$

Problems

8–28 The exponential distribution applies to lifetimes of a certain component. Its failure rate λ is unknown. Find the probability that the component will survive past 5 years assuming: (a) $\lambda = .5/\text{year}$, (b) $\lambda = 1/\text{year}$, and (c) $\lambda = 2/\text{year}$.

8–29 A particular capacitor has a mean time between failures of 5 years. Assuming that the exponential distribution applies, determine the probability that a capacitor placed in continuous operation will survive beyond (a) 2 years, (b) 5 years, (c) 10 years, and (d) 20 years.

8–30 Consider the reliability engineer's problem in the example of fuse lifetimes. Change β to .20, then answer the following.
(a) Recompute the survival probabilities for (1) 5, (2) 10, and (3) 15 years.
(b) Recompute the failure rates at each of these times.

8–31 Repeat Problem 8–30 using $\lambda = .01$ per year while keeping $\beta = .10$, the original level.

8–32 You are conducting a reliability analysis for a new product. Based on prior testing with a similar product, you believe the Weibull failure-time distribution, with parameter $\lambda = .20$ per year, applies. But you have no basis for establishing β. Compute at a time span of 5 years (1) the failure rate and (2) the survival probability assuming that (a) $\beta = .5$, (b) $\beta = .8$, (c) $\beta = 1.2$, and (d) $\beta = 2.0$.

8–33 A solar-powered foghorn is attached to a buoy moored offshore from a remote coast. Five solar collectors independently feed the power supply system. As long as at least 2 collectors remain functional, the buoy will operate satisfactorily. Each collector has an exponential failure-time distribution with a mean failure rate of 1 per year. Find the probability that the buoy will continue to operate without interruption beyond 3 years.

8–34 The failure-time distribution for an item is represented by a normal distribution with mean $\mu = 2{,}000$ hours and standard deviation $\sigma = 500$ (so that negative values for t have negligible probabilities). Find the probability that the item survives beyond (a) 1,800 hours, (b) 2,500 hours, (c) 3,000 hours, and (d) 3,500 hours.

8–35 Consider the system in Figure 8–8. Suppose that each component has a Weibull failure-time distribution with $\beta = .5$ and $\lambda = .01$/hour. Find the system reliability for a span of (a) $t = 100$ hours, (b) $t = 50$, and (c) $t = 10$.

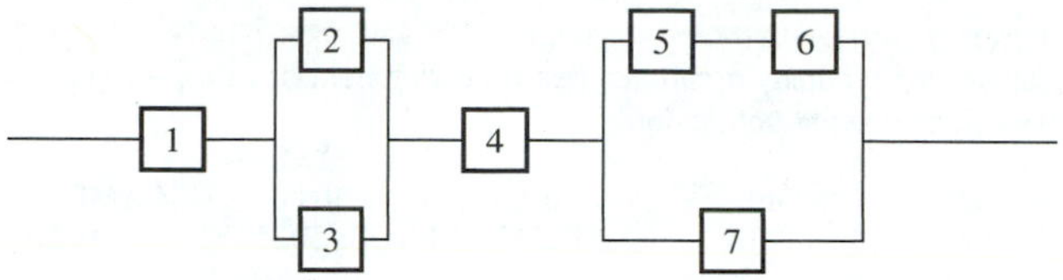

Figure 8–8

8–36 A series system consists of 100 independent units, each with exponential distribution with $\lambda = .005$. Find the system reliability over a span of (a) $t = 10$, (b) $t = 20$, and (c) $t = 50$.

8–37 A parallel system consists of five independent components, each having common exponential failure-time distribution with $\lambda = .10$. Find the system reliability over a span of (a) $t = 2$, (b) $t = 3$, and (c) $t = 5$.

8–5 The Hypergeometric Distribution

We now consider one of the more important discrete probability distributions encountered in statistical sampling applications. The **hypergeometric distribution** provides probabilities for the number of sample observations of a particular category that can be obtained. In this respect it plays the same role as the binomial distribution. But recall that the binomial requires independence between trials. The independence requirement makes that distribution unsuitable for evaluating sampling investigations of small-sized populations, unless sampling is done *with* replacement.

The hypergeometric distribution does not require independence between trials and therefore can be applied when sampling *without* replacement from populations of small size. Thus, it may provide probabilities for the number of defective "black boxes" found in a sample of test items removed from a shipment containing 100, 50, or even 20 items altogether. Since electronic components often must be destroyed during quality assurance testing, sampling must be done without replacement.

***Illustration:** Testing a Shipment of Transformers*

To detail the characteristics of the hypergeometric distribution, it will be helpful to look at the sampling process in its basic form. Consider a shipment of 25 transformers from which a sample of 4 will be subjected to a thorough environmental test. (The test items themselves will end up as burned-out hulks with fused coils.) Each transformer in the shipment will ultimately be satisfactory (S) or defective (D). Figure 8–9 shows the probability tree diagram applicable when the shipment is presumed to have 20% defectives (5 transformers from the entire shipment). Of course, this fact is unknown to the quality assurance department, which must decide to accept or reject the shipment on the basis of the number of defective items R it actually finds in the sample.

Every successive observation is represented in the tree by a fork having one satisfactory and one defective branch. Each path through the tree represents a sequence of observed sample results, which we may view as the elementary events for the complete sampling experiment. Applying basic probability concepts, we can readily find the probability that the sample contains $R = 1$ defective. Referring to the tree, we see that there are four elementary events involving exactly 1 defective. Each of these outcomes is represented on the tree by a heavy-line path.

Figure 8–9

Probability tree diagram for a quality assurance test on a shipment of transformers. Sampling is done without replacement from a small population.

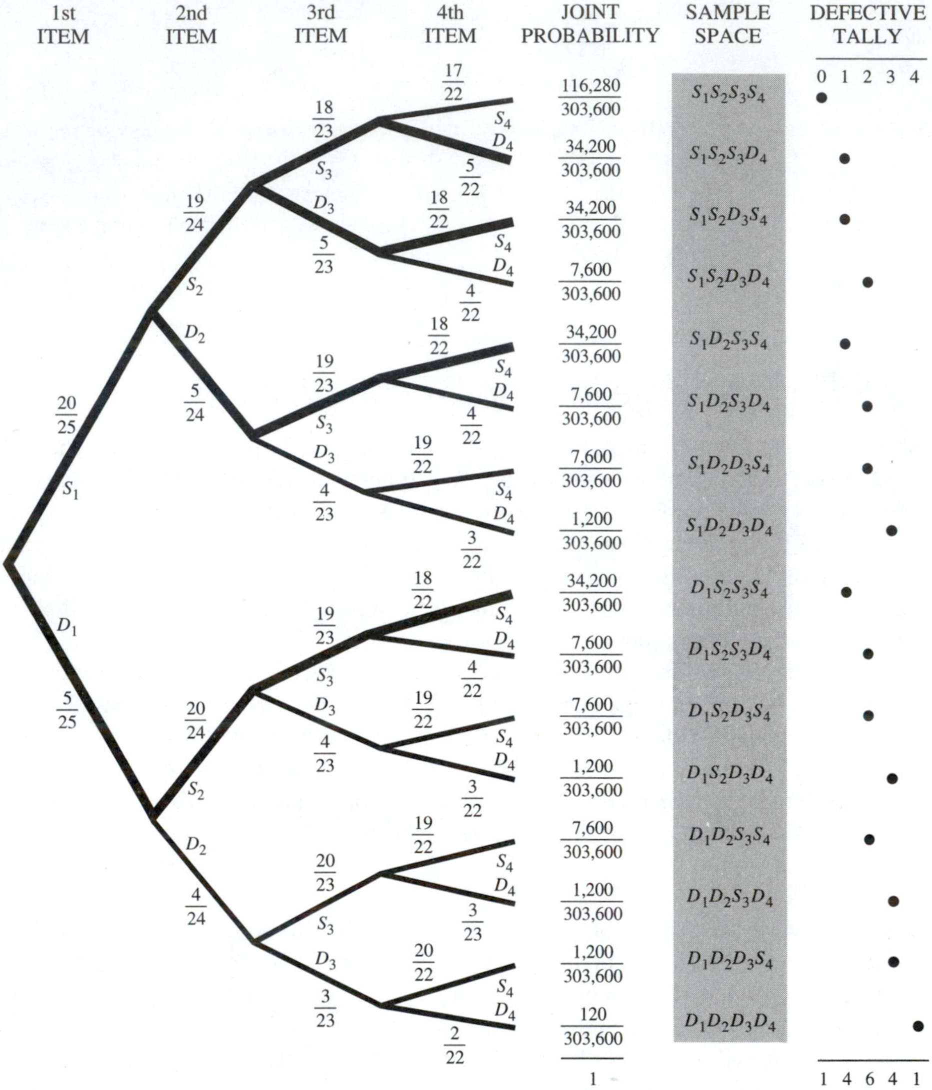

Consider the topmost of these paths, leading to $S_1 S_2 S_3 D_4$. The branches on this path each have a probability based on what item types were obtained in earlier selections. For the initial observation, 20 out of 25 items are satisfactory, and $\Pr[S_1] = \frac{20}{25}$. Given this event, only 19 satisfactories are left out of 24 remaining items to be selected for the second observation; thus, $\Pr[S_2 \mid S_1] = \frac{19}{24}$. Similarly, $\Pr[S_3 \mid S_1 \textit{ and } S_2] = \frac{18}{23}$, since 18 satisfactories will be available for the third observation taken from the 23 remaining items. This leaves 5 defectives in the final 22 items available for the fourth observation, and

$$\Pr[D_4 \mid S_1 \text{ and } S_2 \text{ and } S_3] = \frac{5}{22}$$

The multiplication law allows us to take the product of these branch probabilities in calculating the joint probability for the final outcome:

$$\Pr[S_1 S_2 S_3 D_4] = \frac{20}{25} \times \frac{19}{24} \times \frac{18}{23} \times \frac{5}{22} = \frac{34,200}{303,600}$$

An identical result is obtained for the three other 1-defective outcomes, so that

$$\Pr[R = 1] = \Pr[S_1 S_2 S_3 D_4] + \Pr[S_1 S_2 D_3 S_4] + \Pr[S_1 D_2 S_3 S_4] + \Pr[D_1 S_2 S_3 S_4]$$
$$= 4\left(\frac{34,200}{303,600}\right) = \frac{136,800}{303,600} = .45059$$

This same probability value may be reached by a somewhat different line of reasoning using event possibility counting methods.

Finding Hypergeometric Probabilities

The sample space in Figure 8–9 does not consider which particular items are selected each time. Envision a more detailed sample space that treats each transformer as a separate entity (identified by its serial number). The number of possible results in that detailed accounting equals the number of *combinations* of 4 specific items from the entire shipment of 25,

$$C_4^{25} = \frac{25!}{4!21!} = \frac{25 \times 24 \times 23 \times 22 \times 21!}{4! \times 21!} = \frac{303,600}{4!} = 12,650$$

We can readily determine how many of the above outcomes involve exactly 1 defective transformer. Consider first which one of the five defective transformers will be chosen for the sample; this is the number of ways of picking 1 object from 5, or C_1^5. The other three satisfactory items will be removed from the 20 satisfactories. The number of possibilities for that is C_3^{20}, the total combinations of 3 objects taken from 20. Applying the principle of multiplication, the number of elementary events in the expanded sample space that involve exactly 1 defective is the product of the two combinatorials:

$$C_1^5 C_3^{20} = \left(\frac{5!}{1!4!}\right)\left(\frac{20!}{3!17!}\right)$$
$$= \frac{5 \times 4! \times 20 \times 19 \times 18 \times 17!}{1! \times 4! \times 3! \times 17!} = \frac{5 \times 20 \times 19 \times 18}{3!}$$
$$= \frac{34,200}{3!} = 5,700$$

Counting and dividing, we have

$$\Pr[R = 1] = \frac{C_1^5 C_3^{20}}{C_4^{25}} = \frac{\frac{34,200}{3!}}{\frac{303,600}{4!}}$$
$$= 4\left(\frac{34,200}{303,600}\right) = \frac{136,800}{303,600} = .45059$$

which is the same value found by evaluating the probability tree diagram.

This evaluation suggests a general approach for finding similar probabilities for the number of successes R. Letting N represent the size of the population to be sampled, n the size of the sample, and π the proportion of successes in the population, we use the following expression.

PROBABILITY MASS FUNCTION FOR A HYPERGEOMETRIC DISTRIBUTION

$$h(r; n, \pi, N) = \Pr[R = r] = \frac{C_r^{\pi N} C_{n-r}^{(1-\pi)N}}{C_n^N} \qquad r = 0, 1, \ldots, n \text{ or } \pi N$$

Here the possibilities for the number of successes are all the nonnegative integers up through the smaller of n or πN.

Continuing with the transformer illustration, we have $N = 25$ items in the shipment population. A sample of $n = 4$ is selected for testing, and the proportion of defectives in the population is $\pi = .20$. We may use the above expression to evaluate the probability that exactly 2 defectives are found in testing.

$$\Pr[R = 2] = h(2; 4, .2, 25) = \frac{C_2^{.2(25)} C_{4-2}^{.8(25)}}{C_4^{25}}$$

$$= \frac{C_2^5 C_2^{20}}{C_4^{25}} = \frac{\left(\frac{5!}{2!3!}\right)\left(\frac{20!}{2!18!}\right)}{\frac{25!}{4!21!}}$$

$$= \frac{45{,}600}{303{,}600} = .15020$$

(You may verify this by duplicating the earlier probability tree evaluation for the case of $R = 2$.) The remaining probabilities are:

$$h(0) = \frac{C_0^5 C_4^{20}}{C_4^{25}} = \frac{116{,}280}{303{,}600} = .38300$$

$$h(1) = \frac{C_1^5 C_3^{20}}{C_4^{25}} = \frac{136{,}800}{303{,}600} = .45059$$

$$h(3) = \frac{C_3^5 C_1^{20}}{C_4^{25}} = \frac{4{,}800}{303{,}600} = .01581$$

$$h(4) = \frac{C_4^5 C_0^{20}}{C_4^{25}} = \frac{120}{303{,}600} = .00040$$

We may use the following expression for the

HYPERGEOMETRIC CUMULATIVE PROBABILITY DISTRIBUTION FUNCTION

$$F(r) = \Pr[R \le r] = \sum_{x=0}^{r} h(x; n, \pi, N)$$

For example, for the above illustration we compute

$$F(3) = \Pr[R \leq 3] = .38300 + .45059 + .15020 + .01581 \\ = .99960$$

Expected Value and Variance

As with all the probability distributions discussed, the expected value and variance of the number of successes R may be computed directly from the parameter values that apply. We use the following expressions.

EXPECTED VALUE AND VARIANCE FOR A HYPERGEOMETRIC DISTRIBUTION

$$E(R) = n\pi$$

$$\text{Var}(R) = n\pi(1 - \pi)\left(\frac{N - n}{N - 1}\right)$$

Notice that the expected value is identical to that for the binomial distribution. Thus, on the average, successes will occur in the same proportion in the sample as they do in the population. The following results apply to the earlier quality assurance illustration:

$$E(R) = 4(.2) = .8$$

$$\text{Var}(R) = 4(.2)(1 - .2)\left(\frac{25 - 4}{25 - 1}\right) = .56$$

Thus, only .8 defective is expected in the sample, and the variance is .56.

The expression for the variance is also similar to that for the binomial. It will be helpful to compare the variance expressions. Recall that

$$\text{Var}(R) = n\pi(1 - \pi) \qquad \text{(binomial)}$$

which differs from Var(R) for the hypergeometric distribution by the term $(N - n)/(N - 1)$. This quantity is called the **finite population correction factor**. Notice that when the sample size n is near the population size N, the factor may be considerably below 1. It indicates that a smaller variance applies for R when sampling without replacement from a finite population of fixed size N than when sampling from a continuous process wherein the binomial distribution is appropriate and there is no theoretical limit on population size. Using the parameters from the quality assurance illustration,

$$\frac{N - n}{N - 1} = \frac{25 - 4}{25 - 1} = .875$$

we see that the variance is only 87.5% as large as that for the binomial distribution with the same parameters, $4(.2)(1 - .2) = .64$.

The practical significance of the finite population correction arises when the hypergeometric distribution is being approximated by the normal distribution. We next consider the binomial approximation.

Binomial Approximation to Hypergeometric Distribution

Recall that the binomial distribution also provides probabilities for the number of successes in n trials. That distribution may be used in a probability evaluation for sampling directly from a continuous process, such as an assembly line, when the quality of

Table 8–1
Comparison of Binomial and Hypergeometric Probabilities for the Number of Defective Transformers (when $n = 4$, $\pi = .20$)

Possible Number of Defectives r	Approximate Binomial Probability $b(r; n, \pi)$	Exact Hypergeometric Probability $h(r; n, \pi, N)$ $N = 25$	$N = 100$
0	.4096	.38300	.40333
1	.4096	.45059	.41905
2	.1536	.15020	.15312
3	.0256	.01581	.02326
4	.0016	.00040	.00124

successive items can be assumed independent. It is also used when the population N is large in relation to the sample size n, although that is not exactly correct unless sampling is done with replacement. And, even when observed items are not replaced, the binomial distribution is still sometimes used. In this case, the binomial is used as an *approximation* to the hypergeometric distribution.

The reason for using the binomial distribution at all is that it is easier to compute values for $b(r; n, \pi)$ than for $h(r; n, \pi, N)$, which involves more factorials and has an additional parameter N. But with modern computational aids (even pocket computers) the advantages from the binomial approximation are not substantial. Users working with probability tables instead of computers will prefer approximating with binomial tables, which take far less space than hypergeometric tables would require.

How good is the binomial approximation? This depends on the settings of the parameters. Generally, the larger the population size N is in relation to the sample size n, the better the approximation becomes. Table 8–1 shows a comparison of the approximate binomial probabilities applicable to the earlier transformer quality assurance illustration. Notice that the binomial probabilities are considerably closer to the true ones when $N = 100$ instead of 25. The overall effect of the larger N is summarized by the finite population correction factor, $(N - n)/N - 1)$, which is .875 when $N = 25$ but .970 when $N = 100$. (When the factor is close to 1, the variance of R is nearly the same level with the binomial and hypergeometric distributions.) The levels for π and n will, to a lesser extent, affect the quality of the approximation.

Problems

8–38 A computer instructor gives his students a 100-line Visual Basic program. Altogether 10% of the lines have bugs. What is the probability that there will be exactly 5 erroneous lines in a group of 20 lines selected at random?

8–39 The freshman class at an engineering school contains 300 students. Ten percent of them have reserved space in the civil engineering option. What is the probability that there will be 10% or fewer civil engineering majors within the first 20 names from an alphabetical roster?

8–40 Determine for each of the following situations (1) the expected value for the number of successes, (2) the finite population correction factor, and (3) the variance:

	(a)	(b)	(c)	(d)	(e)
N	100	200	50	100	1,000
π	.05	.20	.50	.10	.10
n	10	25	10	5	100

8–41 A shipment contains 100 printed circuit boards. A sample of 10 boards will be tested. If 2 or fewer defectives are found, the shipment will be accepted. Assuming that 10% of the boards in the shipment are defective, find the probability it will be accepted.

8–42 Find the probability for getting a heart flush in a 5-card poker hand dealt from a complete shuffled deck of ordinary playing cards.

8–43 Repeat Problem 8–42, but consider the probability of exactly 3 queens.

8–44 A quality-control inspector accepts shipments whenever a sample of size 5 contains no defectives, and she rejects otherwise.

(a) Determine the probability that she will accept a poor shipment of 50 items in which 20% are defective.

(b) Determine the probability that she will reject a good shipment of 100 items in which only 2% are defective.

8–45 A professor selects a sample of 5 engineering students from the 25 in her class to state their preferences concerning course content. Each will indicate whether he or she wants more theory or more applications. It is assumed that 40% of the class wants more theory.

Determine the probabilities for all possible levels for the number of students R preferring more theory. (Use five places of accuracy.)

8–46 Determine the probability distribution for the number of defectives in a sample of 5 items taken from a shipment of 500 having 10% defective. (Use five places of accuracy.)

(a) Use exact hypergeometric probabilities.

(b) Use the binomial approximation.

Review Problems

8–47 A sample of 150 items is inspected from a population of 1,000. It is believed that 23% of the population is overweight. Assume that the sample items are selected randomly with replacement.

Determine (1) the expected number of overweights to be found, (2) the variance for that quantity, and (3) the standard deviation.

8–48 Repeat Problem 8–47, assuming that sampling is instead performed without replacement.

8–49 A project engineer is planning to simulate a PERT network for the installation of a data-processing system. She requires a probability distribution for the time taken to complete a government approval activity. This is an uncertain quantity of undetermined probability distribution. Regardless of the distribution form, the expected activity completion time has been established at 20 working hours. She wishes to find the probability that approval will take between 15 and 30 hours.

(a) Approval could be perfunctory—so that it is most likely to require practically no time at all—or it could stretch out considerably for a seemingly indefinite time. What probability distribution would then be appropriate? Use it to find the desired probability.

(b) The expected value and modal value might be the same, with values close to that figure more likely than more extreme approval times. Approximately 68% of the times fall within ± 4 hours of the central value. What probability distribution would then be appropriate? Use it to find the desired probability.

8–50 A parallel-component system is to be designed. The system will remain operational as long as there is at least one working component. Assume that each component has a 5% chance of remaining functional at the start of each day, that they cannot be replaced, and that no component is repairable. Assume also that the status of each component is independent of each preceding one and of what happens on the preceding day.

(a) Find the probability that a 2-component system will survive past 2 days.

(b) What daily survival probability must a single component have in order to provide the identical probability to that found in (a)?

8–51 One popular rule of thumb is to use the binomial approximation to the hypergeometric distribution only when the sample size is less than or equal to 10% of the population size. In each of the following situations where this rule allows the approximations, determine (1) the exact hypergeometric probability for $R = 2$ successes and (2) the approximate binomial probability.

(a)	(b)	(c)	(d)
$N = 100$	$N = 50$	$N = 200$	$N = 100$
$n = 10$	$n = 3$	$n = 10$	$n = 5$
$\pi = .50$	$\pi = .20$	$\pi = .05$	$\pi = .10$

8–52 A data clerk makes entries into a central data bank at a mean rate of 120 per hour. Suppose that 2.5% of all entries are in error.

(a) What distribution is appropriate for finding the probability of exactly 30 errors in 750 entries? Find that value.

(b) What distribution is appropriate for determining the probability that the clerk will make no errors in the next minute? Find that value.

8–53 The logic wafers in a large computer fail according to a Poisson process with mean rate of 1 every 150 hours.

(a) Determine the probability that the next wafer failure will occur between 90 and 180 hours after the last one.

(b) The average time between events to be experienced in 9 failures may be closely approximated by a normal distribution having the same expected value as the above and a standard deviation of 50 hours. Find the probability that this average falls between the same limits as in (a).

8–54 Power cell failures in a satellite have exponentially distributed times with a mean rate of $\lambda = .01$ per day. Presently only 2 cells remain functioning. They are arranged in parallel and have independent lives, so that the satellite may function as long as at least 1 power cell works.

(a) Find the probability that a particular cell will survive for 200 or more days.

(b) Find the probability that the satellite continues to function for 200 or more days.

CHAPTER NINE

Sampling Distributions

Modern statistics involves using known sample information for making *inferences* regarding populations whose true characteristics are unknown. But the procedures of inferential statistics rest on *deductive statistics*, which is concerned with a probabilistic evaluation of the sampling process. Deductive statistics treats the population as the known entity and the sample as the unknown.

The basic thrust of inferential statistics is drawing conclusions regarding the levels of population parameters, such as the mean μ and standard deviation σ. These conclusions can be based directly on the values of the counterpart sample statistics $\bar{X}$ and s. Before the sample data are in, $\bar{X}$, s, and other sample statistics are uncertain quantities; that is, each is a random variable having its own probability distribution. As a special class, the probability distributions for sample statistics are referred to as *sampling distributions*.

Our discussion of these begins with the sampling distribution for the sample mean $\bar{X}$, the most commonly used sample statistic.

9–1 The Sampling Distribution of the Mean

We illustrate the sampling distribution concept using the population in Table 9–1. These data represent the midterm examination results for the 5 students attending a senior metallurgy seminar at an engineering college. The population consists of the exam grade points, with an *A* corresponding to 4 points, a *B* to 3 points, and a *C* to 2 points. The mean and standard deviation of this population are $\mu = 2.8$ and $\sigma = .7483$.

A simple random sample of $n = 2$ student grade points will be selected. Although such a tiny sample is unusual, and statistical investigations ordinarily involve much larger populations, the miniature scale of this example will simplify our presentation. We will now establish the sampling distribution of the sample mean grade points.

Table 9–2 shows the probability distribution for the random variable $\bar{X}$. The probability values for this sampling distribution were found by identifying those combinations of students giving rise to each possible level for $\bar{X}$. For instance, $\bar{X} = 3.5$ when one of the selected students has an *A* (4 points) and the other a *B* (3 points). The following pairs provide this outcome:

(Dan, Gil) (Fran, Gil)

Table 9–1 *Grade-Point Population for Students in Metallurgy Seminar*

Name	Examination Grade	Grade Points
Dan	B	3
Ed	C	2
Fran	B	3
Gil	A	4
Ida	C	2
	$\mu = 2.8$	$\sigma = \sqrt{.56} = .7483$

Since there are $C_2^5 = 10$ equally likely combinations of student pairs, each having probability .1, it follows that $\Pr[\overline{X} = 3.5] = .2$.

Expected Value and Variance

Like any random variable, $\overline{X}$ has an expected value and variance. We have

$$E(\overline{X}) = \sum_{\text{all } \bar{x}} \bar{x} \Pr[\overline{X} = \bar{x}]$$

$$= 2.0(.1) + 2.5(.4) + 3.0(.3) + 3.5(.2) = 2.8$$

Notice that this value is equal to the mean of the population, $\mu = 2.8$. The sample mean for any random sample is expected to be equal to the population mean, so we have the following.

PROPERTY OF THE SAMPLE MEAN

$$E(\overline{X}) = \mu$$

The sample mean has expected value equal to the population mean. Thus, when μ is known, we never actually have to compute $E(\overline{X})$.

The variance of $\overline{X}$ may also be found. Using

$$\text{Var}(\overline{X}) = E(\overline{X}^2) - E(\overline{X})^2$$

Table 9–2 *Sampling Distribution for Sample Mean in Grade Point Values When Selection Is Done without Replacement*

Possible Mean $\bar{x}$	Applicable Student Combinations	$\Pr[\overline{X} = \bar{x}]$
2.0	(Ed, Ida)	.1
2.5	(Dan, Ed) (Dan, Ida) (Ed, Fran) (Fran, Ida)	.4
3.0	(Dan, Fran) (Ed, Gil) (Gil, Ida)	.3
3.5	(Dan, Gil) (Fran, Gil)	.2

we have

$$E(\overline{X}^2) = \sum_{\text{all } \bar{x}} \bar{x}^2 \Pr[\overline{X} = \bar{x}]$$

$$= (2.0)^2(.1) + (2.5)^2(.4) + (3.0)^2(.3) + (3.5)^2(.2) = 8.05$$

so that

$$\text{Var}(\overline{X}) = 8.05 - (2.8)^2 = .21$$

and the standard deviation of $\overline{X}$ is the square root of this:

$$SD(\overline{X}) = \sqrt{.21} = .458$$

Standard Error of $\overline{X}$

During the planning stage, a sample statistic must be treated as a random variable. Its standard deviation is usually referred to as the **standard error**. This terminology emphasizes that data have not yet been collected, so the statistic is an uncertain quantity that might assume a variety of values according to its sampling distribution. Probability limits may be placed on how far a statistic may fall above or below its expected level. These limits are established by a random variable's standard deviation, which thus serves as an index for expressing how much error a statistic may achieve in an attempt to hit its expected value target.

The standard error of $\overline{X}$ is so frequently encountered in statistics that a special symbol is used.

STANDARD ERROR OF THE SAMPLE MEAN

$$\sigma_{\overline{X}} = SD(\overline{X})$$

We have, for the student grade-point illustration,

$$\sigma_{\overline{X}} = .458$$

Just as we can find $E(\overline{X})$ from the level of the population mean μ, the value of $\sigma_{\overline{X}}$ may be obtained directly from the population standard deviation σ (when it is known). This can be a significant computational advantage.

There are two cases, depending on whether or not the sample observations are independent. Independence is guaranteed when sampling is done *with* replacement. When sampling is instead done *without* replacement (the usual case), successive observations will be statistically dependent. That dependence has slight effect on probability values, however, when the population is large in relation to the sample size. As a practical matter, the two cases are (1) *large population* (when, as a good approximation to reality, independence is assumed) versus (2) *small population* (when accuracy demands that the sample observations be treated as dependent). The following expression may be used to evaluate the latter case.

STANDARD ERROR OF $\overline{X}$ WHEN A SMALL POPULATION APPLIES

$$\sigma_{\overline{X}} = \frac{\sigma}{\sqrt{n}}\sqrt{\frac{N-n}{N-1}}$$

where σ is the population standard deviation, N is the population size, and n is the sample size. You may recognize the term under the last radical as the finite population correction factor encountered in Chapter 8.

Returning to our grade-point illustration, the population standard deviation is $\sigma = .7483$. With $N = 5$ and $n = 2$, we find

$$\sigma_{\overline{X}} = \frac{.7483}{\sqrt{2}}\sqrt{\frac{5-2}{5-1}} = .458$$

which is identical to the value computed earlier directly from the sampling distribution.

The finite population correction factor, $(N - n)/(N - 1)$, is a fraction that approaches 1 when the population size N is large in relation to the sample size n. In those cases we may use the following expression.

STANDARD ERROR OF $\overline{X}$ WHEN A LARGE POPULATION APPLIES

$$\sigma_{\overline{X}} = \frac{\sigma}{\sqrt{n}}$$

Notice that the standard error of $\overline{X}$ is inversely proportional to the square root of the sample size. This indicates that raising n will lower the value of $\sigma_{\overline{X}}$. The likely deviations in the values of $\overline{X}$ from μ will therefore be smaller. This very important concept is fundamental in statistical planning and is useful in selecting an appropriate sample size.

Of course, $\sigma_{\overline{X}}$ is equal to $\sigma/\sqrt{n}$ when actually sampling *with replacement*—regardless of the population size. In those cases, the sample observations are *independent* random variables.

Theoretical Justification for Results

When samples are taken at random from populations of large size or selected with replacement, each sample value is a random variable having an identical distribution to the rest, with values of successive observations being statistically independent events. The expected value of a single sample observation X equals the mean of the population from which it comes:

$$E(X) = \mu$$

The justification for this is easy, since the expected value for the sample observation is the long-run average result from repetitions of the sampling experiment. That value must equal the population mean. Likewise, the variance of a single sample observation equals the population variance:

$$\text{Var}(X) = \sigma^2$$

We distinguish each successive sample observation by using subscripts: $X_1, X_2, \ldots, X_n$. Although a detailed discussion is beyond the scope of this book, probability theory extends to higher dimensions, and there is a *joint* probability distribution applicable collectively to the combined sample result. The concept of expected value applies to joint probability distributions as well, with the sample mean considered as a function of the individual random variables. When the expected value concepts of Chapter 7 are extended to higher dimensions, one consequence is that *the expected values are additive* in the following way:

$$\begin{aligned} E(X_1 + X_2 + \cdots + X_n) &= E(X_1) + E(X_2) + \cdots + E(X_n) \\ &= \mu + \mu + \cdots + \mu \\ &= n\mu \end{aligned}$$

Since

$$\overline{X} = \frac{X_1 + X_2 + \ldots + X_n}{n} = \left(\frac{1}{n}\right)(X_1 + X_2 + \ldots + X_n)$$

it follows that

$$E(\overline{X}) = \left(\frac{1}{n}\right) E(X_1 + X_2 + \ldots + X_n)$$
$$= \left(\frac{1}{n}\right)(n\mu) = \mu$$

Additivity applies to the variance as well:

$$\begin{aligned}\mathrm{Var}(X_1 + X_2 + \cdots + X_n) &= \mathrm{Var}(X_1) + \mathrm{Var}(X_2) + \cdots + \mathrm{Var}(X_n)\\ &= \sigma^2 + \sigma^2 + \cdots + \sigma^2\\ &= n\sigma^2\end{aligned}$$

It follows that

$$\mathrm{Var}\,(n\overline{X}) = \mathrm{Var}(X_1 + X_2 + \cdots + X_n) = n\sigma^2$$

Using a property of the variance,

$$\mathrm{Var}(n\overline{X}) = n^2\mathrm{Var}(\overline{X})$$

it follows that

$$\mathrm{Var}(\overline{X}) = \left(\frac{1}{n^2}\right)(n\sigma^2) = \frac{\sigma^2}{n}$$

and, thus,

$$\sigma_{\overline{X}} = SD(\overline{X}) = \sqrt{\frac{\sigma^2}{n}} = \frac{\sigma}{\sqrt{n}}$$

Problems

9–1 Compute the standard error of $\overline{X}$ for the following situations where the population is large:

(a)	(b)	(c)	(d)
$\sigma = 50$ lb $n = 100$	$\sigma = \$16$ $n = 64$	$\sigma = .02$ m $n = 25$	$\sigma = 5$ kHz $n = 36$

9–2 Compute the standard error of $\overline{X}$ for the following situations, where the populations are small and selection is without replacement:

(a)	(b)	(c)	(d)
$\sigma = 5''$ $N = 500$ $n = 100$	$\sigma = .1'$ $N = 1{,}000$ $n = 100$	$\sigma = \$39$ $N = 1{,}000$ $n = 169$	$\sigma = 3$ grams $N = 200$ $n = 25$

9–3 The midterm examination scores for a population of 6 students are as follows:

Student	Score
T. C.	70
C. E.	80
V. L.	60
M. N.	90
N. S.	80
W. T.	70

(a) Compute μ and σ for the population.

(b) For a sample of size 2 selected without replacement, find $E(\overline{X})$ and $\sigma_{\overline{X}}$.

9–4 Construct a table providing the complete sampling distribution for $\overline{X}$ when $n = 2$ random observations are taken without replacement from the population in Problem 9–3.

9–5 Construct a table providing the complete sampling distribution for $\overline{X}$ when $n = 3$ random observations are taken without replacement from the population in Problem 9–3.

9–6 Consider again the student grade-point population in Table 9–1.

(a) Construct a table providing the complete sampling distribution for $\overline{X}$ when $n = 2$ and the selection is made *with* replacement.

(b) Compute from your table (1) $E(\overline{X})$ and (2) $\sigma_{\overline{X}}$.

9–7 Construct a table providing the complete sampling distribution for $\overline{X}$ when $n = 2$ random observations are taken *with* replacement from the population in Problem 9–3.

9–2 Sampling Distribution of $\overline{X}$ When Population Is Normal

We have seen how the sample mean $\overline{X}$ may be treated as a random variable. The sampling distribution for $\overline{X}$ provides probabilities for the possible levels of this variable. When the characteristics of a population are known, we may find that sampling distribution by exercising basic probability concepts. With a small population such as the student grade points in Section 9–1, this involves counting and grouping possible sample outcomes according to the possible level for $\overline{X}$. A different approach should be taken for large populations, however. In this section we consider sampling from a population whose frequency distribution is characterized by a normal curve. Our evaluation of a normally distributed population will provide the necessary background for finding sampling distributions under more general circumstances.

The Normal Distribution for $\overline{X}$

Property of Sample Mean

When the sample mean is computed for a random sample of size n taken from a *normally distributed population* having parameters μ and σ, the sampling distribution of $\overline{X}$ is *also* normally distributed with mean μ and standard deviation $\sigma_{\overline{X}} = \sigma/\sqrt{n}$.

As we have seen, many physical dimensions for both people and objects are closely represented by the normal distribution. In these cases, knowing μ and σ establishes the sampling distribution for the mean of any sample.

Illustration: *Reach and Hand Length*

As an illustration, consider again the anthropometric data for the reach of adult males first described on page 225. The population may be assumed to be normally distributed, and the following parameters apply:

$$\mu = 32.33''$$
$$\sigma = 1.63''$$

The top normal curve in Figure 9–1 represents the frequency distribution of this population.

Since the above parameters are known, it would be unnecessary for an investigator actually to collect sample data on reach. Samples are necessary only when the population has unknown characteristics. But we may use this population to illustrate the concept of the sampling distribution for $\overline{X}$.

Suppose that a random sample of $n = 100$ men will be selected and the reach measured for each. The sampling distribution for the sample mean is normally distributed, with the same mean as the population's, $\mu = 32.33''$, and standard deviation

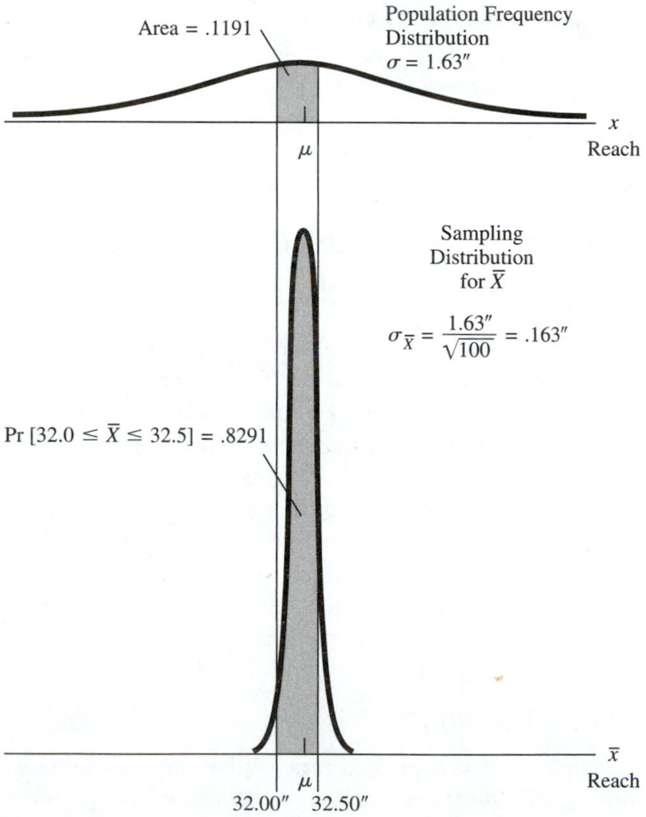

Figure 9–1 *Population frequency curve and sampling distribution for sample mean reach of adult males.*

$$\sigma_{\overline{X}} = \frac{\sigma}{\sqrt{n}} = \frac{1.63''}{\sqrt{100}} = .163''$$

Notice that this standard deviation is only one-tenth as large as that of the population itself. The sampling distribution for $\overline{X}$ exhibits less dispersion than the population's. This is reflected in the normal curve for $\overline{X}$ at the bottom of Figure 9–1, which has its mass concentrated more tightly about μ.

The shaded areas under each curve further illustrate this concept. Each area represents the frequency of reaches between 32.00″ and 32.50″. The proportion of men in the population whose reaches fall within this range is

$$\text{Area} = \Phi\left(\frac{32.50 - 32.33}{1.63}\right) - \Phi\left(\frac{32.00 - 32.33}{1.63}\right)$$
$$= .5398 - .4207 = .1191$$

But the probability is much higher that the sample mean reach of 100 men will fall inside the same limits:

$$\Pr[32.00'' \leq \overline{X} \leq 32.50''] = \Phi\left(\frac{32.50 - 32.33}{.163}\right) - \Phi\left(\frac{32.00 - 32.33}{.163}\right)$$
$$= .8508 - .0217 = .8291$$

One helpful interpretation of this fact is to imagine several different samples, each of size n, taken from the same population. The respective computed values for $\overline{X}$ will tend to be alike, and their average should be very near to μ. The sample means will be much closer together in value than would a similar collection of individual population values. (We know this because $\sigma_{\overline{X}}$ is smaller than σ.)

The Role of the Standard Error

As we have seen, the standard error of $\overline{X}$ determines how closely to μ the computed sample mean will be likely to fall. A sample involving a small level for $\sigma_{\overline{X}}$ should provide a computed $\overline{X}$ that more faithfully represents the population center than would a sample where $\sigma_{\overline{X}}$ is large. In other words, sampling error decreases as $\sigma_{\overline{X}}$ does.

For normally distributed populations there are two determinants of the level for $\sigma_{\overline{X}}$. These are the population standard deviation σ and the sample size n. (The population size is not a factor, since N is theoretically infinite with normal populations.)

Consider first the effect of σ on the level of the standard error. Since $\sigma_{\overline{X}}$ is equal to $\sigma/\sqrt{n}$, the standard error of $\overline{X}$ is directly proportional to the population standard deviation. In absolute terms, the amount of error in the computed value for $\overline{X}$, relative to the true population mean, will be greater when sampling from a population having a large standard deviation than from one where σ is small.

For example, anthropometric data indicate that adult males have a mean hand length of $\mu = 7.49''$, with a standard deviation of $\sigma = .34''$. The standard deviation for this population is smaller than the earlier one for reach, where $\sigma = 1.63''$. The respective values of $\sigma_{\overline{X}}$ will reflect this difference in σ size. Suppose that a random sample of $n = 25$ men is measured both for hand length and for reach. The respective standard errors are

$$\sigma_{\bar{X}} = \frac{.34''}{\sqrt{25}} = .068'' \qquad \text{(hand length)}$$

$$\sigma_{\bar{X}} = \frac{1.63''}{\sqrt{25}} = .326'' \qquad \text{(reach)}$$

Figure 9–2 shows the respective normal curves. Notice that it is virtually certain that $\bar{X}$ (hand length) will fall within $\mu \pm .25''$, but that there is less than a 56% chance for $\bar{X}$ (reach) falling within .25″ of the mean of the reach population. In absolute terms, the smaller σ for hand length leads to less sampling error, with a computed $\bar{X}$ that is more likely to be close to its μ target (in absolute terms).

Figure 9–2
Normal curves for adult male anthropometric data.

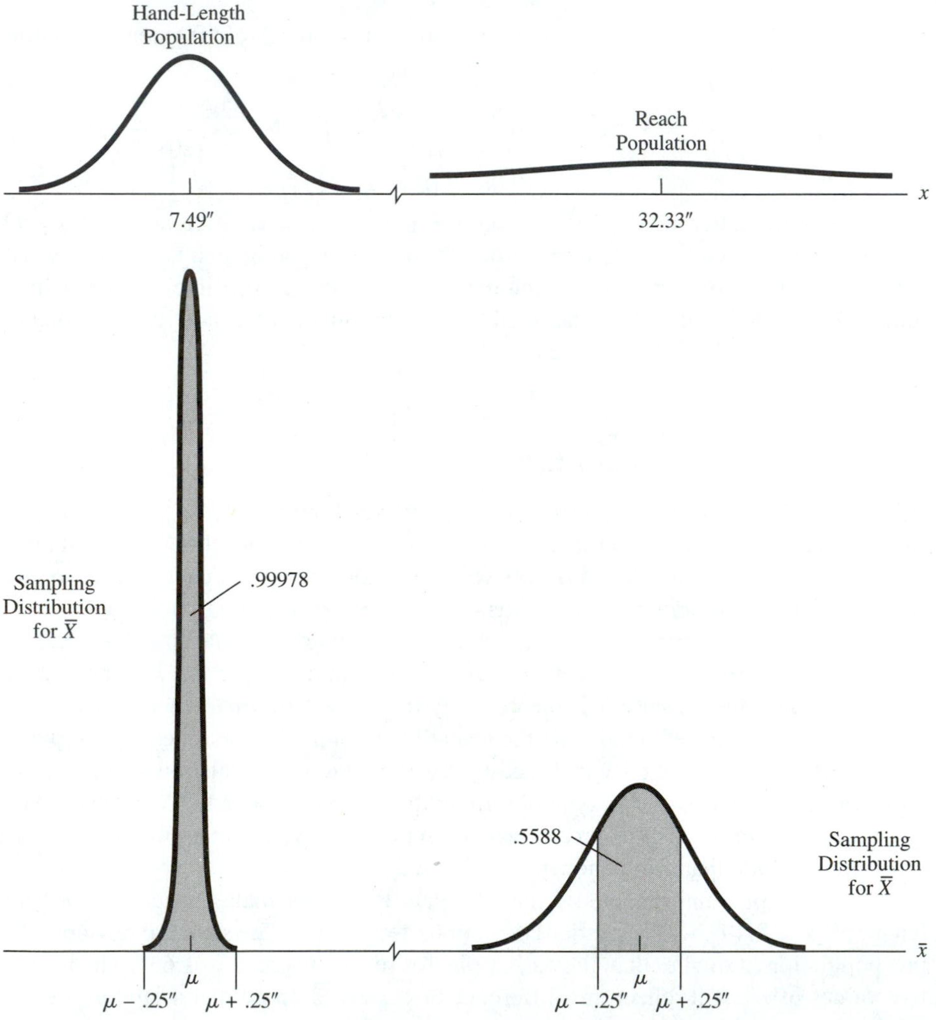

The hand-length population exhibits less variability than the reach population. This is reflected by the smaller standard error of $\bar{x}$ for a sample of men's hand lengths and the tighter clustering of possible values for that mean.

Example: *Distortion Frequencies in Radiotelescopes*

An engineer is evaluating the design for a new radiotelescope amplifier. Prototype testing will determine whether the new device can operate distortion free at higher frequencies than the present unit, where the mean maximum distortion frequency is 50 kHz and the standard deviation is 5 kHz.

The engineer is, of course, uncertain about μ, the mean maximum distortion frequency of the proposed unit. He will recommend his new design only if the sample counterpart figure from $n = 100$ test shots falls above 50 kHz. Assuming that the new amplifier is clearly superior and the true level for μ is 51 kHz, what is the probability that the computed $\overline{X}$ will be so atypically small that the design will fail this test?

The probability will depend on σ, which, like μ, is unknown. Suppose $\sigma = 5$ kHz (the same as on the present system). Suppose also that the population of maximum distortion frequencies on test shots is normally distributed. Then

$$\sigma_{\overline{X}} = \frac{5}{\sqrt{100}} = .5 \text{ kHz}$$

and

$$\Pr[\overline{X} \leq 50 \mid \mu = 51 \text{ kHz}] = \Phi\left(\frac{50 - 51}{.5}\right) = .0228$$

Should the population standard deviation be greater, say $\sigma = 10$ kHz, then the probability that the amplifier design will fail the test (even though truly better) is greater:

$$\sigma_{\overline{X}} = \frac{10}{\sqrt{100}} = 1.0 \text{ kHz}$$

$$\Pr[\overline{X} \leq 50 \mid \mu = 51 \text{ kHz}] = \Phi\left(\frac{50 - 51}{1}\right) = .1587$$

A larger σ gives rise to a larger level for $\sigma_{\overline{X}}$. This increases the chance of obtaining an erroneous sample result.

The second influence on $\sigma_{\overline{X}}$ is the sample size. The standard error of $\overline{X}$ is inversely proportional to the square root of n. Greater fidelity in $\overline{X}$ will therefore be achieved as n becomes larger (which makes $\sigma_{\overline{X}}$ smaller).

Figure 9–3 shows the sampling distributions for mean reach $\overline{X}$ when $n = 25$ and when $n = 100$. Notice that the normal curve for the larger sample size (and smaller standard error) exibits less dispersion, with possible levels of $\overline{X}$ clustering more tightly

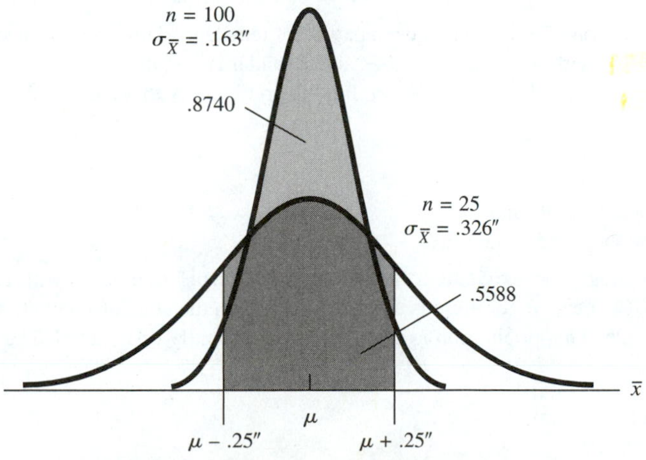

Figure 9–3 *Sampling distributions for mean reach $\overline{X}$ when $n = 25$ and $n = 100$.*

about μ. When $n = 100$, the probability that $\overline{X}$ falls within $\mu \pm .25''$ is .8740, compared with only .5588 for the like event when $n = 25$.

Example: *Probabilities for Mean Task Completion Times*

An industrial engineer wishes to estimate the mean time taken to complete an assembly task. She believes that the standard deviation for normally distributed individual completion times is $\sigma = 25$ seconds, a value obtained from earlier experiments with similar tasks. She plans to make $n = 30$ sample observations. The standard deviation for the random variable $\overline{X}$ is

$$\sigma_{\overline{X}} = \frac{\sigma}{\sqrt{n}} = \frac{25}{\sqrt{30}} = 4.56 \text{ seconds}$$

There is probability $\Phi(z) - \Phi(-z)$ that $\overline{X}$ will fall within the interval $\overline{X} = \mu \pm z\sigma_{\overline{X}} = \mu \pm 4.56z$. Thus, when $z = 2$,

$$\Pr[\overline{X} = \mu \pm 4.56(2) = \mu \pm 9.12] = \Phi(2) - \Phi(-2)$$
$$= .9772 - .0228 = .9554$$

From these calculations, the engineer can see that there is about a 95% chance that her sample result will be within around 9 seconds of the true level for the population mean.

An even larger sample size of $n = 100$ would improve the accuracy. Using the same standard deviation,

$$\sigma_{\overline{X}} = \frac{\sigma}{\sqrt{n}} = \frac{25}{\sqrt{100}} = 2.50 \text{ seconds}$$

There is a .9554 probability that the computed value for $\overline{X}$ will lie within $\mu \pm 2(2.50)$, an accuracy of ± 5 seconds.

Problems

9–8 The height of adult males in a particular city is normally distributed, with $\mu = 69.5''$ and standard deviation $\sigma = 2.65''$. Find (1) the standard error of $\overline{X}$ and (2) the probability that $\overline{X}$ falls within $\mu \pm .5''$ when a random sample is taken of size
(a) $n = 10$ (b) $n = 25$ (c) $n = 50$ (d) $n = 100$

9–9 The operator reaction time to simulated nuclear-reactor cooling-system failure events is believed to be normally distributed, with a mean $\mu = 45$ seconds and standard deviation $\sigma = 8$ seconds. Find (1) the standard error of $\overline{X}$ and (2) the probability that $\overline{X}$ falls within $\mu \pm 1$ second when a random sample is taken of size
(a) $n = 25$ (b) $n = 100$ (c) $n = 200$ (d) $n = 500$

Table D

9–10 The elongation of a steel bar under a particular tensile load may be assumed to be normally distributed, with a mean $\mu = .06''$ and standard deviation $\sigma = .008''$. A sample of $n = 100$ bars is subjected to the test. Find the probability that the sample mean elongation falls
(a) above .062″
(b) below .0505″
(c) between .059″ and .061″
(d) between .0585″ and .0605″

9–11 A sufonating process yields a final output whose pH fluctuates with each successive barrel. The mean μ for these is unknown, although the standard deviation is assumed to be $\sigma = .25$. This population is assumed to be normally distributed. The chief chemical engineer orders replacement of the control valves whenever the sample mean pH of $n = 25$ barrels falls below 6.90 or above 7.10.

(a) Calculate the standard error of $\overline{X}$.

(b) Determine the probability

(1) of replacing the valves when the process is operating as intended and $\mu = 7.0$

(2) of replacing the valves when the process is slightly off target and the mean pH is $\mu = 7.02$

(3) of failing to replace the valves when the process is too alkaline and the mean pH is $\mu = 7.15$

(4) of failing to replace the valves when the process is too acidic and the mean pH is $\mu = 6.80$

9–12 A ceramics engineer will be testing two epoxy substances to determine the average setting time until bonding. The population of individual setting times is believed to be normally distributed with some unknown mean μ and standard deviation σ. The test involves $n = 25$ trials. Find the probability that the sample mean setting time deviates from the unknown μ by no more than 1 hour, assuming that the standard deviation (in hours) has value

(a) $\sigma = 2$ (b) $\sigma = 4$ (c) $\sigma = 8$ (d) $\sigma = 16$

9–13 Consider again the radiotelescope design example. Suppose that the new design is actually inferior, so that the mean maximum distortion frequency is 49 kHz. The same decision criterion will be applied. Find the probability that the new design will be correctly not recommended, assuming

(a) $\sigma = 6$ kHz (b) $\sigma = 7$ kHz (c) $\sigma = 10$ kHz

9–14 Refer to the assembly time example on page 270.

(a) Find the probability when $n = 25$ that the sample mean $\overline{X}$ will fall within the following limits:

(1) $\mu \pm 1.5\sigma_{\overline{X}}$ (2) $\mu \pm 2.5\sigma_{\overline{X}}$ (3) $\mu \pm 3\sigma_{\overline{X}}$ (4) $\mu \pm 4\sigma_{\overline{X}}$

(b) What can you conclude about the accuracy of using $\overline{X}$ to estimate μ and the probability that the resulting estimate will be truthful?

9–15 Refer to Problem 9-14.

(a) Find the level of accuracy (with $z = 2$) obtained when the sample size is: (1) $n = 10$, (2) $n = 50$, (3) $n = 200$, (4) $n = 1{,}000$.

(b) What can you conclude about the accuracy of using $\overline{X}$ to estimate μ and the size of the sample?

9–16 Refer to Problem 9-14.

(a) Find the level of accuracy (with $z = 2$ and $n = 100$) assuming that the standard deviation in assembly times is (1) $\sigma = 5$ seconds, (2) $\sigma = 10$, (3) $\sigma = 15$, (4) $\sigma = 30$.

(b) What can you conclude about the accuracy of using $\overline{X}$ to estimate μ and the level of the population standard deviation?

9–3 Sampling Distribution of $\overline{X}$ for a General Population

One reason the normal distribution is important in statistics is that so many population frequency distributions may be represented by normal curves. But keep in mind that the normal curve is a theoretical construct that provides a convenient approximation to reality. (Remember, it assigns some probability to the mile-tall man!)

Figure 9–4
Illustration of the central limit theorem, showing the tendency toward normality in the sampling distribution of $\overline{X}$ as n increases, for various populations.

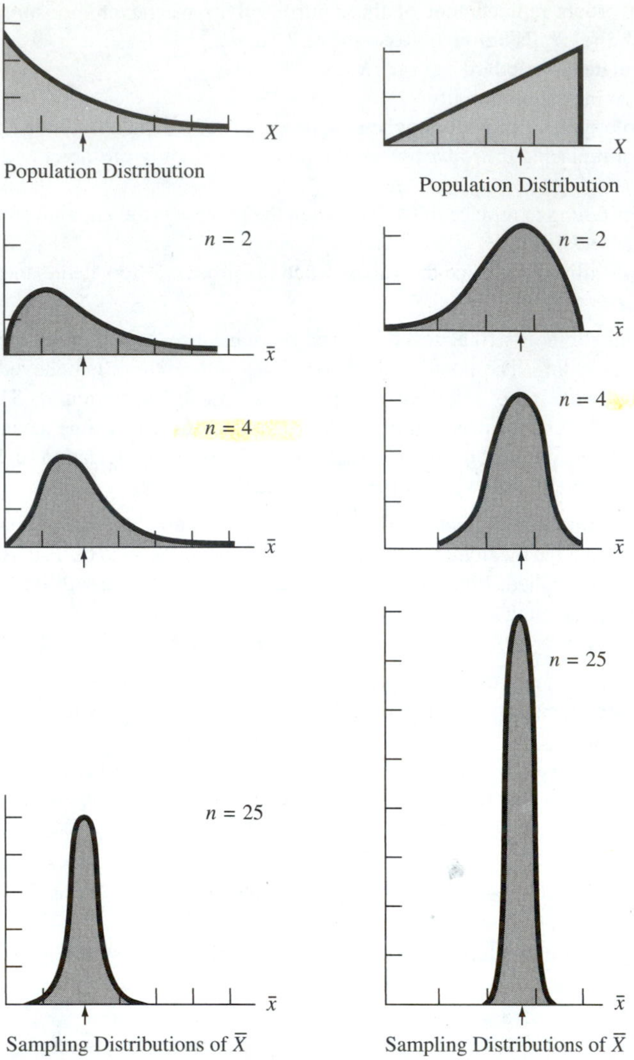

Source: Lawrence L. Lapin, *Statistics for Modern Business Decisions*, 6th ed. © 1993 Dryden Press, Ft. Worth, Texas.

But what makes the normal distribution crucial to statistics is the role it plays in representing the sampling distribution for $\overline{X}$. We have just seen that when the population is normally distributed, then the sampling distribution for $\overline{X}$ must *also* be a normal distribution (with the same mean, but with standard deviation $\sigma_{\overline{X}}$). A surprising fact is that this same property can apply approximately even when the population is *not* normally distributed. This fact is established by the **central limit theorem**.

Central Limit Theorem

Because so many statistical procedures are based on the sample mean, the following theorem exhibits one of the most useful properties in statistics.

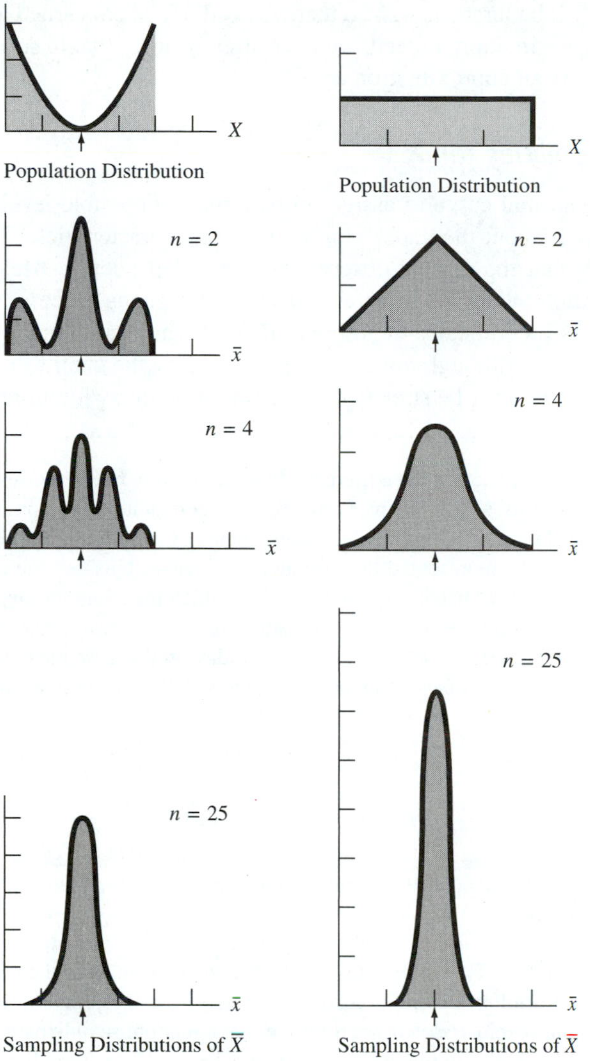

Figure 9–4
(continued)

Central Limit Theorem

Consider a population having mean μ and finite standard deviation σ. Let $\overline{X}$ represent the mean of n independent random observations from this population. Then, as n becomes large, the sampling distribution of $\overline{X}$ tends toward a normal distribution with mean μ and standard deviation $\sigma_{\overline{X}} = \sigma/\sqrt{n}$.

This theorem permits us to *approximate* the sampling distribution for $\overline{X}$ by an appropriate normal curve—*regardless of the form of the population frequency distribution.*

Figure 9–4 illustrates this concept for 4 underlying populations having frequency curves of radically different shapes. The exact sampling distributions for $\overline{X}$ are pictured in each case for 3 sample sizes. When $n = 2$, the shapes of the respective sampling distributions differ considerably. They become rounder and more symmetrical at $n = 4$, although each case still has a distinctive shape. But from all populations, a smooth, nearly symmetrical bell-shaped curve applies for $\overline{X}$ when $n = 25$. In most statistical applications the sample size will be large enough so that we need not be concerned about the quality of the normal approximation. Indeed, we have already noted that in some sense the normal curve is *always* an approximation anyway.

Finding Probabilities for $\bar{X}$

We may use the normal curve to assign probabilities to possible levels for $\bar{X}$ without worrying too much about the underlying population characteristics.* The central limit theorem requires that the sample observations be independent—a feature guaranteed when the population values are generated from a continuous process (so that the sampled population is theoretically of infinite size) or when sampling with replacement from any population. Although not strictly permitted by the central limit theorem, the normal approximation may be stretched to include sampling without replacement.

Example:
Mean Time between Failures

Electronic components often have lifetimes that empirically have been shown to fit an exponential distribution (that is, over time, failures constitute a Poisson process). The underlying population of lifetimes of all components of a particular type being manufactured may then be represented by a frequency curve such as the one at the top left corner in Figure 9–4. Consider one such item for which the mean failure rate λ is unknown. Although it might take on many possible values, the mean rate can be assumed to be a particular value, such as $\lambda = .01$ item per day. In that case the expected time X until an individual item fails is given below, along with the standard deviation:

$$E(X) = \frac{1}{\lambda} = 100 \text{ days}$$

$$SD(X) = \frac{1}{\lambda} = 100 \text{ days}$$

These values may be taken as the population parameters for the lifetimes of all similar items: $\mu = 100$ and $\sigma = 100$ days.

Sample lifetimes might be used to establish μ from the computed mean time between failures, or MTBF, $\bar{X}$. How close will such an estimate lie to the true μ?

From the sampling distribution for $\bar{X}$ we may find the probabilities of events, such as $\bar{X}$ falling within $\mu \pm 5$ days. Suppose that a random sample of $n = 100$ items is to be used and that we continue to assume that $\lambda = .01$. The central limit theorem indicates that the sampling distribution for $\bar{X}$ may be approximated by a normal curve with mean $\mu = 100$ days and standard deviation $\sigma_{\bar{X}} = 100/\sqrt{100} = 10$ days. Thus,

$$\begin{aligned}\Pr[95 \leq \bar{X} \leq 105] &= \Phi\left(\frac{105-100}{10}\right) - \Phi\left(\frac{95-100}{10}\right)\\ &= .6915 - .3085\\ &= .3830\end{aligned}$$

A different probability would apply if the true failure rate were assumed to be some other value, say $\lambda = .05$, so that $\mu = \sigma = 1/.05 = 20$. Then $\sigma_{\bar{X}} = 20/\sqrt{100} = 2$ days, and

$$\begin{aligned}\Pr[15 \leq \bar{X} \leq 25] &= \Phi\left(\frac{25-20}{2}\right) - \Phi\left(\frac{15-20}{2}\right)\\ &= .9938 - .0062\\ &= .9876\end{aligned}$$

---

*A finite variance is assumed to apply.

This example shows that to establish probabilities for $\overline{X}$, we must often make detailed assumptions about the underlying population. If those assumptions are incorrect, the sampling distribution will reflect those errors. The exponential population distribution used in the following example is particularly volatile in this respect because it is fully described by a single parameter λ (so that $\mu = \sigma = 1/\lambda$), and the mean and standard deviation are always the same value.

Finding Probabilities with an Assumed σ

In more general cases, we may precisely compute the probability that $\overline{X}$ falls within μ plus or minus some error by knowing σ only. (Thus, there is no need to make possibly erroneous assumptions regarding μ or the shape of the population frequency curve.) Although the true value of σ is also generally unknown when μ is uncertain, a "ballpark" level for σ may be easily found from a related population for which sample data have already been collected.

For example, consider using $\overline{X}$ to estimate the mean instep length of adult males. Suppose that anthropometric data are available only for above-the-waist measurements. Since instep and hand lengths are nearly the same for most people, we might assume that the two populations also have similar variability. During the planning phase for an anthropometric study involving below-the-waist measurements, we might then use the value $\sigma = .34''$, which is known to apply for hand length, to represent the standard deviation for instep length.

Example: *Starting Salaries of New Petroleum Engineers*

The editor of a professional engineering journal wishes to estimate the mean starting salaries of newly graduated engineers in various specialties. Because of that field's growth, petroleum engineers will be categorized separately for the first time. (In previous years, they had been included with chemical engineers or lumped together with mineral engineers.)

Although a large-sample salary database will be compiled and refined from university and industry sources, that will take a year to complete. A more current (and smaller) sample of starting salaries is needed for immediate publication—and some sampling error must therefore be tolerated. A sample of $n = 25$ petroleum seniors has been selected, and in June a reporter will contact each to determine his or her starting salary. From these responses, the mean starting salary $\overline{X}$ will be obtained. The editor wishes to establish the probability that the sample mean will fall within ±$500 of the true population mean salary.

Historically, salaries have exhibited a positively *skewed* distribution so that the frequency curves have presented thin upper tails extending tens of thousands of dollars above the mean (and short lower tails touching only thousands to the left of μ). The exact shape of the frequency curve for petroleum engineers' salaries would be similar. The central limit theorem indicates that $\overline{X}$ will have a normal curve regardless, and that the shape of this sampling distribution depends only on the population σ. The standard deviation for petroleum engineers' salaries is of course also unknown, but the editor believes that it will be close to last year's value of $1,057 for the broad mineral-engineering category. For planning purposes a value of $1,100 is used.

Under this assumption, the standard error of $\bar{X}$ is $\sigma_{\bar{X}} = \$1{,}100/\sqrt{25} = \$220$, and the desired probability is

$$\begin{aligned}\Pr[\mu - \$500 \le \bar{X} \le \mu + \$500] &= \Phi\left(\frac{\$500}{\$220}\right) - \Phi\left(\frac{-\$500}{\$220}\right)\\ &= .9884 - .0116\\ &= .9768\end{aligned}$$

The editor is satisfied with the high reliability that this indicates will be achieved from his sample of petroleum engineers.

Sampling from Small Populations

As noted, the sampling distribution of $\bar{X}$ may be closely approximated by the normal curve even when sampling without replacement from a finite population. In those cases, the following expression applies for the standard error:

$$\sigma_{\bar{X}} = \frac{\sigma}{\sqrt{n}}\sqrt{\frac{N-n}{N-1}}$$

which involves the finite population correction. The latter may usually be ignored when N is large. As a practical rule of thumb, *the finite population correction should be used whenever* n *exceeds 10% of the population size*.

For example, 183 students will be granted bachelor's degrees at a particular engineering school. Of these, only $N = 154$ will enter the job market. A sample of $n = 25$ of these graduates will be polled to determine the starting salaries of the entire class. Assume the population standard deviation is $\bar{X}$ is $\sigma = \$900$. The standard error of $\bar{X}$ is

$$\sigma_{\bar{X}} = \frac{\$900}{\sqrt{25}}\sqrt{\frac{154-25}{154-1}} = \$165.28$$

The probability that $\bar{X}$ falls within \$500 of μ is

$$\begin{aligned}\Pr[\mu - \$500 \le \bar{X} \le \mu + \$500] &= \Phi\left(\frac{\$500}{\$165.28}\right) - \Phi\left(\frac{-\$500}{\$165.28}\right)\\ &= .9988 - .0012\\ &= .9976\end{aligned}$$

Although the size N of the population and how it relates to the sample size n is important, keep in mind that N is not within the investigator's control. Only the size for n can be chosen. And, since its level will affect the quality of the sample results, n must be chosen carefully. As we shall see, larger n's are more reliable than smaller n's—no matter what N happens to be.

Problems

9–17 Suppose that the failure rate of the electrical components in the example on page 274 is $\lambda = .02$ per day. Find the probability that from a sample size of $n = 25$, the computed $\bar{X}$ will fall

(a) above 55 days
(b) below 40 days
(c) between 45 and 60 days
(d) either below 35 days or above 65 days

9–18 Suppose that the standard deviation applicable to the petroleum engineers' salaries in the text example should be larger, owing to increasing worldwide oil exploration activity. Using $n = 25$, find the probability that $\bar{X}$ falls within $\mu \pm \$500$ when
(a) $\sigma = \$1{,}500$ (b) $\sigma = \$2{,}000$ (c) $\sigma = \$2{,}500$

9–19 The time that each plane waits for service at an aircraft repair facility may be assumed to be exponentially distributed. A sample of 25 aircraft waiting times will be collected. Assuming that planes are removed from the holding area at a mean rate of 2 per day, determine the probability that the sample mean waiting time
(a) does not exceed .6 day (c) equals $.5 \pm .1$ day
(b) is greater than .7 day (d) is between .3 and .7 day

9–20 A computer generates pseudorandom decimals. Each decimal has an expected value of .5 and a standard deviation of $1/\sqrt{12}$. Consider a particular sequence of 12 numbers.
(a) Determine the standard error for the mean of those 12 random decimals.
(b) Find the probability that the mean of the 12 random decimals is
(1) $\leq .4$ (2) $> .65$ (3) $< .35$ (4) $\leq .75$

9–21 An engineer wishes to evaluate a high-frequency sonar transducer. The maximum effective echo-ranging distance at a particular frequency should be several miles. The mean effective distance is unknown, although the same standard deviation may be assumed as that of an existing transducer, for which $\sigma = 5$ nautical miles (NM).

An experiment will be conducted with n shots at a series of buoys spaced at various depths and distances throughout a test range. Find the probability that the sample mean maximum effective distance will lie within ± 1 NM of the unknown population mean when the sample size is
(a) 25 (b) 64 (c) 100 (d) 625

9–22 A quality-control inspector accepts shipments of 500 precision $\frac{1}{2}''$ steel rods if the mean diameter of a sample of $n = 100$ falls between $.4995''$ and $.5005''$. Previous evaluations have established that the standard deviation for individual rod diameters is $.003''$.
(a) What is the probability that the inspector will accept an out-of-tolerance shipment having $\mu = .5003''$?
(b) What is the probability that the inspector will reject a near-perfect shipment having $\mu = .4999''$?

9–4 The Student t Distribution

So far our discussion of sampling distributions has focused on $\bar{X}$, which is proper because that statistic is the one most commonly used. When the sample size is large enough, $\bar{X}$ may be represented by a normal curve with mean μ and standard deviation $\sigma_{\bar{X}}$, and this is ordinarily true regardless of the underlying population frequency distribution. Should the population itself be normally distributed, this relation is exact no matter how small n is. In either case, probabilities for $\bar{X}$ may be obtained from areas under the standard normal curve, using the following transformation into the standard normal random variable:

$$Z = \frac{\bar{X} - \mu}{\sigma_{\bar{X}}} = \frac{\bar{X} - \mu}{\sigma/\sqrt{n}}$$

In this section we are still concerned with $\bar{X}$, but now we want to find the sampling distribution for $\bar{X}$ by estimating σ from the sample results (rather than using an assumed value for σ, as we did before). In order to do this, we should use the Student t distribution instead of the normal distribution.

The Student t Statistic

In many statistical applications, the population parameters are estimated from their sample counterparts: $\overline{X}$ for μ and s for σ. This suggests a way around our lack of knowledge about σ—substitute s for σ in the previous expression for Z. Doing this, we obtain

$$\frac{\overline{X} - \mu}{s/\sqrt{n}}$$

Before data are collected, $\overline{X}$ and s are random variables, so this expression is a composite random variable that is itself a function of those two quantities.

Can we use this new random variable to assign probabilities in the same fashion that we used Z?

This question was answered by the pioneer statistician W.S. Gosset, who published his results under the pen name "Student." (He used a pseudonym because he worked for a brewery that prohibited employees from publishing their findings.) So that his new random variable would not be confused with Z, Gosset used the letter t, and in his honor, the above quantity is referred to as the **Student *t* statistic**.

STUDENT *t* STATISTIC

$$t = \frac{\overline{X} - \mu}{s/\sqrt{n}}$$

The Student t is a continuous random variable whose probability distribution is completely specified by a single parameter, referred to as **the number of degrees of freedom** (df). For the Student t distribution this parameter is determined by the sample size, and here df $= n - 1$. The density curve when df $= 4$ is shown in Figure 9–5.

The degrees of freedom are used by statisticians to indicate the minimal number of free terms in a collection of variables that must sum to a fixed total. The Student t statistic is computed from n sample observations $X_1, X_2, \ldots, X_n$. However, when the level of $\overline{X}$ is fixed, there is freedom to specify the values of $n - 1$ individual X_k's, and the final X_k must balance them all so that $\sum X_k/n = \overline{X}$. In other words, there are only $n - 1$ degrees of freedom for computing $\overline{X}$ (and also s), and hence t.

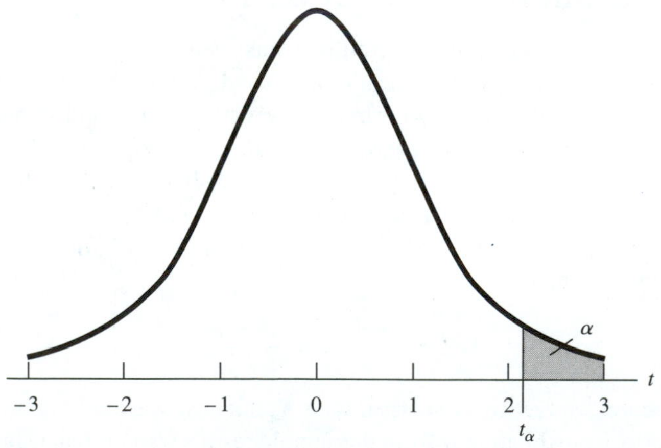

Figure 9–5 *Density curve for the Student t distribution for 4 degrees of freedom.*

The probability distribution for t was derived by assuming that each sample observation is selected randomly from a *normally distributed population* having some mean μ and finite standard deviation σ (although the values of these parameters need not be specified). This is a very restrictive condition that narrows the use of the Student t distribution to situations in which the sampled population has a frequency distribution that resembles a bell-shaped normal curve and is nearly symmetrical. Despite this limitation, the Student t distribution may be used in a wide variety of statistical applications.

The algebraic expression for the Student t density is cumbersome, and areas under the curve must be found through numerical integration. Appendix Table G allows us to evaluate upper-tail probabilities,

$$\alpha = \Pr[t > t_\alpha]$$

For any one of several levels of α, a **critical value** t_α may be read from the table. For example, when df = 4 and the upper-tail probability is $\alpha = .05$, the critical value is read from Table G as $t_{.05} = 2.132$, and thus

$$.05 = \Pr[t > 2.132] \quad (\text{df} = 4)$$

Similarly, when the upper-tail probability is specified at $\alpha = .01$, the tabled critical value is $t_{.01} = 3.747$, and

$$.01 = \Pr[t > 3.747] \quad (\text{df} = 4)$$

Appendix Table G has a strange layout compared with the familiar table for areas under the normal curve (Table D). This is partly a space-saving measure, since a separate probability distribution applies for each number of degrees of freedom. Critical values can be listed for only a few levels of α. Also, most statistical applications involve looking up a critical value that matches a particular tail area (rather than matching an α to a particular level of t). There is a separate row of critical values in Table G for each df. For example, when df = 30, then for $\alpha = .01$, we read $t_{.01} = 2.457$ and

$$.01 = \Pr[t > 2.457] \quad (\text{df} = 30)$$

Like the normal distribution, the Student t is *symmetrical*, so the area above any level t_α must be equal to the area below $-t_\alpha$. For example,

$$.01 = \Pr[t < -2.457] \quad (\text{df} = 30)$$

Thus, probabilities may be established for any range of values. For instance, there is a .90 probability that the computed t will fall within $\pm t_{.05}$ and a .99 chance that t lies within $\pm t_{.005}$.

The Student t and Normal Curves

Recall that t plays the same role as the standard normal random variable Z, except that Z involves a specified σ whereas t is based on the yet-to-be-computed s. Let's compare these two probability distributions. Figure 9–6 shows the density curve for both distributions. The shapes are similar, although the curve for Student t has thicker tails, so the random variable exhibits greater variability than Z does. This is easy to justify intuitively. Z reflects less uncertainty about where $\overline{X}$ will fall because more population information is available (σ is specified). That same $\overline{X}$ uncertainty is magnified in t because the population variability is also uncertain (σ is unknown).

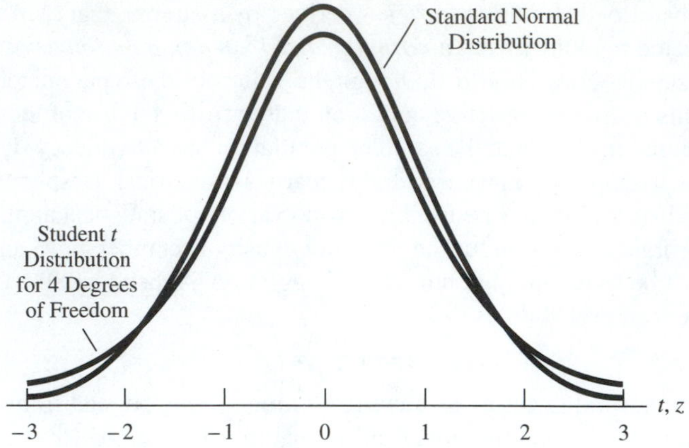

Figure 9–6
Density curves for standard normal and Student t distributions.

However, the density curve for t approaches the shape of the standard normal curve as the number of degrees of freedom becomes large. This tends to occur rapidly with large sample sizes.

Example: *Metal Fatigue and Aircraft Accidents*

A Federal Aviation Agency statistician is interested in those aircraft crashes in which metal fatigue in structural members is an indicated cause. He has collected wing segments from 8 different downed aircraft of a particular model and will be performing fatigue tests that measure the elongation of strips under various weight loads. He theorizes that the metal from the downed planes will have weakened with time and stress. Although a larger sample size than 8 would be ideal, the investigation is limited by the actual number of crashes.

Suppose that the mean elongation of new metal of the same type under a particular load is .05″. The statistician will conclude that μ for wing segments in crashed aircraft is larger only if the sample measurements are found to violate that hypothesis significantly. How can he determine whether his test results are significant?

The statistician establishes a rule providing a probability of only 1% of incorrectly concluding that metal strips taken from crash members experience greater distortion than new strips when in fact the two metals have identical characteristics so that $\mu = .05''$. In terms of the Student t statistic, we see that for $\text{df} = n - 1 = 8 - 1 = 7$,

$$.01 = \Pr[t > t_{.01}]$$

when $t_{.01} = 2.998$ (from Table G). The following rule therefore meets the stated criterion, as applied to the t computed from the sample data:

Conclude crash and new metals are identical if $t \leq 2.998$.

Conclude crash metals have greater distortion if $t > 2.998$.

The test results provide $\bar{X} = .13''$ with $s = .04''$. Thus,

$$t = \frac{\bar{X} - \mu}{s/\sqrt{n}} = \frac{.13 - .05}{.04/\sqrt{8}} = 5.66$$

Since the computed value exceeds 2.998, the statistician may safely conclude that structural members from crashed aircraft experience significantly greater distortion than new metals.

9–23 Using the Student *t* distribution, find the critical values for the following tail areas: *Problems*

(a)	(b)	(c)	(d)
$\alpha = .05$	$\alpha = .10$	$\alpha = .005$	$\alpha = .0005$
df = 11	df = 6	df = 25	df = 30

9-25

a) $Pr[t > 1.711] = 0.05$

9–24 The drying times of 15 sample epoxy substances are assumed to be normally distributed with unknown mean and standard deviation. Find the critical value above which the probability is α that the computed *t* will fall when

(a) $\alpha = .05$ (b) $\alpha = .01$ (c) $\alpha = .005$ (d) $\alpha = .0005$

b) df=25-1 =24

9–25 Establish the probability that the computed *t* from a sample of size $n = 25$ will fall

(a) above 1.711 *0.05*   (d) above 3.091
(b) below .685 *0.75*   (e) between .256 and 3.745
(c) below 2.797 *0.995*

b) $Pr[t > .685] = 0.25$
$Pr[t < .685] = 1 - 0.25$
$= 0.75$

9–26 Suppose that the FAA statistician in the example obtained the following results under different loadings, each with 8 sample metal segments from the crashed aircraft. Compute *t* and determine for each sample which conclusion should be made, using the rule given in the example.

e) $Pr[t > 3.745] = 0.0005$
$Pr[t > .256] = 0.4$
$Pr[t > .256] - Pr[t > 3.745]$
$= 0.4 - 0.0005$
$= 0.3995$

Loading	Mean Elongation for New Metal		Computed Values	
(a) Light	$\mu = .01''$	$\bar{X} = .008''$	$s = .002''$	
(b) Moderate	$\mu = .03''$	$\bar{X} = .07''$	$s = .02''$	
(c) Heavy	$\mu = .10''$	$\bar{X} = .18''$	$s = .07''$	
(d) Very Heavy	$\mu = .20''$	$\bar{X} = .43''$	$s = .12''$	

9–27 The texture index for a chemical product has a nominal mean value of $\mu = 100$. On the basis of 25 sample readings, the following decision rule has been established for determining when to replace a kiln thermostat:

Replace whenever $t < -2.492$.

Leave alone otherwise.

For each of the following texture results, indicate whether or not the thermostat should be replaced:

(a)	(b)	(c)	(d)
$\bar{X} = 98$	$\bar{X} = 101$	$\bar{X} = 97$	$\bar{X} = 99$
$s = 4.1$	$s = 5.2$	$s = 3.7$	$s = 4.1$

9–28 A chemical engineer has computed the following results for the active ingredient yields from 20 pilot batches processed under a retorting procedure:

$$\bar{X} = 31.4 \text{ grams/liter}$$
$$s = 2.85$$

Determine the approximate probability for getting a result this rare or rarer if the true mean yield is 29 grams per liter.

9–5 Sampling Distribution of the Proportion

In investigations involving qualitative populations the *proportion* π of units exhibiting a particular attribute or characteristic is the parameter of main interest. As with using $\bar{X}$ to generalize about μ, statisticians usually rely on the computed value for the sample proportion P to draw conclusions regarding the unknown π. The sample proportion for an attribute is found by dividing the number of sample observations R having that attribute by the sample size:

$$P = \frac{R}{n}$$

In Chapter 7 we established that when sample observations are random and independent, R is a random variable with a binomial distribution. That distribution applies as long as the sampled population results from a continuous process or sampling is done with replacement. If sampling is performed without replacement from a finite population, the hypergeometric distribution applies instead. Probabilities from the sample proportion may be obtained from either distribution by matching the possible level of P with the probability for the corresponding level of R.

In either case, it is usually convenient to approximate those probability distributions by an appropriate normal curve, and such an approximation is usually satisfactory when π is near .5 or when n is large.

Indeed, the central limit theorem may be invoked to establish that a normal curve is appropriate for P when sampling meets the conditions of a Bernoulli process. Then each sample observation, or *trial*, may be represented by a variable having two values: $X = 0$ (if a failure) or $X = 1$ (if a success). The probabilities for the respective trial outcomes are $1 - \pi$ and π, and individual Xs have the **Bernoulli distribution**. The binomial random variable R can represent the sum of n Bernoulli Xs, and the proportion of successes P is equal to the sample mean $\bar{X}$ of those same observations.

Probabilities for P Using Normal Approximation

In Chapter 7 we saw that the tables ordinarily used in finding exact binomial probabilities are limited to just a few parameter levels. The normal approximation works well for a wide variety of levels for n and π and is often used in statistical applications involving the proportion. From the binomial distribution for R we get the following.

EXPECTED VALUE AND STANDARD DEVIATION FOR P

$$E(P) = \pi$$

$$SD(P) = \sigma_P = \sqrt{\frac{\pi(1-\pi)}{n}}$$ (2)

where in keeping with our earlier notation, we use σ_P to denote the standard error of P. When the population size N is small (and sampling is without replacement), σ_P must be modified, as we shall see shortly.

These expressions serve as the parameters for approximating the normal distribution. A continuity correction of .5 is used in finding cumulative probabilities. Applying this, we may use the following expression.

CUMULATIVE PROBABILITY FOR P UNDER NORMAL APPROXIMATION

r is # items found with particular attribute in sample n

$$\Pr\left[P \le \frac{r}{n}\right] = \Phi\left(\frac{(r+.5)/n - \pi}{\sigma_P}\right)$$ (3)

As an illustration, suppose that the silicon wafers used in making a particular microcircuit have a final chip yield that contains 9% defectives. A sample of $n = 40$ chips is taken. What is the probability that the sample proportion of defective chips is less than or equal to .15?

Exact binomial probabilities are not tabled for $n = 40$ or $\pi = .09$, so it is convenient to use the normal approximation. The standard error for P is

$$\sigma_P = \sqrt{\frac{.09(.91)}{40}} = .0452$$

The level .15 corresponds to $r = .15(40) = 6$ defective items. Using the normal approximation, the desired probability is found:

$$\Pr[P \le .15] = \Phi\left(\frac{(6+.5)/40 - .09}{.0452}\right)$$
$$= \Phi(1.60) = .9452$$

Example: When to Replace Circuit Boards

The transmission errors in a teleprocessing system occur randomly over time. The actual proportion is uncertain, but the supervising engineer has established a policy that the transmitter circuit boards must be replaced whenever the sample proportion of erroneous bits in a test transmission is greater than .000025. For this purpose a test burst of 100,000 bits is used; these are checked by the receiver to see if they match the established test pattern. The supervisor is interested in checking this policy to see if it results in too many unnecessary replacements.

The nominal system specification allows the proportion of erroneous bits to be 2 per 100,000 ($\pi = .00002$). Assuming that this applies for the entire population of transmitted bits, what is the probability that the boards will be replaced unnecessarily after a particular test?

We have

$$\sigma_P = \sqrt{\frac{.00002(.99998)}{100{,}000}}$$
$$= .000014142$$

so that the normal approximation provides

$$\Pr[\text{Unnecessary replacement}] = \Pr[P > .000025] = \Pr\left[P > \frac{2.5}{100{,}000}\right]$$
$$= 1 - \Pr\left[P \leq \frac{2.5}{100{,}000}\right]$$
$$= 1 - \Phi\left(\frac{(2.5 + .5)/100{,}000 - .00002}{.000014142}\right)$$
$$= 1 - \Phi(.71)$$
$$= 1 - .7612 = .2388$$

This probability was judged by the engineer to be too large. He therefore revised the cutoff level of the sample proportion for replacement from the original .000025 to .00004, for which the following applies:

$$\Pr[P > .00004] = \Pr\left[P > \frac{4}{100{,}000}\right]$$
$$= 1 - \Pr\left[P \leq \frac{4}{100{,}000}\right]$$
$$= 1 - \Phi\left(\frac{(4 + .5)/100{,}000 - .00002}{.000014142}\right)$$
$$= 1 - .9616 = .0384$$

The new cutoff value was accepted by management, although it provides less protection than before against erroneously failing to replace the circuit boards when the true proportion of errors is actually higher than specifications allow.

Sampling from Small Populations

The normal approximation may also be used as the sampling distribution of P when samples are taken from small populations. As noted, the value of σ_P will be computed differently. For this purpose we use the following expression.

STANDARD ERROR FOR P
(SMALL POPULATIONS)

$$\sigma_P = \sqrt{\frac{\pi(1-\pi)}{n}}\sqrt{\frac{N-n}{N-1}}$$

As noted, in this case the normal distribution approximates the hypergeometric distribution, and σ_P includes the finite population correction factor.

Problems

9–29 Referring to the circuit board example, suppose that the true proportion of erroneous bits is higher than the specification level. Of course, the supervising engineer does not know this and must rely on the $P > .00004$ rule to determine whether the boards should be replaced. Using the same sample size as before, calculate the probability that for a particular test burst the circuit boards will *not* be replaced when

(a) $\pi = .00003$ (b) $\pi = .00004$ (c) $\pi = .00005$

9–30 Repeat Problem 9–29, using a sample size of $n = 200{,}000$ bits instead.

9–31 A sample of 100 parts is to be tested from a shipment containing 500 items altogether. Although the inspector does not know the proportion defective in the shipment, assume that this quantity is .03. Determine the following probabilities for the sample proportion defective:
(a) $P \leq .05$ (b) $P \leq .02$ (c) $P > .04$ (d) $.02 \leq P \leq .03$

9–32 A quality-control inspector accepts shipments whenever a sample of size 20 contains 10% or fewer defectives. She rejects otherwise.
(a) Determine the probability that she will accept a poor shipment of 100 items in which the actual proportion of defectives is 12%.
(b) Determine the probability that she will reject a good shipment containing 150 items, only 8% of which are defective.

9–33 A welding robot needs overhauling if it misses 4.5% of its welds. It is judged to be operating satisfactorily if it misses only .8% of its welds.

A test is performed involving 50 sample welds. If the proportion of missed sample welds is greater than .02, the robot will be overhauled. Otherwise, it will continue to operate. Assuming that in-between percentages may be ignored:
(a) What is the probability that a satisfactory robot will be overhauled unnecessarily?
(b) What is the probability that a poorly functioning robot will be left in operation?

9–34 A computer manufacturer rejects any shipment of printed circuit boards in which the proportion of defectives in a sample of 25 boards tested exceeds .16. For each of the following shipments, determine the probability of rejection:

(a)	**(b)**	**(c)**	**(d)**
$N = 200$	$N = 150$	$N = 250$	$N = 75$
$\pi = .10$	$\pi = .12$	$\pi = .15$	$\pi = .18$

9–6 Sampling Distribution of the Variance: The Chi-Square and F Distributions

Although the sample variance s^2 is computed in most sampling investigations involving quantitative data, it is ordinarily an adjunct statistic for procedures that focus on the sample mean. In those cases, direct inferences concern μ only, and no conclusions are made about the population variance (standard deviation). However, there are instances when the major concern is with the value of σ^2 (or σ) instead of μ. In those cases, *variability* rather than central tendency is the major concern.

For instance, in chemical processing a method having slower reaction times that are consistent may be more desirable than one that is on the average faster but has more varied individual times. Alternative methods would then be compared in terms of the respective reaction-time population standard deviations, with the mean times being of secondary interest. Similarly, although customers are very concerned with their waiting time in a queue (waiting line), they may be more upset about the unpredictability of the amount of wasted time than about its average duration. That is, σ^2 for the waiting time population is often more important than μ.

As we have seen with inferences regarding means and proportions, it is appropriate to use the counterpart sample statistic in making conclusions about population variability. Statisticians use the computed value of s^2 to make generalizations about the uncertain level of σ^2. Our present concern is with finding probabilities for the possible levels of s^2.

The Chi-Square Distribution

The sampling distribution of sample variance s^2 is simplest to obtain by transforming the random variable into the chi-square statistic.

CHI-SQUARE STATISTIC

$$\chi^2 = \frac{(n-1)s^2}{\sigma^2}$$

where the symbol χ is the Greek lowercase *chi* (pronounced "kie") and χ^2 is read "chi-square." This statistic (in the planning stage) is a random variable having the **chi-square distribution**.

Figure 9–7 shows several density curves for the special chi-square random variable used in most statistical applications. Like the Student *t*, the chi-square distribution is specified by a single parameter, the number of degrees of freedom (df). When it is applied to making probability statements about s^2, df $= n - 1$. (There are only $n - 1$

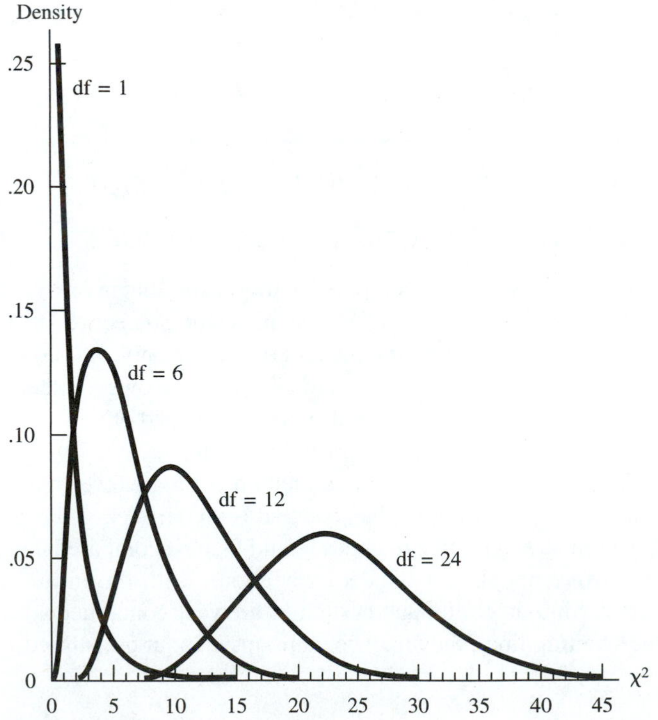

Figure 9–7 *Chi-square density curves for several degrees of freedom.*

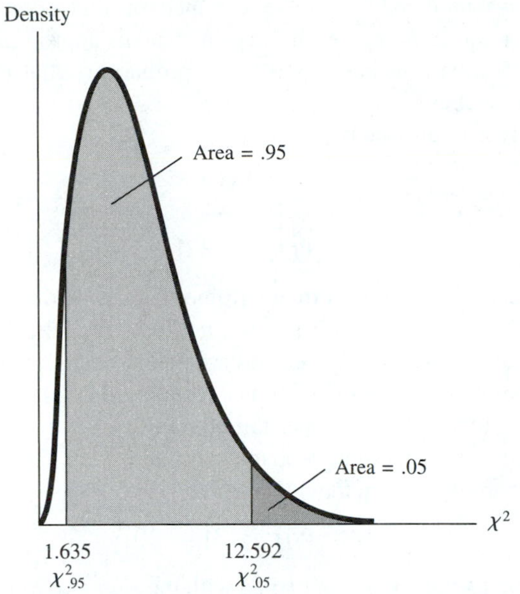

Figure 9–8
Chi-square density curve for 6 degrees of freedom.

degrees of freedom because in addition to the n observed Xs, $\overline{X}$ is used in calculating s^2, and 1 degree of freedom is lost in specifying the level for $\overline{X}$.)

Notice that the chi-square densities are positively skewed, with upper tails extending indefinitely, but the degree of skew becomes less pronounced as the degrees of freedom increase. (As with the Student t, the normal curve may be used as an approximation when n is large.) For each upper-tail probability α there corresponds a critical value χ^2_α for the chi-square random variable such that

$$\alpha = \Pr[\chi^2 > \chi^2_\alpha]$$

Figure 9–8 shows the chi-square density curve for 6 degrees of freedom. Two upper-tail areas are given. The respective critical values for these are $\chi^2_{.05} = 12.592$ and $\chi^2_{.95} = 1.635$, so that

$$\begin{aligned} .05 &= \Pr[\chi^2 > 12.592] \\ .95 &= \Pr[\chi^2 > 1.635] \end{aligned}$$

and there is a $.95 - .05 = .90$ probability that χ^2 will lie between 1.635 and 12.592. These values were read from Appendix Table H, which provides critical values of the chi-square statistic for several settings and 30 df levels. Table H resembles the layout for the Student t table. To accommodate the lack of symmetry in chi-square curves, several α levels greater than .5 are listed.

Probabilities for the Sample Variance

The probability of any s^2 event may be found by identifying the corresponding χ^2 event and determining the probability for the latter.

For example, suppose that a human-factors engineer will be collecting a random sample of $n = 25$ of the effective reach radii for seated men. The population is assumed to be normally distributed. Although the variance of this population is unknown, anthropometric data for standing reach suggest that an appropriate value would be $\sigma^2 = (1.5'')^2 = 2.25$ square inches. What is the probability that the sample variance will exceed 3 square inches?

This probability is expressed by

$$\begin{aligned}\alpha = \Pr[s^2 > 3] &= \Pr\left[\frac{(n-1)s^2}{\sigma^2} > \frac{(25-1)3}{2.25}\right] \\ &= \Pr[\chi^2 > 32]\end{aligned}$$

To evaluate this probability, we need to determine the area under the chi-square curve that lies above 32. For a sample size of $n = 25$, the chi-square density with df $= 25 - 1 = 24$ applies. Only a few entries are provided in Table H, and we see that the critical value coming closest to 32 without exceeding that quantity is 29.553, a level for χ^2 corresponding to an $\alpha = .20$ upper-tail probability. The next larger critical value is 33.196, representing an upper-tail probability of $\alpha = .10$. Thus, we may conclude that the desired probability falls in the interval,

$$.10 < \Pr[s^2 > 3] < .20$$

Investigators are usually more concerned with the *standard deviation* than the variance. Instead of dealing with a separate sampling distribution, one can obtain the probabilities for the sample standard deviation s by applying the chi-square distribution in evaluating the corresponding s^2 event. Thus, $s > 5$ is equivalent to $s^2 > 25$ and $s = 10$ to $s^2 = 100$.

Example:* *Selecting a Laser Crystal for Infrared Spectroscopy

One criterion used in laser-based infrared spectroscopy is the dislocation density of the laser crystal. An engineer conducted a test to measure this quantity in crystals grown from an experimental composition. He was more concerned with the consistency in dislocation density from crystal to crystal than with the absolute level. Thus, he used the computed displacement statistics from $n = 10$ crystals,

$$\begin{aligned}\bar{X} &= 530/\text{cm}^2 \\ s &= 98\end{aligned}$$

to make a comparison to an existing crystal, in which the standard deviation in dislocation density was assumed to be $\sigma = 200/\text{cm}^2$.

Assuming that the population standard deviations (variances) are the same for the existing and experimental crystals, what is the probability that results as extreme as the foregoing, or more so, could have been obtained?

We must find

$$\begin{aligned}\Pr[s \le 98 \mid \sigma = 200] &= \Pr\left[\frac{(n-1)s^2}{\sigma^2} \le \frac{(10-1)(98)^2}{(200)^2}\right] \\ &= \Pr[\chi^2 \le 2.161]\end{aligned}$$

For $10 - 1 = 9$ degrees of freedom, Appendix Table H provides the nearest bracketing critical values of $\chi^2_{.99} = 2.088$ and $\chi^2_{.98} = 2.532$, representing upper-tail probabilities of .99 and .98. The desired event is a lower-tailed one, so that the complementary probabilities apply, and the following conclusion is reached:

$$.01 < \Pr[s \le 98 \mid \sigma = 200] < .02$$

Assumptions of the Chi-Square Distribution

The chi-square distribution represents a random variable that is the sum of the squares of several independent normal random variables. This means that using the chi-square distribution to represent the sampling distribution for s^2 is theoretically valid only when the individual sample observations are taken from a *normally distributed population.* This assumption of normality is identical to the one underlying the Student t and may often be violated in certain applications, so that the chi-square does not technically apply. Nevertheless, the chi-square often serves as a satisfactory approximation to the true sampling distribution even when the population is not normal.

Expected Value, Variance, and the Normal Approximation

It may be established that the expected value and variance of the chi-square distribution summarized in Table H are, for df $= n - 1$:

$$E(\chi^2) = n - 1$$
$$\text{Var}(\chi^2) = 2(n - 1)$$

As noted earlier, the chi-square densities resemble a bell-shaped curve when the sample size (df) is large. As a practical matter, *the normal distribution is used to approximate the chi-square whenever* $n > 30$. The above moments may be used to convert χ^2 into a standard normal random variable by subtracting the expected value and dividing by the standard deviation. The following expression is used.

NORMAL DEVIATE FOR APPROXIMATING CHI-SQUARE

$$z = \frac{\chi^2 - (n - 1)}{\sqrt{2(n - 1)}}$$

Returning to the problem of determining the sitting reach of men, suppose that a sample size of $n = 100$ were used instead. What is the probability that the sample variance will exceed 3 square inches?

Here we use the normal approximation in evaluating the following because $n > 30$ and df $= 100 - 1 = 99$ falls outside of Table H. We have

$$\begin{aligned} \Pr[s^2 > 3] &= \Pr\left[\frac{(n-1)s^2}{\sigma^2} > \frac{(100-1)3}{2.25}\right] \\ &= \Pr[\chi^2 > 132] \\ &\approx 1 - \Phi\left(\frac{132 - (100 - 1)}{\sqrt{2(100 - 1)}}\right) \\ &= 1 - \Phi(2.35) = 1 - .9906 \\ &= .0094 \end{aligned}$$

The F Distribution

A second sampling distribution—**the *F* distribution**—is also associated with the sample variance. This distribution is useful for comparing the variances of *two* samples, which is done by finding their *ratio*.

Consider two normally distributed populations, A and B, each with identical variances. One random sample is selected from each. We let s_A^2 and s_B^2 denote the respective sample variances and n_A and n_B the sample sizes and get the following expression.

F STATISTIC

$$F = \frac{s_A^2}{s_B^2}$$

This random variable has the F distribution, which is completely described by two parameters, $n_A - 1$ and $n_B - 1$. These parameters are referred to as the *degrees of freedom* for the *numerator* and the *denominator*, respectively.

Figure 9–9 illustrates the density curves for the F distribution for three different pairs of degrees of freedom. Notice that the F distribution is positively skewed with possible values ranging upward from zero. As with the Student t distribution, critical values may be determined for a specified upper-tail area α, and these are provided in Appendix Table J. In keeping with the earlier notation, the respective critical values are denoted by F_α, so that

$$\alpha = \Pr[F \geq F_\alpha]$$

Because there is a different F distribution for each degrees-of-freedom pair, entries are provided for only two levels, $\alpha = .01$ or $\alpha = .05$. Values for $F_{.05}$ are shown in lightface type, while those for $F_{.01}$ are shown in boldface. When the numerator and denominator degrees of freedom are each 4, the critical values are read from row 4 and column 4 of Table J: $F_{.01} = 15.98$ and $F_{.05} = 6.39$. The latter point is the critical value above which the area under the corresponding density curve is .05; that area is shaded in Figure 9–9. The tail area of $F_{.01}$ lies off the page. These values indicate that under that distribution, values taken by the random variable F will be greater than 6.39 only 5% of the time and will exceed 15.98 just 1% of the time.

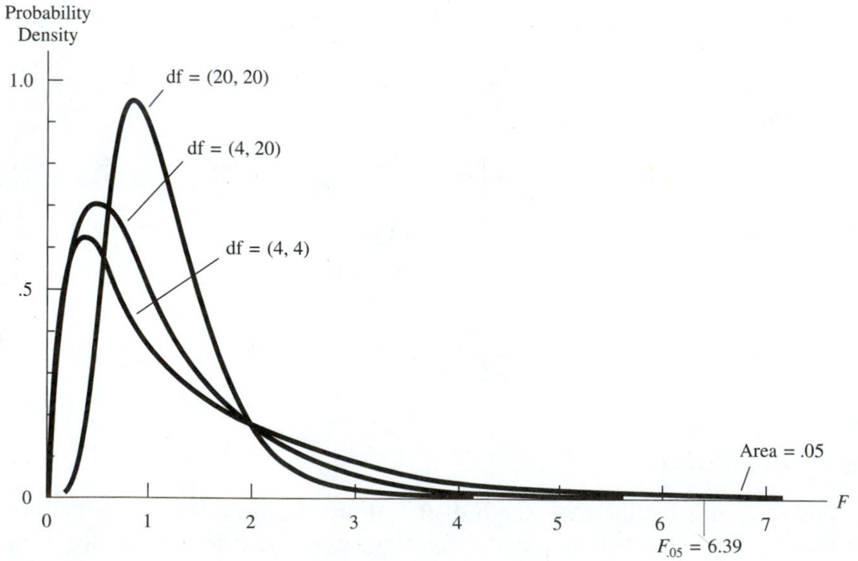

Figure 9–9
Density curves for F distribution.

To illustrate using the F distribution, suppose that a computer science student plans to make two sets of information retrievals from a database. She will record how long it takes to complete each retrieval. Set A involves $n_A = 5$ retrievals made at lunchtime, between noon and 1 P.M. Set B involves $n_B = 10$ retrievals made in the early morning, between midnight and 1 A.M. She plans to compute the two sample variances in completion time. From Table J, using $5 - 1 = 4$ degrees of freedom for the numerator and $10 - 1 = 9$ for the denominator, we see that there is a .05 probability that s_A^2 will be at least 3.63 times as large as s_B^2:

$$\Pr\left[\frac{s_A^2}{s_B^2} \geq 3.63\right] = .05$$

while there is a .01 probability that s_A^2 will be at least 6.42 times as large as s_B^2:

$$\Pr\left[\frac{s_A^2}{s_B^2} \geq 6.42\right] = .01$$

The theoretical requirement for the F distribution is that the parent *population* be normally distributed. This same assumption underlies both the Student t and chi-square distributions. Although the F distribution may still be applied when populations are nonnormal, doing so involves potential inaccuracies. In those instances we must acknowledge that the F distribution only approximates the true distribution of the sample variance ratio.

Problems

9–35 Determine the upper-tail critical values for the chi-square statistic in the following cases:

(a)	(b)	(c)	(d)
$\alpha = .01$	$\alpha = .001$	$\alpha = .50$	$\alpha = .95$
df $= 10$	df $= 25$	df $= 2$	df $= 12$

9–36 Compute the chi-square statistic for the following sample results under the assumed σ. Then determine the nearest upper-tail area and the corresponding critical value such that the computed chi-square does not exceed that quantity: $\alpha = 0.95$

(a)	(b)	(c)	(d)
$s = 10$	$s = 8$	$s = 215$	$s = 53$
$n = 25$	$n = 10$	$n = 20$	$n = 15$
$\sigma = 9$	$\sigma = 15$	$\sigma = 300$	$\sigma = 35$

Table H
$$\chi^2 = \frac{(n-1)s^2}{\sigma^2}$$

degrees of freedom $\Rightarrow n-1$

9–37 The waiting time of time-sharing jobs for access to a central processing unit is believed to be normally distributed with unknown mean and standard deviation. Find the following percentiles for the sample standard deviation s, when a sample of size $n = 25$ is taken and assuming that the true value of the population standard deviation is $\sigma = 1.5$ minutes:
(a) 1st (b) 10th (c) 50th (d) 90th (e) 99th

9–38 Compute the chi-square statistic for the following sample results under the assumed σ. Because of the large sample sizes, the normal approximation must be used. Determine the normal deviate that corresponds and establish the probability for getting a value for s as great or greater:

(a)	(b)	(c)	(d)
$s = 7$	$s = 12$	$s = 325$	$s = 33$
$n = 65$	$n = 90$	$n = 100$	$n = 225$
$\sigma = 5$	$\sigma = 15$	$\sigma = 300$	$\sigma = 30$

9–39 The mean sitting height of adult males may be assumed to be normally distributed, with mean 36″ and standard deviation 1.3″. For a sample size of $n = 100$ men, determine the probabilities for the following possible levels of the sample standard deviation in sitting height:
(a) $s \le 1.00''$ (b) $s > 1.5''$ (c) $s \le 1.4''$ (d) $1.25'' \le s \le 1.35''$

9–40 The customer waiting times at a certain post office are assumed to be normally distributed, with unknown mean and standard deviation. An industrial engineering student will be monitoring 25 noontime customers, timing their arrivals and service with a watch. As closely as you can, find an interval bracketing the probability that she will observe a standard deviation in waiting time exceeding 5 minutes when the true value of the population standard deviation is:
(a) $\sigma = 4.3$ minutes
(b) $\sigma = 5.4$ minutes
(c) $\sigma = 6.0$ minutes
(d) $\sigma = 4.7$ minutes

9–41 Consider two random samples of size $n_A = 15$ and $n_B = 20$. Which of the following pairs of variances provides ratios extreme enough to fall into a range that occurs only 5% of the time when the underlying populations have identical variances?

(a)	(b)	(c)	(d)
$s_A^2 = 10.1$	$s_A^2 = 25.4$	$s_A^2 = 50.5$	$s_A^2 = 100.8$
$s_B^2 = 5.3$	$s_B^2 = 9.6$	$s_B^2 = 17.3$	
$s_B^2 = 235.6$			

9–42 Find the critical values $F_{.05}$ and $F_{.01}$ for the following sampling situations:

(a)	(b)	(c)	(d)
$n_A = 10$	$n_A = 25$	$n_A = 5$	$n_A = 101$
$n_B = 10$	$n_B = 5$	$n_B = 25$	$n_B = 101$

9–43 Denoting by υ_A and υ_B the numerator and denominator degrees of freedom, let $F_\alpha(\upsilon_A, \upsilon_B)$ denote the critical value for F such that

$$\Pr\left[\frac{s_A^2}{s_B^2} \geq F_\alpha(\upsilon_A, \upsilon_B)\right] = \alpha$$

Show that $F_{1-\alpha}(\upsilon_A, \upsilon_B) = 1/F_\alpha(\upsilon_B, \upsilon_A)$.

Review Problems

9–44 The height of adult males in a particular city is normally distributed, with $\mu = 69.5''$ and standard deviation $\sigma = 2.65''$. A sample of $n = 10$ men is to be measured. Find (as closely as possible) the probabilities that (a) $\bar{X}$ falls within $\mu \pm .75''$ and (b) s falls within $\sigma \pm .1''$.

9–45 Refer to the student grade-point population in Table 9–1. Suppose a sample of 2 is taken without replacement. Construct a table showing the complete sampling distribution of

(a) the sample proportion of *B*s

(b) the sample range (absolute value for difference in observed values)

9–46 Repeat Problem 9–45, but assume instead that the observations are made *with* replacement.

9–47 An engineering society is preparing for a sampling investigation of salaries of ceramics engineers. A random sample of 25 engineers will be selected and their annual incomes determined.

(a) The degree of error in using $\bar{X}$ to estimate μ cannot be known, since the standard error of $\bar{X}$ depends on the unknown population standard deviation. Assuming that this quantity will be the same as that for another income population for which σ was found to be \$1,500, determine the 99th percentile for the sample standard deviation from the present investigation. (There will be a 1% chance that the computed s will exceed this quantity.)

(b) Under the assumption that the true value of the population standard deviation equals the 99th percentile for s found in (a), determine the probability that the sample mean income falls within \$500 of the true population mean.

9–48 An automotive engineer will be conducting crash tests to determine the standard repair costs for cars damaged by hitting a barrier at 5 mph. A sample of $n = 9$ crashes is all her budget allows. She knows neither the mean nor the standard deviation for the population of costs, although she believes the frequency distribution to be nearly symmetrical.

Suppose that the engineer will request a front-end redesign study if the computed sample mean exceeds the historical figure for past models by more than some yet-to-be-determined amount.

(a) On what sample statistic should the engineer base her decision?

(b) What level of the quantity identified in (a) provides a 5% chance of requesting a study when the true mean repair cost is identical to the historical one?

(c) Assuming she uses the critical value found in (b), would she request a redesign study if $\bar{X} = \$242$, $s = \$45$, and the historical mean is $\mu = \$195$?

9–49 Suppose that the engineer in Problem 9–48 will order an extension of crash testing if the sample standard deviation exceeds the historical level by some amount.

(a) On what sample statistic should the engineer base her decision?

(b) What level of the quantity identified in (a) provides a 1% chance of ordering further crashes when the true standard deviation in repair cost is identical to the historical figure?

(c) Assuming she uses the critical value found in (b), would she order an extension if $s = \$45$ and the historical standard deviation is $\sigma = \$30$?

9–50 A plant foreman shuts down a process to replace control valves whenever more than 5% of the items in a sample of 100 are found to be out of tolerance. Use the normal approximation to find the probabilities for the following events:

(a) He fails to replace the control valves when the process is actually yielding 10% off-size items.

(b) He replaces the valves unnecessarily when the process is performing satisfactorily and only 1% of production is out of tolerance.

9–51 Repeat Problem 9–50, but use the binomial distribution to find the exact probabilities.

CHAPTER TEN

Statistical Estimation

SCIENCE AND ENGINEERING rely heavily on sample information. Nearly all physical constants, such as $g = 32$ ft/sec^2, are based on empirical observations and are really statistical estimates. In large measure the pioneering efforts at evaluating such parameters may be characterized as sampling studies. Practically all modern measurements of physical objects culminate in some sort of estimate—from the level of resistance in a particular substance used in a printed circuit board to the strength reported for structural members of a particular shape, size, and composition.

Mean values and proportions are perhaps the most familiar quantities. The unknown population parameter value μ or π must usually be estimated from sample data. The resulting values are generally based on the computed level for the respective sample statistic, with $\bar{X}$ being used to estimate μ and P serving as the estimator of π. Other statistics might be used in estimating these parameters. For instance, the mean of a symmetrical population might just as well be estimated by the sample median. Although all estimators exhibit the potential for sampling error, some are more accurate than others. Several criteria exist for gauging the suitability of each candidate statistic for estimating a particular population parameter.

Once a particular statistic has been selected, such as $\bar{X}$ for estimating μ, questions remain regarding the form of the estimate itself. Newspaper readers are accustomed to reading reports in which means and percentages are listed as single quantities. By now, we know there are pitfalls in using such statistics for any serious purpose—largely because of the potential for sampling error. That error must be dealt with in a systematic way.

10–1 Estimators and Estimates

The sample statistic employed in estimating a population parameter is called an **estimator**. An important part of the statistical art is deciding what particular statistic to employ. Several criteria guide in this selection. These help explain why particular statistics are favored in drawing inferences. There are two principal forms a statistical estimate may take.

The Estimation Process

The estimation process begins in the planning stage of the statistical investigation. The population has already been identified and the parameter to be estimated has been determined. An early part of the planning process is deciding which statistic will be the estimator. We shall see how the sampling distribution of the chosen statistic may provide some idea of the precision and reliability of the final estimate. These considerations guide the final selection of the sample size.

The form of the estimate itself will depend largely on the purpose of the study. In a formal report of the study results, the favored form is the **interval estimate**. For example, the mean waiting time μ in a particular queueing situation might be expressed as

$$35 \leq \mu \leq 45 \text{ seconds}$$

The reported value for μ is inexact, and the limits in this inequality provide the end points for the interval estimate of μ, [35, 45]. Interval estimates are usually reported in the inequality form shown. An equivalent expression is

$$\mu = 40 \pm 5 \text{ seconds}$$

The central value (40, here) is often the computed value of the sample statistic serving as the estimator.

The limits of the interval estimate indicate the degree of **precision** involved. A more precise estimate of the mean waiting time would be

$$\mu = 40 \pm 1 \qquad \text{or} \qquad 39 \leq \mu \leq 41 \text{ seconds}$$

The quality of a statistical estimate is measured in two dimensions. Precision or accuracy is one. The other is the **reliability** of the result. The reliability of an estimate is simply the probability that it is correct.

Reliability and precision are competing ends. One can always be improved at the expense of the other. For example, suppose we used sample data to estimate the mean height of male engineering majors. On the basis of a sample measurement of one randomly chosen man we could conclude that the mean height of all men in the population is that person's height plus or minus $.01''$, say

$$\mu = 5'8.5'' \pm .01'' \qquad \text{or} \qquad 5'8.49'' \leq \mu \leq 5'8.51''$$

a very precise outcome indeed. But how reliable would such a result be? Most would agree on a small probability that the above estimate is valid. With little chance of being correct, an overly precise estimate is useless.

Alternatively, using the same data, we can achieve the following very reliable estimate:

$$\mu = 5'8.5'' \pm 1.00' \qquad \text{or} \qquad 4'8.5'' \leq \mu \leq 6'8.5''$$

an outcome virtually certain to be true (so that the reliability is 100%). But this second estimate is too imprecise to be of any use whatsoever.

Better reliability and satisfactory precision can be achieved by raising the probability that the estimator's value will be close to its target parameter. To achieve this, statisticians seek estimators with sampling distributions that cluster ever more tightly about a central value as n becomes larger. By paying the price of increasing the number of sample observations, an investigator using such an estimator can simultaneously improve precision and reliability.

The second form of estimation involves a **point estimate**. This is a single quantity, such as 5′9.5″ for the mean height of a population of adult males or .64 for the proportion of voters approving of the president. Unlike interval estimates, point estimates give no hint of the sampling error inherent in their generation. They are particularly dangerous when publicly disseminated, because the general public too often assumes such values to be the true parameter levels, without even realizing that sampling was involved.

Selecting an Estimator

The estimator used in making statistical inferences from sample data will be a sample statistic, and its value will be computed from the observed data. The following are typical estimators:

$$\hat{\mu} = \overline{X} \qquad \hat{\sigma}^2 = s^2 \qquad \hat{\pi} = P = \frac{R}{n}$$

(Traditional statistical notation denotes the estimator for a population parameter by placing a caret over the symbol for that parameter.) The estimators used in statistical methods are carefully chosen because they exhibit various desirable properties.

Although there is not sufficient space in this book for the mathematical arguments, theoretical rationales have been important for justifying the choice of the above estimators. Intuitively, it is reasonable to use $\overline{X}$ to estimate μ, since the population mean would be computed in the same way as the sample mean—were all the population data available for doing so. One property of the sample mean or sample proportion is that they are *maximum likelihood estimators*. This means that their computed values establish levels for the respective population parameters that are most likely to have generated the particular sample results obtained.

Criteria for Statistics Used as Estimators

Several criteria are employed in assessing the worthiness of a particular statistic as an estimator. We will describe the more common ones.

Unbiased Estimators An estimator is **unbiased** if its expected value equals the parameter being estimated. There are no guarantees that any sample estimate will be correct, but those determined from unbiased estimators will, on the average, hit their targets. In Chapter 9 we noted for the sample mean that

$$E(\overline{X}) = \mu$$

which establishes that $\overline{X}$ is an unbiased estimator of μ. Similarly,

$$E(P) = \pi$$

and P is an unbiased estimator of π.

This property is not a universal one. Consider the sample statistic

$$\frac{\sum_{i=1}^{n}(X_i - \overline{X})^2}{n}$$

which resembles the sample variance s^2. However, the divisor in this expression is n, not $n - 1$ as we have defined that quantity for s^2.

Like σ^2, s^2 expresses the mean of the squared observation deviations (from their own mean). The divisor in computing σ^2 is the population size N. Why didn't we use the entire sample size n as the divisor for s^2, as in the foregoing equation?

The answer is that s^2 as we have been using it ($n - 1$ divisor) is an unbiased estimator of σ^2. That is, $E(s^2) = \sigma^2$, so that on the average, values computed for s^2 for successive samples taken from the same population will be equal to σ^2. This does not apply to the statistic having n as the divisor. Values computed from that biased statistic will tend to undershoot the σ^2 target.

Like $\bar{X}$, the sample median is an unbiased estimator for μ—but only when the population is symmetrical. Unless we know that the population frequency distribution is not skewed, the median will consistently overshoot or undershoot μ.

Example: *Estimating the Gold Requirements of Logic Circuit Wafers*

An industrial engineer for a microcircuit manufacturer is estimating production costs for a new logic chip. Hundreds of chips are produced on a single silicon sheet that is like a wafer cake with as many as 50 layers. The most expensive layers involve gold, which is applied in an aerosol mist. The actual amount of gold consumed cannot be predicted exactly, since the layer consistency will vary from wafer to wafer and the unrecoverable gold loss fluctuates. From a pilot run of 15 batches, the following data were obtained for the gold consumption per wafer:

Amount of Gold (oz) X	$(X - \bar{X})$	$(X - \bar{X})^2$
.15	−.01	.0001
.17	.01	.0001
.13	−.03	.0009
.18	.02	.0004
.15	−.01	.0001
.16	.00	.0000
.17	.01	.0001
.15	−.01	.0001
.17	.01	.0001
.15	−.01	.0001
.19	.03	.0009
.18	.02	.0004
.17	.01	.0001
.15	−.01	.0001
.13	−.03	.0009
$2.40 = \sum X$		$.0044 = \sum(X - \bar{X})^2$

$\bar{X} = 2.40/15 = .16$ $\quad s^2 = .0044/(15 - 1) = .000314$

The engineer uses this sample mean to estimate that the true mean gold consumption per wafer is .16 ounce. If each wafer yields 100 marketable chips, the amount of gold used to make each chip is $.16/100 = .0016$ ounce. At a current price of \$500 per ounce, the mean value of the gold in each chip is $\$500(.0016) = \$.80$.

Consistent Estimators An estimator is **consistent** if the precision and reliability of the estimate improve as the sample size is increased. This will be the case if the standard error of the estimating statistic becomes smaller as n is increased. It should be obvious that $\overline{X}$ and P are consistent estimators, since

$$\sigma_{\overline{X}} = \frac{\sigma}{\sqrt{n}} \qquad \text{and} \qquad \sigma_P = \sqrt{\frac{\pi(1-\pi)}{n}}$$

both become smaller as n increases.

Efficient Estimators An estimator is more **efficient** than another if for the same sample size it will provide greater sampling precision and reliability. This will be achieved by the estimator whose sampling distribution clusters more tightly about the parameter being estimated, so the more efficient statistic will be the one having the smaller standard error. The sample mean $\overline{X}$ is more efficient in estimating μ than is the sample median, whose standard error is 1.25 times as large as $\sigma_{\overline{X}}$.

A substantial portion of statistical theory is devoted to applying these criteria in selecting estimates and in expressing their suitability. However, meeting these criteria does not in itself guarantee that a particular statistic will be a good estimator. As a practical matter, an estimator must be easy to compute and must have a sampling distribution that can be readily determined. A workable procedure for evaluating sample results must then be built around that statistic.

Commonly Used Estimators

The sample mean $\overline{X}$ is a favored estimator of μ because it is unbiased, consistent, and more efficient than many alternative statistics. The sampling distribution of $\overline{X}$ is particularly convenient. For large sample sizes taken from most commonly encountered populations, the central limit theorem indicates that a normal curve may be assumed for $\overline{X}$. In this chapter we will see how this feature can be used in quantifying estimates based on $\overline{X}$. We will also see how to choose a sample size that guarantees meeting the investigator's precision and reliability goals.

In addition to being unbiased, the sample variance s^2 is a consistent estimator of the population variance σ^2. Its square root may be used in estimating the population standard deviation. In the preceding example, the sample standard deviation for wafer gold content is $s = \sqrt{.000314} = .018$ ounce. This figure may be used to estimate σ, the standard deviation in the population of gold content.

The sample proportion P is an unbiased and consistent estimator of the population proportion π. The following example provides an interesting illustration of using P as an estimator.

Example: Work Sampling Eliminates Stopwatch "Tyranny"

Work measurement is a fundamental task of industrial engineers in production management. The information thereby gained may be used to group tasks into comprehensive job packages by finding the mean completion time of each task. Pioneers in the field of work measurement found these durations by timing workers as they performed each step. This was done with a timepiece (stopwatch) and required one watcher for each worker being timed. Not only was this an expensive way to gather data, but it was resented by the workers.

Modern work measurement rarely involves the stopwatch or any kind of continuous monitoring. Rather, the test subject may be observed for a few seconds at random times. A tally is then compiled, showing how often each particular chore was observed. From these data the proportion of time a worker spends on each step or task can be estimated. For example, one industrial engineer's clerk collected the following data for one worker in an 8-hour day:

Task	Number of Observations
Mounting	25
Clamping	9
Adjusting	105
Assembling	58
	197

The proportion of observed times during which the worker was adjusting is $P = \frac{105}{197} = .533$. Assuming that the day's work provides a representative sample, for this worker .533 may be used to estimate the proportion of all similar working times spent adjusting items.

This procedure is called **work sampling**, and it can be extended to estimating mean times spent on each task. Continuing with the above, the worker completed 48 parts, and out of the 480-minute shift he was idle 87 minutes. He finished his parts in $480 - 87 = 393$ minutes, averaging $\frac{393}{48} = 8.2$ minutes per part. It is estimated that he spent 53.3% of his productive time adjusting, so that the mean time spent on that task is

$$.533(8.2) = 4.4 \text{ minutes}$$

Work sampling has removed workers from the tyranny of the stopwatch while allowing a many-fold increase in the efficiency of work measurement. (The clerk in this example was able to monitor 17 other workers in the same day. That would have been impossible under continuous time monitoring.)

Problems

10–1 For each of the following situations, indicate whether or not it is possible to take a sample from a population in order to make an estimate within the next few days. Explain your answers.

(a) The mean annual earnings are to be determined for electrical engineers graduating this coming June.

(b) The preferred majors of the next year's freshmen engineering students are to be found.

(c) The lifetimes of cathode ray tubes with an experimental phosphor coating are to be determined.

(d) A power company must estimate the electricity its customers will require 10 years from now.

10–2 For each of the following parameters, state whether a point or an interval estimate would be more appropriate:

(a) A ballistic coefficient is desired for a recoverable rocket engine to be reused in future satellite launches.

(b) The mean strength of a new structural member is to be reported in an engineering journal.

(c) A forecast value for the cost per unit of a particular ingredient in gasoline must be established for planning purposes.

(d) An engineering journal is publishing mean salaries for engineers in various categories.

10–3 The sample proportion of observations of a particular characteristic is P. Show that NP is an unbiased estimator of the number of items in a population of size N that have that characteristic.

10–4 For a class project, a graduate industrial engineering student observed a coffee shop waitress during 50 random times when she was on station over a 40-hour time span. The following tally summary was obtained:

Task	Number of Observations
Taking orders	6
Walking	12
Serving food	8
Cleaning tables	5
Preparing orders	10
Collecting cash	9

An investigation of her checks shows that she served 480 customers during that period. She spent 4.5 of the working hours resting off station.

(a) Estimate the mean time per customer this employee spends serving food.

(b) Estimate the mean time per customer she walked on official business during the study period. If her average pace was 2.5 mph, how far is it estimated she walked during the study period?

10–5 Show that the median M of a random sample of size 3, taken without replacement from the following population, is an unbiased estimator of μ.

105 110 115 120 125 130 135

10–6 Using the information from Problem 10–5, compute $SD(M)$. Then establish the value of $SD(\bar{X})$ (also using $n = 3$). Which estimator, M or $\bar{X}$, is more efficient?

10–7 Find the sampling distribution of the range R (greatest minus smallest values) for a random sample of size 2 taken without replacement from the following student grade-point averages:

2.0 3.0 3.5 4.0 2.5 3.0

Show that R is a biased estimator of the population range.

10–2 Interval Estimates of the Mean

As we have seen, the interval estimate is the form favored by statisticians because it provides some acknowledgment of the error inherent in the sampling process. An interval estimate reflects sampling error in two dimensions. First, the end points demonstrate that the sample results do not precisely yield the true mean. Second, because the sampling distribution of $\bar{X}$ may usually be approximated by the normal curve, it is possible to extend probability concepts and quantify their sampling procedure in terms of its reliability.

Precision and reliability goals may be established at the outset, before the data are collected. Our present concern is with the evaluation stage of the sampling study, after the sample results are known. At that time a computed value for $\bar{X}$ is known, and there is no longer any uncertainty about the sample. That computed $\bar{X}$ serves as the estimate of the unknown population mean μ. It defines the **center** of the interval estimate. How do we establish the end points?

Assuming that the population has a finite standard deviation σ, then for a sample of size n, the standard error of $\bar{X}$ is $\sigma_{\bar{X}} = \sigma/\sqrt{n}$. *Before* the data are collected, the normal curve is applicable. At that time, the probability is .95 that $\bar{X}$ will fall within ± 1.96 standard deviations of the mean:

$$\Pr[\mu - 1.96\sigma_{\bar{X}} \leq \bar{X} \leq \mu + 1.96\sigma_{\bar{X}}] = .95$$

This expression can be translated into an equivalent form,

$$\Pr[\bar{X} - 1.96\sigma_{\bar{X}} \leq \mu \leq \bar{X} + 1.96\sigma_{\bar{X}}] = .95$$

so that μ is placed inside the end points.

We may generalize the foregoing. Denoting the interval probability for $\bar{X}$ by $1 - \alpha$, the area under that portion of the normal curve over the prescribed interval is also $1 - \alpha$, leaving areas of $\alpha/2$ remaining in each tail. The interval itself may be expressed in terms of the corresponding **critical normal deviate**, which we denote by $z_{\alpha/2}$, above which the upper-tail area under the standard normal curve is $\alpha/2$. With $.95 = 1 - \alpha$, we have $\alpha = .05$ and $\alpha/2 = .025$. Thus,

$$z_{\alpha/2} = z_{.025} = 1.96$$

In general, the $1 - \alpha$ probability interval inside the probability brackets may be expressed as

$$\mu = \bar{X} \pm z_{\alpha/2}\frac{\sigma}{\sqrt{n}} \quad \text{or} \quad \bar{X} - z_{\alpha/2}\frac{\sigma}{\sqrt{n}} \leq \mu \leq \bar{X} + z_{\alpha/2}\frac{\sigma}{\sqrt{n}}$$

The critical normal deviate for many specified upper-tail areas may be read from Appendix Table E. For instance, $z_{.01} = 2.33$ and $z_{.005} = 2.57$.

The above interval is centered at $\bar{X}$, and its end points depend partly on σ, an often unknown quantity that must be specified before probabilities can be found for $\bar{X}$. They also are affected by the level for $1 - \alpha$ (and hence $z_{\alpha/2}$) and by n, all of which can be established by the investigator. The inequality suggests an appropriate procedure for making an interval estimate of the population mean.

Confidence and Meaning of the Interval Estimate

Again, it is important to establish our perspective. Estimates are made *after* the sample data have been collected. Although there is a probability that any particular sample will place μ inside $\bar{X} \pm z_{\alpha/2}\sigma_{\bar{X}}$, once the actual sample results are available, $\bar{X}$ is a calculated and known quantity.

A probability value is no longer applicable to $\bar{X}$. Once its value has been computed, statisticians employ a related concept to express remaining lack of certainty regarding the still unknown μ. This concept is the **confidence level**.

Definition

Suppose that a sample experiment is repeated many times, each time with a different collection of n observations, and that an interval estimate is obtained from each. The **confidence level** is the percentage of those estimates providing intervals that actually would contain the true value of the population parameter being estimated.

Figure 10–1 illustrates the confidence level concept. A procedure yielding an interval estimate of the mean with a 95% confidence level can be expected to provide

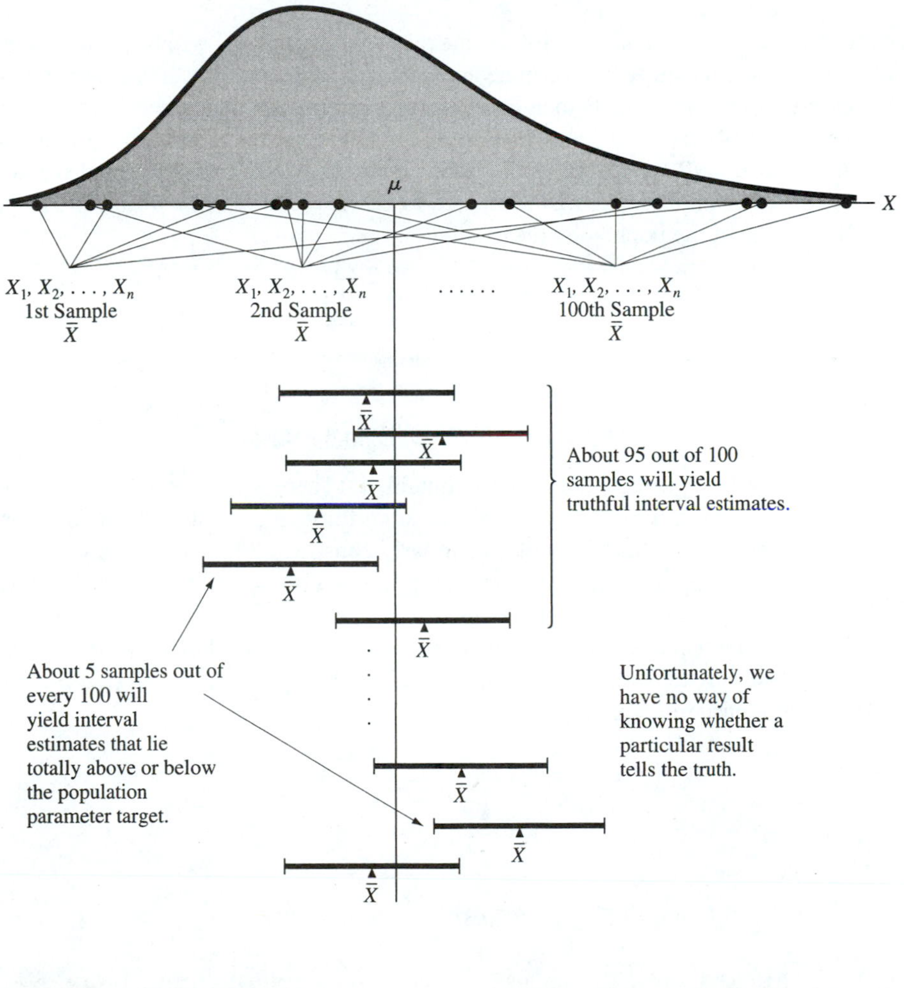

μ
X
$X_1, X_2, \ldots, X_n$
1st Sample
$\bar{X}$
$X_1, X_2, \ldots, X_n$
2nd Sample
$\bar{X}$
. . . . . .
$X_1, X_2, \ldots, X_n$
100th Sample
$\bar{X}$
About 95 out of 100 samples will yield truthful interval estimates.
About 5 samples out of every 100 will yield interval estimates that lie totally above or below the population parameter target.
Unfortunately, we have no way of knowing whether a particular result tells the truth.

Figure 10–1 *Conceptual illustration of 95% confidence level in making interval estimates of the mean.*

correct results only 95% of the time. Were such a procedure repeated 100 times, about 95 of the intervals obtained would contain μ and the remainder—the entire interval lying above or below μ—would be totally off. Of course, only one sample of size n is ordinarily taken, and *there is no way to know whether the estimate obtained is a true one*. Such is the unavoidable price we must pay for using sample evidence. The only way to improve the situation is to raise the confidence level, using the bigger z (and thus reducing the precision), or to increase the sample size.

Confidence Interval for Mean When σ Is Known

The interval estimate itself is called a **confidence interval**. The simplest confidence interval is when the value of σ is known (say, from an earlier investigation or a repeated population). We use the following expression to compute the $100(1-\alpha)\%$

CONFIDENCE INTERVAL ESTIMATE OF THE MEAN WHEN σ IS KNOWN

$$\mu = \bar{X} \pm z_{\alpha/2}\frac{\sigma}{\sqrt{n}} \quad \text{or} \quad \bar{X} - z_{\alpha/2}\frac{\sigma}{\sqrt{n}} \leq \mu \leq \bar{X} + z_{\alpha/2}\frac{\sigma}{\sqrt{n}}$$

Illustration: *Heights of Pilots and Submariners*

To illustrate, consider a sample of 100 airline pilots whose heights will be measured for a human-engineering study to determine aircraft cockpit parameters. It is assumed that the standard deviation in pilots' heights is identical to that for the general population, wherein $\sigma = 2.44''$. A $100(1-\alpha)\% = 90\%$ confidence level is chosen so that the upper-tail area is $\alpha/2 = .05$. Appendix Table E provides the critical normal deviate $z_{.05} = 1.64$. The sample mean was computed to be $\bar{X} = 69.75''$. The 90% confidence interval estimate of the population mean height is

$$\mu = 69.75'' \pm 1.64\frac{2.44''}{\sqrt{100}}$$
$$= 69.75'' \pm .40''$$

or

$$69.35'' \leq \mu \leq 70.15'' \qquad \text{(pilots, 90\%)}$$

This interval may or may not actually contain μ. (There is no way to know for certain unless all pilots in the population are measured through a census.) A proper interpretation for the above is that 90% of all similarly constructed intervals obtained from separate random samples, each involving 100 pilots, will bracket the true value of μ, but 10% will not—lying totally above or below μ.

Should the human engineers find such a result too nebulous, they may increase the confidence level, say to $100(1-\alpha) = 99\%$. In that case $\alpha/2 = .005$ and $z_{.005} = 2.57$. The 99% confidence interval is

$$\mu = 69.75'' \pm 2.57\frac{2.44''}{\sqrt{100}}$$
$$= 69.75'' \pm .63''$$

or

$$69.12'' \leq \mu \leq 70.38'' \qquad \text{(pilots, 99\%)}$$

The price paid for the increased confidence *level* is a wider, less precise confidence *interval*.

Should the lack of precision in this estimate also be undesirable, the engineers would have to take a larger sample. Suppose that in a related investigation of submariners' heights they used a sample of 500 measurements. Assuming the same standard deviation as before, with a computed sample mean of $\bar{X} = 68.39''$, the following 99% confidence interval is obtained:

$$\mu = 68.39'' \pm 2.57 \frac{2.44''}{\sqrt{500}}$$

$$= 68.39'' \pm .28''$$

or

$$68.11'' \leq \mu \leq 68.67'' \qquad \text{(submariners, 99\%)}$$

Notice that this confidence interval is quite narrow and considerably more precise than the last one found for the pilots.

Both pilots and submariners are selected partly by their physical size. Very tall people or very short people are not allowed into these groups, and so the tails of the general population are underrepresented. That would make it inappropriate for the same standard deviations to be used for pilot and submariner heights as for men in the general population, and the true level for σ in each case may be considerably different than 2.44″. Ordinarily, in a sampling study involving an unknown μ, σ is also unknown and must also be estimated from the sample results.

Confidence Interval Estimate of Mean When σ is Unknown

The normal curve is inappropriate when the population standard deviation σ is unknown. In those cases estimating μ involves procedures based on the Student *t* distribution. This distribution applies to the statistic

$$t = \frac{\bar{X} - \mu}{s/\sqrt{n}}$$

which reflects that σ is estimated by its sample counterpart *s*.

The following expression is used to calculate the $100(1 - \alpha)\%$ confidence interval.

CONFIDENCE INTERVAL ESTIMATE OF THE MEAN WHEN σ IS UNKNOWN

$$\mu = \bar{X} \pm t_{\alpha/2} \frac{s}{\sqrt{n}} \qquad \text{or} \qquad \bar{X} - t_{\alpha/2} \frac{s}{\sqrt{n}} \leq \mu \leq \bar{X} + t_{\alpha/2} \frac{s}{\sqrt{n}}$$

where, as before, two tail areas are excluded and $t_{\alpha/2}$ is the critical value at $n - 1$ degrees of freedom corresponding to an upper-tail area of $\alpha/2$. For example, when the sample size is $n = 25$ and the confidence level is 95%, the value $t_{.025} = 2.064$ is read from Appendix Table G when $\text{df} = 25 - 1$.

We may illustrate this using the example from Section 10–1 for the gold required in the manufacture of a particular logic circuit wafer. The industrial engineer in that example computed the following from a sample of $n = 15$:

$$\bar{X} = .16 \text{ ounce} \qquad s = \sqrt{.000314} = .018 \text{ ounce}$$

A 95% confidence interval is constructed. Using $\alpha/2 = .025$, $t_{.025} = 2.145$ is read from Appendix Table G for df $= 15 - 1$. Thus,

$$\mu = .16 \pm 2.145 \frac{.018}{\sqrt{15}}$$
$$= .16 \pm .010 \text{ ounce}$$

or

$$.150 \le \mu \le .170 \text{ ounce}$$

This interval does not give a very precise indication of μ. But unless the manager has a budget that can absorb further sampling expense, it will have to suffice.

***Example:** Evaluating a Physical Therapy Apparatus*

A mechanical engineer on the staff of a physical therapy research team made a sampling study to evaluate his new design for an exerciser. The device is intended to strengthen the muscles in persons suffering from chronic lower back pain. The availability of only one test device limited the number of test subjects that could be accommodated. A random sample of 12 patients was treated with the exerciser, and the following recovery times (days) were obtained:

15	23	32
18	16	22
41	29	25
27	30	18

The mean recovery time is $\bar{X} = 24.7$ days, and the sample standard deviation is $s = 7.6$ days.

An estimate was made of the mean recovery time for all patients who might receive similar therapy. Using a 95% level of confidence, the engineer used the critical value $t_{.025} = 2.201$ to calculate the following interval estimate:

$$\mu = 24.7 \pm 2.201 \frac{7.6}{\sqrt{12}}$$
$$= 24.7 \pm 4.8 \text{ days}$$

or

$$19.9 \le \mu \le 29.5 \text{ days}$$

These results were encouraging, although the device could not be recommended for general use without a controlled experiment that rigorously compared its therapy with more traditional methods. Further sampling investigations were necessary.

Example: Californium Isotope Production by Irradiating Berkelium

The data in Table 10–1 show the results of $n = 10$ production runs for producing isotopes. In each run the source material, berkelium (^{249}Bk) is irradiated and chemically treated to provide a californium isotope (^{250}Cf). From the sample results, a 95% confidence interval estimate can be constructed for the mean percentage yield of californium μ. Applying the critical value $t_{.025} = 2.262$ (from Table G) and the sample statistics, $\bar{X} = 16.92\%$ and $s = 8.692\%$, we obtain:

$$\mu = \bar{X} \pm t_{.025} \frac{s}{\sqrt{n}}$$

$$= 16.92 \pm 6.22\%$$

or

$$10.70\% \leq \mu \leq 23.14\%$$

This interval is quite wide, reflecting the high degree of variability in isotope yields from run to run.

Table 10–1

Results of Experimental Production of Californium Isotopes

Berkelium Source (mg)	Californium Product (mg)	Isotope Yield (%)
1.5	.160	10.7
1.5	.260	17.3
1.0	.135	13.5
1.5	.204	13.6
1.5	.079	5.3
3.0	1.140	38.0
3.0	.540	18.0
1.2	.235	19.6
3.1	.637	20.5
3.3	.420	12.7
	$\bar{X} = 16.92$	$s = 8.692$

Source: Knauer et al., *Transactions of American Nuclear Society,* Vol. 55, November 15–19, 1987, pp. 239–240.

Large Sample Sizes The critical values for the Student t distribution given in Appendix Table G are available for all degrees of freedom from 1 to 30. Above 30 degrees of freedom, it may be necessary to use linear interpolation. Consider the following example.

Table G gives critical values when df = 120. The next tabled values are for df = ∞, so that whenever the degrees of freedom fall above 120, the critical values may be taken from the df = ∞ row.

The values from that row are identical to the corresponding normal deviates. This reflects that the density for the Student t approaches the normal curve for large n's. Thus, for large sample sizes the critical normal deviate $z_{\alpha/2}$ can be used in place of $t_{\alpha/2}$.

***Example:* Estimating the Mean Time Between Failures (MTBF) of Floppy Disks**

Floppy disks used in personal computers take a great deal of punishment. A mainframe manufacturer wishes to establish specifications for producers who supply disks to computer users. The desired end is higher standards for the industry. The specifications will depend partly on testing with 3 prototypes, designated as *A*, *B*, and *C*, which differ in composition and density.

Separate random samples of 100 disks of each type have been selected for endurance testing. Entries are made on and read from each disk by the same program on a specially constructed disk drive until disk-caused I/O errors occur. Periodically, a robot removes each test disk to a chamber where temperatures are changed according to a program, fluctuating between 60 and 80 degrees Fahrenheit. Another robot gives the disks an occasional rubdown with a lightly oiled cloth to simulate human handling. Each disk is subjected throughout testing to slight electromagnetic disturbances. The following data were obtained for the observed cumulative drive time until first error (failure):

Disk *A*	Disk *B*	Disk *C*
$\bar{X} = 49$ hours	$\bar{X} = 55$ hours	$\bar{X} = 53$ hours
$s = 8.2$	$s = 10.1$	$s = 7.5$

With $n = 100$, the degrees of freedom for the Student t distribution are $n - 1 = 99$. Using a 95% confidence level, Appendix Table G provides the following critical values:

$$\text{df} = 60 \qquad t_{.025} = 2.000$$
$$\text{df} = 120 \qquad t_{.025} = 1.980$$

Linear interpolation provides the value

$$\begin{aligned} t_{.025} &= 2.000 + \left(\frac{99 - 60}{120 - 60}\right)(1.980 - 2.000) \\ &= 1.987 \end{aligned}$$

The following 95% confidence intervals were obtained:

$$\begin{aligned} \mu &= 49 \pm 1.987 \frac{8.2}{\sqrt{100}} \\ &= 49 \pm 1.63 \text{ hours} \end{aligned}$$

or

$$47.37 \leq \mu \leq 50.63 \text{ hours} \qquad (\text{Disk } A)$$
$$52.99 \leq \mu \leq 57.01 \text{ hours} \qquad (\text{Disk } B)$$
$$51.51 \leq \mu \leq 54.49 \text{ hours} \qquad (\text{Disk } C)$$

Notice that the MTBF confidence interval for Disk *A* lies totally below the intervals for the other two disks. The evidence is strong that *A* exhibits performance inferior to *B* or *C*. The relative performance standings of *B* and *C* are clouded, as reflected by the respective overlapping confidence intervals. Further testing with more floppy disks will be needed to determine which type of disk has the greater true MTBF.

Theoretical Assumptions Although a theoretical assumption of the Student t is that the *population* be normally distributed, it is usually appropriate for all populations except those having frequency distributions with a pronounced skew.

Confidence Interval When Population Is Small

Once the confidence level is chosen and the corresponding value for $z_{\alpha/2}$ is found, the width of the interval estimate of μ depends on $\sigma_{\overline{X}}$. The standard error of $\overline{X}$ is affected by both n and the population standard deviation σ (which, when unknown, may be estimated by s). When sampling is performed without replacement (the usual case) from a small population, the level for $\sigma_{\overline{X}}$ must reflect the population size N. As we have seen, the finite population correction factor is applied in computing $\sigma_{\overline{X}}$. We may analogously amend the procedure for finding confidence intervals, using the following expressions:

$$\mu = \overline{X} \pm z_{\alpha/2}\frac{\sigma}{\sqrt{n}}\sqrt{\frac{N-n}{N-1}} \quad (\sigma \text{ known})$$

and

$$\mu = \overline{X} \pm t_{\alpha/2}\frac{s}{\sqrt{n}}\sqrt{\frac{N-n}{N-1}} \quad (\sigma \text{ unknown})$$

Determining the Required Sample Size

The quality of a statistical estimate has two dimensions, *precision* and *reliability*. Although the desired level of one of these is always attained at the expense of the other, an investigator can achieve high levels of both precision and reliability by increasing the sample size. The following expression may be used to calculate the required sample size for estimating the mean.

REQUIRED SAMPLE SIZE FOR ESTIMATING THE MEAN

$$n = \frac{z^2_{\alpha/2}\sigma^2}{d^2}$$

where

d = desired precision (or maximum error)

$z_{\alpha/2}$ = critical normal deviate for specified reliability $1 - \alpha$

σ = assumed population standard deviation

and where we define

$$\text{Reliability} = \Pr[|\,\overline{X} - \mu\,| \leq d]$$

To illustrate, suppose that an investigator wishing to estimate the mean stretching force required to permanently distort reinforcing bars of a particular type desires precision of $d = 5$ pounds, with a reliability probability of $1 - \alpha = .95$. Previous studies of similar rods indicate that σ should be about 50 pounds. Appendix Table E provides the critical normal deviate $z_{.025} = 1.96$. Thus, the required sample size is

$$n = \frac{(1.96)^2(50)^2}{5^2} = 385 \quad \text{(rounded up)}$$

Example: *Estimating Height and Hand Length*

Imagine a study in which population means for men's heights and hand lengths are to be estimated. For planning purposes, we could use the existing anthropometric tables for Air Force flight crews,* getting

$$\sigma = 2.44'' \quad \text{(height)} \qquad \text{and} \qquad \sigma = .34'' \quad \text{(hand length)}$$

Since a different target population applies, a new sampling study will be performed. Although this may be unrealistic, for purposes of illustration we will suppose that heights and hand lengths will be measured independently in different samples. In each case the desired precision is $d = .10''$ and the required reliability is $1 - \alpha = .99$. What are the respective required sample sizes?

Appendix Table E provides $z_{.005} = 2.57$, and we have

Height	Hand Length
$n = \dfrac{(2.57)^2(2.44'')^2}{(.10'')^2}$	$n = \dfrac{(2.57)^2(.34'')^2}{(.10'')^2}$
$= 3{,}933$	$= 77$

Notice that the n needed for estimating mean height is over 49 times as great as the sample size needed to estimate mean hand length. This is because, in absolute terms, heights are considerably more varied than hand lengths, with σ for the former being over 7 times as great. Our model therefore requires that n be over $(7)^2 = 49$ times as great for heights as for hands.

The sample size model does not reflect that the two samples should probably have different d's to account for the differences in the relative sizes of the observed units. For instance, precision of $\pm.10''$ may be adequate for establishing the height of a cockpit passageway (and $\pm1.0''$ or less accuracy might suffice). But that degree of precision would hardly be satisfactory for a glove manufacturer in sizing cutting dies, where $\pm.01''$ might be more appropriate.

*H. T. E. Hertzberg, G. S. Daniels, and E. Churchill, *Anthropometry of Flying Personnel*, WADC Technical Report 52–321, U.S.A.F. (September 1954).

This example graphically illustrates the crucial role that σ plays in establishing the required sample size. But like μ itself, σ may not be known. We may guess its value with the help of the empirical rule, which tells us that approximately 99.7% of the population values will lie within $\pm3\sigma$ of the mean. About 1 out of 1,000 will fall below $\mu - 3\sigma$, with a similar fraction falling above $\mu + 3\sigma$. This suggests that two percentiles, the .1 and the 99.9, span a range of 6 standard deviations. Knowing these two quantities, we may compute the assumed population standard deviation from

$$\text{Planning } \sigma = \frac{Q_{.999} - Q_{.001}}{6}$$

Of course, there is no more reason to know the above percentiles than σ itself—but those quantities may be guessed with some confidence. A chemical engineer should have some idea about an absolute minimal processing time (say, 1 hour) and about some maximum amount that it is almost inconceivable to exceed (say, 100 hours). A crude value for σ would then be $(100 - 1)/6 = 16.5$ hours.

10-8 Suppose that a fourth type of floppy disk (D) is evaluated in the investigation described in the example. The following results apply: $\bar{X} = 57.4$ hours and $s = 11.1$. Construct a 99% confidence interval estimate for the true MTBF of this disk.

Problems

10-9 A sample of $n = 100$ observations is taken from a population of unknown mean wherein the standard deviation is assumed to be $\sigma = 5$ grams. The computed value of the sample mean is $\bar{X} = 29.4$ grams. Construct confidence interval estimates of μ for each of the following levels of confidence:
(a) 90%   (b) 95%   (c) 99%   (d) 99.8%

b) $z_{0.025}$ c) $z_{0.005}$ d) $z_{0.001}$

10-9) $n = 100$, $\bar{X} = 29.4$
a) Con. Int = 90%.
$100(1-\alpha) = 90$, $\frac{\alpha}{2} = 0.05$
$\mu = x \pm z_{\alpha/2} \frac{\sigma}{\sqrt{n}}$
$= 29.4 \pm 1.64 \left(\frac{5}{\sqrt{100}}\right)$
$28.58 \leq \mu \leq 30.32$

Table E, $z_{0.05} = 1.64$

10-10 A sample investigation of $n = 15$ waiting times at a tool cage provided $\bar{X} = 5.3$ minutes with $s = 1.2$ minutes. Construct confidence interval estimates of μ for each of the following confidence levels:
(a) 90%   (b) 95%   (c) 99%   (d) 99.5%

No σ, use t!

10-11 The precision of an estimate depends largely on the sample size. In each of the following cases, construct a 95% confidence interval estimate of μ, assuming that the population standard deviation is known to be $\sigma = \$14.50$:

10-10) $n = 15$, $s = 1.2$
$\bar{X} = 5.3$, $df = 15-1 = 14$

(a)	(b)	(c)	(d)
$n = 36$	$n = 100$	$n = 500$	$n = 1,000$
$\bar{X} = \$100.25$	$\bar{X} = \$99.75$	$\bar{X} = \$100.75$	$\bar{X} = \$100.50$

$\mu = x \pm t_{\alpha/2} \frac{s}{\sqrt{n}} = 5.3 \pm 1.761 \left(\frac{1.2}{15}\right)$
5.3 ± 0.55
$4.75 \leq \mu \leq 5.85$

10-12 Construct 95% confidence intervals for each of the following sample results:

$t_{\alpha/2}$ in table G

(a)	(b)	(c)	(d)
$\bar{X} = \$5.55$	$\bar{X} = 10.1$ lb	$\bar{X} = 10.8$ k bits	$\bar{X} = 98.1$ kHz
$s = \$1.29$	$s = .5$	$s = 2.3$	$s = .5$
$n = 25$	$n = 10$	$n = 22$	$n = 17$

10-13 An inspector wishes to estimate the mean weight of the contents in a shipment of 16-ounce cans of corn. The shipment contains 1,000 cans. A sample of 200 cans is selected, and the contents of each are weighed. The sample mean and standard deviation were computed to be $\bar{X} = 15.9$ ounces and $s = .3$ ounce. Construct a 99% confidence interval estimate of the population mean.

10-14 One student measured the heights of 35 male colleagues in her surveying class. It may be assumed that her data constitute a representative random sample of the heights of all men attending the university. She computed $\bar{X} = 70.2''$. Assuming that the population standard deviation is the same as used in the text illustrations, what is her 95% confidence interval estimate of the mean height of university men?

10-15 The industrial engineer evaluating logic wafers was also interested in the amount of silver required in the photographic stages. For the same 15 batches, the following net usages (ounces per wafer) were determined:

.25	.18	.24	.19	.20
.23	.27	.21	.23	.21
.19	.22	.20	.25	.25

Construct a 95% confidence interval estimate of the mean silver usage per wafer.

10–16 The mechanical engineer who designed the physical therapy device discussed in the example collected the following data for the amount of time (hours) spent by the test patients using his machine:

8	12	26	10	23	21
16	22	18	17	36	9

Construct a 99% confidence interval estimate for the mean time spent until recovery by all back patients who use the same therapy.

10–17 The student in Problem 10–14 computed the sample standard deviation $s = 3.1''$.
(a) Do you think it would be more appropriate for her to use this standard deviation in constructing interval estimates of μ? Explain.
(b) Find her 95% confidence interval estimate of the mean height using $s = 3.1''$ as the standard deviation.

10–18 A car manufacturer obtained the following sample results for the mileage traveled before a transmission overhaul was required in each of $n = 100$ cars tested: $\bar{X} = 76{,}400$ miles and $s = 5{,}250$ miles.
(a) Construct a 95% confidence interval estimate for the mean mileage until a transmission overhaul is needed.
(b) Interpret the estimate obtained in (a).

10–19 A product designer is testing a new consumer battery. A test was conducted in which the new battery was discharged in parallel with the existing model. The following results express the number of hours by which the new battery outlasted the old one:

5	−4	10	15	11	25	−5	17
−5	0	8	12	14	8	1	5

Construct a 95% confidence interval estimate of the mean lifetime advantage of the new battery.

10–20 The data in Table 10–2 provide the exposure rates to various radioactive elements, as measured in randomly selected sites in Riyadh, Saudi Arabia. Construct 95% confidence interval estimates for the mean exposure rates for (a) uranium, (b) thorium, (c) potassium, and (d) cesium.

10–21 With a .95 specified reliability, determine the required sample size for estimating the mean in each of the following situations:

Table 10–2
Exposure Rates ($\times 10^{-5}$ Gy/yr) to Radioactive Elements in Riyadh, Saudi Arabia

Site	Uranium ^{232}U	Thorium ^{232}Th	Potassium ^{40}K	Cesium ^{137}Cs
1	6.2	8.4	7.7	1.1
2	7.2	7.9	10.3	—
3	7.8	10.7	12.0	1.2
4	8.9	11.3	11.8	1.8
5	8.5	8.0	7.1	1.0
6	5.3	7.2	6.6	.4
7	8.7	12.3	7.8	2.1
8	8.3	10.0	9.4	.5
9	5.6	7.2	8.1	.4
10	7.9	10.1	12.0	.3
11	5.7	6.8	7.2	.4
12	7.4	10.1	10.6	3.6
13	8.1	7.2	8.3	.5
14	7.0	9.8	11.6	2.0
15	6.8	6.1	7.6	2.3
16	8.6	6.7	7.6	1.6
17	6.4	7.4	7.0	.8
18	12.2	13.5	9.3	1.2
19	7.0	9.8	11.6	2.0
20	6.0	8.2	8.0	.98
21	7.4	8.2	8.6	8.0

Source: Tawfik et al., *Transactions of American Nuclear Society,* Vol. 55, November 15–19, 1987, pp. 87–88.

(a)	(b)	(c)	(d)
$d = \$100$ $\sigma = \$1{,}000$	$d = .15''$ $\sigma = 20''$	$d = .40$ pounds $\sigma = 5.6$ pounds	$d = 1{,}000$ kHz $\sigma = 5{,}000$ kHz

10–22 With a desired precision of .5 gram, determine the required sample size for estimating the mean of a population having an assumed standard deviation of 10 grams for each of the following reliability levels:
(a) .90 (b) .95 (c) .99 (d) .999

10–23 A statistical analyst wishes to determine the sample size needed to estimate the mean maximum pressure that can be withstood by an airlock seal. Desired precision is ± 100 psi with reliability .95. She has no feel for an assumed population standard deviation; however, the designer believes that it is almost certain that an individual seal will withstand a test pressure of 5,000 psi, whereas it is extremely unlikely that any seals will survive 6,500 psi.
(a) What value should the analyst use as her assumed standard deviation?
(b) Using your answer to (a), what is the required sample size?

10–3 Interval Estimates of the Proportion

Procedures for estimating the population proportion parallel those for the mean. The process is somewhat simplified because only one parameter, π, is involved. Rather than basing procedures on the true sampling distribution for P, which we know is either the binomial (sampling from a large population) or the hypergeometric (small population), when n is large enough, we employ the normal approximation.

Estimates When Sampling from Large Populations

For sufficiently large sample sizes, the true binomial sampling distribution for P is closely approximated by the normal distribution with mean π and standard deviation

$$\sigma_P = \sqrt{\frac{\pi(1-\pi)}{n}}$$

In constructing the interval estimate of π, the computed value of P is used in the above to estimate σ_P. The following expression is used to determine the $100(1-\alpha)\%$ confidence interval.

CONFIDENCE INTERVAL ESTIMATE OF THE PROPORTION WHEN THE POPULATION IS LARGE

$$\pi = P \pm z_{\alpha/2}\sqrt{\frac{P(1-P)}{n}}$$

or

$$P - z_{\alpha/2}\sqrt{\frac{P(1-P)}{n}} \leq \pi \leq P + z_{\alpha/2}\sqrt{\frac{P(1-P)}{n}}$$

As an illustration, suppose that an estimate is made of the proportion of operating time spent on unscheduled maintenance for computer mainframes of a particular make and model. A random sample of 100 terminal log entries indicated the computers were down 5 times, so that $P = .05$. A 95% confidence interval is constructed for the population proportion of unscheduled downtime, using $z_{.025} = 1.96$ from Appendix Table E:

$$\pi = .05 \pm 1.96\sqrt{\frac{.05(1-.05)}{100}} = .05 \pm .043$$

or

$$.007 \leq \pi \leq .093$$

An interpretation similar to the one used for the mean applies to the above. Were the procedure repeated many times, using a different random sample of 100 log entries each time, about 95% of the interval estimates would correctly bracket the true level of π, but the remaining intervals would lie totally above or below π.

When to Use Normal Approximation Keep in mind that the foregoing procedure is based on the normal approximation to the binomial. That approximation works best for large sample sizes or levels of π that are not extreme. When in doubt, use the following rule.

RULE FOR MAKING NORMAL APPROXIMATION
Use the normal approximation whenever *both* of the following hold:

$$n\pi \geq 5$$
$$n(1-\pi) \geq 5$$

In applying this rule, the level of π must be estimated from the sample results by the computed level of P. When n does not satisfy these inequalities, confidence intervals may instead be constructed using exact binomial probabilities.

Estimating the Proportion When Sampling from Small Populations

When sampling without replacement from a small population (usually one in which n is more than 10% of the population size), the standard error of P involves the finite population correction factor:

$$\sigma_P = \sqrt{\frac{\pi(1-\pi)}{n}}\sqrt{\frac{N-n}{N-1}}$$

Employing this equation, we obtain slightly narrower interval estimates.

The following expression is used to compute the $100(1-\alpha)\%$ confidence interval.

CONFIDENCE INTERVAL ESTIMATE OF THE PROPORTION WHEN THE POPULATION IS SMALL

$$\pi = P \pm z_{\alpha/2}\sqrt{\frac{P(1-P)}{n}}\sqrt{\frac{N-n}{N-1}}$$

As an illustration, suppose an inspector opens a sample of 50 cartons out of a shipment of 300, noting that 6 cartons are underweight. What is the 95% confidence interval estimate for the proportion of underweight cartons in the entire shipment?

The sample proportion of underweight cartons is $P = \frac{6}{50} = .12$. Using $z_{.025} = 1.96$, $n = 50$, and $N = 300$, the following is obtained:

$$\pi = .12 \pm 1.96\sqrt{\frac{.12(1-.12)}{50}}\sqrt{\frac{300-50}{300-1}} = .12 \pm .082$$

or

$$.038 \leq \pi \leq .202$$

The inspector is 95% confident that the true proportion of underweight cartons will lie between .038 and .202. Of course, π could lie outside these limits, since the sample result might not have been very representative of the whole. There is no way for the inspector to know for sure without opening all cartons in the shipment.

When sampling without replacement from a small population, the same restrictions given earlier apply to the normal approximation. When n is too small to justify that approximation, exact *hypergeometric* probabilities may be used to construct confidence limits. (Because those probabilities usually have to be computed from scratch, the hypergeometric is often approximated by the binomial.)

Required Sample Size

A procedure similar to the one described for the mean is used for finding the required sample size for estimating the *population proportion*. Sampling situations in which choice of n is an important issue will ordinarily involve a number of observations large enough so that the normal curve may be assumed as the sampling distribution. We use the following expression to find the

REQUIRED SAMPLE SIZE FOR ESTIMATING THE PROPORTION

$$n = \frac{z_{\alpha/2}^2 \pi(1-\pi)}{d^2}$$

where

$$d = \text{desired precision (maximum error)}$$
$$z_{\alpha/2} = \text{critical normal deviate for specified reliability } 1-\alpha$$
$$\pi = \text{assumed population proportion}$$

The precision is in the same units as the proportion and will be a value such as .05, .01, or .005. Notice that the formula incorporates π. Thus, paradoxically, to find the proper size for n, we must have a pretty good idea about the level of π, the quantity being estimated.

As an illustration, the sample size required to estimate the proportion of defectives π produced by a process believed to yield as many as 10% defectives, to a precision of $d = .02$ with 95% reliability ($z_{.025} = 1.96$) would be:

$$n = \frac{(1.96)^2(.10)(.90)}{(.02)^2}$$
$$= 865$$

In making this determination, it was necessary to assume a value for π of .10. Of course, the true value of π could differ, perhaps considerably. Should the true π be greater—say, .15—the required sample size would have to be larger,

$$n = \frac{(1.96)^2(.15)(.85)}{(.02)^2}$$
$$= 1{,}225$$

or else the desired precision or reliability would not be met. On the other hand, the true π might be smaller and more than the desired precision might be obtained.

Over- or undersampling results whenever the assumed π turns out to be smaller or larger than the true level. When estimating the proportion, it is always possible to prevent undersampling, because the product $\pi(1-\pi)$ achieves a maximum when $\pi = .5$. Using that level in computing the required n guarantees that for the stated reliability the desired precision will be achieved.

Problems

10–24 Construct 95% confidence intervals for the following sample results:

(a)	(b)	(c)	(d)
$n = 100$	$n = 25$	$n = 500$	$n = 1,000$
$P = .23$	$P = .25$	$P = .05$	$P = .005$

10–25 The proportion of defectives found in a sample of 100 items taken at random from a production line is .10. Construct interval estimates of the true process proportion defective for each of the following levels of confidence:
(a) 90% (b) 95% (c) 99% (d) 99.5%

10–26 For each of the following situations, construct 95% confidence intervals for the population proportion:

(a)	(b)	(c)	(d)
$N = 500$	$N = 1,000$	$N = 5,000$	$N = 1,000$
$n = 100$	$n = 200$	$n = 1,000$	$n = 150$
$P = .15$	$P = .05$	$P = .075$	$P = .875$

10–27 An automobile parts distributor found 34 packages containing defective brake linings in a sample of 200 taken at random from a shipment containing 1,000 packages. Construct a 99% confidence interval estimate of the proportion of packages in the entire shipment that contain defective linings.

10–28 The controls in a brewery need adjustment whenever the proportion π of underfilled cans is .015 or greater. There is no way of knowing the true proportion, however. Periodically, a sample of 100 cans is selected and the contents are measured.

(a) For one sample, 4 underfilled cans were found. Construct the resulting 95% confidence interval estimate of π.

(b) What is the probability of getting as many or more underfilled cans as in (a) when in fact π is only .01?

10–29 Suppose that the proportion of voters approving the president's performance is to be estimated. Two pollsters found that (1) in May, 75 approved out of 130 persons polled, and (2) in October, 642 approved out of 1,056 persons polled.

(a) For each of these results construct a 95% confidence interval estimate of the true proportion approving.

(b) The difference in confidence limits indicates the amount of precision in an interval estimate. Determine this quantity for each of the intervals you found in (a).

10–30 An industrial engineer's assistant made 50 random observations of the upholstery installation team in an automobile assembly plant. During 12 of the observations the workers were arranging materials beside their workstation.

(a) Construct a 95% confidence interval estimate of the proportion of time installers spend arranging materials.

(b) A total of 495 cars passed upholstery installers during the 8-hour shift, during which time the line was in operation 434 minutes. Using your answer to (a), determine an interval estimate for the mean time per car spent by the installation team just arranging materials.

10–31 Determine the sample size required to estimate the population proportion when the specified reliability is .99 and the following apply:

(a)	(b)	(c)	(d)
$\pi = .20$ $d = .02$	$\pi = .50$ $d = .05$	$\pi = .05$ $d = .005$	$\pi = .01$ $d = .001$

10–4 Interval Estimates of the Variance

Like the mean and proportion, the variance (or standard deviation) is an important parameter often estimated from its sample counterpart. In Chapter 9 we saw that the sampling distribution for s^2 could be represented by a standard version of the *chi-square distribution.*

Making Estimates Using the Chi-Square Distribution

The chi-square distribution also serves as the basis for constructing confidence intervals for σ^2 (or σ). The following expression is used to construct the $100(1 - \alpha)\%$ confidence interval.

CONFIDENCE INTERVAL ESTIMATE OF THE VARIANCE

$$\frac{(n-1)s^2}{\chi^2_{\alpha/2}} \leq \sigma^2 \leq \frac{(n-1)s^2}{\chi^2_{1-\alpha/2}}$$

Notice that owing to the asymmetry of the chi-square distribution there are two critical values, $\chi^2_{\alpha/2}$ and $\chi^2_{1-\alpha/2}$, which must be read from Appendix Table H using $n - 1$ degrees of freedom.

As an illustration, consider the population of waiting times experienced by message segments buffered in a telecommunications satellite until retransmission. Sample results provide a sample standard deviation of $s = 10.4$ microseconds for $n = 25$ messages. A 90% confidence interval estimate of the variance in waiting time for all similar message transmissions is constructed below. The critical values for $\text{df} = 25 - 1$ are $\chi^2_{.05} = 36.415$ and $\chi^2_{1-.05} = \chi^2_{.95} = 13.848$. Thus,

$$\frac{(25-1)(10.4)^2}{36.415} \leq \sigma^2 \leq \frac{(25-1)(10.4)^2}{13.848}$$

or

$$71.28 \leq \sigma^2 \leq 187.45$$

Taking the square roots of the end points, we obtain a 90% confidence interval for the population standard deviation:

$$8.4 \leq \sigma \leq 13.7 \quad \text{microseconds}$$

This inequality can be interpreted similarly to the earlier interval estimates. In repeated samples, only about 90% of intervals constructed in this way will actually bracket σ^2 (or σ); the remainder will lie totally above or below the true parameter value.

This procedure assumes that the chi-square distribution is suitable. One of the theoretical assumptions of that distribution is that the *population* be normally distributed. As a practical matter, however, the chi-square works satisfactorily for most populations having frequency distributions not highly skewed. The procedure is in those cases a "robust" one.

Problems

10–32 Construct 90% confidence intervals for (1) the population variance and (2) the population standard deviation in each of the following cases:

(a)	(b)	(c)	(d)
$s = .15$	$s = 34.2$	$s = 1.01$	$s = 335.1$
$n = 25$	$n = 15$	$n = 20$	$n = 10$

10–33 The sample standard deviation for $n = 20$ observations was computed to be $s = 12.6$. Construct interval estimates of the population standard deviation for each of the following levels of confidence:
(a) 80% (b) 90% (c) 98%

10–34 Using the silver-content data in Problem 10–15, construct a 90% confidence interval estimate of the standard deviation in wafer silver consumption.

10–35 Using the time data in Problem 10–16, construct a 98% confidence interval estimate of the standard deviation in patient time on the exercise machine.

10–5 Confidence Intervals for the Difference between Means

The most common type of estimate made in two-population investigations is the magnitude of *difference* between the respective means. As noted earlier, the common designation refers to one group as population A and the second as population B. Using the respective letter subscripts, the following designations apply:

$$\mu_A = \text{Mean of population A}$$
$$\mu_B = \text{Mean of population B}$$

The designation of A and B is completely arbitrary, although the A population is often believed to have the larger values.

The following example will be used to illustrate several techniques for drawing inferences about the means of two populations.

Illustration: *The Effectiveness of Personal Computers in Engineering Education*

Faced with the prospect of upgrading the computer labs at his school, one engineering dean came to the conclusion that funds would be better spent by providing every student with a personal computer. Two random samples of students were chosen to participate in a two-year experiment to measure the effectiveness of owning personal computers. The control group consisted of 40 new sophomores who were not given personal computers and who were left to fend for themselves on computer assignments through the existing campus computer center. The experimental group consisted of a like number of students with the same background but who were permitted to check out a personal computer on long-term loan. Each student in that group received a modest budget for procuring software and supplies. The effectiveness of personal computers in engineering education was to be gauged by the respective grade-point averages achieved by the populations represented by the two groups over the two-year period.

The sample results will be described shortly.

Independent Samples

The simplest procedure for estimating the difference between means involves two independent sample groups. We will denote the sample statistics computed for the respective groups by the subscripts A and B. In making inferences regarding the difference $\mu_A - \mu_B$, we use the following composite statistic.

DIFFERENCE BETWEEN SAMPLE MEANS

$$D = \bar{X}_A - \bar{X}_B$$

This equation serves as an unbiased estimator of the difference in population means. The respective sample sizes for the two groups are denoted as n_A and n_B. A different sample size may apply to each group.

The two sample means $\bar{X}_A$ and $\bar{X}_B$ are independent random variables and, by the central limit theorem, they are approximately normally distributed. Denoting by σ_A and σ_B the respective population standard deviations, the standard errors for the respective sample means are $\sigma_A/\sqrt{n_A}$ and $\sigma_B/\sqrt{n_B}$. The difference D will also be approximately normally distributed, with mean $\mu_A - \mu_B$ and variance equal to the sum of the individual variances,

$$\sigma_D^2 = \frac{\sigma_A^2}{n_A} + \frac{\sigma_B^2}{n_B}$$

The values of σ_A and σ_B are ordinarily unknown. Two parallel procedures apply for estimating $\mu_A - \mu_B$, depending on the sample sizes.

Large Sample Sizes When n_A and n_B are both greater than 30, the following estimate is used for the standard error for D:

$$s_D = \sqrt{\frac{s_A^2}{n_A} + \frac{s_B^2}{n_B}}$$

The $100(1-\alpha)\%$ confidence interval estimate for the difference in population means may be determined in the usual manner:

$$\mu_A - \mu_B = D \pm z_{\alpha/2} s_D$$

This may be stated more conveniently by the following equivalent expression.

CONFIDENCE INTERVAL FOR THE DIFFERENCE BETWEEN MEANS USING INDEPENDENT SAMPLES (LARGE *n*'s)

$$\mu_A - \mu_B = (\bar{X}_A - \bar{X}_B) \pm z_{\alpha/2} s_D$$

As an illustration, consider the sample results in Table 10–3 for the personal computer investigation. We denote the experimental group by the letter A and the control group by B. These data will be used to estimate the difference between sophomore and junior grade-point averages (GPAs), $\mu_A - \mu_B$. Although 40 students were originally included in each group, each has suffered a reduction by 5 students, so that $n_A = n_B = 35$. The following statistics were computed:

$$\bar{X}_A = 3.38 \qquad \bar{X}_B = 3.26 \qquad s_A = .47 \qquad s_B = .46$$

The standard error for D is

$$s_D = \sqrt{\frac{(.47)^2}{35} + \frac{(.46)^2}{35}} = .111$$

The 95% confidence interval estimate of the difference in population means, using $z_{.025} = 1.96$, is

$$\begin{aligned} \mu_A - \mu_B &= (3.38 - 3.26) \pm 1.96(.111) \\ &= .12 \pm .22 \end{aligned}$$

so that

$$-.10 \le \mu_A - \mu_B \le .34$$

This indicates that the mean GPA for all students who receive free personal computers exceeds the corresponding figure for all students not given computers by some amount between $-.10$ (a disadvantage of a tenth of a grade point) and $+.34$ (a more than three-tenths advantage).

A proper interpretation of this interval is that if the sampling experiment were repeated several times, about 95% of all intervals computed in a similar fashion would contain the true $\mu_A - \mu_B$ difference. The engineering dean would probably not find these results particularly convincing, as the interval end points span a considerable range.

Small Sample Sizes—Case of Equal Variances When either $n_A \le 30$ or $n_B \le 30$, the procedure employs the Student t distribution. The key theoretical assumption is that a common variance σ^2 applies for the two populations, so that $\sigma_A^2 = \sigma_B^2 = \sigma^2$. The standard error of D may then be expressed by the following:

$$\sigma_D = \sqrt{\frac{\sigma_A^2}{n_A} + \frac{\sigma_B^2}{n_B}} = \sigma\sqrt{\frac{1}{n_A} + \frac{1}{n_B}}$$

The sample variances s_A^2 and s_B^2 each serve as unbiased estimators of σ^2. By pooling the sample results, the estimate of σ^2 may be improved. Thus, we may use a weighted

Table 10–3
Sample Results for Personal Computer Evaluation

Experimental Group (A) Free Personal Computers		Control Group (B) Computers Not Provided	
Student	Sophomore-Junior GPA X_A	Student	Sophomore-Junior GPA X_B
R. A.	3.05	M. A.	3.10
W. A.	3.05	N. A.	3.45
B. B.	3.80	T. A.	3.10
A. C.	3.60	E. B.	3.50
T. C.	3.70	F. D.	3.80
A. D.	3.10	K. D.	2.30
P. D.	2.45	S. E.	3.05
E. E.	3.75	K. F.	3.60
H. E.	2.80	M. F.	3.55
Q. E.	3.60	N. J.	3.10
R. F.	4.00	B. K.	3.75
D. G.	2.45	A. K.	3.70
B. H.	3.30	J. K.	2.85
G. H.	3.60	D. L.	2.40
D. J.	3.35	L. L.	3.70
H. J.	3.45	R. L.	3.35
R. J.	4.00	C. M.	3.90
L. K.	2.55	E. M.	3.40
J. L.	3.80	J. M.	3.75
R. L.	4.00	W. M.	3.90
L. M.	4.00	F. P.	3.20
M. M.	3.60	D. Q.	2.60
M. N.	2.75	M. R.	3.30
L. P.	2.55	L. S.	2.65
O. P.	3.25	T. S.	3.50
M. R.	3.40	T. T.	2.85
S. R.	3.20	C. U.	3.65
C. S.	3.25	D. U.	2.90
E. S.	2.95	T. V.	2.25
D. T.	3.95	V. V.	3.20
S. T.	3.80	T. W.	3.75
R. U.	3.50	W. W.	3.45
S. V.	3.25	R. Y.	3.30
A. Z.	3.85	W. Y.	3.45
W. Z.	3.60	D. Z.	2.70

average of the sample variances as an unbiased estimator of σ^2. That will also improve the estimator of σ_D, which may be determined from the following expression:

STANDARD ERROR FOR DIFFERENCE BETWEEN SAMPLE MEANS (SMALL n's)—EQUAL VARIANCES:

$$s_D = \sqrt{\frac{(n_A - 1)s_A^2 + (n_B - 1)s_B^2}{n_A + n_B - 2}}\sqrt{\frac{1}{n_A} + \frac{1}{n_B}}$$

Two sample means are used in computing the sample variances, and one degree of freedom is lost for each. In order for the weighted average of the sample variances to be an unbiased estimator of σ^2, the weights must each be one less than the respective sample size. Their sum serves as the denominator in the first fraction and equals the combined sample size, reduced by two.

The following expression is used to compute the $100(1 - \alpha)\%$ confidence interval.

CONFIDENCE INTERVAL FOR THE DIFFERENCE BETWEEN MEANS USING SMALL INDEPENDENT SAMPLES (EQUAL VARIANCES)

$$\mu_A - \mu_B = (\bar{X}_A - \bar{X}_B) \pm t_{\alpha/2}s_D$$

The critical value $t_{\alpha/2}$ is read from Appendix Table G for $n_A + n_B - 2$ degrees of freedom.

The following illustration demonstrates the procedure.

Illustration: Comparing Curing Times for Ceramic Coated Fan Jet Blades

A mechanical engineer compared the mean time of two methods for curing the ceramic coating of fan jet blades. Method A uses a static chamber containing high-temperature, high-humidity air. Method B uses a slowly moving air stream having low temperature and humidity. The following data were collected.

Curing Time (hours) Using Method A (high temperature and humidity)			**Curing Time (hours) Using Method B (low temperature and humidity)**		
15.2	13.7		14.4	14.5	
13.3	15.5		13.2	12.7	
16.4	14.4		12.3		
14.6					
$n_A = 7$	$\bar{X}_A = 14.73$	$s_A = 1.07$	$n_B = 5$	$\bar{X}_B = 13.42$	$s_B = .99$

The standard error is computed as

$$s_D = \sqrt{\frac{(7-1)(1.07)^2 + (5-1)(.99)^2}{7+5-2}}\sqrt{\frac{1}{7} + \frac{1}{5}} = .61$$

The engineer used the above to construct a 95% confidence interval estimate. With $7 + 5 - 2 = 10$ degrees of freedom, Appendix Table G provides the critical value

$t_{.025} = 2.228$. Substituting the sample means and standard error into the earlier expression, the 95% confidence interval estimate is:

$$\mu_A - \mu_B = (14.73 - 13.42) \pm 2.228(.61) = 1.31 \pm 1.36$$

or

$$-.05 \leq \mu_A - \mu_B \leq 2.67$$

The difference in mean curing time is estimated to lie somewhere between $-.05$ hour and 2.67 hours.

As with the single-sample applications involving the Student t statistic, this procedure rests on the assumption that the underlying populations have *normal* frequency distributions. In spite of that theoretical condition, the procedure generally provides satisfactory results as long as the population frequency curves are assumed to be unimodal and two-tailed. But the assumption of equal population variances is a more sensitive issue. When those are believed to be unequal, a different procedure must be used.

Small Sample Sizes—Case of Unequal Variances When the populations have different variances, the Student t may still be used to provide *approximate* results. The confidence interval is computed just like before, but the standard error for D and the number of degrees of freedom differ.

STANDARD ERROR FOR D AND DEGREES OF FREEDOM FOR SMALL INDEPENDENT SAMPLES (UNEQUAL VARIANCES)

$$s_D = \sqrt{\frac{s_A^2}{n_A} + \frac{s_B^2}{n_B}}$$

$$\text{df} = \frac{(s_A^2/n_A + s_B^2/n_B)^2}{(s_A^2/n_A)^2/(n_A - 1) + (s_B^2/n_B)^2/(n_B - 1)}$$

(rounded to nearest integer)

To illustrate, we consider a study comparing two maintenance policies.

Illustration: *Comparing Two Maintenance Policies*

As part of a summer project, an industrial engineering student is investigating maintenance policies that might be adopted by a large trucking firm. One alternative (A) involves a regular schedule of preventive maintenance after every 25,000 miles of operation, when belts, seals, gaskets, and many bearings are automatically replaced, whether or not the item is actually unsatisfactory. This policy will be compared with a radically different one (B), where there is no fixed interval for preventive maintenance and nothing is repaired until there is a breakdown or some obvious problem. Then all of the above items are checked and replaced only if necessary.

Two independent sample groups of trucks were selected to be run under one of the proposed policies for one year, after which the total maintenance cost for that period would be determined. The following data were obtained:

Policy A		
$12,366	$12,575	$13,589
11,950	13,820	12,276
11,786	12,479	13,125
12,659		

Policy B		
$7,024	$11,115	$10,443
18,203	6,450	4,255
12,357	19,204	4,158
23,425	3,718	8,295
6,225	4,870	9,146

The sample sizes are $n_A = 10$ and $n_B = 15$. The following sample statistics were computed:

$$\bar{X}_A = \$12{,}663 \qquad \bar{X}_B = \$9{,}926$$
$$s_A = \$664 \qquad s_B = \$6{,}044$$

From the above, the standard error for D and the number of degrees of freedom are computed.

$$s_D = \sqrt{\frac{(664)^2}{10} + \frac{(6{,}044)^2}{15}} = 1{,}575$$

$$\text{df} = \frac{[(664)^2/10 + (6{,}044)^2/15]^2}{[(664)^2/10]^2/(10-1) + [(6,044)^2/15]^2/(15-1)} = 14.50 \quad \text{or} \quad 15$$

For 95% confidence, with 15 degrees of freedom Appendix Table G provides the critical value $t_{.025} = 2.131$. Using the above and the computed sample means, the 95% confidence interval is constructed.

$$\mu_A - \mu_B = (\$12{,}663 - \$9{,}926) \pm 2.131(1,575)$$
$$= \$2{,}737 \pm \$3,356$$

or

$$-\$619 \le \mu_A - \mu_B \le \$6{,}093$$

The mean cost advantage per truck of fix-only-when-broken (policy B) over scheduled preventive maintenance (policy A) is therefore estimated to fall somewhere between −$619 and $6,093 per year.

Matched-Pairs Samples

A matched-pairs experiment is achieved by matching each sample unit in the experimental group to a counterpart in the control group. Each resulting pair is then evaluated as a single entity. The matching should ordinarily be done before the sample observations are taken. The objective of matching is to obtain units in each pair that are alike in many of the factors that might contribute to the level of the variable being considered.

In the personal computer evaluation, a matched pair should comprise students who are expected to perform equally well. Thus, the students in each pair should be of nearly the same demonstrated ability and achievement level. Table 10–4 shows how the students

Table 10–4
Matching Students Using SAT Scores and Freshman GPA

	Experimental Group			Control Group		
Pair	Student	Freshman GPA	Quantitative SAT	Student	Freshman GPA	Quantitative SAT
1	R. J.	4.00	740	C. M.	3.95	780
2	A. Z.	3.90	760	B. K.	3.90	750
3	S. T.	3.90	720	T. W.	3.90	690
4	L. M.	3.80	790	A. K.	3.85	720
5	B. B.	3.80	760	C. U.	3.80	710
6	R. F.	3.80	690	F. D.	3.75	730
7	R. L.	3.75	720	W. M.	3.75	670
8	Q. E.	3.60	685	L. L.	3.70	650
9	T. C.	3.60	585	J. M.	3.65	620
10	G. H.	3.50	640	W. W.	3.50	590
11	E. E.	3.40	600	K. F.	3.40	650
12	M. M	3.40	560	E. B.	3.40	610
13	J. L.	3.25	635	T. S.	3.35	595
14	D. T.	3.25	625	M. F.	3.30	660
15	R. U.	3.25	615	R. Y.	3.25	640
16	H. J.	3.25	575	E. M.	3.25	605
17	W. Z.	3.20	670	W. Y.	3.20	640
18	A. C.	3.20	610	M. R.	3.20	585
19	O.P.	3.20	590	T. A.	3.20	580
20	S. R.	3.10	710	S. E.	3.20	570
21	M. R.	3.10	690	F. P.	3.10	730
22	B. H.	3.10	660	M. A.	3.10	710
23	R. A.	3.00	685	J. K.	3.10	610
24	D. J.	2.90	675	N. A.	3.00	575
25	S. V.	2.90	650	R. L.	2.90	650
26	A. D.	2.90	620	V. V.	2.90	640
27	C. S.	2.90	580	N. J.	2.85	595
28	W. A.	2.85	610	T. T.	2.80	550
29	H. E.	2.75	590	D. U.	2.70	580
30	E. S.	2.65	610	D. Z.	2.60	725
31	M. N.	2.50	575	L. S.	2.55	640
32	L. K.	2.40	585	D. Q.	2.45	655
33	P. D.	2.30	625	D. L.	2.25	610
34	L. P.	2.20	580	K. D.	2.25	575
35	D. G.	2.10	560	T. V.	2.20	640

in the study were actually paired. That is, they were paired first in terms of their freshman GPA, deemed to be the best predictor of future academic achievement. In the case of ties in GPA, quantitative SAT scores were used to refine the final pairing choice.

Keep in mind that either the matched-pairs procedure or independent sample, but *not both*, will be used in a controlled experiment involving two populations. We have applied both procedures to the same data only to facilitate a comparison of the relative strengths and weaknesses of the two procedures.

Letting i denote the ith pair, with X_{Ai} and X_{Bi} representing the sample observations for that pair from the respective groups, we express the difference as follows.

MATCHED-PAIR DIFFERENCE

$$d_i = X_{Ai} - X_{Bi}$$

Altogether, there are n sample pairs. (We use a lowercase d to distinguish *pairwise* differences from the uppercase D used to represent the difference between two sample *means*.)

We denote the sample mean for the matched-pair differences by $\bar{d}$. The sample standard deviation is denoted by s_d. Both of these statistics are computed in the usual manner. The sample results for the personal computer illustration are shown in Table 10–5. From these we obtain

$$\bar{d} = \frac{\sum d_i}{n} = \frac{4.30}{35} = .123$$

and

$$s_d = \sqrt{\frac{\sum(d_i - \bar{d})^2}{n-1}} = .130$$

The mean matched-pair difference $\bar{d}$ is an unbiased estimator of $\mu_A - \mu_B$.

Although two populations, A and B, generate the respective samples, we consider the single *population of paired differences*. Theoretically, inferences regarding that population may be made in exactly the same way as for any single population. When the standard deviation of the paired-difference population is unknown, the Student t distribution may be applied.

The following expression is used to compute the $100(1 - \alpha)\%$ confidence interval.

CONFIDENCE INTERVAL FOR THE DIFFERENCE BETWEEN MEANS USING MATCHED PAIRS

$$\mu_A - \mu_B = \bar{d} \pm t_{\alpha/2}\frac{s_d}{\sqrt{n}}$$

where the critical value $t_{\alpha/2}$ is found from Appendix Table G for $n - 1$ degrees of freedom.

The above may be used to find a 95% confidence interval for the difference in GPAs between the experimental and control groups in the personal computer investigation. From Appendix Table G, for $n - 1 = 34$ degrees of freedom, we find $t_{.025} = 2.034$ (using linear interpolation). From this we have

$$\mu_A - \mu_B = .123 \pm 2.034\left(\frac{.130}{\sqrt{35}}\right)$$

$$= .123 \pm .045$$

Table 10–5
Matched-Pairs Differences for the Personal Computer Experiment

	Student Pair		Sophomore-Junior GPA		
i	Group A Student	Group B Student	Group A X_{Ai}	Group B X_{Bi}	Difference d_i
1	R. J.	C. M.	4.00	3.90	.10
2	A. Z.	B. K.	3.85	3.75	.10
3	S. T.	T. W.	3.80	3.75	.05
4	L. M.	A. K.	4.00	3.70	.30
5	B. B.	C. U.	3.80	3.65	.15
6	R. F.	F. D.	4.00	3.80	.20
7	R. L.	W. M.	4.00	3.90	.10
8	Q. E.	L. L	3.60	3.70	−.10
9	T. C.	J. M.	3.70	3.75	−.05
10	G. H.	W. W.	3.60	3.45	.15
11	E. E.	K. F.	3.75	3.60	.15
12	M. M.	E. B.	3.60	3.50	.10
13	J. L.	T. S.	3.80	3.50	.30
14	D. T.	M. F.	3.95	3.55	.40
15	R. U.	R. Y.	3.50	3.30	.20
16	H. J.	E. M.	3.45	3.40	.05
17	W. Z.	W. Y.	3.60	3.45	.15
18	A. C.	M. R.	3.60	3.30	.30
19	O. P.	T. A.	3.25	3.10	.15
20	S. R.	S. E.	3.20	3.05	.15
21	M. R.	F. P.	3.40	3.20	.20
22	B. H.	M. A.	3.30	3.10	.20
23	R. A.	J. K.	3.05	2.85	.20
24	D. J.	N. A.	3.35	3.45	−.10
25	S. V.	R. L.	3.25	3.35	−.10
26	A. D.	V. V.	3.10	3.20	−.10
27	C. S.	N. J.	3.25	3.10	.15
28	W. A.	T. T.	3.05	2.85	.20
29	H. E.	D. U.	2.80	2.90	−.10
30	E. S.	D. Z.	2.95	2.70	.25
31	M. N.	L. S.	2.75	2.65	.10
32	L. K.	D. Q.	2.55	2.60	−.05
33	P. D.	D. L.	2.45	2.40	.05
34	L. P.	K. D.	2.55	2.30	.25
35	D. G.	T. V.	2.45	2.25	.20

so that

$$.078 \leq \mu_A - \mu_B \leq .168$$

This confidence interval is considerably tighter than the one found for the same data when the independent-sample procedure was applied.

Example:* *Evaluation of Computer-Aided Design

The chief engineer in a machine parts manufacturing company is comparing CAD (computer-aided design) (A) to the traditional method (B). The two procedures are compared in terms of the mean time from start until production drawings and specifications are ready. A random sample of 20 parts has been selected, each to be designed twice, once by an engineer using borrowed time on the CAD system at a nearby facility and again by an engineer in-house working in the traditional manner. The two engineers designing each sample part have been matched in terms of the quality of their past performance. The engineers in the CAD group have each just completed an after-hours training program and have been judged proficient in the new system.

The following completion times (days) have been obtained:

	Design Time			Design Time	
Part	**CAD**	**Tradit.**	**Part**	**CAD**	**Tradit.**
1	8.2	11.4	11	4.3	6.1
2	15.4	25.2	12	2.7	3.5
3	4.6	4.0	13	5.2	4.8
4	5.5	6.0	14	13.8	16.9
5	8.0	11.5	15	22.5	30.0
6	10.6	9.4	16	15.3	18.4
7	5.3	8.1	17	7.5	9.0
8	5.0	6.4	18	12.7	14.4
9	19.3	20.1	19	14.5	13.5
10	7.7	9.3	20	6.3	8.1

From these data the following sample results have been obtained:

$$\bar{d} = -2.085 \qquad s_d = 2.67$$

A 95% confidence interval for the difference in mean time may be constructed from the above. Appendix Table G provides that for $20 - 1 = 19$ degrees of freedom the critical value $t_{.025} = 2.093$. The difference in mean design times is thus estimated to be

$$\begin{aligned}\mu_A - \mu_B &= -2.085 \pm 2.093\frac{2.67}{\sqrt{20}}\\ &= -2.085 \pm 1.250\end{aligned}$$

or

$$-3.335 \le \mu_A - \mu_B \le -.835 \text{ days}$$

This interval indicates that there is a mean time advantage in using computer-assisted design of about 1 to 3 days per part.

Matched Pairs Compared to Independent Samples

Suppose that the two groups have the same sample sizes so that $n_A = n_B$, and the two populations have nearly equal variances. It may be proved that, for any given reliability or confidence level and degree of precision, matched-pairs sampling will require a substantially smaller sample size than independent sampling.

It is natural to wonder at this point why independent samples are used at all. Keep in mind that matched-pairs sampling is the more complicated procedure, and each sample observation using such a design can be an order of magnitude more costly to make. Also, there may be no suitable matching criterion to apply, and even if there is, there may be no useful database for exercising it. In addition, separate samples may be obtained from the two populations by different investigators or at different times, so the entire experiment would have to be repeated if matched pairs were to be used.

Problems

10–36 An engineering society wishes to determine by how much, if at all, the mean outcome of practicing electrical engineers exceeds that of mechanical engineers. Two independent random samples provided the following data:

Electrical Engineers	Mechanical Engineers
$n_A = 76$	$n_B = 58$
$\bar{X}_A = \$37,246$	$\bar{X}_B = \$36,412$
$s_A = \$8,371$	$s_B = \$8,856$

Construct a 95% confidence interval estimate for the difference in population means.

10–37 Construct a 95% confidence interval estimate of the difference between the MTBFs of two types of power cells, assuming that the following results have been obtained for independent samples. Assume equal population variances.

	Cell A	Cell B
1	535 days	499 days
2	448	434
3	615	642
4	728	607
5	269	368
6	542	474
7	521	406
8	750	575
9	716	582
10	614	365
11	680	494
12	388	336

10–38 An electrical engineer wishes to find the difference between the mean time between failures for transformers obtained from two different vendors. The following data were obtained from independent high-temperature testing:

Vendor A	Vendor B
$n_A = 76$	$n_B = 225$
$\bar{X}_A = 1{,}246$ hr	$\bar{X}_B = 1{,}347$ hr
$s_A = 157$	$s_B = 217$

Construct a 99% confidence interval for the difference between MTBF for the two vendors.

10–39 Suppose that the observations in Problem 10–37 have been matched according to type of application, environmental conditions, and power demands. The pairs are as listed. Construct a 95% confidence interval estimate for the difference in MTBF between the two types of power cell.

10–40 A chief engineer for construction wishes to determine how electrical subcontractors differ in their ability to complete jobs quickly. From past records he has found $n = 43$ jobs done by each firm that may be considered to be representative random samples. These have been paired in terms of size, overall difficulty, general environment, time of year, and several other factors. For each pair the logged completion duration for subcontractor B was subtracted from the corresponding time taken by subcontractor A. The following data were determined:

$$\bar{d} = .5 \text{ day} \qquad s_d = 1.4 \text{ days}$$

Construct a 95% confidence interval estimate for the difference between mean completion times for the two contractors.

10–41 A heavy-equipment foreman for a regional construction company wants to compare maintenance costs for two makes of diesel engines. He has selected a random sample of 50 vehicles equipped with brand A engines. Each of these has been paired with a counterpart brand B engine, also randomly selected. Each match was based on age and type of vehicle and on anticipated usage. The following sample results summarize the difference in annual maintenance cost for each engine pair:

$$\bar{d} = -\$113.67 \qquad s_d = \$58.42$$

Construct a 99% confidence interval estimate for the difference in mean annual maintenance cost for the two makes of engine.

10–42 The following sample data have been collected independently for the drying times (in hours) of two brands of paint:

Paint A				Paint B			
12.7	13.4	14.5	11.7	9.8	10.4	12.6	13.7
10.6	11.4	12.2	13.7	12.3	11.7	12.1	10.8
14.1	13.3	12.6	12.2	12.6	11.9	10.1	9.9
11.3	12.5	12.3	13.7	12.3	12.1	11.6	10.8

Paint A				Paint B			
15.1	13.7	12.5	14.4	13.1	11.5	10.9	11.4
12.5	13.3	13.3	14.5	10.2	10.4	12.7	12.6
12.5	13.3	13.5	12.5	11.2	11.7	12.4	13.1
10.7	10.5	12.4	11.9	11.0	12.0	13.1	12.0
12.0	13.5	14.1					

Construct a 95% confidence interval estimate for the difference in mean drying times for the two paints. Assume unequal population variances and small sample sizes.

10–43 The following grade-point data have been obtained from matched pairs taken randomly from two populations of engineering students. Group A consists of cooperative students who regularly rotate between school and industry. Group B consists of regular full-time students.

Coop.	Reg.	Coop.	Reg	Coop.	Reg.	Coop.	Reg.
3.65	3.71	2.84	2.75	3.22	3.15	3.61	3.40
3.25	2.96	2.86	2.91	3.11	2.95	3.55	3.35
3.33	3.17	3.55	3.51	4.00	3.86	3.74	3.55
2.54	2.27	2.85	2.62	3.45	3.28	3.61	3.48
3.44	3.42	3.65	3.51	3.12	2.88	3.05	3.12
3.14	3.02	3.67	3.50	3.24	3.18	3.15	2.91
3.29	3.17	3.64	3.52	3.78	3.42	3.20	2.95
2.65	2.90	2.85	2.69	3.53	3.37	3.15	3.11
3.22	2.84	3.55	3.65	3.52	3.44		

Construct a 99% confidence interval for the difference in mean GPA for all students in the respective populations.

10–6 Bootstrapping Estimation

The estimation procedure described earlier in the chapter for estimating the mean using the Student t distribution is based on a critical assumption regarding the underlying population: *the parent population must be normally distributed.* That same condition applies to the chi-square distribution used in estimating the variance. When the original population deviates from having a normal distribution, the confidence intervals can distort reality.

Estimating the Population Mean with Resampling

Let's consider the simple case of drawing conclusions about the true level for the population mean μ. As we have seen, samples will ordinarily deviate from the parent population. As an alternative to the confidence interval, which relies on the results from a *single* sample, statistical bootstrapping generates *several* samples from which a pattern can be determined regarding how the possible sample results differ from each other. This procedure uses the raw power of the computer to generate those samples through a process called *resampling*. The procedure uses no mathematical theory, freeing it from the dangers of drawing any conclusions that rest on assumptions that might be invalid.

The basic premise is that a random sample can be expected to mirror the population from which it was drawn. Resampling goes a step further and uses the actual sample observations themselves as a *surrogate* for the underlying population, whose true characteristics remain unknown. As its name implies, resampling involves taking a succession of samples from this surrogate population. In essence, the sampling process is *simulated* over and over again with the objective of drawing meaningful conclusions regarding the true population. It is because the resampling procedures build upon the original sample data only that the term *statistical bootstrapping* is used. Analysts using these methods must "lift themselves by their bootstraps."

The Resampling Procedure

To illustrate, we use the following sample data representing the drying times (in minutes) of moldings made from an experimental epoxy substance.

145 227 178 165 188 239 205 265 194 217

The process is extremely simple. Suppose that for each of the $n = 10$ observations we write its value onto a million slips of paper. We then place those n million slips of paper into a large box. Those values constitute a surrogate population for the epoxy drying times that would be achieved with 10 million individual moldings.

Suppose we now randomly select from the box some of those slips and record the sample values. We would in effect be mimicking how the original sample values were obtained from the actual population of all molding drying times that would come into existence if the epoxy were to be used in the future. To avoid confusion with the original data, we will now call the new values a **resample**. Many separate resamples, each of size n, must ordinarily be selected in reaching a meaningful conclusion.

It would not be practical to actually create millions of slips of paper in the above fashion. Instead, we may paperlessly duplicate key elements of the process by using the computer to conduct a *Monte Carlo simulation*. Computer-generated random numbers can then determine which values constitute each resample. That is done in such a way that every original sample observation has an equal chance of being selected each time.

Only the n original values are involved. These are selected randomly *with replacement*, so that the resulting resamples will have identical characteristics to what would have been obtained by making random selections using n million paper slips and values. We start by assigning a random digit to each of the original data values.

Original sample value:	145	227	178	165	188	239	205	265	194	217
Represented by random digit:	1	2	3	4	5	6	7	8	9	0

A resample is built by taking ten random digits read from a list of random numbers, using the above assignment to match the data value to the successive random number.

The following results are obtained for the first resample.

Random digit:	4	7	6	6	5	9	8	7	2	2
Resample value:	165	205	239	239	188	194	265	205	227	227

$$\bar{X} = (165 + 205 + 239 + 239 + 188 + 194 + 265 + 205 + 227 + 227)/10$$

$$= 215.4$$

Notice that random numbers 7, 6, and 2 all appear twice, while no 0, 1, or 3 is obtained. The data values 205, 239, and 227 each occur twice in the resample, while 217, 145, and 178 do not.

The next 10 random numbers give a different set of values for the second resample.

Random digit:	4	1	6	3	5	9	4	0	7	3
Resample value:	165	145	239	178	188	194	165	217	205	178

$$\bar{X} = (165 + 145 + 239 + 178 + 188 + 194 + 165 + 217 + 205 + 178)/10$$

$$= 187.4$$

Successive resamples will vary, following the particular random numbers determining them. Any two resamples are no more likely to resemble each other than each is to be identical to the original data themselves. We would expect such variability with repeated physical random samples taken from the original population, each involving application of epoxy in a mold. But it would be far cheaper to let the collection of resamples similarly exhibit essential characteristics of the underlying population.

The computer makes this process fàst and easy.

Resampling with the Computer

In just a few seconds of personal computer time, 1,000 resamples each having size $n = 10$, were generated. These are summarized in the computer printout shown in Figure 10–2. The computer run was made using *Resampling Stats*, a PC program requiring a short set of command instructions for processing the original sample data values. The top portion of the printout gives the instructions, telling the program to repeat the series of commands 1,000 times. (The boldface annotations have been added to the original printout.) In each iteration a random resample is created from the original values, and the mean is computed. That set of $\bar{X}$s can be summarized in several ways, including a complete histogram.

We may use the computed resample means to make inferences about the population of all potential epoxy drying times. The inherent nature of sampling error is obvious from the histogram, which shows that many different levels of the sample mean will be obtained from repeated samples taken from the same surrogate population. Nearly all sample means will differ from any particular value, most notably the true level for the population mean μ. That would be the case even if the sample data had been physically generated from fresh experiments instead of resampling.

Figure 10–2
Details of Resampling Stats *run for epoxy drying times.*

INSTRUCTIONS	EXPLANATION
`COPY (145 227 178 165 188 239 205 265 194 217) A`	**Store data in array A.**
`REPEAT 1000`	**Generate 1,000 resamples:**
`  SAMPLE 10 A B`	**from A, each of size 10; store in array B;**
`  MEAN B C`	**calculate mean for each; store in C;**
`  SCORE C Z`	**keep score by counting frequencies for C.**
`  END`	
`PERCENTILE Z (2.5 97.5) C95`	**Compute 2.5 and 97.5 percentiles of means.**
`PRINT C95`	**Print values.**
`HISTOGRAM Z`	**Construct histogram of means.**

RESULTS

```
C95        =        182        222.7

    100+
       +
       +
F      +
r      +
e    75+                     * *
q      +                     * *
u      +                   *** ****
e      +                   ********
n      +                  *********
c    50+                 **********
y      +                 **********
       +               * *************
*      +               ***************
       +              ****************
Z    25+              ****************
       +             ********************
       +             ********************
       +          * **********************
       +          **************************  *
      0+-----------------------------------------
         |^^^^^^^^^|^^^^^^^^^|^^^^^^^^^|^^^^^^^^^|
        160       180       200       220       240
```

Mean of Resample

The Resampling Interval Estimate

Collectively, the resampling results tell us much more about the mean of the underlying population than could be generated from a single sample. One way of doing this, while at the same time acknowledging the presence of sampling error and giving an indication of its magnitude, is from an *interval estimate*. We accomplish this by finding the limits within which a stated percentage of the 1,000 computed resample means happens to fall.

The instructions in Figure 10–2 request that two percentile values be found. These are the 2.5 and the 97.5 percentiles, $Q_{.025}$ and $Q_{.975}$, which the program finds by sifting through the 1,000 resample means generated. The computer printout provides $Q_{.025} = 182$ and $Q_{.975} = 222.7$.

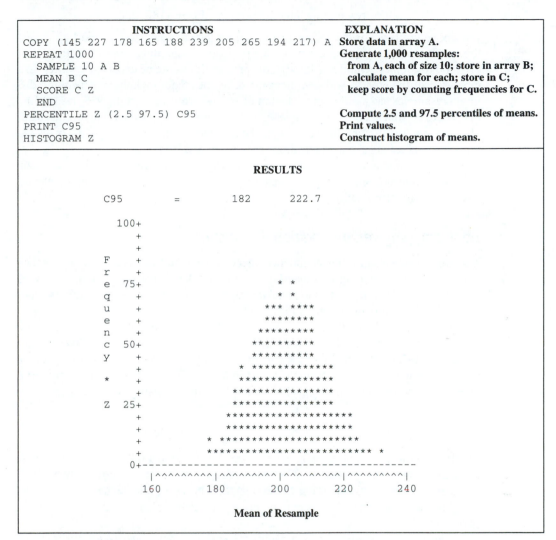

The two percentiles establish limits within which 95% of the computed means fall. Together the limits define the

$$95\% \text{ resampling interval estimate of } \mu = (182, 222.7)$$

The above limits tell us the approximate level for the unknown μ. By reporting the estimate over a *range*, the above conveys the presence of sampling error. It has the power of 1,000 resamples behind it. Although the precise limits can be expected to differ, in a repetition of the process with ten thousand different random numbers, generating different resamples and a different collection of $\bar{X}$s, we would expect the 95% resampling interval to be similar and to have closely matching limits. And, even if the original sample values were different, the process can be expected to deliver a very similar resampling interval estimate, but whose limits may be quite different.

The resampling interval estimate may be used as an alternative to the traditional confidence interval. Since it reports what was actually achieved, the resampling interval has a concrete interpretation.

Bootstrapping versus Traditional Statistics

The primary advantage of the confidence intervals obtained from resampling is that they assume nothing regarding the underlying nature of the parent population. It may be helpful if we use the same data to compare the two procedures when the assumption of population normality might be invalid.

Figure 10–3 shows the results of a run made using *Resampling Stats*. This 1,000 resample run was made using the sample data given on page 306 for the recovery times of patients using a physical therapy apparatus. The histogram summarizes the sample means computed from each resample. Using individual $\bar{X}$s as estimators of μ, the computer run provides the following 95% resampling interval estimate of the population mean:

$$20.958 \leq \mu \leq 28.958 \qquad \text{(resampling with recovery times)}$$

The lower limit for this interval is the .025 fractile from the 1,000 computed resample means, and the upper limit is the .975 fractile. A comparison of the above with the 95% confidence interval estimate for μ found on page 306 shows it to be narrower by about 1 day on each end. This discrepancy might be attributed to nonnormality of the parent population.

The discrepancies between the traditional and resampling interval estimates can be substantial. A second example is provided by the computer run in Figure 10–4 showing the original sample of $n = 12$ light bulb lifetimes (hours of operation). The printout lists the original data and computed values of $\bar{X} = 267.76$ and $s = 338.77$ and presents a sample histogram of the individual values. The data do not present the bell shape we associate with the normal curve and are instead consistent with a single-tailed distribution, such as the exponential, commonly used with lifetime data.

Figure 10–5 shows the results of 1,000 resamples made using the original sample light bulb lifetimes to establish the surrogate population. The 95% resampling interval estimate obtained for the mean is

$$147.10 \leq \mu \leq 420.54 \qquad \text{(resampling with light bulb lifetimes)}$$

Figure 10–3

Computer run with physical therapy recovery time data using Resampling Stats.

INSTRUCTIONS	EXPLANATION
`COPY (15 18 41 27 23 16 29 30 32 22 25 18) B`	**Store data in array B.**
`REPEAT 1000`	**Generate 1,000 resamples:**
`  SAMPLE 12 B D`	**from B, each of size 12; store in array D;**
`  MEAN D DM`	**calculate mean for each; store in DM;**
`  SCORE DM ZM`	**keep score by counting frequencies for DM.**
`  END`	
`PERCENTILE ZM (2.5 97.5) C95`	**Compute 2.5 and 97.5 percentiles of means.**
`PRINT C95`	**Print values.**
`HISTOGRAM ZM`	**Construct histogram of means.**

RESULTS

```
C95      =      20.958      28.958
```

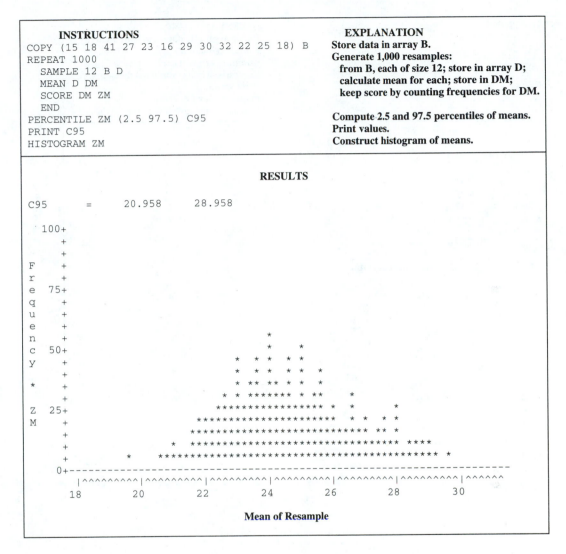

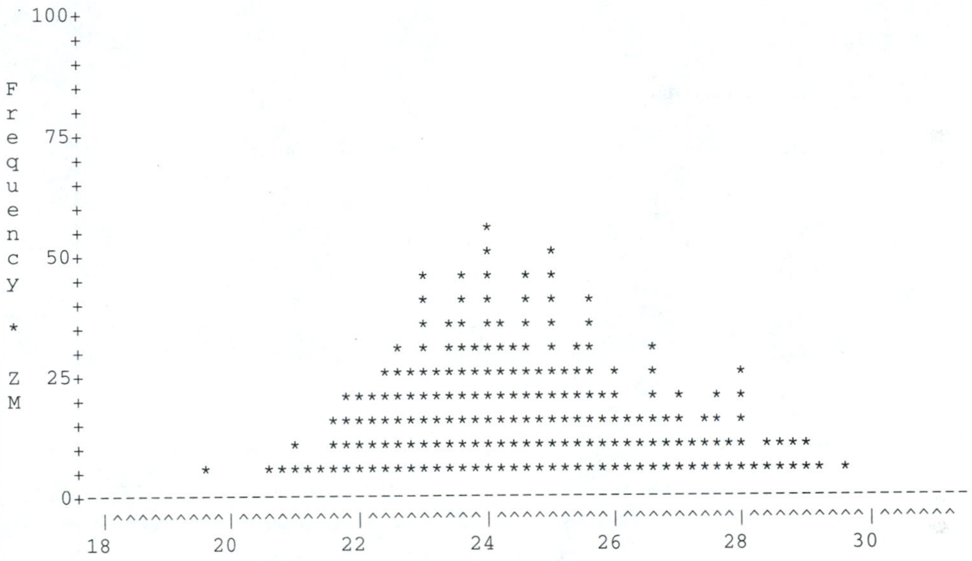

Mean of Resample

Using the original sample data with $t_{.025} = 2.064$, the traditional 95% confidence interval is computed:

$$\mu = 267.76 \pm 2.064(338.77)/\sqrt{(25 - 1)} = 267.76 \pm 139.84$$

or

$$127.92 \leq \mu \leq 407.60 \qquad \text{(traditional with light bulb lifetimes)}$$

The two intervals coincide in spite of the evidence that the population is not normally distributed.

That is *not* the case with the variance. Figure 10–5 provides a 95% resampling interval estimate of σ^2:

$$49{,}216 \leq \sigma^2 \leq 167{,}040 \qquad \text{(resampling with light bulb lifetimes)}$$

Figure 10–4
Computer run using Resampling Stats *showing original light bulb lifetime sample data.*

INSTRUCTIONS	EXPLANATION
`READ FILE "LIFETIME.DAT" T`	**Retrieve sample data and store in array B.**
`PRINT T`	**Print original data.**
`MEAN T TM`	**Compute original sample mean.**
`STDEV T TS`	**Compute original standard deviation.**
`PRINT TM`	**Print mean.**
`PRINT TS`	**Print standard deviation.**
`HISTOGRAM T`	**Construct histogram for original data.**

Original Data

```
T        =        3       7      20       1     151     399
                406     995      57      93     554       1
                387     908      31       8      21      55
                149       2     421     887      50     137
                951

TM       =   267.76   (mean)

TS       =   338.77   (standard deviation)

    20+
      +
      +
F     +
r     +
e   15+
q     +
u     +
e     +
n     +
c   10+
y     +
      +
*     +
      + *
T    5+ *
      + *
      + *  *
      + ** *   *            *
      + **** * **          ***      *               **  * *
     0+-----------------------------------------------------
        |^^^^^^^^^|^^^^^^^^^|^^^^^^^^^|^^^^^^^^^|^^^^^^^^^|
        0        200       400       600       800      1000
```

Original Sample Lifetime

Applying the traditional method, we use $\chi^2_{.025} = 12.30$ and $\chi^2_{.975} = 39.63$ obtained from Appendix Table H with linear interpolation, the following 95% confidence interval is obtained for the variance:

$$(25 - 1)(338.77)^2/(39.63) \leq \sigma^2 \leq (25 - 1)(338.77)^2/(12.30)$$

or

$$69{,}502 \leq \sigma^2 \leq 223{,}932 \qquad \text{(traditional with light bulb lifetimes)}$$

Figure 10–5

Computer run using Resampling Stats *showing resampling results with light bulb lifetime data.*

INSTRUCTIONS	EXPLANATION
`READ FILE "LIFETIME.DAT" T`	**Retrieve sample data and store in array T.**
`REPEAT 1000`	**Generate 1,000 resamples:**
`SAMPLE 25 T RT`	**from BT each of size 25; store in array RT;**
`MEAN RT RTM`	**calculate mean for each; store in RTM;**
`VARIANCE RT RTV`	**calculate variance for each; store in RTV;**
`SCORE RTM ZM`	**keep score by counting frequencies for DM.**
`SCORE RTV ZV`	**keep score by counting frequencies for RTV.**
`END`	
`PERCENTILE ZM (2.5 97.5) C95`	**Compute 2.5 and 97.5 percentiles of means.**
`PRINT C95`	**Print values.**
`HISTOGRAM ZM`	**Construct histogram of variances.**
`PERCENTILE ZV (2.5 97.5) CV95`	**Compute 2.5 and 97.5 percentiles of variances.**
`PRINT CV95`	**Print values.**
`HISTOGRAM ZV`	**Construct histogram of variances.**

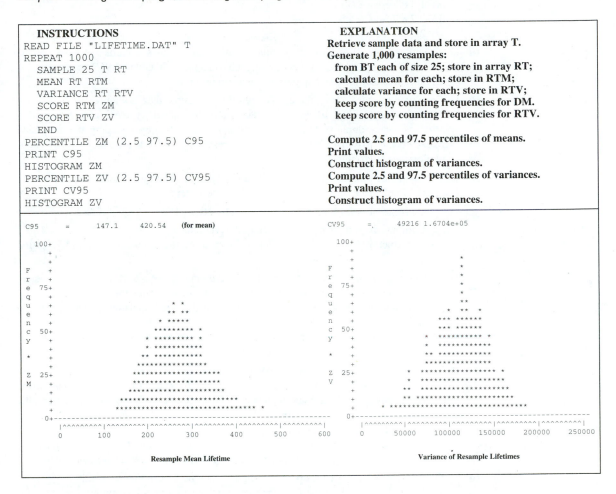

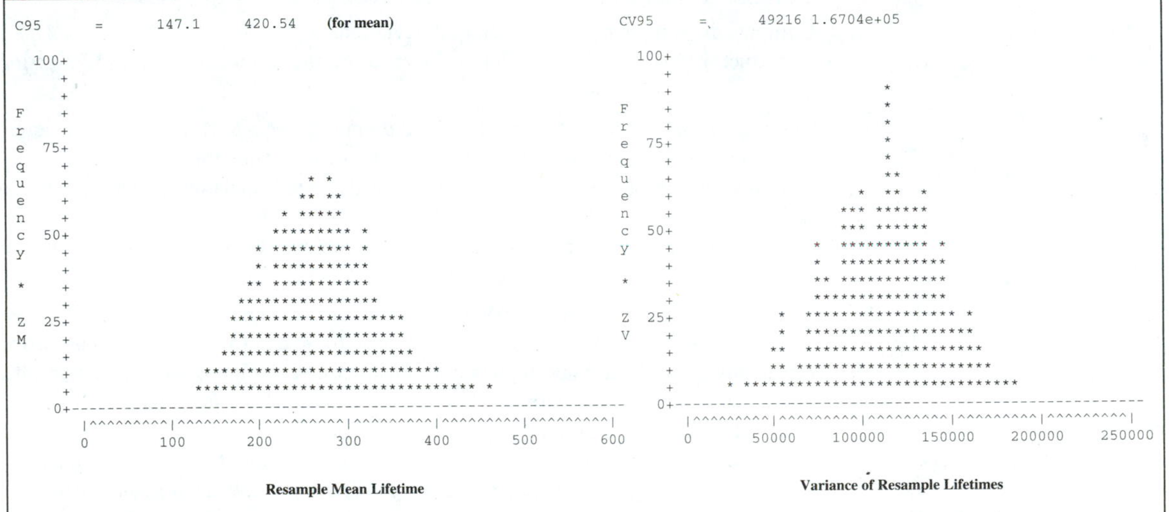

The resampling interval disagrees considerably with the traditional interval and is narrower too.

The information provided by the resampling interval is free of any restrictive assumptions and might therefore be superior for analytical purposes. The traditional interval for σ^2, however, must be rejected (even though it overlaps with the resampling interval) because it rests on a foundation built with the faulty assumption that the parent population has a normal distribution.

Problems

10–44 The following sample data were obtained for the curing time (hours) of an experimental adhesive.

31.5 28.7 42.5 39.3 29.8 31.0 36.3 33.4 35.5 37.2

Using a random number assignment such that 1 corresponds to the lowest possible sample time, 2 to the second lowest, and so on up to 0 for the highest, determine (by hand) the resample values for each of the following sets of random numbers and compute the resample means.

(a) 1 4 0 7 5 6 8 8 0 1

(b) 9 7 8 7 3 4 6 6 4 0

(c) 2 5 5 1 8 5 6 9 6 1

10–45 *Computer Exercise.* Refer to the sample adhesive curing times in Problem 10–44. Generate 1,000 resamples of size 10, computing the mean for each. Obtain a 95% resampling interval estimate of the true population mean curing time.

10–46 *Computer Exercise.* Using the epoxy drying time sample data in Figure 10–2, generate 1,000 resamples of size 10, computing the *median* for each. (If you are using *Resampling Stats*, you may duplicate the instructions in Figure 10–2 by replacing the command MEAN B C by MEDIAN B C.) Obtain a 95% resampling interval estimate of the true population median drying time.

10–47 The following sample data were obtained for the digitization time (milliseconds) for images of Earth made by a ground mapping satellite system.

14.5 50.4 23.5 19.3 49.7 35.2 16.3 152.3 76.1 47.0

(a) Compute the sample mean and standard deviation.

(b) Construct a 95% confidence interval estimate of the population mean digitization time.

10–48 *Computer Exercise.* Refer to Problem 10–47 and your answers to that exercise. Generate 1,000 resamples of size 10, computing the mean digitization time for each.

(a) Obtain a 95% resampling interval estimate of the true population mean digitization time.

(b) Compare the above to the interval obtained in Problem 10–47 (b); do the results coincide closely? If not, briefly discuss why.

10–49 *Computer Exercise.* Refer to the epoxy drying time sample data in Figure 10–2.

(a) Generate 5,000 resamples of size 10, computing the mean for each. (If you are using *Resampling Stats*, you may duplicate the instructions in Figure 10–2, replacing the command REPEAT 1000 with REPEAT 5000.) Obtain a histogram and a 95% resampling interval estimate of the true population mean salary.

(b) Compare this interval to the one obtained in the text. Should they differ by much?

(c) Compare the histogram obtained to the one in Figure 10–2. What do you notice?

10–50 *Computer Exercise.* Use the epoxy drying time sample data in Figure 10–2.

(a) Generate for each of the following cases 1,000 resamples, computing the mean for each. Obtain in each case a 95% resampling interval estimate of the true population mean drying time and a histogram.

  (1) Use a sample size of $n = 25$. (The original data will still be the initial 10 data values. If you are using *Resampling Stats*, change the SAMPLE 10 A B command in Figure 10–2 to SAMPLE 25 A B.)

  (2) Use instead a sample size of $n = 100$.

(b) What are the important differences between (1) and (2)? What may you conclude regarding sample size and estimating precision?

10–51 Refer to the digitization data in Problem 10–47. Construct a 90% confidence interval estimate of the population variance.

10–52 *Computer Exercise.* Refer to Problem 10–51 and your answers to that exercise. Generate 1,000 resamples of size 10, computing the variance in digitization time for each.

(a) Obtain a 90% resampling interval estimate of the true population variance for digitization time.

(b) Compare the above to the interval obtained in Problem 10–51; do the results coincide closely? If not, briefly discuss why.

Review Problems

10–53 An estimate must be made of the mean quantity of scrap sheetmetal consumed in fabricating home-furnace housings. A random sample of 100 furnaces were monitored throughout the complete production cycle. All raw sheetmetal for individual furnaces was allotted in advance and color coded. All scraps were set aside and their dimensions noted. The following summarize these amounts: $\bar{X} = 55.3$ square feet and $s = 5.4$ square feet. Construct a 95% confidence interval estimate for the mean amount of scrap per furnace.

10–54 The following costs per 1,000-liter batch were obtained for pilot runs of a technical chemical:

\$105	\$110	\$108	\$97	\$ 99
104	112	103	98	105

(a) Find the value of an efficient, unbiased, and consistent estimator of the mean cost.
(b) Find the value of an unbiased and consistent estimator of the variance in cost.
(c) Find the value of an unbiased estimator of the proportion of costs exceeding \$100.

10–55 Using the data in Problem 10–54, construct 90% confidence interval estimates of the following:
(a) the mean cost per batch
(b) the variance in cost per batch
(c) the standard deviation in cost per batch

10–56 The mean drying time of an epoxy coating is to be estimated. Assuming a standard deviation value of 1.5 hours, determine the required sample size under the following conditions:
(a) The reliability for being in error at most .25 hour in either direction is .99.
(b) The reliability for being in error at most .25 hour in either direction is .95.
(c) The reality for precision of $\pm.5$ hour is .99. How does the resulting sample size requirement compare with your answer to (a)?

10–57 An electrical engineer wishes to estimate the mean signal-to-noise ratio for a radar operating in the Doppler-shift mode while following aircraft at a nominal range and altitude. A minimum value of 10 is anticipated, and the maximum ratio believed possible is 40. How many observations are necessary to estimate the mean signal-to-noise ratio with a precision of ± 2 at a reliability of .95?

10–58 The following sample data were obtained for the time (in minutes) required to hand-assemble a throttle:

11.2	18.5	8.7	12.4	13.5
9.9	12.9	15.4	12.6	16.7
10.2	10.5	14.4	17.7	15.5

(a) Construct a 95% confidence interval estimate of the mean throttle-assembly time.
(b) Construct a 96% confidence interval estimate of the standard deviation in throttle-assembly time.

10–59 For a class project an industrial engineering student used work sampling to estimate the mean times taken by campus bookstore checkers to make change. In 100 random observations, change was being made 23 times. Over the observed span, 154 customers checked out, and 258 employee minutes were logged at the checkstands.

Assuming the total checkout time per customer does not vary from the above, construct a 95% confidence interval estimate of the mean time to make change.

10–60 A civil engineer wishes to find the energy consumption of a proposed sports complex that incorporates new design concepts. The major concern is natural-gas consumption, which will be simulated using Monte Carlo methods.

(a) For planning purposes the engineer must establish a value for the standard deviation in annual gas requirements. It is virtually certain that 10,000 mcf (thousand cubic feet) will be consumed, even in the warmest of years; it is inconceivable that gas requirements in any given year will exceed 100,000 mcf. What value should be used for the standard deviation in annual gas consumption?

(b) Assuming precision requirements of ±500 mcf for estimating mean annual gas requirements with .99 reliability, how many years should be simulated?

10–61 Using the data in Problem 10–60, consider the following. Because the civil engineer could not afford the computer time needed to simulate the ideal number of years, only 200 years were simulated. The gas consumption summary statistics are $\bar{X} = 55{,}300$ mcf and $s = 9{,}250$ mcf.

(a) Construct a 95% confidence interval estimate of the mean annual gas consumption.

(b) The engineer wants to see how much the simulation undersampled. How many more simulated years would be needed to meet precision of ±500 mcf with .99 reliability, assuming that the standard deviation of the foregoing sample is a satisfactory estimate of the population standard deviation?

10–62 An electrical engineer wishes to use the following sample lifetime data to compare two different brands of relay fuses in terms of their mean times to failure.

Brand A (hours)			Brand B (hours)			
1,138	1,401	578	1,062	1,169	10	1,237
2,622	881	205	149	2	557	1
169	678	344	1,164	302	737	1,688
1,222	2,550	235	1,411	2,719	330	225
797	1,205	1,209	828	321	558	1,272

Construct a 95% confidence interval for the difference in mean times to failure. Assume unequal variances.

10–63 A materials researcher wishes to use the following sample curing-time data to compare two methods for applying an adhesive.

Method A (hours)		Method B (hours)		
14	13	8	7	7
16	14	9	10	7
13	14	12	5	10
15	15	6	14	6
16	11	7	10	7

Construct a 95% confidence interval for the difference in mean curing times. Assume unequal variances.

CHAPTER ELEVEN

Statistical Testing

LIKE ESTIMATION, **statistical testing** is one of the basic forms of statistical inference. This broad area of statistics usually culminates in an immediate decision, making it the more dynamic form of statistical inference. Testing is based on two complementary assumptions, or *hypotheses*, regarding unknown populations. Taking various forms, the most common hypotheses involve assumed levels for a population parameter.

This chapter begins with tests for the unknown value of the population mean. These procedures are useful for evaluating decisions involving two complementary actions when the unknown level of μ is the pivotal factor. For example, the decision whether or not to repair equipment might depend on how much the mean reading deviates from some nominal figure. Statistical inferences are needed in such evaluations because μ can be evaluated only indirectly from sample data. For this purpose the sample mean serves as the *test statistic*. Thus, repairs might be requested only if the computed level of $\bar{X}$ exceeds a specified level.

Of course, samples may give false readings of reality. The potential for such sampling error can be controlled but never eliminated. Sampling error may be managed by establishing a *decision rule* that achieves some optimal balance among the probabilities for taking incorrect actions. For example, the incidence of unnecessary repairs must be kept small, while failure to make needed repairs must be avoided. Hypothesis testing procedures focus on such errors of commission and omission, which may be controlled by judicious determination of the decision rule or by selecting a sample large enough to keep the error probabilities acceptably small.

11–1 Basic Concepts of Hypothesis Testing

To introduce hypothesis testing, we will consider a decision involving the population mean. The following example illustrates the underlying concepts.

***Illustration**: Bacterial Leaching in Ore Processing*

A mining engineer is studying ways to increase the production of metal from a large copper deposit. A substantial amount of copper-bearing ore is presently bypassed because its quality is too low for economical processing at current world prices. The engineer is interested in a new process based on bacterial leaching. This technique isolates key minerals in poor ores through a chemical reaction caused by strains of the bacterium *Thiobacillus*. These microbes convert the iron in the ore from a ferrous form into ferric iron, which then oxidizes the insoluble copper sulfides present in the ore's pyrites. The process culminates in collection of a liquid solution from which copper may be recovered.

The effectiveness of bacterial leaching depends on the particular strain of bacteria used and the composition of the ore. A statistical sampling study will be performed to determine whether bacterial leaching should be adopted on a major scale. A sample of 100 loads of ore will be removed from random coordinates throughout the unworkable deposit site and processed by a promising strain of bacteria at a test leaching dump. The final yield, in pounds of copper per ton of ore, will then be determined for each sample load. These data will represent the yields obtainable from bacterial leaching of the more than 100 million tons of target ore.

The success of large-scale bacterial leaching will depend on the level of the mean copper yield μ that will ultimately be achieved from the entire deposit. On the basis of current prices and present processing costs, the break-even level has been established at 36 pounds of recoverable copper per ton of presently unworkable ore. This quantity is the pivotal level for the statistical investigation.

The Structure of a Hypothesis Test

Statistical testing involves two complementary hypotheses. Here these are defined in terms of levels for the unknown population mean μ. The engineer has established the following hypotheses:

Null hypothesis (process is uneconomical)

$$H_0 : \mu \leq 36 \text{ lb/ton}$$

Alternative hypothesis (process is economical)

$$H_1 : \mu > 36 \text{ lb/ton}$$

Although the adjective "null" has historically been applied to the hypothesis representing no change, this designation is arbitrary. Most often, *the null hypothesis will be the one for which erroneous rejection is the more serious consequence*. For notational convenience, the null hypothesis is represented by H_0 and the opposite or alternative hypothesis by H_1.

Hypothesis testing involves two complementary actions or choices. Our mining engineer must decide whether to adopt full-scale bacterial leaching or to abandon the process for now (and perhaps do further testing). These two actions may be expressed in terms of the foregoing hypotheses as

Accept H_0 (abandon bacterial leaching for now)

Reject H_0 (adopt bacterial leaching)

The decision must be made without knowing which of the two hypotheses is true. A determination must be made whether the sample evidence tends to favor the null hypothesis or refute it. This will be based on the following.

TEST STATISTIC

Sample mean copper yield $\overline{X}$ (lb/ton)

A large level for $\overline{X}$ (say, 50 lb/ton) will give the support to the efficacy of bacterial leaching, whereas a small value (such as 10) will deny it. Some critical level must serve as the point of demarcation between adopting the process or abandoning it.

After the sample data are obtained, the engineer will compute the test statistic and apply the following.

DECISION RULE

Accept H_0 (abandon bacterial leaching for now)
if $\overline{X} \leq 40$ lb/ton

Reject H_0 (adopt bacterial leaching)
if $\overline{X} > 40$ lb/ton

The pivotal value of 40 lb/ton serves as the **critical value** of the test statistic. It is sometimes convenient to portray the decision rule in terms of **acceptance** and **rejection regions**, as in the following:

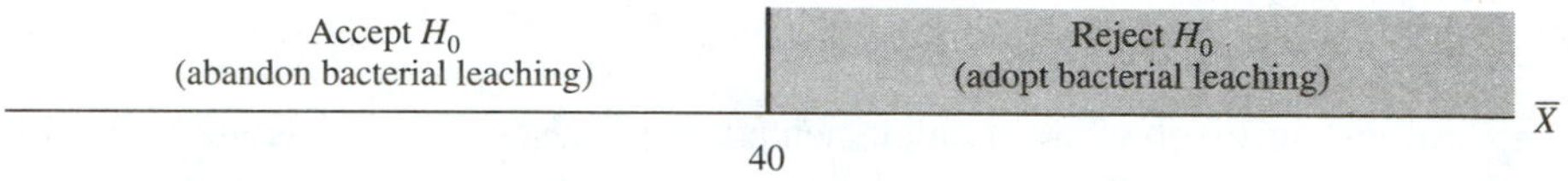

The critical value of $\overline{X}$ lies above the 36 lb/ton break-even level for μ that the engineer used to establish the hypotheses. The test statistic's critical value will ordinarily lie outside the range of the null hypothesis. This provides some leeway and reduces the chance that an atypical sample result will cause the wrong decision.

At this point it may be helpful to summarize the structure of the mining engineer's investigation. Table 11–1 arranges the problem in terms of a decision table. There is a column for each of the decision-maker's two choices or acts and a row for the two uncertain population events, representing the truth or falsity of the null hypothesis. Notice that each act-event combination culminates in a particular outcome. Two of these are correct decisions:

To accept H_0 when H_0 is true
(abandon an uneconomical process)
To reject H_0 when H_0 is false
(adopt an economical process)

Two other outcomes result in **errors**. These are:

Type I error: To reject H_0 when it is true
(adopt an uneconomical process)
Type II error: To accept H_0 when it is false
(abandon an economical process)

Table 11–1
Decision Table for Bacterial Leaching Illustration

Events	Acts	
	Accept H_0 (abandon process) $\overline{X} \leq 40$	**Reject H_0** (adopt process) $\overline{X} > 40$
H_0 True (process uneconomical) $\mu \leq 36 \quad (= \mu_0)$	**Correct Decision** (abandon uneconomical process)	**Type I Error** (adopt uneconomical process) probability α
H_0 False (process economical) $\mu > 36 \quad (= \mu_0)$	**Type II Error** (abandon economical process) probability β	**Correct Decision** (adopt economical process)

Finding the Error Probabilities

A major concern in hypothesis testing is controlling the incidence of the two kinds of errors. The degree of such control may be summarized by the error probabilities:

$$\alpha = \text{Pr [type I error]} = \text{Pr [reject } H_0 \mid H_0 \text{ true]}$$
$$\beta = \text{Pr [type II error]} = \text{Pr [accept } H_0 \mid H_0 \text{ false]}$$

Although both errors are undesirable, rejecting a true null hypothesis is typically the more serious. For this reason the type I error probability α is also referred to as the **significance level**.

The sampling distribution of $\overline{X}$ may be assumed to be normally distributed with mean μ and standard error $\sigma_{\overline{X}} = \sigma/\sqrt{n}$. Appropriate parameters must be established for the particular normal curve used in finding the error probabilities.

Using the decision rule established earlier for the bacterial leaching study, we may evaluate the error probabilities. Under the null hypothesis $\mu \leq 36$, so a range of values applies. Any value 36 or less is permitted under H_0, such as $\mu = 12.3$, $\mu = 27.5$, or $\mu = 34.99$. Type I error probabilities are usually determined for the worst case, when μ falls exactly at the H_0 limit, which is denoted by μ_0. In the present illustration $\mu_0 = 36$ lb/ton. In computing α, it is assumed that $\mu = \mu_0$.

Although σ is not precisely known, the mining engineer has sufficient experience to arrive at a "ballpark" guess. A yield almost certain to be exceeded, 1 pound per ton (which is what would be expected by just washing the ore with water and processing the rinse waste), was subtracted from the present yield of higher-quality, conventionally processed ores—150 lb/ton. Using the procedure of Section 10–2, a range of 6 standard deviations was thereby established. Thus, for planning purposes he found

$$\sigma = \frac{Q_{.999} - Q_{.001}}{6} = \frac{150 - 1}{6} = 24.83$$

which for convenience was rounded to 25 lb/ton. (A better estimate of σ may be obtained from a related set of data or once the sample results are obtained.) Using this quantity, we have $\sigma_{\overline{X}} = 25/\sqrt{100} = 2.5$.

We use the engineer's decision rule to compute the type I error probability

$$\begin{aligned}\alpha &= \Pr\,[\text{reject } H_0 \,|\, H_0 \text{ true}]\\ &= \Pr\,[\overline{X} > 40 \,|\, \mu = 36\ (= \mu_0)]\\ &= 1 - \Phi\left(\frac{40-36}{2.5}\right) = 1 - .9452 = .0548\end{aligned}$$

Thus, there is a better than 5% chance that the engineer will get a sample result so untypically great that he will adopt an uneconomical bacterial leaching program. (This is actually a worst-case figure, since when $\mu = \mu_0 = 36$, the adopted program will exactly break even; for any lower level, such as $\mu = 20$, the chance of incorrect rejection of H_0 will be more remote.)

This appears to be very good. But what about that other error, accepting H_0 when it is false?

Similar computations provide type II error probabilities. There are many ways for the null hypothesis to be false—this will happen in the present illustration whenever the true μ exceeds μ_0. Suppose we assume that $\mu = 42$ lb/ton. We then have

$$\begin{aligned}\beta &= \Pr\,[\text{accept } H_0 \,|\, H_0 \text{ false}]\\ &= \Pr\,[\overline{X} \le 40 \,|\, \mu = 42]\\ &= \Phi\left(\frac{40-42}{2.5}\right) = .2119\end{aligned}$$

There is a substantial probability that the engineer will get an untypically small level for $\overline{X}$ when indeed the bacterial leaching process would be quite economical.

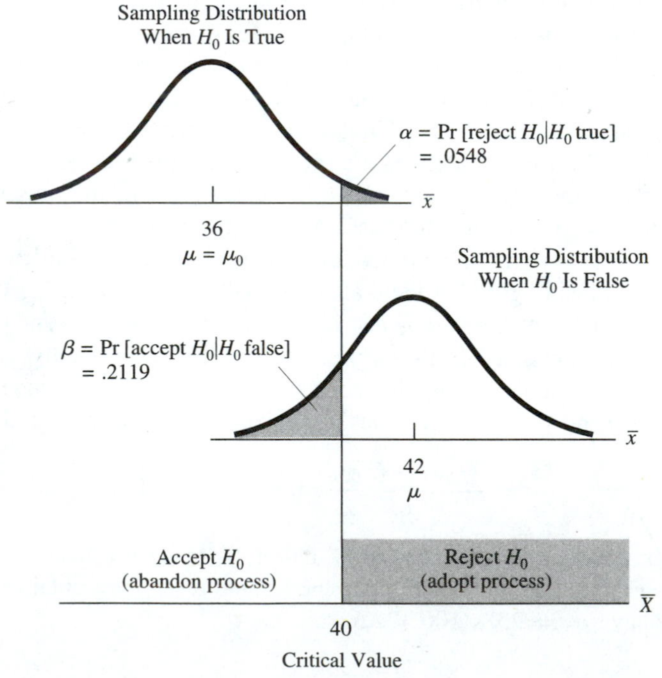

Figure 11–1 *Illustration of hypothesis testing for bacterial leaching decision.*

Figure 11–1 shows the two separate normal curves that apply for the type I and II error probabilities we have just computed. Notice that α is represented by an upper-tail area in the top curve, whereas β is represented by a lower-tail area in the bottom curve.

Similar type II error probabilities could be obtained for other economical levels for population mean copper yield. Figure 11–2 shows several cases, including $\mu = 50$ (which gives $\beta = .00003$) and $\mu = 39$ (giving $\beta = .6554$).

The dilemma of hypothesis testing—achieving an acceptable balance between α and the spectrum of possible β's—may be resolved by selecting an appropriate decision rule. And, as in statistical estimation, where increasing the sample size will provide greater precision and reliability, a larger n will reduce both the type I and II error probabilities.

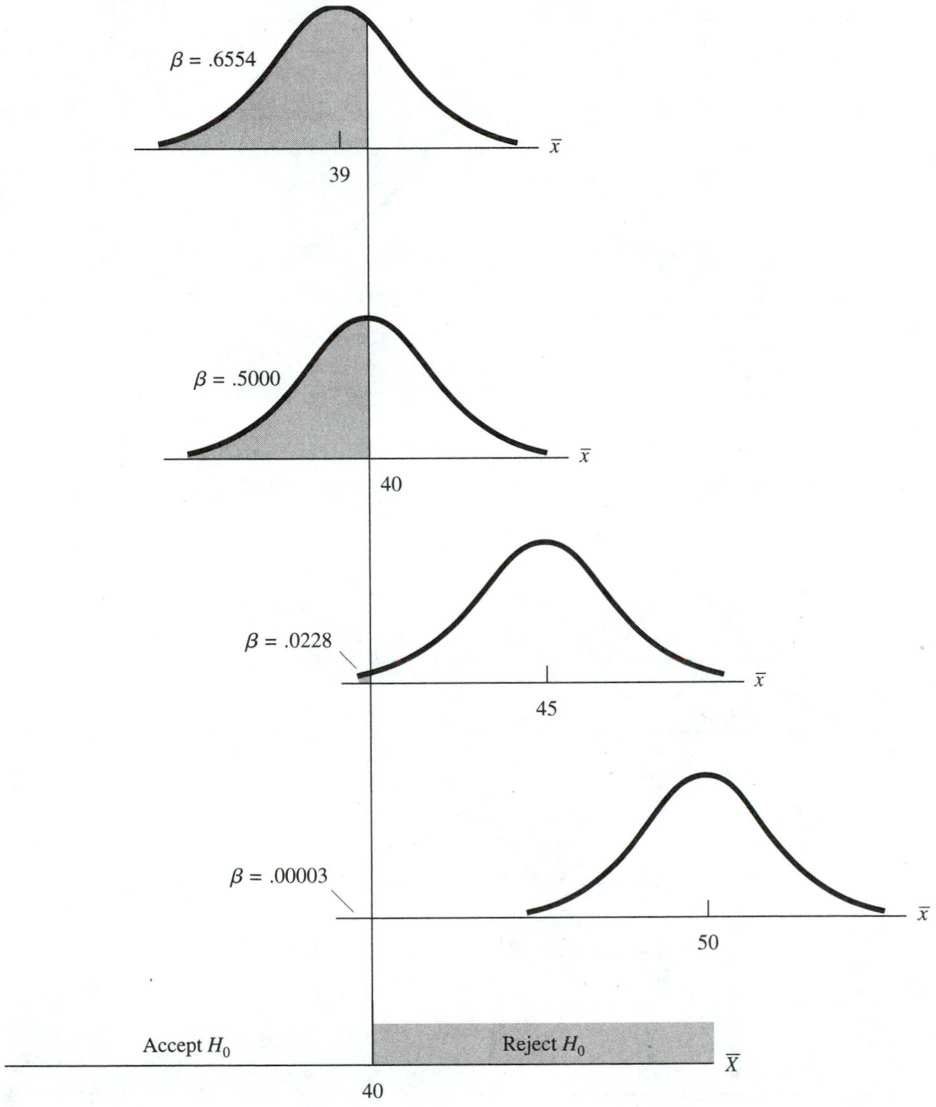

Figure 11–2 *Type II error probabilities for accepting H_0 when it is false, for several possible means.*

Determining an Appropriate Decision Rule

A new decision rule may be found simply by shifting the critical value for $\overline{X}$ to a new position. Two new decision rules for the bacterial leaching decision are illustrated in Figure 11–3.

Figure 11–3
Error probabilities as they change with a shift in critical value for bacterial leaching decision.

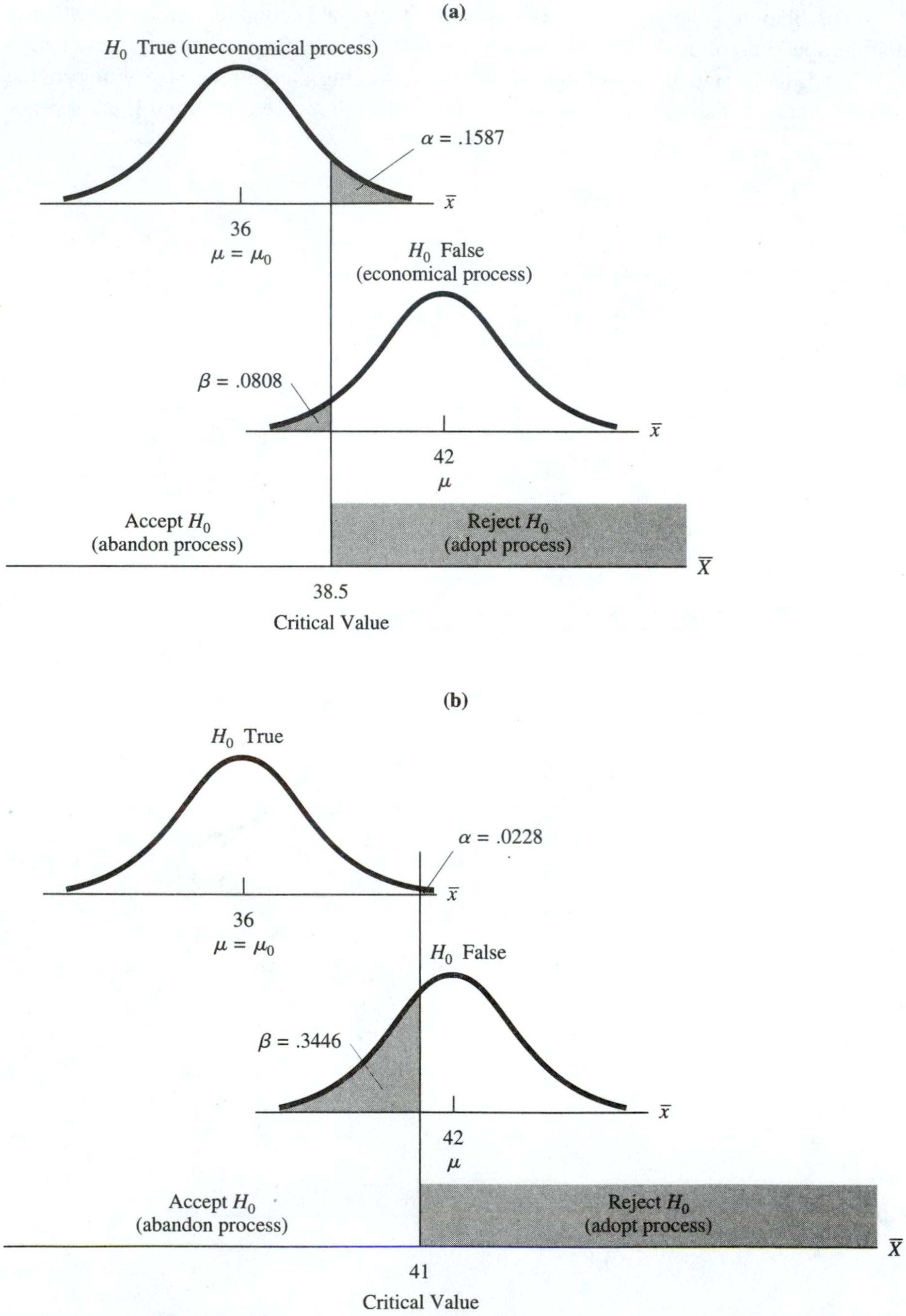

In (a) the critical value has been reduced to 38.5 lb/ton. Notice under this rule that it will be more likely to reject a true H_0 (adopt an uneconomical process) with $\alpha = .1587$. But there is less chance of accepting a false H_0 (abandoning an economical process) with $\beta = .0808$. In (b) the critical value for $\bar{X}$ is at 41 lb/ton, with reverse changes providing a lower α (.0228) and a higher β (.3446).

Either of these rules might be an improvement over the original one. That will depend on the attitudes of the engineer and mining company management. The null hypothesis was formulated in such a way that the type I error is the more serious one of adopting an uneconomical bacterial leaching process. This is typical of statistical testing, where a target level (such as 5%) for the probability of this outcome is often prescribed in advance; that in turn establishes a unique critical value and decision rule. Unfortunately, a decision-maker must live with whatever βs the chosen decision rule brings. As we have seen, when the engineer's critical value is 40, there will be a considerable chance that the bacterial leaching process will be abandoned when it might indeed be economical—perhaps considerably so.

Choice of critical value is really a matter of deciding where to position the "fulcrum" so that a proper balance is achieved between α and the βs. If none of the equilibria prove satisfactory, the decision-maker must improve the "leverage," which can be achieved only by increasing the sample size.

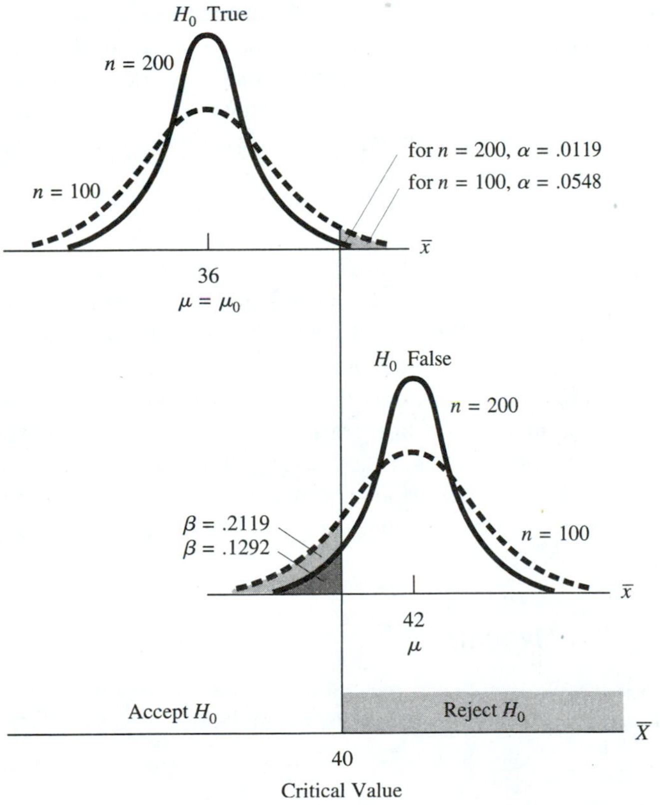

Figure 11–4
Both error probabilities are reduced when the sample size is increased from 100 to 200.

The Effect of Sample Size

Like precision and reliability in estimation, the type I and type II error probabilities of hypothesis testing are competing ends, so that *one* cannot be improved except at the expense of the other. The only way to improve *both* α and β is to reduce potential sampling error by increasing the sample size. Figure 11–4 shows how the mining engineer's situation improves when n is raised from 100 to 200 while the original decision rule is retained. Notice that the normal curves cluster more tightly about the mean, with new levels of $\alpha = .0119$ (when $\mu = \mu_0 = 36$) and $\beta = .1292$ (when $\mu = 42$).

Making the Decision

The preceding discussion applies entirely to the planning stage of a hypothesis testing study. At that time—before data are collected—the hypotheses are formulated, the sample size is established (usually by budgetary considerations), and the decision rule is established (often being dictated by the chosen significance level α). The sample data are then collected. The actual decision is more or less automatic, depending on the computed level of the test statistic.

We consider two hypothetical scenarios that might result after the data are in.

Scenario One Suppose that the mean yield per ton from the 100 test loads brought to the leaching dump turns out to be $\bar{X} = 47.6$ lb/ton. Since this quantity falls in the rejection region, the null hypothesis that $\mu \leq 36$ must be rejected; the sample results are statistically significant. The engineer concludes that large-scale bacterial leaching will be economical and recommends to management that it be adopted. The engineer takes comfort in the magnitude by which the computed mean exceeds the break-even level of 36 lb/ton. This quantity exceeds that hypothetical level for μ by more than 4 times $\sigma_{\bar{X}}$.

Scenario Two Suppose instead that the resulting mean copper yield from the test batches is only $\bar{X} = 32.1$ lb/ton. This value falls in the acceptance region, and the engineer concludes that bacterial leaching is not economical at this time. (Indeed, under the established decision rule the engineer would take the same action even if the computed $\bar{X}$ were much closer to the critical value of 40 lb/ton—say 38, or even 39.9.) In accepting H_0, the test results are found not to be statistically significant. This does not mean the findings are unimportant, only that H_0 cannot be rejected. The engineer realizes that there is a substantial probability that this decision is incorrect because μ might indeed be greater than the postulated break-even point. Unfortunately, there is no way to know. The case for bacterial leaching can always be reopened should conditions change, perhaps because of rising copper prices or development of an improved strain of bacteria.

Formulating the Hypotheses

As noted, H_0 is usually designated in such a way that the type I error is the most important one to avoid. Consider the following example.

***Example:* Does a New Type of Battery Last Longer?**

A design engineer for consumer products is evaluating a new battery substance based on different chemistry than is presently in use. It gives batteries a mean lifetime of approximately 100 hours. The new battery may last at least as long, on the average, as the present one—or it may die more rapidly. A sampling study will be used, and a decision whether or not to change to the new battery will be based on the observed mean time between failures (MTBF) of test cells composed under the new chemistry. Either of the following assumptions regarding the mean lifetime μ of the new battery might serve as the null hypothesis:

$$\mu \leq 100 \text{ hours} \qquad \text{or} \qquad \mu \geq 100 \text{ hours}$$

Since the company wants to keep its reputation for innovation, the more serious error was judged to be keeping the present battery when the new one lasts longer. (Certainly, it would be undesirable to introduce a new battery that doesn't have a longer life, but this was judged far less damaging.) Thus, the following designation was made:

$H_0 : \mu \geq 100$ hours (new battery is at least as good)

$H_1 : \mu < 100$ hours (new battery has a shorter lifetime)

A decision rule based on $\overline{X}$ was established so that H_0 would be rejected for shorter computed MTBFs and accepted for longer ones:

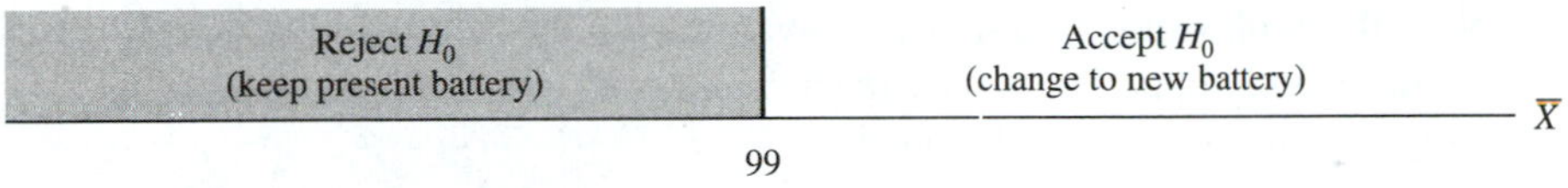

The test described in the battery example is a **lower-tailed test**, since the rejection region coincides with the lower tail of the normal curve for $\overline{X}$ that applies under the null hypothesis. The reversed rejection region in the bacterial leaching illustration indicates that an **upper-tailed test** applies in that application. Some tests involve hypotheses in which the rejection region is split in two, with an intervening acceptance region. The following is an example of such a **two-sided test**.

***Example:* Meeting Labeling Requirements**

Packaging items for public consumption can be an exacting task because of regulations or trade practices regarding labeling of the volume or weight of the contents. Too much ingredient is as much to be avoided as too little. Sampling is often used in quality-control investigations concerning weights and measures. Typically, the null hypothesis takes the form that the labeled mean is being met exactly, with the opposite alternative. For example,

$H_0 : \mu = 32$ grams (labeling specifications are met)

$H_1 : \mu \neq 32$ grams (labeling specifications are violated)

There are two ways for the null hypothesis to be untrue—if there is overfilling ($\mu > 32$ grams) or if there is underfilling ($\mu < 32$ grams). The following form is used for the decision rule:

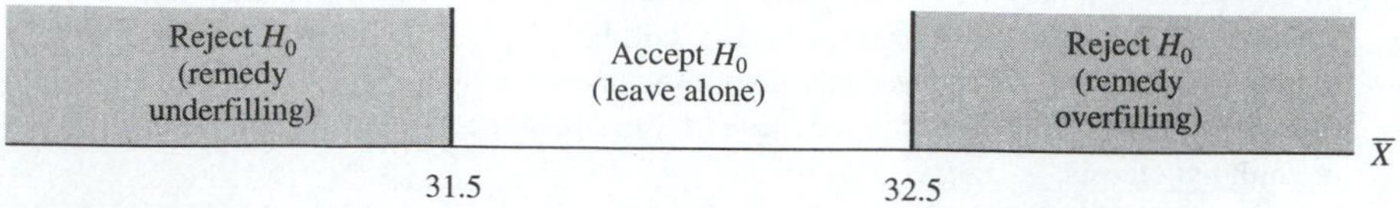

Drawing Conclusions with Prob Values

Sometimes there is no basis for judging the relative seriousness of errors, making meaningless the balancing act between the type I and type II errors. Also, a predetermined decision rule would be impractical when different people are going to make their own assessments from the same data. In such cases a less formal approach may be taken in drawing conclusions from sample data. A final determination may then be based on the *prob value* (sometimes referred to as the *p*-value).

Definition

Assuming that the null hypothesis is true and the pivotal parameter level applies, the **prob value** for a statistical evaluation is the probability of getting a sample result opposite to one expected under H_0 and that is as extreme or more so than the one actually obtained.

To demonstrate, we continue with the bacterial leaching illustration. The prob values for the two scenarios are computed below.

Scenario One $\bar{X} = 47.6$ is the sample result:

$$\begin{aligned} \text{prob} &= \Pr[\bar{X} \geq 47.6 \,|\, \mu = 36] \\ &= 1 - \Phi\left(\frac{47.6 - 36}{2.5}\right) = 1 - \Phi(4.64) = 0.0000 \end{aligned}$$

The above indicates the extreme rarity of getting such a sample result when the null hypothesis is true. The evidence in this case would provide a very strong refutation of H_0, which nearly any investigator would under the circumstances reject.

Scenario Two $\bar{X} = 32.1$ is the sample result:

$$\begin{aligned} \text{prob} &= \Pr[\bar{X} \geq 32.1 \,|\, \mu = 36] \\ &= 1 - \Phi\left(\frac{32.1 - 36}{2.5}\right) = 1 - \Phi(-1.56) = .9406 \end{aligned}$$

A sample mean of 32.1 or greater is quite likely under the null hypothesis. The sample evidence in this case is consistent with a true H_0, which in this case ought to be accepted.

How a prob value is actually employed in reaching conclusions can be very informal. Often statistical results are simply listed, along with the relevant prob value, in a report or published article. The authors may not even reach their own conclusions, instead choosing to leave any inferences up to the individual readers themselves. Indeed, many such reportings will not even state the null hypothesis per se, it being implicit from the general discussion accompanying the data.

Problems

11–1 Suppose that copper prices have dropped so that the break-even level for the mean yield from bacterial leaching rises to 37 lb/ton. Complete the following:
(a) Reformulate the engineer's hypotheses.
(b) Using the critical value of 40, determine (1) the type I error probability α and (2) the type II error probability β when $\mu = 42$.

11–2 Suppose a mining engineer must decide whether to use a new steel alloy instead of the one presently used to reinforce the mineshaft linings. The null hypothesis is that the present bars are at least as strong as the new ones. He will either accept this assumption, retaining the present reinforcing bars, or reject it and replace them.

State in words (a) the type I error, and (b) the type II error.

11–3 Consider each of the hypothesis-testing applications below. Indicate for each outcome whether it is a correct decision, a type I error, or a type II error:
(a) H_0: New power cell's lifetime does not exceed old one's.
  (1) Change to new when old lasts as long or longer.
  (2) Keep old when new lasts longer.
  (3) Keep old when old lasts as long or longer.
  (4) Change to new when new lasts longer.
(b) H_0: Memory chips are satisfactory.
  (1) Reject satisfactory shipment.
  (2) Accept satisfactory shipment.
  (3) Reject unsatisfactory shipment.
  (4) Accept unsatisfactory shipment.
(c) H_0: New design is safe.
  (1) Approve an unsafe design.
  (2) Disapprove an unsafe design.
  (3) Disapprove a safe design.
  (4) Approve a safe design.

11–4 The following hypotheses are to be tested, with $\mu_0 = 100$:

$$H_0 : \mu \leq \mu_0 \qquad H_1 : \mu > \mu_0$$

Assume that the population standard deviation is $\sigma = 28$ and the sample size is $n = 100$. The following decision rule applies:

$$\text{Accept } H_0 \text{ if } \overline{X} \leq 104$$
$$\text{Reject } H_0 \text{ if } \overline{X} > 104$$

Determine the type I error probability α when $\mu = 100$ and the type II error probability β when $\mu = 110$.

11–5 Repeat Problem 11–4 using 106 instead as the critical value. Is the new α larger or smaller? Is the new β larger or smaller?

11–6 Repeat Problem 11–4 using a larger sample size of $n = 150$. Is the new α larger or smaller? Is the new β larger or smaller?

11–7 The following hypotheses are to be tested, with $\mu_0 = 900$:

$$H_0 : \mu \geq \mu_0 \qquad H_1 : \mu < \mu_0$$

Assume that the population standard deviation is $\sigma = 50$ and the sample size is $n = 100$. The following decision rule applies:

$$\text{Accept } H_0 \text{ if } \overline{X} \geq 885$$
$$\text{Reject } H_0 \text{ if } \overline{X} < 885$$

Determine the type I error probability α when $\mu = 900$ and the type II error probability β when $\mu = 875$.

11–8 Refer to the decision rule in Problem 11–4. For each of the following cases (1) find the prob value and (2) indicate whether it appears to support the null hypothesis, to refute it, or neither.

(a) $\bar{X} = 97.5$ (b) $\bar{X} = 106.5$ (c) $\bar{X} = 102.1$ (d) $\bar{X} = 106.2$

11–9 Refer to the decision rule in Problem 11–7. For each of the following cases (1) find the prob value and (2) indicate whether it appears to support the null hypothesis, to refute it, or neither.

(a) $\bar{X} = 882$ (b) $\bar{X} = 907$ (c) $\bar{X} = 869$ (d) $\bar{X} = 889$

11–2 Procedures for Testing the Mean

We are now ready to apply the basic concepts of the preceding section to testing the mean. A general procedure has been developed for this application that may be adapted to a wide variety of hypothesis-testing situations. Some of these pertain to other parameters than the mean and will be described later in this chapter.

The Hypothesis-Testing Steps

All complete statistical tests involve the same basic steps.

Step 1. *Formulate the hypotheses and specify the possible choices.* In all cases involving the mean, the decision parameter is the unknown μ. A limiting value μ_0 must be identified for defining H_0. In the bacterial leaching illustration μ_0 was the break-even point for the procedure. Although the choice of μ_0 is often an economic consideration, other reasons may prevail. The level of μ_0 may be prescribed, as in a labeling specification for package contents. Or, μ_0 might be a benchmark level that corresponds to a present procedure, process, material, or part; we saw such an example in the preceding battery example.

One-Sided Tests Error considerations usually determine which side of μ_0 to include under H_0. The null hypothesis will be selected from one of the two forms,

$$\mu \leq \mu_0 \qquad \text{or} \qquad \mu \geq \mu_0$$

The two errors are:

1. taking the wrong action when in fact $\mu \leq \mu_0$
2. taking the wrong action when in fact $\mu \geq \mu_0$

The null hypothesis will be $H_0: \mu \leq \mu_0$ if (1) is more serious and $H_0: \mu \geq \mu_0$ if (2) is more serious.

Two-Sided Tests In a two-sided test the null hypothesis takes the form

$$H_0: \mu = \mu_0$$

Since H_0 is untrue both when $\mu < \mu_0$ and when $\mu > \mu_0$, the investigation of the above null hypothesis is a two-sided test.

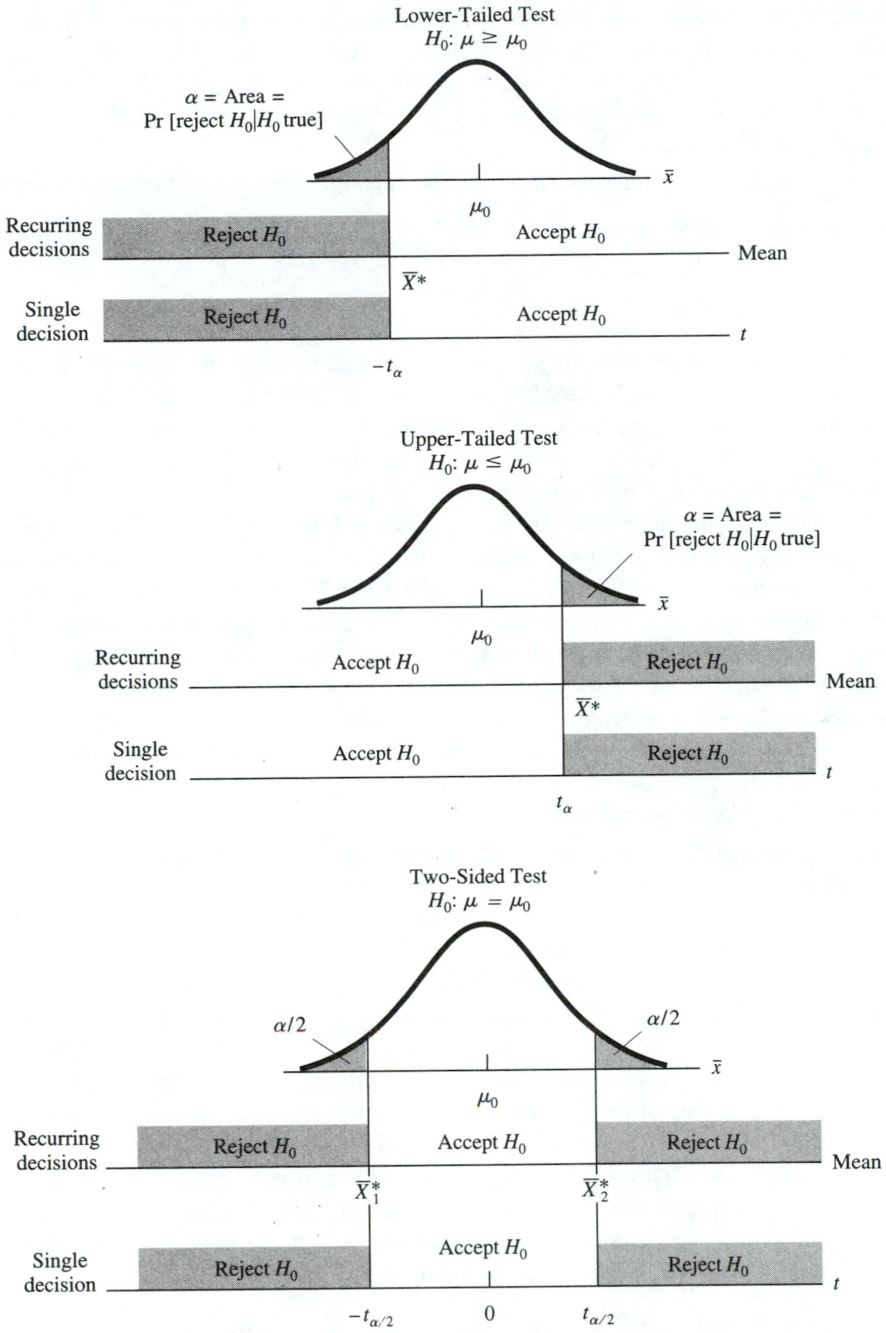

Figure 11–5 *Three forms of hypothesis tests for mean.*

Step 2. *Select the test procedure and test statistic.* In testing the mean, $\bar{X}$ is the natural test statistic. By the central limit theorem we may ordinarily use the normal curve to represent the sampling distribution of $\bar{X}$. When the population standard deviation is known, $\sigma_{\bar{X}}$ may be computed, and for any given level of μ we may establish probabilities for possible values of $\bar{X}$.

But when σ is unknown, testing procedures employ the following instead.

STUDENT t TEST STATISTIC (σ UNKNOWN)

$$t = \frac{\bar{X} - \mu_0}{s/\sqrt{n}}$$

There are two types of testing procedures, depending on the nature of the decisions to be made. With the **single-decision procedure**, H_0 will be accepted or rejected just once. With a **recurring-decisions procedure**, a series of decisions will be made, and H_0 will be accepted or rejected each time. This dichotomy is useful in selecting the test statistic and determining the form of the decision rule.

Under recurring decisions, one decision rule will be used over and over, possibly being applied by different persons and in several locations. The test statistic should be easy to compute and readily understood by all participants. In a recurring-decisions procedure involving means, the computed mean $\bar{X}$ is ordinarily used and compared to a single critical value $\bar{X}^*$ or a pair of critical values $\bar{X}_1^*$ and $\bar{X}_2^*$. Simplicity of procedure generally requires that the population standard deviation value be stipulated for all test applications, so that σ is treated as a *known* quantity.

Under a single-decision procedure, no stipulation on the level of the population standard deviation is required, and σ may remain *unknown*. The Student t statistic then applies, and the decision rule will be based on it.

Step 3. *Establish the significance level and the acceptance and rejection regions for the decision rule.* As we have seen, the common practice in statistical testing is to base the procedure on avoidance of the more serious type I error. The probability α of this, also referred to as the *significance level*, may then be prescribed in advance. With one-sided applications, the resulting experiment will be a lower- or upper-tailed test, depending on whether small or large values of the test statistic refute H_0. In a two-sided test there are two ways in which H_0 can be rejected. The type I error probability α is split evenly between the two cases. Figure 11–5 summarizes the various situations.

The critical value of the test statistic defines the acceptance and rejection regions. In testing the mean when σ is known, these regions are determined by the **critical normal deviate** z_α (or $z_{\alpha/2}$), for which the normal curve upper-tail area is α(or $\alpha/2$). If σ is unknown, critical values for the Student t statistic t_α(or $t_{\alpha/2}$) are used instead.

When a recurring-decisions procedure applies, so that σ is known and $\bar{X}$ serves directly as the test statistic, the critical normal deviate is used.

CRITICAL VALUES FOR THE SAMPLE MEAN
(σ KNOWN)

Lower-Tailed Test $(H_0: \mu \geq \mu_0)$:

$$\bar{X}^* = \mu_0 - z_\alpha \sigma_{\bar{X}}$$

Upper-Tailed Test $(H_0: \mu \leq \mu_0)$:

$$\bar{X}^* = \mu_0 + z_\alpha \sigma_{\bar{X}}$$

Figure 11–6
Capsule summary of final hypothesis-testing steps for mean.

Step 4 Collect data and compute $\overline{X}$, s.
Then calculate

$$t = \frac{\overline{X} - \mu_0}{s/\sqrt{n}}$$

Only one of these is needed.

Step 5 Make decision.

Lower-Tailed (H_0: $\mu \geq \mu_0$)	**Upper-Tailed** (H_0: $\mu \leq \mu_0$)
Accept H_0 if $t \geq -t_\alpha$	*Accept* H_0 if $t \leq t_\alpha$
Reject H_0 if $t < -t_\alpha$	*Reject* H_0 if $t > t_\alpha$

Two-Sided (H_0: $\mu = \mu_0$)

Accept H_0 if $-t_{\alpha/2} \leq t \leq t_{\alpha/2}$
Reject H_0 if $t < -t_{\alpha/2}$ or if $t > t_{\alpha/2}$

or

Compute the $100(1-\alpha)\%$ confidence interval:

$$\overline{X} - t_{\alpha/2}\frac{s}{\sqrt{n}} \leq \mu \leq \overline{X} + t_{\alpha/2}\frac{s}{\sqrt{n}}$$

Accept H_0 if μ_0 falls inside.
Otherwise, *Reject* H_0.

SINGLE DECISION (σ unknown)

Identify procedure

RECURRING DECISIONS (σ known)

Step 3 Find critical values for the decision rule.

Lower-Tailed (H_0: $\mu \geq \mu_0$)	**Upper-Tailed** (H_0: $\mu \leq \mu_0$)
$\overline{X}^* = \mu_0 - z_\alpha \sigma_{\overline{X}}$	$\overline{X}^* = \mu_0 + z_\alpha \sigma_{\overline{X}}$
Accept H_0 if $\overline{X} \geq \overline{X}^*$	*Accept* H_0 if $\overline{X} \leq \overline{X}^*$
Reject H_0 if $\overline{X} < \overline{X}^*$	*Reject* H_0 if $\overline{X} > \overline{X}^*$

Two-Sided (H_0: $\mu = \mu_0$)

$$\overline{X}_1^* = \mu_0 - z_{\alpha/2}\sigma_{\overline{X}} \qquad \overline{X}_2^* = \mu_0 + z_{\alpha/2}\sigma_{\overline{X}}$$

Accept H_0 if $\overline{X}_1^* \leq \overline{X} \leq \overline{X}_2^*$
Reject H_0 if $\overline{X} < \overline{X}_1^*$ or if $\overline{X} > \overline{X}_2^*$

When $n \leq .10N$:

$$\sigma_{\overline{X}} = \frac{\sigma}{\sqrt{n}}$$

When $n > .10N$:

$$\sigma_{\overline{X}} = \frac{\sigma}{\sqrt{n}}\sqrt{\frac{N-n}{N-1}}$$

Steps 4 and 5 Collect data, compute $\overline{X}$, and decide.

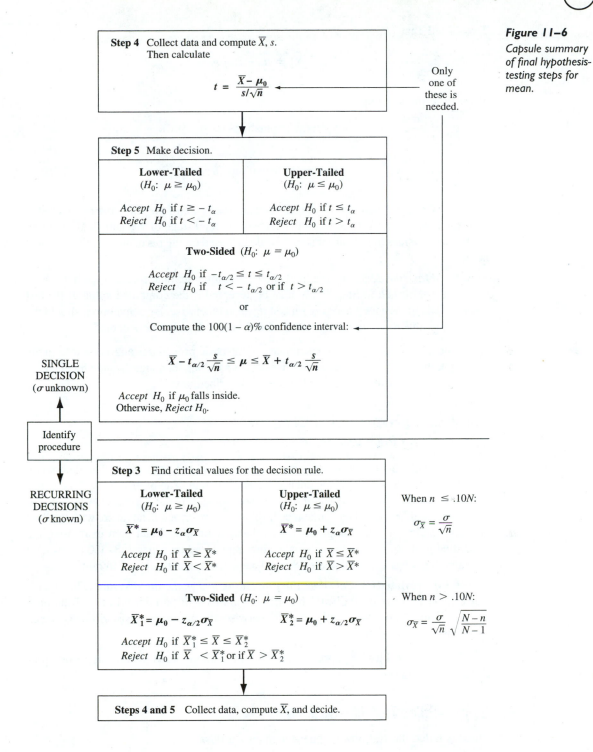

Two-Sided Test ($H_0 : \mu = \mu_0$).

$$\bar{X}_1^* = \mu_0 - z_{\alpha/2}\sigma_{\bar{X}}$$
$$\bar{X}_2^* = \mu_0 + z_{\alpha/2}\sigma_{\bar{X}}$$

Note: In computing the above,

$$\sigma_{\bar{X}} = \frac{\sigma}{\sqrt{n}} \qquad \text{for large populations } (n \leq .10N)$$

$$\sigma_{\bar{X}} = \frac{\sigma}{\sqrt{n}}\sqrt{\frac{N-n}{N-1}} \qquad \text{for small populations } (n > .10N)$$

Step 4. *Collect the sample data and compute the value of the test statistic.* This step incorporates the time-consuming and expensive second stage of the statistical sampling study. No revisions to the earlier steps should be based on the particular results obtained here.

Step 5. *Make the decision.* The final choice is automatic, in accordance with the decision rule established in Step 3. H_0 must be accepted if the computed value of the test statistic falls in the acceptance region and rejected otherwise. By convention, should the computed value round precisely to the critical value, H_0 must be accepted.

Figure 11–6 gives a capsule summary of the basic procedures and variations in the final hypothesis-testing steps. Some investigators do not set a significance level (α) in advance and prefer the abbreviated approach described next.

Abbreviated Hypothesis-Testing Steps When Using Prob Values

The abbreviated hypothesis-testing procedure employs the prob value. The dichotomy of recurring versus nonrecurring decision is unimportant. The prob value will be determined by the Student t distribution if σ is unknown and by the normal distribution if σ is known. In either case the significance level is not prescribed in advance, with the final decision made *after* the sample data have been summarized and then examining the computed prob value. The hypothesis-testing steps listed on pages 358–360 are exactly the same for Steps 1, 2, and 4. Step 3 is skipped, and the last step is different.

Step 5 (abbreviated). *Calculate the prob value and make the decision.* The significance level is never prescribed. The null hypothesis will be rejected only if the prob value is judged to be sufficiently small. Otherwise, the null hypothesis will be accepted for the time being.

Figure 11–7 shows a capsule summary of the abbreviated hypothesis-testing procedure for the mean.

Upper-Tailed Test Illustrations

Three detailed illustrations of upper-tailed tests follow.

Single-Decision Procedure Consider the following application of stress testing.

Figure 11–7 *Capsule summary of final hypothesis-testing steps for abbreviated testing procedure with the mean.*

Identify procedure

σ Unknown

Step 4 Collect data and compute $\overline{X}$, s.

Step 5 Make the decision based on the level of the prob value, using Student t distribution.

Lower-Tailed Test (H_0: $\mu \geq \mu_0$)	**Upper-Tailed Test** (H_0: $\mu \leq \mu_0$)
$\text{prob} = \Pr\left[t \leq \frac{\overline{X} - \mu_0}{s/\sqrt{n}}\right]$	$\text{prob} = \Pr\left[t \geq \frac{\overline{X} - \mu_0}{s/\sqrt{n}}\right]$

Two-Sided (H_0: $\mu = \mu_0$)

$$\text{prob} = \Pr\left[t \geq \frac{|\overline{X} - \mu_0|}{s/\sqrt{n}}\right]$$

Reject H_0 if the prob value is judged to be sufficiently small. Otherwise, *Accept* H_0.

σ Known

Step 4 Collect data and compute $\overline{X}$ and $\sigma_{\overline{X}}$.

$$\sigma_{\overline{X}} = \frac{\sigma}{\sqrt{n}} \qquad \text{or} \qquad \sigma_{\overline{X}} = \frac{\sigma}{\sqrt{n}}\sqrt{\frac{N - n}{N - 1}}$$

Largen: n > .10N   Smalln: n ≤ .10N

Step 5 Make the decision based on the level of the prob value, using normal distribution.

Lower-Tailed Test (H_0: $\mu \geq \mu_0$)	**Upper-Tailed Test** (H_0: $\mu \leq \mu_0$)
$\text{prob} = \Pr\left[Z \leq \frac{\overline{X} - \mu_o}{\sigma_{\overline{X}}}\right]$	$\text{prob} = \Pr\left[Z \geq \frac{\overline{X} - \mu_o}{\sigma_{\overline{X}}}\right]$

Two-Sided (H_0: $\mu = \mu_0$)

$$\text{prob} = \Pr\left[Z \geq \frac{|\overline{X} - \mu_0|}{\sigma_{\overline{X}}}\right]$$

Reject H_0 if the prob value is judged to be sufficiently small. Otherwise, *Accept* H_0.

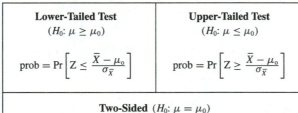

Illustration: *Axial Stress Testing of a New Alloy*

A mechanical engineer is considering a new nickel-chrome-iron alloy. She has ordered seven sample castings, which are to be tested at a materials laboratory for endurance under axial stress. The engineer is seeking a metal strong enough to meet customer specifications for parts in a new stamping machine. These require that the mean number of cycles to failure μ obtained in vibration testing exceed 500,000. The target population represents the endurance measurements that might be obtained if every possible casting—not just the sample items—were subjected to the test procedure. So that quick results may be obtained, a lower endurance figure was specified than is ordinarily found in stress testing. Accordingly, unusually extreme levels were set for the test displacement and force parameters.

The following hypothesis-testing steps were taken.

Step 1. *Formulate the hypotheses and specify the possible choices.* The engineer selects as her hypotheses:

$$H_0: \mu \leq 500{,}000 \qquad \text{(endurance specifications are not exceeded)}$$
$$H_1: \mu > 500{,}000 \qquad \text{(endurance specifications are exceeded)}$$

The possible choices are

Reject H_0 (adopt the new alloy)
Accept H_0 (continue searching for a new alloy)

The designations reflect her strong desire to avoid adopting the alloy when it does not exceed customer specifications. These specifications establish $\mu_0 = 500{,}000$ as the limiting value for the mean number of cycles to failure.

Step 2. *Select the test procedure and test statistic.* The engineer's decision will be made just once. She chooses the Student t as her test statistic.

Step 3. *Establish the significance level and the acceptance and rejection regions for the decision rule.* Because she regards so seriously the type I error of rejecting H_0 when it is true, the engineer chooses a 1% significance level, so that $\alpha = .01$. With $n = 7$ and $\text{df} = 7 - 1 = 6$, Appendix Table G provides the critical value $t_{.01} = 3.143$. Large positive levels for t will refute H_0, and the test is upper-tailed. The decision rule is shown in Figure 11–8.

Figure 11–8

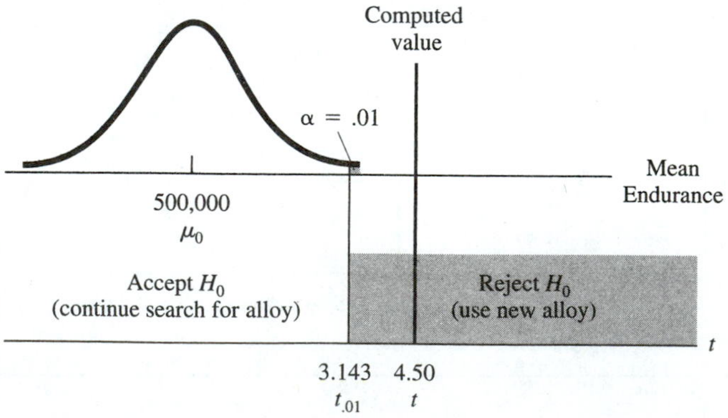

```
MTB > Print c1.

C1
  585500   495600   583500   585700   529500   560100   565400

MTB > Describe c1.

               N     MEAN   MEDIAN   TRMEAN    STDEV   SEMEAN
C1             7   557900   565400   557900    34007    12853

             MIN      MAX       Q1       Q3
C1        495600   585700   529500   585500
```

Figure 11-9 *Minitab printout showing sample data obtained from the alloy test.*

Step 4. *Collect the sample data and compute the value of the test statistic.* The sample data for the test items in cycles to failure are shown in the Minitab printout in Figure 11–9.

The computer printout provides a sample mean of $\bar{X} = 557{,}900$ and a sample standard deviation of $s = 34{,}007$. The computed value of the test statistic is

$$t = \frac{557{,}900 - 500{,}000}{34{,}007/\sqrt{7}} = 4.50$$

Step 5. *Make the decision.* The computed t-value exceeds the critical value and falls in the rejection region. The engineer must therefore reject H_0 and adopt the new alloy in the stamping machine.

Recurring-Decisions Procedure The following illustration details how hypothesis testing can be employed to establish a permanent testing policy.

Illustration: *Deciding When to Calibrate Chemical Controls*

A chemical engineer must set operational policies for the manufacture of a chemical base. At one stage, sample vials will be selected at random time intervals, and the level of impurities measured for each. A tolerable mean level of impurities μ for all potential vials is 150 parts per million. Anything lower would be satisfactory. The engineer needs to find a decision rule that technicians may use to decide whether or not to recalibrate control settings, thereby reducing the incidence of impurities. A new sample of vials will be taken every eight hours, so the calibration decision must be made three times a day. The following hypothesis-testing steps apply.

Step 1. *Formulate the hypotheses and specify the possible choices.* The engineer selects as his hypotheses:

$$H_0: \mu \leq 150 \text{ ppm} \qquad \text{(impurity level is tolerable)}$$

$$H_1: \mu > 150 \text{ ppm} \qquad \text{(impurity level is intolerably high)}$$

The two possible choices are

$$\text{Reject } H_0 \qquad \text{(recalibrate)}$$

$$\text{Accept } H_0 \qquad \text{(don't recalibrate)}$$

The pivotal level for the mean is thus $\mu_0 = 150$. With these hypotheses the type I error is unnecessarily recalibrating when the impurity levels are actually tolerable. The type II error is failing to recalibrate when impurities are intolerably high. Since impurities can be reduced—at a modest cost—in a later stage of processing, and since recalibration is very expensive, the former error is judged to be more serious.

Step 2. *Select the test procedure and test statistic.* The sample mean impurity level $\overline{X}$ (ppm) serves as the test statistic. To establish his decision rule, the engineer uses a value for the population standard deviation of $\sigma = 20$ ppm. This figure was obtained historically from similar processing of related chemicals. The standard deviation is assumed to be stable, even though the mean fluctuates, and is therefore treated as a known quantity.

Step 3. *Establish the significance level and the acceptance and rejection regions for the decision rule.* The engineer decides that a 5% significance level will adequately protect against the type I error. To keep the incidence of the type II error low, he relies on a fairly substantial sample size of $n = 25$ vials.

Because a continuous process is involved, the population is theoretically infinite, and the following applies for the standard error of $\overline{X}$:

$$\sigma_{\overline{X}} = \frac{20}{\sqrt{25}} = 4.0 \text{ ppm}$$

Since *large* $\overline{X}$s will refute the null hypothesis of tolerable impurity levels, the test is *upper-tailed.* Using $\alpha = .05$, the critical normal deviate is found from Table E to be $z_{.05} = 1.64$, and the critical value for the sample mean is

$$\overline{X}^* = 150 + 1.64(4.0) = 156.6 \text{ ppm}$$

The decision rule in Figure 11–10 applies.

Step 4. *Collect the sample data and compute the value of the test statistic.* This step is conducted every eight hours, with the mean impurity level computed for every 25-vial sample.

Step 5. *Make the decision.* As with Step 4, a decision is made every eight hours, in accordance with the decision rule in Figure 11–10.

Figure 11–10

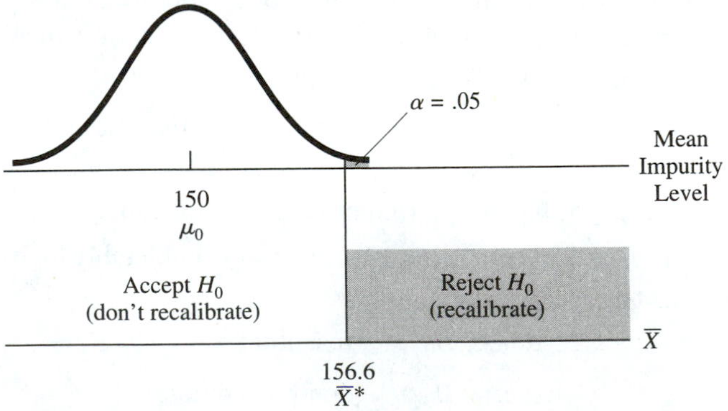

Table 11–2

Experimental Results from Testing Water Spray in Presence of a Flame

Average Droplet Size (microns) X	X^2	Average Droplet Size (microns) X	X^2
7	49	71	5,041
13	169	91	8,281
25	625	33	1,089
62	3,844	45	2,025
63	3,969	51	2,601
37	1,369	62	3,844
49	2,401	38	1,444
50	2,500	697 $=\sum X$	39,251 $=\sum X^2$

$$\bar{X} = 697/15 = 46.47$$

$$s = \sqrt{\frac{39{,}251 - 15(46.47)^2}{15 - 1}} = 22.14$$

Note: These results are consistent with the data in S. C. Li, P. A. Libby, and P. A. Williams, "Spray Impingement on a Hot Surface in Reacting Stagnation Flows," *American Institute of Aeronautics and Astronautics Journal*, June 1995.

Abbreviated Procedure The following illustration describes a test involving sprays of water.

***Illustration**: Investigating Whether a Flame Affects Water Droplet Size*

A researcher for a fire extinguisher manufacturer is studying the effect of a spray of water hitting a surface after traveling through a fire. Of particular interest are the droplet sizes in the stagnation flow. Previous research shows that the mean droplet size does not exceed about 25 microns near the contact surface. She believes that the presence of a flame would increase droplet diameters, but that can be determined only through experimentation. She has no feel for the relative seriousness of the types I and II errors and cannot prescribe a significance level for evaluating the results.

Step 1. *Formulate the hypotheses and specify the possible choices.* Letting μ represent the mean droplet size near the impingement site from a spray of water crossing a flame, the researcher selects as her hypotheses:

$H_0: \mu \leq 25$ microns (the flame has no effect on mean droplet size)

$H_1: \mu > 25$ microns (the flame increases the mean droplet size)

She will either accept H_0 and prepare further experiments, assuming that droplet size is unaffected by the presence of flame, or reject H_0 and allow for increased droplet size when the spray goes through a flame.

Step 2. *Select the test procedure and test statistic*. The researcher has no idea of what the standard deviation in droplet size is. She chooses the Student t as her test statistic.

Step 3. Not applicable.

Step 4. *Collect the sample data and compute the value of the test statistic*. The data in Table 11–2 were obtained from $n = 15$ separate experiments using a special test apparatus.

Step 5 (abbreviated). *Calculate the prob value and make the decision*. The following applies.

$$\text{prob} = \Pr\left[\frac{\bar{X} - \mu_0}{s/\sqrt{n}} \geq \frac{46.47 - 25}{22.14/\sqrt{15}}\right] = \Pr\,[t \geq 3.76]$$

Using $15 - 1 = 14$ degrees of freedom, Appendix Table G gives a smaller critical value, 3.326, for a tail area of .0025. The prob value must be smaller than .0025. The sample data strongly refute the null hypothesis. After examining this result, the researcher rejected H_0 and prepared her further research under the assumption that the presence of flame increases droplet size.

Lower-Tailed Test Illustrations

Procedures with a lower-tailed test are described using the following detailed examples.

Recurring-Decisions Procedure The chemical engineer in the earlier example must set another test policy.

***Illustration**: Deciding When to Purge Separation Tanks*

The chemical engineer finds a decision rule for a later stage of processing the same chemical base. At that stage, the major concern is with the final yield of an active ingredient. As before, a sample of vials will be selected at random time intervals, with the grams per liter of that ingredient determined for each. Company standards are that the mean yield μ should be at least 35 grams per liter. A decision rule will be used by technicians to translate sample results into action. For yields smaller than the critical level, the separation tanks will have to be purged, thereby increasing the efficiency of the reactions. This decision will be faced only once every 24 hours. The following hypothesis-testing steps apply.

Step 1. *Formulate the hypotheses and specify the possible choices*. The engineer selects as his hypotheses:

$$H_0\!: \mu \geq 35 \text{ grams/liter}$$
$$H_1\!: \mu < 35 \text{ grams/liter}$$

The pivotal level for the mean is thus $\mu_0 = 35$. The possible choices are

Reject H_0 (purge tanks)
Accept H_0 (don't purge tanks)

The type I error is unnecessarily purging the tanks when the minimum yield requirements are being met. The type II error is failing to purge the tanks when it is really necessary to do so. The first error is judged more serious since that would involve unnecessary expense. The type II error is less expensive, since low-yield chemicals can be enriched as needed during the final processing stage.

Step 2. *Select the test procedure and test statistic.* The sample mean yield $\bar{X}$ (grams/liter) serves as the test statistic. To establish his decision rule, the engineer uses a known value for the population standard deviation of $\sigma = 1$ gram/liter.

Step 3. *Establish the significance level and the acceptance and rejection regions for the decision rule.* The engineer decides that a 1% significance level will adequately protect against the type I error. A sample size of $n = 100$ vials has been determined to be most cost-effective.

As with the preceding test, the population is theoretically infinite, and the following applies for the standard error of $\bar{X}$:

$$\sigma_{\bar{X}} = \frac{1}{\sqrt{100}} = 0.1 \text{ gram/liter}$$

Since *small* $\bar{X}$s will refute the null hypothesis of acceptable yield level, the test is *lower-tailed.* Using $\alpha = .01$, the critical normal deviate is found from Table E to be $z_{.01} = 2.33$, and the critical value for the sample mean is

$$\bar{X}^* = 35 - 2.33(0.1) = 34.77 \text{ grams/liter}$$

The decision rule in Figure 11–11 applies.

Figure 11-11

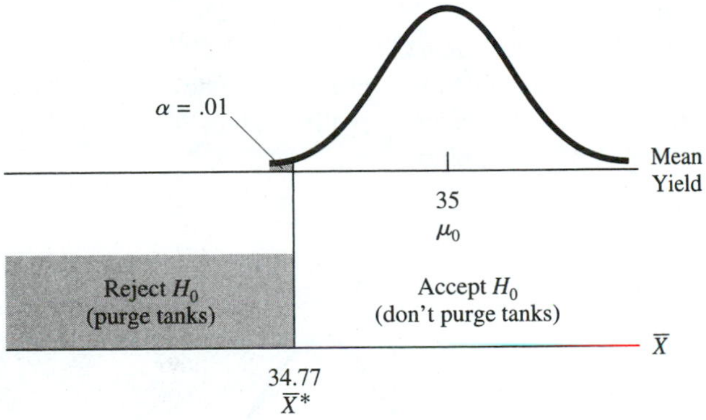

The test is applied over and over again with a new sample each time. A fresh $\bar{X}$ is computed every day, and the tanks are purged whenever $\bar{X}$ falls into the above rejection region.

Single-Decision Procedure The following illustration considers another chemical engineering application where the hypothesis testing is done only once.

***Illustration**: Selecting the Faster Fermentation Temperature*

A chemical engineer is pilot-testing a fermentation process for the manufacture of a pharmaceutical product. All design parameters have been selected except for the temperature at which fermentation itself will take place. Current temperature-control hardware is designed to operate at a low nominal level, although theoretically an environment 10°C higher should provide a faster fermentation rate. Since the higher-temperature operation will require specially designed control devices, that environment will be incorporated into the final design only if sample batches ferment significantly faster than the nominal mean time of 30 hours. A series of 50 pilot batches are to be run at the hotter temperature.

The following hypothesis-testing steps apply.

Step 1. *Formulate the hypotheses and specify the possible choices.* Expressing by μ the mean fermentation time of all future batches that might be processed under high temperatures, the engineer establishes:

$H_0: \mu \geq 30$ hours (higher-temperature fermentation is not faster)

$H_1: \mu < 30$ hours (higher-temperature fermentation is faster)

The limiting case is the nominal processing time $\mu_0 = 30$ hours. The possible choices are:

Reject H_0 (design for high temperature)

Accept H_0 (design for low temperature)

The $\geq$ orientation was selected for H_0 because the more serious error is to choose the higher-temperature process when it is not actually the faster one.

Step 2. *Select the test procedure and test statistic.* Because the population standard deviation is not known, the test statistic is the Student t.

Step 3. *Establish the significance level and the acceptance and rejection regions for the decision rule.* The engineer establishes a significance level of $\alpha = .05$. This level reflects his conservatism regarding the possible type II error (accepting H_0 when H_0 is false). The probabilities (βs) of this second error (not designing the higher-temperature process when in fact it is faster) would be uncomfortably large with a smaller α. The critical value for t is found from Table G to be $t_{.05} = 1.678$ (using linear interpolation). The test is lower-tailed, since a fast mean processing time $\bar{X}$ under the higher-temperature environment will refute H_0. Thus, negative values for t lying below -1.678 will lead to rejection. Figure 11–12 applies.

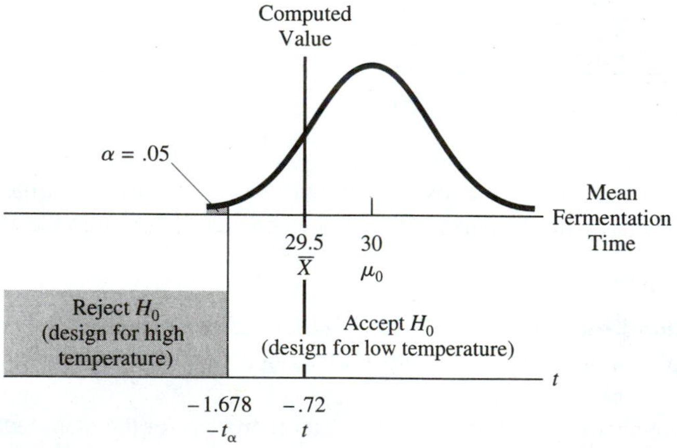

Step 4. *Collect the sample data and compute the value of the test statistic.* The following test results were obtained:

$$\bar{X} = 29.5 \text{ hours}$$
$$s = 4.91$$

This provides the computed Student t statistic

$$t = \frac{29.5 - 30}{4.91/\sqrt{50}} = -.72$$

Step 5. *Make the decision.* This quantity falls inside the acceptance region, so the engineer must accept H_0 and design the process for low fermentation temperatures.

In accepting the null hypothesis that the low-temperature environment results in faster fermentation, there is the danger that the type II error has been committed. It is good statistical procedure to consider the potential for such error.

The engineer might be incorrectly designing for low fermentation temperatures when the actual value of μ is faster than nominal. Of course, that could not be established without further testing. As we have seen, the incidence of such type II errors may be controlled by using a larger α or by increasing the sample size.

Abbreviated Procedure The following illustration involves reaching a conclusion regarding digital compression of photographs.

***Illustration**: Using Fuzzy Logic to Improve Lossy JPEG Digital Compression Technique*

A software engineer for a telecommunications company has read an article on lossy digital compression. The engineer wants to know whether or not the proposed enhancement using fuzzy logic would actually improve existing Lossy JPEG digital compression techniques. That question may be largely resolved by measuring the number of inversion loops necessary to restore the original image. In the article the researchers used different settings of distortion and compression ratios. In each case they used $n = 15$ randomly selected images and took 10 trials each to compute the average number of loops required to inversion.

The software engineer knows for the same types of test images and gray scale settings that his algorithm has a mean inversion rate of about 8 loops.

Step 1. *Formulate the hypotheses and specify the possible choices.* Letting μ represent the mean average number of loops per image required to complete the image inversion, the software engineer selects as his hypotheses:

$H_0 : \mu \geq 8$ loops (the fuzzy logic with lossy JPEG is no better than current experience)

$H_1 : \mu < 8$ loops (the fuzzy logic with lossy JPEG is better)

He will either accept H_0 and retain the present digital compression technique or reject H_0 and switch over to the reported method.

Step 2. *Select the test procedure and test statistic.* The engineer has no idea of what the standard deviation in number of loops per image is. He chooses the Student t as his test statistic.

Step 3. Not applicable.

Step 4. *Collect the sample data and compute the value of the test statistic.* The data in Table 11–3 were reported. The following summary statistics were computed from these data.

$$\bar{X} = 5.38 \qquad s = 1.32$$

Table 11–3
Reported Performance Summary of Fuzzy Logic with Lossy JPEG Digital Image Compression (Target gray scale area = 1.5; Average gray scale area tolerance = .025)

Average Number of Inversion Loops X		
4.8	6.4	6.1
4.4	5.4	4.6
8.8	4.9	4.1
4.7	5.3	7.5
4.9	4.1	4.7

Source: C. J. Wu and A. H. Sung, "The Application of Fuzzy Logic to JPEG, "*IEEE Transactions on Consumer Electronics*. November 1994.

Step 5 (abbreviated). *Calculate the prob value and make the decision.* The following applies:

$$\text{prob} = \Pr\left[\frac{\bar{X} - \mu_0}{s/\sqrt{n}} \leq \frac{5.38 - 8}{1.32/\sqrt{15}}\right] = \Pr[t \leq -7.69]$$

Using 15–1 = 14 degrees of freedom, Appendix Table G gives a smaller critical value for a tail area of .0005. The prob value must therefore be smaller than .0005, and the sample data strongly refute the null hypothesis. After examining this result, the engineer rejected H_0 and adopted the new fuzzy logic and lossy JPEG digital image compression technique.

Two-Sided Test Illustrations

Recurring-Decisions Procedure In the following illustration, a routine maintenance policy is established.

***Illustration**: Replacing a pH Regulator*

The superintendent of a new chemical plant is establishing operation policies, such as when to perform critical maintenance. When practical, those actions will be based on sample data. This is the case for a process in which the pH of an intermediate stage must be maintained at a neutral level. A control valve keeps the mean batch pH at 7.0. Owing to the caustic nature of the fluids flowing through it, the valve must occasionally be replaced. This action will be based on the mean pH of 100 sample vials taken at random times from the separation tank containing the output from that stage.

Hypothesis Steps 1–3 show how the policy for valve replacement was established. Steps 4 and 5 apply to a particular sample outcome.

Step 1. *Formulate the hypotheses and specify the possible choices.* The following hypotheses apply:

$$H_0 : \mu = 7.0 \quad \text{(process is neutral)}$$
$$H_1 : \mu \neq 7.0 \quad \text{(process is not neutral)}$$

The possible choices are

Reject H_0 (shut down for valve replacement)
Accept H_0 (continue the process)

Step 2. *Select the test procedure and test statistic.* Since a permanent policy is being established, a *recurring-decisions procedure* applies. The computed mean pH $\bar{X}$ for the sample vials serves as the test statistic. The standard deviation for individual test vials has long been established at .50 on the pH scale. This establishes the standard error for $\bar{X} : \sigma_{\bar{X}} = .50/\sqrt{100} = .05$.

Step 3. *Establish the significance level and the acceptance and rejection regions for the decision rule.* It is very expensive to shut down the process to replace the control valve, so plant policy permits just a 1% chance of doing this unnecessarily. The significance level for each sample test is therefore $\alpha = .01$. The critical normal deviate is $z_{.005} = 2.57$, and the critical values for the sample mean are

$$\bar{X}_1^* = 7.0 - 2.57(.05) = 6.87 \text{ pH}$$
$$\bar{X}_2^* = 7.0 + 2.57(.05) = 7.13 \text{ pH}$$

The acceptance and rejection regions are shown in Figure 11–13.

Figure 11–13

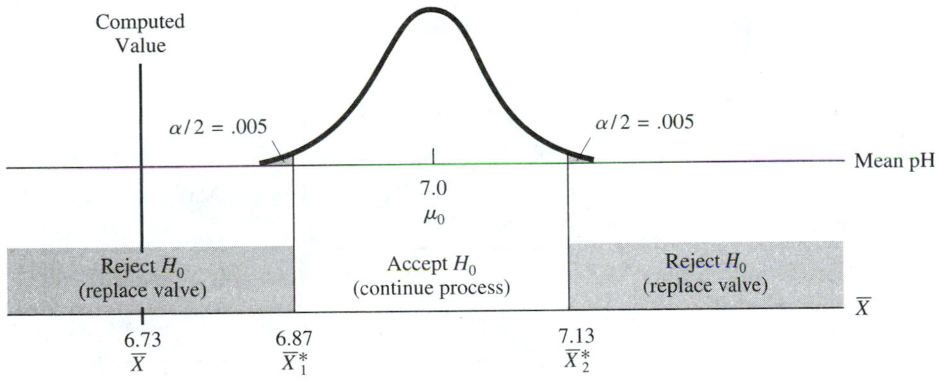

Step 4. *Collect the sample data and compute the value of the test statistic.* The computed mean pH for a particular test turned out to be $\bar{X} = 6.73$.

Step 5. *Make the decision.* Since this value falls in one of the two rejection regions, H_0 must be rejected. At the present time it must be concluded that the process mean pH is not neutral (and is too acidic). The process will be shut down to replace the control valve.

Single-Decision Procedure The following illustration shows how a two-sided test can be used in a single decision.

Illustration: *Placement of Road Signs*

It currently takes a highway department an average of 1.5 hours to determine the placement of each road sign. The director will allow that to be raised to 2.0 hours in return for increased attention to a variety of factors now ignored. From a list of 20 new design factors, 10 were selected and used in siting decisions for a random sample of 15 signs.

Step 1. *Formulate the hypotheses and specify the possible choices.* Letting μ represent the mean placement time for signs, the director tests the following hypotheses:

$$H_0 : \mu = 2.0 \text{ hours}$$
$$H_1 : \mu \neq 2.0$$

The possible choices are

Reject H_0 (add more factors or reduce the number of factors, depending on whether mean times are low or high)

Accept H_0 (keep the number of design factors at 10)

Step 2. *Select the test procedure and test statistic.* This is a single-decision procedure, and the standard deviation in placement times is unknown. The Student t applies as the test statistic.

Step 3. *Establish the significance level and the acceptance and rejection regions for the decision rule.* The director selects an $\alpha = .05$ significance level. This gives the two-sided acceptance and rejection regions portrayed in Figure 11–14.

Figure 11–14

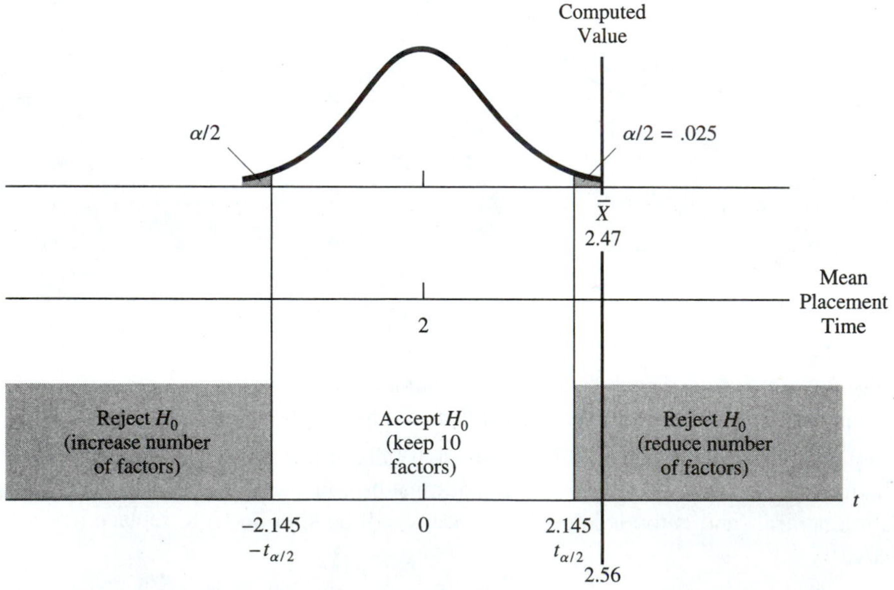

Step 4. *Collect the sample data and compute the value of the test statistic.* The following sample placement times were obtained:

1.7	1.6	2.6
2.3	1.5	3.2
3.1	2.3	1.9
2.8	2.7	1.8
4.0	2.4	3.1

From these, the following summary statistics were computed.

$$\bar{X} = 2.47 \qquad s = .71$$

$$t = \frac{2.47 - 2}{.71/\sqrt{15}} = 2.56$$

Step 5. *Make the decision.* The computed t falls in the upper rejection region. The director must therefore *reject* the null hypothesis and reduce the number of design factors thereby making the sign placement times closer to the desired mean of 2.0 hours.

Abbreviated Procedure The following illustration shows how a two-sided hypothesis test can be used in conjunction with a pure research experiment.

***Illustration**: Fuel-Element Powers—Comparing Diffusion-Theory Calculated and Measured Values*

A researcher performed an experiment with reactor fuel elements that involved gamma rays emitted during decay. For particular reactor core locations (presumed here to be selected randomly), he computed the power produced by the elements, using diffusion theory. The theoretical values were divided by their measured powers to give the ratio of calculated-to-experimental (C/E) values shown in Table 11–4.

We will use the sample data to establish whether the gamma-scan measuring system coincides with diffusion theory. If it does, then the C/E values should average to 1. The following steps apply.

Step 1. *Formulate the hypotheses and specify the possible choices.* Letting μ represent the mean calculated-to-experimental ratio, the following apply:

$$H_0: \mu = 1 \quad \text{(measures coincide with diffusion theory)}$$

$$H_1: \mu \neq 1 \quad \text{(measures deviate from diffusion theory)}$$

He will either accept H_0 and conclude that the results are consistent with diffusion theory or reject H_0 and conclude otherwise.

Step 2. *Select the test procedure and test statistic.* The researcher has no idea of what the standard deviation in calculated-to-experimental ratio is. He chooses the Student t as his test statistic.

Step 3. Not applicable.

Table 11–4
Comparison of Diffusion-Theory Calculated and Measured Fuel-Element Powers

(1) Core Position	(2) Calculated Diffusion Theory C	(3) Measured Gamma Scan E	(4) C/E Ratio X
A–2	.641	.675	.950
A–3	.767	.802	.956
A–4	.924	1.016	.909
A–5	1.074	1.126	.954
A–6	.928	1.034	.897
A–7	.794	.792	1.002
A–8	.605	.634	.954
B–3	.925	1.016	.910
B–5	.987	1.003	.984
B–7	.867	.911	.952
C–2	1.227	1.253	.979
C–4	1.096	1.058	1.036
C–5	1.109	1.178	.941
C–6	1.372	1.378	.996
C–8	1.179	1.245	.947
D–2	.792	.840	.943
D–3	1.318	1.231	1.071
D–5	1.064	1.023	1.040
D–7	1.233	1.134	1.087
D–8	.790	.803	.984
E–2	1.035	.974	1.063
E–4	1.241	1.146	1.083
E–6	1.234	1.130	1.092
E–8	1.067	1.026	1.040
F–3	.758	.717	1.057
F–5	.867	.788	1.100
F–7	.766	.721	1.062

Source: Hobbs, *Transactions of American Nuclear Society* Vol. 55, November 15–19, 1987, pp. 195–197.

Step 4. *Collect the sample data and compute the value of the test statistic*. The data in Table 11–4 were reported. The following summary statistics were computed.

$$\overline{X} = .9996 \qquad s = .0626$$

Step 5 (abbreviated). *Calculate the prob value and make the decision*. The following applies:

$$\begin{aligned}\text{prob} &= \Pr\left[t \geq \frac{|\overline{X} - \mu_0|}{s/\sqrt{n}}\right] \\ &= \Pr\left[t \leq \frac{.9996 - 1}{.0626/\sqrt{27}}\right] + \Pr\left[t \geq -\frac{.9996 - 1}{.0626/\sqrt{27}}\right] \\ &= \Pr[t \leq -.033] + \Pr[t \geq .033] \approx .8\end{aligned}$$

The sample data are quite consistent with the null hypothesis. Because of the very high prob value, the researcher accepted the null hypothesis and concluded that the measures coincide with diffusion theory.

Hypothesis Testing and Confidence Intervals

A two-sided hypothesis test may be explained in terms of statistical estimation. An equivalent procedure for conducting such a test would be to construct a $100(1-\alpha)\%$ confidence interval from the sample results. Then if μ_0 falls inside the interval, H_0 must be accepted; otherwise, it must be rejected. α is the proportion of such intervals that will fail to bracket μ_0 when it is the true mean. In other words, when this procedure is used, the type I error of rejecting H_0 when it is true will occur at relative frequency α.

For the sign placement experiment, $1-\alpha = .95$. Using the sample results from before, we construct a 95% confidence interval estimate for μ:

$$\mu = \bar{X} \pm t_{\alpha/2}\frac{s}{\sqrt{n}} = 2.47 \pm 2.145\frac{.71}{\sqrt{15}} = 2.47 \pm .39$$

or

$$2.08 \leq \mu \leq 2.86$$

Since the hypothesized value of $\mu_0 = 2.0$ lies totally outside, we must reject the null hypothesis and reach the same conclusion as before.

We may interpret using the confidence interval as follows. If the experiment were duplicated several times, and if similar confidence intervals were constructed for each new set of sample values, then we would expect 95% of those intervals to contain μ_0 within their respective limits. But about 5% of them would have limits that lie totally above or below μ_0; for these, H_0 would be rejected. Thus, 5% of the time the type I error would be committed.

Problems

11–10 The mechanical engineer from the stress-testing illustration performs another stress test on a new material, where $\sigma = 600{,}000$ cycles to failure and $H_0 : \mu \leq 10{,}000{,}000$. Suppose a sample size of $n = 50$ castings is tested at a significance level of $\alpha = .05$.

(a) Assuming that the same choices must be made, identify the acceptance and rejection regions.

(b) The computed mean is 10,115,000. What action should the engineer take?

11–11 Suppose that the chemical engineer in the fermentation-temperature illustration had reported identical sample results, but for a larger sample size of $n = 100$ instead. Perform hypothesis-testing Steps 3–5.

11–12 The plant superintendent in the pH regulator illustration found that too much supplemental processing was required because faulty control valves were not replaced often enough. He raised the significance level to $\alpha = .05$.

(a) Perform hypothesis-testing Step 3.

(b) What action should be taken for each of the following sample results?

(1) $\bar{X} = 6.85$ (2) $\bar{X} = 7.24$ (3) $\bar{X} = 6.59$ (4) $\bar{X} = 7.07$

11–13 Refer to the data in the logic circuit wafer example on page 298. Should the industrial engineer accept or reject the null hypothesis that $\mu \geq .20$ ounce? Use $\alpha = .01$.

11–14 Consider a sample size of $n = 100$ taken at random from a small population of size $N = 500$. The null hypothesis is that $\mu \leq 5{,}000$. Suppose the population standard deviation is known to be $\sigma = 150$ and that $\bar{X}$ will serve as the test statistic.

(a) Assuming that a significance level of $\alpha = .01$ is desired, find the critical value for the sample mean and determine the decision rule.

(b) Should H_0 be accepted or rejected if the computed sample mean turns out to be $\bar{X} = 5{,}060$?

11–15 A plant produces rods with a specified 1-cm mean diameter. The standard deviation is .01 cm. A decision rule must be established for determining when to correct for over- or undersized output. That choice will be based on a sample of 100 rods, under the stipulation that there be only a 1% chance of taking corrective action when the mean diameter is exactly on target.

(a) Perform hypothesis-testing Steps 1–3.

(b) What action should be taken if $\bar{X} = .993$ cm?

(c) Suppose that $\bar{X} = 1.0023$ cm. Construct a 99% confidence interval estimate for the mean diameter. Does this contain μ_0? What action should be taken?

11–16 Refer to the data in the physical therapy apparatus example on page 306. Should the mechanical engineer accept or reject the null hypothesis that $\mu \leq 20$ days? Use $\alpha = .05$.

11–17 The plant in Problem 11–15 will soon be producing a .5-cm rod. The standard deviation is unknown. Again, corrective action will be based on a sample of 100 rods, with the same error goal.

(a) Perform hypothesis-testing Steps 1–3.

(b) What act should be taken if $\bar{X} = .497$ cm and $s = .0075$ cm?

(c) Construct a 99% confidence interval for the mean diameter when $\bar{X} = .508$ cm and $s = .013$ cm. Does this contain μ_0? What action should be taken?

11–18 NASA scientists obtained the following data from airborne radiometer scans using the ESTAR method of aperature synthesis.

Soil Moisture Percentage		
19	16	15
16.5	15	14

Source: D. M. LeVine, A. J. Griffis, C. T. Swift, and T. J. Jackson, "ESTAR: A Synthetic Aperature Microwave Radiometer for Remote Sensing Applications," *Proceedings of the IEEE*, December 1994, p. 1795.

These data represent sample measurements taken over the 220–240^0 K brightness temperature range. The theoretical model states that the mean volumetric soil moisture percentage should be 13.5%. Use the abbreviated procedure to conclude whether the sample data support or refute the null hypothesis that the mean ESTAR measurements equal the theoretical level.

11–19 The following data represent a sample of measured coolant pressures in the TOPAZ-II space nuclear reactor.

Coolant System Pressure (kPa)					
86	83	86	83	83	81
83	79	83	78	83	

SOURCE: M. S. El-Genk and D. V. Paramonov, "An Analysis of Disassembling the Radial Reflector of a Thermionic Space Nuclear Reactor Power System," *Nuclear Safety*, January–June 1994, p. 78.

These were made over a range from 80 to 100 Kw of electric power to TFC heaters. Over that range the thermionic transient analysis model (TITAM) predicts a mean cooling system pressure of 88 kPa. Use the abbreviated testing procedure to determine whether the experimental results are consistent with the theoretical prediction.

11–20 An engineering department manager wants to decide whether or not to include a computer-aided design (CAD) software package in his budget. He is skeptical about the vendor's time-savings claims. A test has been conducted using pairs of nearly equally skillful engineers who independently design the same item—one using CAD and the other unassisted. For $n = 8$ parts the following percentage time-savings by CAD over the unassisted design were obtained:

80	10	37	26
45	29	44	5

At the 5% significance level, can the manager conclude that CAD will indeed yield savings in time over present design methods?

11–21 The following sample-failure data (thousands of miles) were obtained for a type of catalytic converter:

62.3	44.4	49.2	63.3	47.6	60.1
37.4	55.8	57.5	58.3	56.2	54.3

(a) Construct a 95% confidence interval estimate for the mean distance to failure.

(b) At the 5% significance level, must you accept or reject the null hypothesis that the catalytic converter will last, on the average, 50 thousand miles?

11–22 A biological reduction process is being evaluated as an alternative to a conventional technique for which the mean yield of active ingredients is $\mu = 10$ grams/liter with a standard deviation of $\sigma = .5$ gram/liter. The biological method will be an effective replacement for the present procedure if it can be assumed to provide a yield of more than 12 grams/liter. A total of 100 batches will be pilot-tested under bacterial reduction.

(a) If test results indicate that significant improvement in yield might be made, the biological procedure will be adopted. There should be only a 5% chance that this will be done when the new technique's yield is not high enough to be effective as a replacement. Assuming the same standard deviation in yield applies whichever procedure is used, perform hypothesis-testing Steps 1–3.

(b) The test results provide a yield of 13.3 grams/liter. What action should be taken?

11–23 A quality-control inspector for a microwave transmitter manufacturer is assessing a shipment of 400 crystal controls. The actual broadcast frequency will depend on the resonant frequency, which will vary slightly from crystal to crystal, but the mean level should achieve the rated target of .55 mHz. The exact standard deviation for the individual crystal frequencies is unknown. A random sample of 50 crystals from the shipment will be tested and the resonant frequency determined for each. The entire shipment will be rejected if the observed mean is significantly above or below the rated level and accepted otherwise. The inspector wants just a 1% chance of rejecting a shipment in which the mean frequency exactly matches the rated level.

(a) Formulate the inspector's hypotheses.

(b) Sample results provide $\bar{X} = .5503$ mHz and $s = 485$ Hz. Construct a 99% confidence interval estimate of the mean resonant frequency for the entire shipment. What action should the inspector take?

11–24 A new test stand might be acquired by the inspection department of GizMo, an instrumentation fabricator, if the mean time to check out a standard "black box" is significantly less than that for the present unit. The new test stand will save enough operating costs to pay for itself only if it can "burn in" all new instrument boxes in 15 hours or less, on the average. The equipment vendor allows GizMo to time-test a sample of 50 boxes. No educated guesses have been made regarding the target population's characteristics.

(a) Assuming that GizMo wants no more than a 1% chance of buying the new test stand when it will not save enough to justify its cost, perform hypothesis-testing Steps 1–3.

(b) The test results provide a mean burn-in time of 13.5 hours with a standard deviation of 3.4 hours. What action should be taken?

11–25 The final stage of a chemical process is sampled and the level of impurities is determined. The final product is recycled if there are too many impurities, and the controls are readjusted if there are too few (which is an indication that too much catalyst is being added). If it is concluded that the mean impurity level is the nominal level of $\mu = .01$ gram/liter, the process is continued without interruption. The standard deviation of impurities has been established to be $\sigma = .005$ gram/liter. A sample of $n = 100$ specimens is measured every 24 hours, and the level of impurities is determined for each.

(a) Perform hypothesis-testing Steps 1–3, assuming a significance level of 5%.

(b) What action should be taken for each of the following sample results for the mean level of impurities?

(1) $\bar{X} = .0095$ (2) $\bar{X} = .0088$ (3) $\bar{X} = .0112$ (4) $\bar{X} = .0104$

(c) Find the type II error probability when the true mean impurity level is

(1) $\mu = .009$ (2) $\mu = .012$ (3) $\mu = .008$ (4) $\mu = .014$

11–26 The world average exposure rate to ^{238}U is12.0($\times 10^{-5}$Gy/yr).* Using the sample results from Table 10–2, should you conclude that the mean exposure rate to this radioactive element in Riyadh, Saudi Arabia, is identical to the worldwide figure? Use $\alpha = .05$.

11–27 The following world exposure rates ($\times 10^{-5}$Gy/yr) apply to two radioactive elements:*

(a)	(b)
^{232}Th	^{40}K
14.0	9.0

Using the sample results from Table 10–2, should you conclude that Riyadh, Saudi Arabia, has an identical exposure rate to the worldwide figure?

11–3 Testing the Proportion

Tests of the proportion are next in importance to those for the mean. The proportion is especially important in statistical quality control where a major concern is whether the level of defectives in a shipment is high or low; similarly, the overall quality of a production process might be measured by the proportion of defective output. Proportions are also fundamental in work measurement, where work sampling is often used instead of stopwatch procedures. The proportion is also a key parameter in specifying a Bernoulli process, and a considerable amount of probability analysis might be based upon an assumed level.

The testing procedures established for the mean are easily extended to the proportion. The sample proportion P is a natural test statistic for experiments regarding π, but sometimes the number of successes R is employed equivalently. The appropriate sampling distribution for P or R is either the binomial or the hypergeometric—the latter applying when sampling without replacement from small populations. A normal curve closely approximates these underlying distributions over the wide range of π's and n's, and tests of the proportion are most often based on that approximation.

Testing the Proportion Using the Normal Approximation

The sample proportion P is a convenient test statistic for testing the value of the population proportion. As with testing the mean, we divide applications into single-decision and recurring-decisions procedures. Figure 11–15 summarizes these. (A continuity correction of $.5/n$ is used in computing critical values for P.)

*A. Towfik et al., *Transactions of American Nuclear Society*, Vol. 55, November 15–19, 1987, p. 89.

Figure 11–15 *Summary of hypothesis-testing procedures for proportion.*

Step 4 Collect data and compute P.
Then calculate

$$z = \frac{P - \pi_0}{\sigma_P}$$

Only one of these is needed.

Step 5 Make decision.

Lower-Tailed	**Upper-Tailed**
$(H_0\colon \pi \geq \pi_0)$	$(H_0\colon \pi \leq \pi_0)$
Accept H_0 if $z \geq -z_\alpha$	*Accept* H_0 if $z \leq z_\alpha$
Reject H_0 if $z < -z_\alpha$	*Reject* H_0 if $z > z_\alpha$

Two-Sided $(H_0\colon \pi = \pi_0)$

Accept H_0 if $-z_{\alpha/2} \leq z \leq z_{\alpha/2}$
Reject H_0 if $z < -z_{\alpha/2}$ or if $z > z_{\alpha/2}$

or

Compute the $100(1 - \alpha)\%$ confidence interval:

$$P - z_{\alpha/2}\sigma_P \leq \pi \leq P + z_{\alpha/2}\sigma_P$$

Accept H_0 if π_0 falls inside.
Otherwise, *Reject* H_0.

When $n \leq .10N$:

$$\sigma_P = \sqrt{\frac{\pi_0(1 - \pi_0)}{n}}$$

When $n > .10N$:

$$\sigma_P = \sqrt{\frac{\pi_0(1 - \pi_0)}{n}} \times \sqrt{\frac{N - n}{N - 1}}$$

Single decision

Identify procedure

Recurring decisions

Step 3 Find critical values for the decision rule. — or Use $R^* = nP^*$.

Lower-Tailed	**Upper-Tailed**
$(H_0\colon \pi \geq \pi_0)$	$H_0\colon \pi \leq \pi_0)$
$P^* = \pi_0 + .5/n - z_\alpha\sigma_P$	$P^* = \pi_0 - .5/n + z_\alpha\sigma_P$
Accept H_0 if $P \geq P^*$	*Accept* H_0 if $P \leq P^*$
Reject H_0 if $P < P^*$	*Reject* H_0 if $P > P^*$

Compare R to R^*

Two-Sided $(H_0\colon \pi = \pi_0)$

$P_1^* = \pi_0 + .5/n - z_{\alpha/2}\sigma_P$ $\qquad$ $P_2^* = \pi_0 - .5/n + z_{\alpha/2}\sigma_P$

Accept H_0 if $P_1^* \leq P \leq P_2^*$
Reject H_0 if $P < P_1^*$ *or* if $P > P_2^*$

or Use
$R_1^* = nP_1^*$
$R_2^* = nP_2^*$

Compare R to R_1^*, R_2^*

Steps 4 and 5 Collect data, compute P, and decide. — or Find R.

The normal approximation may be applied whenever the sample size is sufficiently large. We will use it whenever both of the following are true:

$$n\pi \geq 5$$
$$n(1-\pi) \geq 5$$

A Lower-Tailed Test

A lower-tailed test is described in the following illustration.

***Illustration**: Is Double Data Entry Necessary?*

A computer systems analyst for a large electric utility wondered if it might be practical to cut the tremendous expense of the data-processing staff. Hundreds of data clerks presently encode meter-reader tags for entry into the customer-billing data base. To minimize billing errors, two clerks independently enter the data from each tag, and if the entries do not coincide, the tag is reprocessed. Although the possibility that errors will occur is extremely remote when tags are processed in this manner, the number of clerks required is twice that of a single data entry system. Furthermore, extra data processing is required to make the verification comparisons.

The analyst wonders if significant cost savings might be achieved by abandoning double-entry processing in favor of entering the data just once—obviously a much more error-prone procedure. Under the proposed scheme, large errors would be filtered out by comparing usage with the average of previous months; only those customer tags on which usage is above or below average by a factor of 3 would be recycled once for verification. Any subsequent over- or underbilling would ultimately be rectified by the cumulative readings obtained for a customer in future months.

Because of the expenses that would arise in handling customer complaints for overbilling and owing to the cost of uncollected funds from underbilled accounts, the efficacy of this procedure depends on a fairly low single-entry error rate. Management consensus was that any new-procedure error rate not exceeding .5% would result in savings for the company. The pivotal value for the proportion π of all erroneous entries was therefore set at $\pi_0 = .005$.

A sampling experiment was initiated and performed under conditions that duplicated very closely those anticipated under the new procedure. For the test period the clerks were asked to be especially careful, as their work was not going to be totally verified. To preclude lackadaisical performance, each clerk's work would still be checked on a random basis for accuracy—and the test subjects were so informed. The following steps were performed using a sample of $n = 2{,}000$ meter tags:

Step 1. *Formulate the hypotheses and specify the possible choices*. The following hypotheses were selected by the analyst:

$H_0 : \pi \geq .005\ (\pi_0)$ (new procedure is impracticable)

$H_1 : \pi < .005$ (new procedure may be an improvement)

The analyst will either reject H_0 and recommend the new procedure or accept H_0 and not recommend it.

Step 2. *Select the test procedure and test statistic.* Since there is a single decision to be made, the normal deviate z will serve as the test statistic. This will be computed using the sample proportion of errors P.

Step 3. *Establish the significance level and the acceptance and rejection regions for the decision rule.* The analyst plans to recommend adoption of the new procedure if the observed error rate is sufficiently small. A significance level of $\alpha = .05$ was selected, so the critical normal deviate is $z_\alpha = 1.64$.

The decision rule is summarized in Figure 11–16.

Figure 11–16

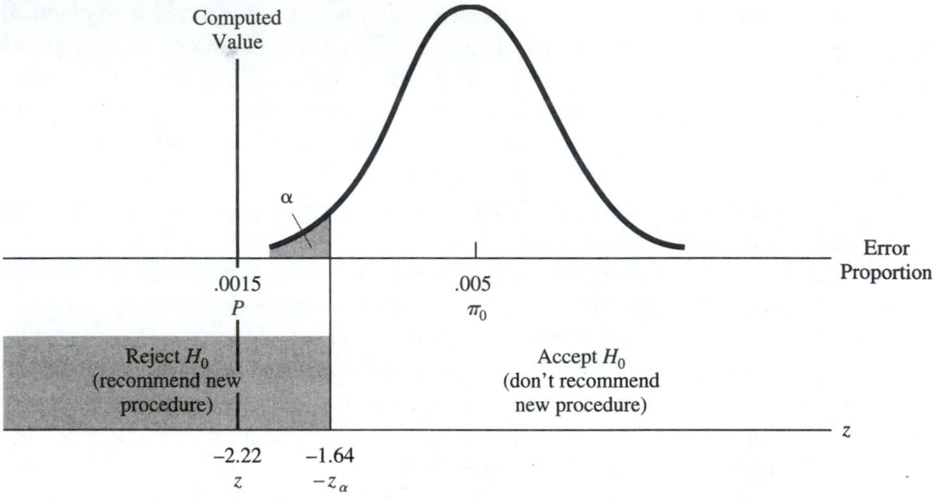

Step 4. *Collect the sample data and compute the value of the test statistic.* Under H_0, the standard error of P is

$$\sigma_P = \sqrt{\frac{.005(.995)}{2{,}000}} = .00158$$

Only 3 of the sample tags were in error, so the computed error proportion was only $P = \frac{3}{2{,}000} = .0015$. The computed value for the normal deviate is

$$z = \frac{.0015 - .005}{.00158} = -2.22$$

Step 5. *Make the decision.* The computed value falls below the critical value, and the analyst must reject the null hypothesis and conclude that the new procedure might provide savings in operational costs. The analyst therefore recommends the new procedure.

An Upper-Tailed Test

To illustrate an upper-tailed test of the proportion, consider the following quality-control illustration.

Illustration*: *Establishing a Policy for Receiving Microcircuit Chips

An electrical engineer for a computer manufacturer wishes to establish the policy for inspecting and receiving shipments of microcircuit chips for installation onto printed circuit boards (pcb's). Presently, all chips are installed without any testing. Defective chips are removed during later-stage testing of the pcb's. He believes that it will be cheaper to first test a sample of chips. Then, if the sample proportion defective is small, the shipment will be *accepted* and be released directly for assembly. If that proportion is high, the shipment will be *rejected*, and all chips will be tested first; no defectives will be included with those assembled onto pcb's.

The engineer needs to establish where to draw the line between accepting and rejecting shipments of chips. This determination is based on a break-even level of .10 for the proportion of defectives in any 1,000-unit shipment. Choosing a sample size of $n = 50$, he uses the hypothesis-testing approach to arrive at the answer.

Step 1. *Formulate the hypotheses and specify the possible choices.* The engineer selects as his hypotheses:

$$H_0: \pi \leq .10 \qquad \text{(shipment quality is good)}$$

$$H_1: \pi > .10 \qquad \text{(shipment quality is poor)}$$

Accepting H_0 is to accept the shipment, while rejecting H_0 is to reject the shipment. The pivotal level of the population (shipment) proportion defective is the break-even level $\pi_0 = .10$. With these hypotheses, the type I error will be to reject a good-quality shipment, while the type II error will be to accept a poor-quality shipment.

Step 2. *Select the test procedure and test statistic.* The sample proportion defective P serves as the engineer's test statistic for this recurring-decisions procedure.

Step 3. *Establish the significance level and the acceptance and rejection regions for the decision rule.* The engineer decides that a 1% significance level will adequately protect against the type I error. Since neither $n\pi_0 = 50(.10)$ nor $n(1 - \pi_0) = 50(.90)$ falls below 5, the sample size is large enough for him to make the normal approximation. The critical normal deviate is $z_{.01} = 2.33$. In finding the critical value P^*, he first computes the standard error of P that applies under H_0:

$$\sigma_P = \sqrt{\frac{.10(1 - .10)}{50}} = .0424$$

(The population size of $N = 1{,}000$ is great enough that the finite population correction may be ignored.) For this *upper-tailed* test, the critical value is computed as follows:

$$\begin{aligned} P^* &= \pi_0 - \frac{.5}{n} + z_{.01}\sigma_P \\ &= .10 - \frac{.5}{50} + 2.33(.0424) \\ &= .189 \end{aligned}$$

Figure 11–17

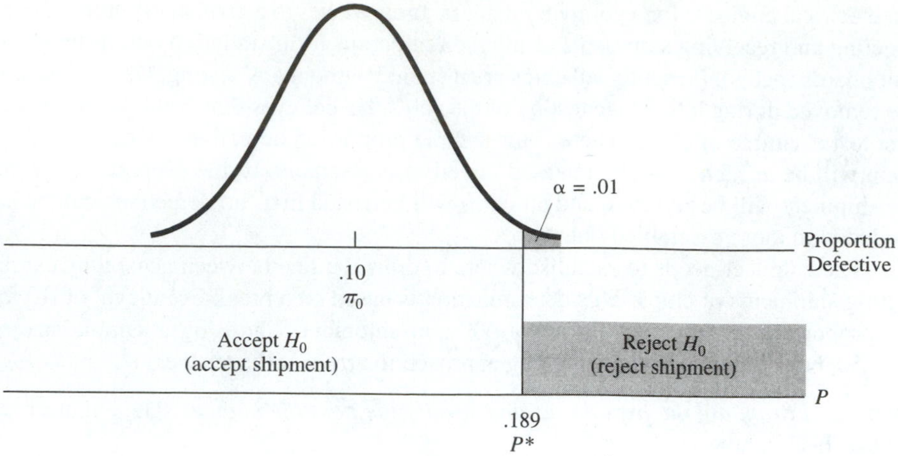

The decision rule in Figure 11–17 applies.

Hypothesis-testing Steps 4 and 5 will be performed for each shipment of chips received. The following results would be achieved for five future shipments:

Number of Defectives R	Proportion Defective $P = R/n$	Action
3	.06	Accept
9	.18	Accept
10	.20	Reject
7	.14	Accept
12	.24	Reject

After reviewing this decision rule, the engineer chooses to simplify the procedure by not requiring the inspectors to compute the sample proportion.

Alternative Procedure for Testing Proportion

When several persons may be involved in repeated applications of the test, it may be easier to streamline the procedure by basing the decision rule on the number of observed successes R as the test statistic. The critical value is found by translating P^* into its number counterpart:

$$R^* = nP^* \quad \text{(integer value)}$$

where R^* is always the nearest *integer* $\geq nP^*$ for a lower-tailed test and $\leq nP^*$ for an upper-tailed test. The form of the decision rule is analogous. For two-sided tests, R_1^* is the nearest integer $\geq nP_1^*$ and R_2^* the nearest integer $\leq nP_2^*$.

Using this alternative procedure, the engineer in the preceding illustration computes

$$nP^* = 50(.189) = 9.45$$

so that rounding down he gets the critical number of sample defectives $R^* = 9$ and arrives at the following decision rule:

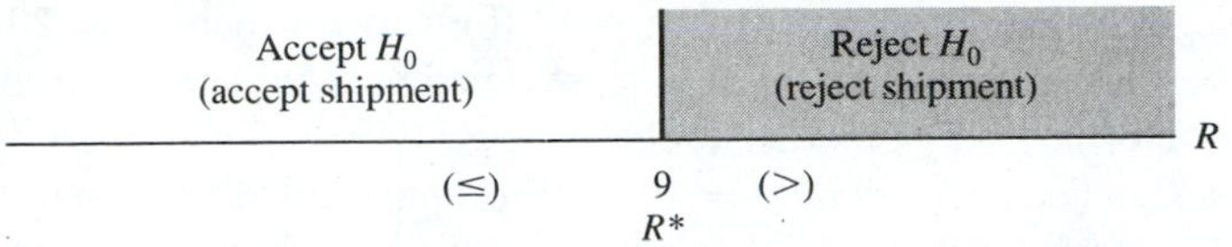

Regardless of whether P or R serves as the test statistic, identical results will be obtained.

Testing with Binomial Probabilities

When n is too small to make the normal approximation, the decision rule may instead be based directly on binomial or hypergeometric probabilities. (The latter always apply when sampling without replacement from a finite population, but can ordinarily be approximated by the binomial when N is large.) Once π_0 and H_0 have been set and n chosen, the decision rule follows directly from the *targeted* significance level α. It will ordinarily take the R form.

Lower-Tailed Tests The critical value R^* is that level of r providing the *smallest* cumulative binomial probability such that

$$P[R \le r \,|\, \pi = \pi_0] > \alpha$$

This gives a type I error probability

$$\text{Pr [reject } H_0 \,|\, \pi = \pi_0] = \text{Pr }[R < R^* \,|\, \pi = \pi_0] \le \alpha$$

which cannot exceed the targeted significance level.

Upper-Tailed Tests The critical value R^* is that level of r providing the *smallest* cumulative binomial probability such that

$$\text{Pr }[R \le r \,|\, \pi = \pi_0] \ge 1 - \alpha$$

Like the above, this guarantees that the type I error probability,

$$\text{Pr [reject} \,|\, \pi = \pi_0] = \text{Pr }[R > R^* \,|\, \pi = \pi_0] \le \alpha$$

will not exceed the targeted significance level.

Two-Sided Tests The lower critical value R_1^* is the level of r providing the *smallest* cumulative binomial probability such that

$$\text{Pr }[R \le r \,|\, \pi = \pi_0] > \frac{\alpha}{2}$$

The upper critical value R_2^* is that level of r providing the *smallest* cumulative binomial probability such that

$$\text{Pr }[R \le r \,|\, \pi = \pi_0] \ge 1 - \frac{\alpha}{2}$$

To illustrate, suppose the same engineer in the preceding illustration wishes to establish a decision rule for a relay module. These come in shipments of $N = 5{,}000$, and he wishes to use $\pi_0 = .05$ to decide how to dispose of these items.

Because testing is so expensive, he can afford to use only $n = 20$. Since $n\pi_0 = 20(.05) = 1$ is less than 5, the normal approximation should not be used, and his decision rule will be based on binomial probabilities.

Using a significance level of $\alpha = .05$ for this upper-tailed test, the critical number of sample defectives has the smallest cumulative binomial probability greater than or equal to $1 - .05 = .95$. From Appendix Table A, we read

$$\Pr[R \leq 3] = .9841$$

as the required value, so that $R^* = 3$. The following decision rule applies to relays:

$n = 20$:

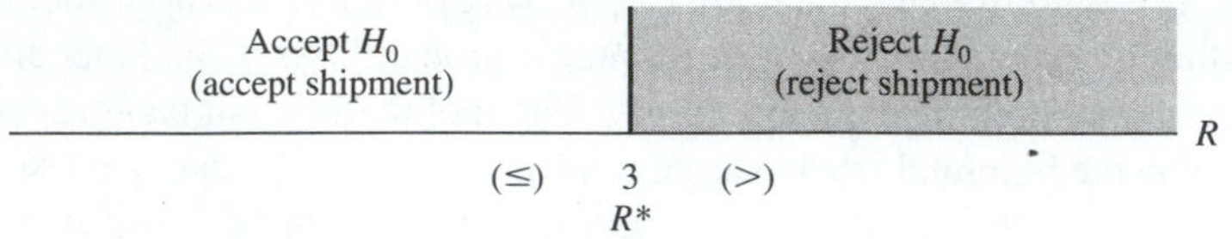

The *achieved* significance level is lower than the *targeted* .05 level and is equal to

$$\Pr[R > 3 \mid \pi = .05] = 1 - .9841 = .0159$$

The engineer was disappointed to find that using this rule results in very high probabilities for accepting shipments that are really bad (committing a type II error). For instance, when the true proportion defective is $\pi = .20$,

$$\Pr[\text{accept}] = \Pr[R \leq 3 \mid \pi = .20] = .4114 \qquad (n = 20)$$

Not wanting to change the target significance level, he chooses instead to increase the sample size to $n = 50$. Returning to Appendix Table A, for that sample size, a new level for r and smallest cumulative probability is found:

$$\Pr[R \leq 5 \mid \pi = .05] = .9622 \qquad (n = 50)$$

so that $R^* = 5$. The new decision rule is as follows:

$n = 50$:

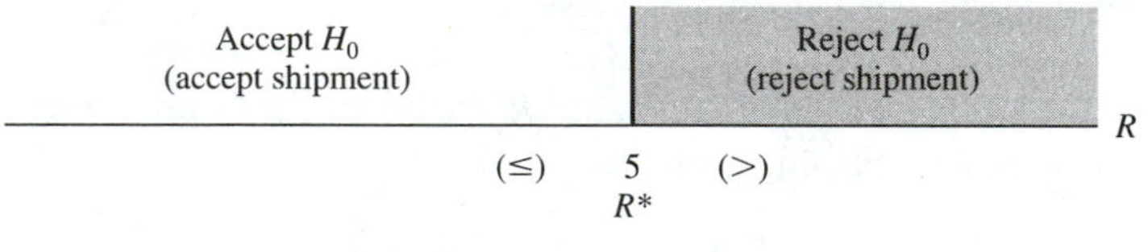

Although this rule raises the achieved significance level slightly to $1 - .9622 = .0378$, there is now better protection against the type II error of accepting bad shipments. In the case of $\pi = .20$, the new sample size and decision rule provide

$$\Pr[\text{accept}] = \Pr[R \leq 5 \mid \pi = .20] = .0480 \qquad (n = 50)$$

Abbreviated Tests with the Proportion

When evaluating another investigator's data—the typical case when reviewing the literature—an abbreviated approach may be used. A conclusion is then reached after computing a prob value. The hypothesis-testing steps on pages 381–382 are modified by skipping Step 3 and computing a prob value in Step 5. This is the probability, assuming the null hypothesis is true, for getting a result more extreme than the one actually obtained. It is a judgment call whether to then accept or reject the null hypothesis.

The following illustration uses the binomial distribution directly to demonstrate the abbreviated procedure.

***Illustration**: Are Too Many Bridges Needing Structural Repairs?*

A reporter finds that 5 out of 100 randomly selected highway bridges have serious structural decay. A professor of civil engineering wants to determine if that finding is consistent with the highway department's claim that no more than 1 percent of the bridges are in need of emergency structural repairs. Is he correct?

Step 1. *Formulate the hypotheses and specify the possible choices.* Letting π represent the true proportion of highway bridges needing structural repairs, we will test the following hypotheses:

$$H_0 : \pi \leq .01 \qquad \text{(the highway department is correct)}$$

$$H_1 : \pi > .01 \qquad \text{(the highway department is not correct)}$$

Accepting H_0 is to accept the highway department claim, while rejecting H_0 is to reject that claim.

Step 2. *Select the test procedure and test statistic.* The binomial distribution applies. The observed number of bridges needing repair R is the test statistic.

Step 3. Not applicable.

Step 4. *Collect the sample data and compute the value of the test statistic.* The reporter's data, $R = 5$ and $n = 100$, are assumed to be correct.

Step 5 (abbreviated). *Calculate the prob value and make the decision.* Using the binomial distribution directly, with $\pi = .01$ as the pivotal level, Appendix table A provides

$$\Pr[R \geq 5] = 1 - \Pr[R \leq 4] = 1 - .9966 = .0034$$

The reported result would be highly unlikely if the null hypothesis were true. But the professor could reject the highway department's claim only if he believed the reporter's data. He decided to audit those findings before suggesting to the highway department that they update their evaluations and begin taking remedial action.

The following illustration uses the normal approximation to the binomial to demonstrate the abbreviated procedure.

***Illustration**: Does an Engineering Certification Exam Need Overhauling?*

A multiple choice exam is used by a state to certify engineers. A passing score for that segment requires at least 60% correct. A psychometrist, also knowledgeable in the subject, notices that the answers are so loose that in nearly every case two out of the four possible choices can be eliminated on the basis of consistency alone—no knowledge of the subject required. She hypothesizes that some person with no training at all might be able to pass the test by eliminating illogical responses and then randomly selecting one of the remaining choices as his answer.

Has the psychometrist reached the correct conclusion?

Step 1. *Formulate the hypotheses and specify the possible choices.* Letting π represent the proportion of answers that will be correct, the following hypotheses apply:

$$H_0 : \pi = .50 \qquad \text{(the psychometrist is correct)}$$
$$H_1 : \pi \neq .50 \qquad \text{(the psychometrist is not correct)}$$

Accepting H_0 is to accept the psychometrist's claim, while rejecting H_0 is to reject that claim.

Step 2. *Select the test procedure and test statistic.* The normal approximation to the binomial distribution will be used. The proportion of questions P answered correctly is the test statistic.

Step 3. Not applicable.

Step 4. *Collect the sample data and compute the value of the test statistic.* The test section has $n = 20$ questions. A passer needs to get $P = .60$ or more correct.

Step 5 (abbreviated). *Calculate the prob value and make the decision.* Using the normal approximation to the binomial (ignoring the continuity correction), with $\pi = .5$ as the pivotal level, we find

$$\begin{aligned}
\sigma_P &= \sqrt{\frac{.5(1-.5)}{20}} = .111 \\
\text{prob} &= 2 \Pr\left[P \geq .60 \mid \pi = .5\right] \\
&= 2 \Pr\left[\frac{P-\pi_0}{\sigma_P} \geq \frac{.60-.50}{.111}\right] = 2[1-\Phi(.91)] = 2[1-.8152] = 2(.1841) \\
&= .3862
\end{aligned}$$

The above result is consistent with the null hypothesis. An unqualified person might very well be able to pass the test using good test-taking techniques—even if he knows nothing about the subject.

Two remedies are possible:

1. Increase the number of questions.
2. Reword the possible answers to make it impossible to use pure logic to eliminate some choices.

Doubling the number of questions to $n = 40$ yields

$$\begin{aligned}
\sigma_P &= \sqrt{\frac{.5(1-.5)}{40}} = .079 \\
\text{prob} &= 2 \Pr\left[\frac{P-\pi_0}{\sigma_P} \geq \frac{.60-.50}{.079}\right] = 2[1 - \Phi(1.27)] = 2[1-.8980] = (.1020) \\
&= .2040
\end{aligned}$$

The probability of a fraud's passing is still uncomfortably high.

If no answers can be eliminated, then a random chooser would face the revised hypotheses:

$$H_0 : \pi = .25$$
$$H_1 : \pi \neq .25$$

For the above, we recompute the prob value:

$$\sigma_P = \sqrt{\frac{.25(1-.25)}{20}} = .097$$

$$\text{prob} = 2\Pr[P \geq .60 \mid \pi = .25]$$

$$= 2\Pr\left[\frac{P-\pi_0}{\sigma_P} \geq \frac{60-.20}{.097}\right] = 2[1-\Phi(3.61)] = 2(1-.99984) = 2(.00016)$$

$$= .00032$$

A passing score is not at all consistent with the null hypothesis. Remedy 2 provides the better improvement.

11–28 Suppose that the systems analyst described in the data-entry illustration used $\pi_0 = .0035$ and a 1% significance level instead. Assuming the same sample results, what action should the analyst take?

11–29 For the data-entry illustration in the text, suppose that $\pi_0 = .007$ had been used with a 10% significance level.

(a) Determine the critical value and resulting decision rule.

(b) What decision will be reached if the number of sample defectives is

(1) 25? (2) 6? (3) 10? (4) 31?

11–30 Consider again the microcircuit illustration on page 383. Assume that the given decision rule with $n = 50$ will be used. Using the normal approximation, determine the type II error probability for incorrectly accepting a shipment when the proportion of defectives is

(a) .12 (b) .15 (c) .17 (d) .19

11–31 A quality-control inspector must decide about a very large shipment of 10-ohm amplifier resistors. A sample of $n = 100$ resistors will be tested and the number of defectives determined. Depending on that result, the entire shipment will be purchased or returned. The inspector wants at most a 5% chance of purchasing a poor-quality batch (having 10% or more defectives).

(a) Using the number of defectives in the sample as the test statistic, perform hypothesis-testing Steps 1 and 3. Use the binomial distribution directly instead of the normal approximation.

(b) What action should be taken when the number of sample defectives is

(1) 4? (2) 6? (3) 5? (4) 7?

11–32 A statistician is testing the null hypothesis that exactly half of all engineers will still be in the profession 10 years after receiving their bachelor's. He took a random sample of 200 graduates from the class of 1979 and determined their occupations in 1989. He found that 111 persons were still employed primarily as engineers.

(a) Construct a 95% confidence interval estimate for the proportion of engineers remaining in the profession. (Ignore the continuity correction.)

(b) At the 5% significance level, should the statistician accept or reject his hypothesis?

11–33 Consider fuses of a particular type that are inspected by both the producer and the consumer.

(a) The manufacturer defines as marketable any batch of fuses in which the proportion of defectives does not exceed .07. The inspection rules are based on the requirement that there should be no more than a 1% chance of scrapping any marketable batch. This outcome is referred to as the **producer's risk**. A sample of 500 fuses will be tested from a production run of 4,000. Perform hypothesis-testing Steps 1–3. What action should be taken if the sample is found to contain 8% defectives?

(b) One of the users of the fuses is a laboratory, where the policy is that any batch of fuses of a particular type is satisfactory if it contains less than 5% defectives. The inspection rules stipulate that there should be only a 5% chance of accepting an unsatisfactory batch. This outcome is referred to as the **consumer's risk**. A sample of 100 fuses is to be tested from a shipment of 500. Perform hypothesis-testing Steps 1–3. What action should be taken if the sample is found to contain 2 defectives?

11–34 The number of switches placed on line in a communications system depends on the level of traffic. At 20 random times over a brief interval it is determined what proportion of the time all switches are busy. An additional switch is then placed on line if this quantity exceeds a critical level. A similar rule is applied to determine when to remove one of the switches, which is done whenever the sample proportion falls below a smaller critical value.

(a) Assume that there should be only a 5% chance of unnecessarily adding a switch when all the switches are busy no more than 50% of the time.
  (1) Perform hypothesis-testing Steps 1–3.
  (2) What action should be taken if all switches are busy in 65% of the observed sample times?

(b) Assume that there should be only a 10% chance of incorrectly removing a switch when all the switches are busy at least 30% of the time.
  (1) Perform hypothesis-testing Steps 1–3.
  (2) What action should be taken if all switches are busy in 25% of the observed sample times?

11–35 You want to establish a policy for accepting or rejecting shipments of parts. Each lot contains 1,000 units. Using a pivotal level of .01 for the lot proportion defective, answer the following:

(a) Is the normal approximation appropriate?
(b) Using $n = 20$ and an α target of .05, determine the critical value for the number of sample defectives. Express the corresponding decision rule.
(c) Find the type II error probability given that the true lot proportion defective is .10. Do you think that the rule found in (b) is satisfactory? Comment on how you might improve the test.
(d) Increase the sample size to 100.
  (1) Find the new critical value.
  (2) Express the decision rule.
  (3) Determine the type II error probability when the true lot proportion defective is .10.

11–36 Consider the dilemma faced by a workstation manufacturer receiving shipments of keyboards from a vendor. These arrive in lots of 50 each. Based on the number of sample defectives, a decision rule for accepting or rejecting shipments needs to be established so that there is at most a 5% chance of rejecting a good shipment having a maximum of 10% defective keyboards.

(a) For a sample of size 5, construct a table of cumulative hypergeometric probabilities.
(b) Using the above, find the critical value and express the decision rule.
(c) Find the critical value when binomial probabilities are used to approximate the hypergeometric. Is there any change from (b)?

11–37 An engineer wishes to establish a policy for deciding when to accept or reject shipments. He wants to have just a 5% chance of rejecting a good shipment having only .06 defectives and just a 10% chance of accepting a really bad shipment having .25 defectives. Answer the following assuming that the normal approximation is appropriate.

(a) Find an expression when H_0 is true that treats n and P^* as unknown quantities and z_α and π_0 as givens.
(b) Let z_β represent the normal deviate that corresponds to the accept probability with P^* when $\pi = \pi_1$. Find an expression when H_1 is true that treats n and P^* as unknown quantities and z_β and π_1 as givens.
(c) Assuming that the normal approximation is appropriate, find the sample size n that will achieve the engineer's objective. (Ignore the continuity correction.)
(d) Using that sample size, find P^* and express the engineer's decision rule.

11–4 Hypothesis Tests for Comparing Two Means

We are now ready to establish the framework for testing hypotheses when the decision is based on the relative levels of two population means. A typical engineering application of such a test would be a decision whether or not to adopt a new method, material, procedure, or vendor. The final action is based on data collected from two populations, one sample representing the status quo, the other reflecting the proposed population. The decision rule determines whether or not the sample results are significant in signaling that a difference exists between true levels of the population means. An appropriate null hypothesis must then be accepted or rejected.

That null hypothesis will take one of the following forms:

$$H_0 : \mu_A - \mu_B \geq D_0 \quad \text{(lower-tailed test)}$$
$$H_0 : \mu_A - \mu_B \leq D_0 \quad \text{(upper-tailed test)}$$
$$H_0 : \mu_A - \mu_B = D_0 \quad \text{(two-sided test)}$$

These may be tested using either independent samples or matched pairs.

Independent Samples

The null hypothesis (in whatever form) is usually evaluated from independent samples using the difference in sample means, $D = \bar{X}_A - \bar{X}_B$, as the test statistic. As with estimating $\mu_A - \mu_B$, there are three methods for testing H_0, depending on whether the sample sizes are large or small.

Large Sample Sizes When $n_A > 30$ and $n_B > 30$, the normal deviate serves as the test statistic. The following expression is used.

NORMAL DEVIATE USING INDEPENDENT SAMPLES (LARGE n's)

$$z = \frac{(\bar{X}_A - \bar{X}_B) - D_0}{s_D}$$

using

$$s_D = \sqrt{\frac{s_A^2}{n_A} + \frac{s_B^2}{n_B}}$$

Illustration*: *Evaluation of Test Stands

Consider a decision by a quality-control department regarding testing methods.

The quality-assurance director in an electronics fabricating facility wants to determine if new test stands might be effective in increasing the flow of production units. A sampling investigation on a borrowed unit will determine the final choice whether to acquire the new stands or to retain the present manual checking procedure. Sample data have already been obtained on the present procedure's testing times, so a further independent sample of times will be obtained using a borrowed test stand. The latter sample is designated by the letter A (an arbitrary choice). A total of $n_A = 50$ observations will be made. The manual testing method involves $n_B = 100$ observations.

The same basic five hypothesis-testing steps apply as before.

Step 1. *Formulate the hypotheses and specify the possible choices.* We let μ_A denote the mean testing time using the test stands and μ_B the mean testing time under the manual method. The director wants to switch only if it can be demonstrated that using the test stands reduces the mean testing time by at least 5 minutes.

$H_0 : \mu_A - \mu_B \geq -5$ min (test stand inspection is not sufficiently faster than manual inspection)

$H_1 : \mu_A - \mu_B < -5$ min (test stand inspection is sufficiently faster)

The pivotal level for the differences in means is $D_0 = -5$. The possible choices are

Reject H_0	(acquire the test stands)
Accept H_0	(retain manual testing)

This designation reflects her strong desire to avoid adopting the test stands when they actually take longer than manual testing.

Step 2. *Select the test procedure and test statistic.* Although the difference D between sample means is the basis for the test, it will be transformed into the normal deviate expressed earlier, so that z will serve as the test statistic.

Figure 11–18

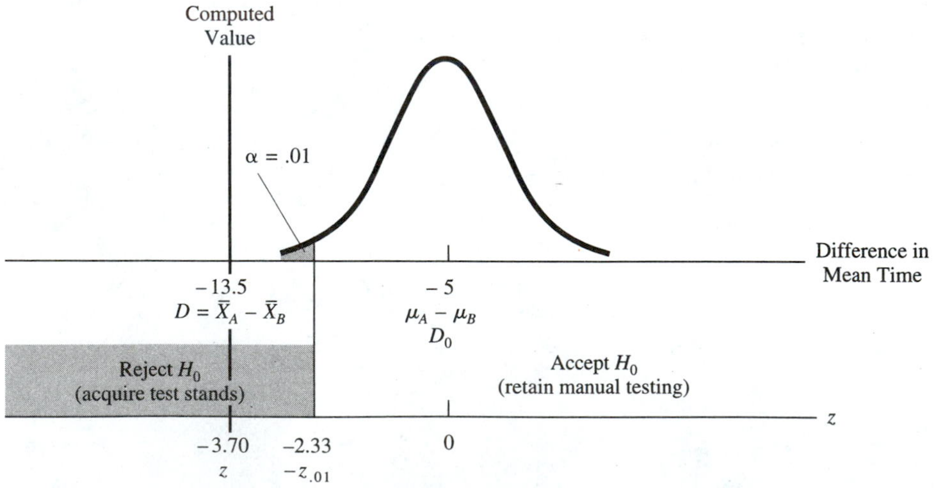

Step 3. *Establish the significance level and the acceptance and rejection regions for the decision rule.* Because the director regards so seriously the type I error of rejecting H_0 when it is true, she chooses a 1% significance level, so that $\alpha = .01$. From Appendix Table E this corresponds to a critical normal deviate of $z_{.01} = 2.33$. Extreme negative levels for z will refute H_0, and the test is lower-tailed. The decision rule is shown in Figure 11–18.

Step 4. *Collect the sample data and compute the value of the test statistic.* The sample results for the testing procedure evaluation are given in Table 11–5. The normal deviate z serves as the test statistic. The computed value of the standard error is

$$s_D = \sqrt{\frac{(12.6)^2}{50} + \frac{(14.5)^2}{100}} = 2.297$$

from which the normal deviate is computed:

$$z = \frac{(46.5 - 60.0) - (-5)}{2.297} = -3.70$$

Table 11–5

Sample Data for Evaluating Quality-Control Testing Methods

Test Stand *A* Testing Times (min.)			Manual Testing *B* Testing Times (min.)				
58.1	60.1	47.2	73.3	80.6	54.7	75.7	81.9
42.9	46.5	48.8	50.2	47.7	76.9	51.1	79.8
51.0	49.3	34.4	55.2	88.6	85.9	73.3	70.5
57.9	43.9	41.7	56.9	68.2	40.5	47.0	72.0
49.4	56.1	32.8	58.4	49.6	69.7	83.5	34.1
49.5	50.7	47.0	47.4	80.3	65.6	64.0	47.1
47.9	51.7	32.5	40.7	34.7	29.0	54.5	75.4
50.5	41.1	48.5	52.4	33.7	54.8	51.2	34.2
46.4	27.1	70.9	68.5	67.7	68.8	71.3	53.1
36.6	38.2	42.0	70.7	35.8	47.5	73.8	68.3
46.6	42.0		69.7	73.2	54.4	33.6	55.0
35.2	14.2		70.5	52.3	71.1	74.5	56.2
24.5	66.4		52.6	80.9	70.2	69.6	61.7
51.2	50.4		49.8	73.5	51.6	54.1	76.8
44.1	69.8		54.1	53.2	38.1	34.2	44.9
44.9	42.4		52.6	84.9	69.9	57.3	53.0
73.5	43.7		58.1	74.0	40.8	57.4	65.9
14.5	41.8		65.0	68.5	77.5	24.2	66.8
34.6	52.3		49.1	72.1	45.1	73.0	59.2
64.5	65.8		60.8	48.4	62.7	60.8	60.6
$n_A = 50$			$n_B = 100$				
$\bar{X}_A = 46.5$			$\bar{X}_B = 60.0$				
$s_A = 12.6$			$s_B = 14.5$				

Step 5. *Make the decision.* The computed normal deviate falls inside the rejection region. The director must therefore reject H_0 and acquire the new test stands.

This illustration involves a lower-tailed test. An upper-tailed test with $H_0\colon \mu_A - \mu_B \leq D_0$ will have a rejection region encompassing large positive values for z. A two-sided test, when $H_0\colon \mu_A - \mu_B = D_0$, will involve a double rejection region.

When either $n_A \leq 30$ or $n_B \leq 30$, the testing procedure is based on the Student t distribution. Two cases apply, depending on the levels of the respective variances.

Small Sample Sizes—Case of Equal Variances The following expression is used.

t STATISTIC FOR SMALL INDEPENDENT SAMPLES (EQUAL VARIANCES)

$$t = \frac{(\overline{X}_A - \overline{X}_B) - D_0}{s_D}$$

using

$$s_D = \sqrt{\frac{(n_A - 1)s_A^2 + (n_B - 1)s_B^2}{n_A + n_B - 2}}\sqrt{\frac{1}{n_A} + \frac{1}{n_B}}$$

with df $= n_A + n_B - 2$.

Illustration*: *Alloys for Armor Plating

Consider the following test performed by an engineer evaluating two candidate alloys for armor plating. One candidate is an all-metal alloy (A); the other alloy (B) is a high-quality steel with embedded grains of Teflon. The test procedure involves shooting uranium darts at sample plates from a gun at successively higher muzzle velocities until penetration. The basis for comparison is the mean velocity at penetration. A total of $n_A = 15$ all-metal and $n_B = 10$ Teflon alloy plates were used in the experiment.

Step 1. *Formulate the hypotheses and specify the possible choices.* Denoting the respective mean penetration velocities as μ_A and μ_B, the engineer chooses as his hypotheses

$H_0: \mu_A \geq \mu_B$ or $\mu_A - \mu_B \geq 0$ (the all-metal alloy offers at least as much resistance as the Teflo alloy)

$H_1: \mu_A < \mu_B$ or $\mu_A - \mu_B < 0$ (the all-metal alloy offers less resis tance than the Teflon alloy)

The pivotal level for the differences in means is $D_0 = 0$. The possible choices are

Reject H_0 (recommend the Teflon alloy)

Accept H_0 (don't recommend Teflon)

This designation reflects his strong desire to avoid recommending Teflon when this unconventional material actually provides inferior resistance.

Step 2. *Select the test procedure and test statistic.* The Student t statistic computed for the difference in sample means is the basis for the test.

Step 3. *Establish the significance level and the acceptance and rejection regions for the decision rule.* The engineer chooses a type I error probability of .01. From Appendix Table G this corresponds to a critical value of $t_{.01} = 2.500$, with $15 + 10 - 2 = 23$ degrees of

Figure 11-19

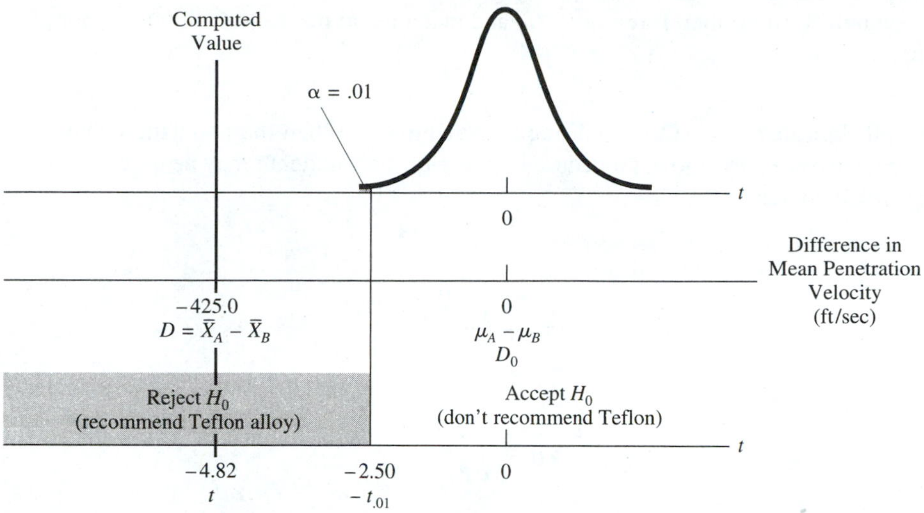

freedom. Extreme negative levels for t will refute H_0, and the test is lower-tailed. The decision rule is shown in Figure 11–19.

Step 4. *Collect the sample data and compute the value of the test statistic.* The sample results are given in Table 11–6. The standard error for D is computed:

$$s_D = \sqrt{\frac{14(207.4)^2 + 9(228.6)^2)}{15+10-2}}\sqrt{\frac{1}{15}+\frac{1}{10}} = 88.16$$

which is used to compute the value of the test statistic:

$$t = \frac{(4193.2 - 4{,}618.3) - (0)}{88.16} = -4.82$$

Table 11–6
Sample Data for Armor-Plating Experiment

All-Metal Alloy *A* Piercing Velocity (ft/sec)			Teflon Alloy *B* Piercing Velocity (ft/sec)	
4,405	4,036	4,244	4,291	4,573
4,324	3,855	4,122	4,691	4,209
4,289	4,674	4,320	4,720	4,935
4,023	3,979	4,222	4,895	4,606
3,993	4,309	4,103	4,616	4,647
$n_A = 15$			$n_B = 10$	
$\bar{X}_A = 4{,}193.2$			$\bar{X}_B = 4{,}618.3$	
$s_A = 207.4$			$s_B = 228.6$	

Step 5. *Make the decision*. The computed value for t falls inside the rejection region. The engineer must therefore reject H_0 and recommend the Teflon alloy as armor plating.

Small Sample Sizes—Case of Unequal Variances When the populations have different variances, the above procedure is incorrect. The Student t may be used to provide *approximate* results.

t STATISTIC FOR SMALL INDEPENDENT SAMPLES (UNEQUAL VARIANCES)

$$t = \frac{(\bar{X}_A - \bar{X}_B) - D_0}{s_D}$$

using

$$s_D = \sqrt{\frac{s_A^2}{n_A} + \frac{s_B^2}{n_B}}$$

with

$$\text{df} = \frac{(s_A^2/n_A + s_B^2/n_B)^2}{(s_A^2/n_A)^2/(n_A - 1) + (s_B^2/n_B)^2/(n_B - 1)}$$

(rounded to nearest integer)

To illustrate, an abbreviated hypothesis test is performed.

***Illustration**: Comparing Plankton Abilities to Create Bioluminescent Flow Visualization*

A researcher replicated an experiment employing two species of plankton to create flow visualization from the bioluminescence. She compared the mean intensities of Pyrocystis frusiforous (B, 1 cell ml^{-1}) and Gonyaulax polyedra (A, 6 cells ml^{-1}).* Photomultiplier light intensity readings were obtained in a tank with a wall stress of 2 dynes cm^{-2}.

Step 1. *Formulate the hypotheses and specify the possible choices*. We let μ_A denote the mean photomultiplier light intensity reading from plankton A and μ_B the mean reading from plankton B.

The following hypotheses apply.

$H_0: \mu_A \le \mu_B$ or $\mu_A - \mu_B \le 0$ (plankton A gives mean flow visualization not exceeding that of B)

$H_1: \mu_A > \mu_B$ or $\mu_A - \mu_B > 0$ (plankton A gives greater mean flow visualization than type B)

The pivotal level for the differences in means is $D_0 = 0$. Rejecting H_0 is to conclude that plankton A yields greater mean flow visualization than plankton B, while accepting H_0 is to conclude the opposite.

Step 2. *Select the test procedure and test statistic*. The Student t statistic is used.

Step 3. Not applicable.

*M. I. Latz, J. Rohr, and J. Hoyt, "A Novel Flow Visualization Technique Using Bioluminescent Marine Plankton," *IEEE Journal of Oceanic Engineering*, April 1995, pp. 144–47.

Step 4. *Collect the sample data and compute the value of the test statistic*. The following readings (counts $\sec^{-1}$) were obtained.

Gonyaulax (A):	546	241	346	435	443	384
Pyrocystis (B):	314	355	325	343		

The respective sample sizes are $n_A = 6$ and $n_B = 4$, and the sample statistics are

$$\overline{X}_A = 399.2 \qquad \overline{X}_B = 334.3$$

$$s_A = 102.80 \qquad s_B = 18.3$$

The above standard deviations are compelling evidence that the population variances are unequal, with plankton A yielding greater variability in light intensity. The standard error is

$$s_D = \sqrt{\frac{(102.80)^2}{6} + \frac{(18.3)^2}{4}} = 42.95$$

The test statistic is computed to be

$$t = \frac{(399.2 - 334.3) - (0)}{42.95} = 1.511$$

with

$$\text{df} = \frac{[(102.80)^2/6 + (18.3)^2/4]^2}{[(102.80)^2/6]^2/(6-1) + [(18.3)^2/4]^2/(4-1)} = 5.47 \qquad \text{or} \qquad 5$$

Step 5 (abbreviated). *Calculate the prob value and make the decision*. Appendix Table G provides the nearest critical value of $t_{.10} = 1.476$. Data yielding a test statistic as great or greater should occur in about 10 percent of all similar sampling experiments when the null hypothesis is true. This suggests that the results do not strongly refute the null hypothesis.

Two-Sided Tests Using Confidence Intervals A two-sided test may be conducted in conjunction with a confidence interval for $\mu_A - \mu_B$. For example, the following $100(1-\alpha)\% = 95\%$ confidence interval was determined for the difference in mean design time between computer-aided design (A) and traditional methods (B):

$$-3.335 \leq \mu_A - \mu_B \leq -.835$$

Because the above interval does not contain 0, the null hypothesis $H_0: \mu_A = \mu_B$, (or, equivalently, $H_0: \mu_A - \mu_B = 0 \neq D_0$) may be rejected at the $\alpha = .05$ significance level. Should the corresponding $100(1 - \alpha)\%$ confidence interval instead contain D_0, the null hypothesis should be accepted at the α significance level.

Summary of Procedures There are altogether three basic forms for using independent samples to test the difference in means: large sample, small samples (equal variances), and small samples (unequal variances). Each of these cases may involve lower-tailed, upper-tailed, or two-sided tests. The regular or abbreviated procedures may be used in any of those $3 \times 3 = 9$ cases, raising the number of possible forms for these tests to $2 \times 9 = 18$. And, there are three more possible forms when confidence intervals are used with $H_0: \mu_A - \mu_B = D_0$. Space does not permit us to illustrate all 21 different forms. All of these possibilities are summarized in Figure 11–20.

Figure 11-20 *Capsule summary of final hypothesis-testing steps for comparing two means with independent samples.*

LARGE *n*'s (general populations)

Step 4 Collect data and compute z.

$$z = \frac{(\bar{X}_A - \bar{X}_B) - D_0}{s_D} \quad \text{with } s_D = \sqrt{\frac{s_A^2}{n_A} + \frac{s_B^2}{n_B}}$$

Step 5 Make the decision.

Using Decision Rule:

Lower-Tailed (H_0: $\mu_A - \mu_B \geq D_0$)	**Upper-Tailed** (H_0: $\mu_A - \mu_B \leq D_0$)
Accept H_0 if $z \geq -z_\alpha$	*Accept* H_0 if $z \leq z_\alpha$
Reject H_0 if $z < -z_\alpha$	*Reject* H_0 if $z > z_\alpha$

Two-Sided (H_0: $\mu_A - \mu_B = D_0$)

Accept H_0 if $-z_{\alpha/2} \leq z \leq z_{\alpha/2}$
Reject H_0 if $z < -z_{\alpha/2}$ or if $z > z_{\alpha/2}$

or

use the $100(1 - \alpha)\%$ confidence interval:

$$\mu_A - \mu_B = (\bar{X}_A - \bar{X}_B) \pm z_{\alpha/2} s_D$$

Accept H_0 if D_0 is inside; otherwise, *reject*.

Or Using Abbreviated Procedure:

Prob $= \Pr[Z \leq -z_\alpha]$ for lower-tailed test
Prob $= \Pr[Z \geq z_\alpha]$ for upper-tailed test
Prob $= \Pr[Z \geq |z_{\alpha/2}|]$ for two-sided test

Reject H_0 if deemed sufficiently low.

SMALL *n*'s (normally distributed populations)

Step 4 Collect data and compute t.

$$t = \frac{(\bar{X}_A - \bar{X}_B) - D_0}{s_D}$$

using when $\sigma_A^2 = \sigma_B^2$:

$$s_D = \sqrt{\frac{(n_A - 1)s_A^2 + (n_B - 1)s_B^2}{n_A + n_B - 2}} \sqrt{\frac{1}{n_A} + \frac{1}{n_B}}$$

$$\text{df} = n_A + n_B - 2$$

or using when $\sigma_A^2 \neq \sigma_B^2$:

$$s_D = \sqrt{\frac{s_A^2}{n_A} + \frac{s_B^2}{n_B}}$$

$$\text{df} = \frac{(s_A^2/n_A + s_B^2/n_B)^2}{(s_A^2/n_A)^2/(n_A - 1) + (s_B^2/n_B)^2/(n_B - 1)}$$

Step 5 Make the decision.

Same using t_α instead.

Same using $t_{\alpha/2}$ instead.

Find the nearest corresponding tail areas for t using Appendix Table G according to the above degrees of freedom.

Notes: No finite population adjustments are made. It is unimportant whether or not a decision is recurring.

Matched Pairs Samples

What matched pairs are used, the same types of hypotheses as before may be tested using for each of the n sample pairs the difference

$$d_i = X_{A_i} - X_{B_i}$$

from which the mean and standard deviation are computed:

$$\bar{d} = \frac{\sum d_i}{n} \qquad s_d = \sqrt{\frac{\sum(d_i - \bar{d})^2}{n-1}}$$

These are then the basis for the test, which regardless of sample size, is based on the Stu dent t distribution. The following test statistic is used.

t STATISTIC USING MATCHED PAIRS

$$t = \frac{\bar{d}}{s_d/\sqrt{n}}$$

The critical value corresponds to the desired significance level when the number of degrees of freedom is $n-1$.

***Illustration**: Cost-Effectiveness of Electronic Mail*

A government agency evaluated the effectiveness of electronic mail to replace interoffice memos. Although rapid and increased communication is the major benefit of the new technology, significant operating-cost reductions are needed to substantiate procurement of the new hardware. A three-month test was performed using electronic mail in a sample of 10 district offices. Each office in this experimental group was matched with a counterpart where all interoffice memos are generated on hard copy. The monthly cost for the office using electronic mail (B) includes equipment rental, less any salary recovery from a smaller clerical staff. That amount will be subtracted from the clerical communications cost of the matching office where the standard stenographic procedure (A) is used.

The following steps apply:

Step 1. *Formulate the hypotheses and specify the possible choices.* Letting μ_A and μ_B denote the respective mean monthly mailing cost, we have

$$H_0: \mu_A \leq \mu_B \quad \text{or} \quad H_0: \mu_A - \mu_B \leq 0 \quad \text{(electronic mail is not cheaper)}$$

$$H_1: \mu_A > \mu_B \quad \text{or} \quad H_1: \mu_A - \mu_B > 0 \quad \text{(electronic mail is cheaper)}$$

Here, $D_0 = 0$. Accepting H_0 would be to not recommend E-mail, while rejecting H_0 would be to recommend it. The H_0 designation was chosen to control the error of recommending electronic mail when it will not in fact reduce clerical costs.

Step 2. *Select the test procedure and test statistic.* The Student t statistic computed from the matched-pair differences is used.

Step 3. *Establish the significance level and the acceptance and rejection regions for the decision rule.* The chosen significance level is $\alpha = .05$. Using $10-1 = 9$ degrees of freedom from Appendix Table G, this corresponds to a critical value of $t_{.05} = 1.833$.

Figure 11-11

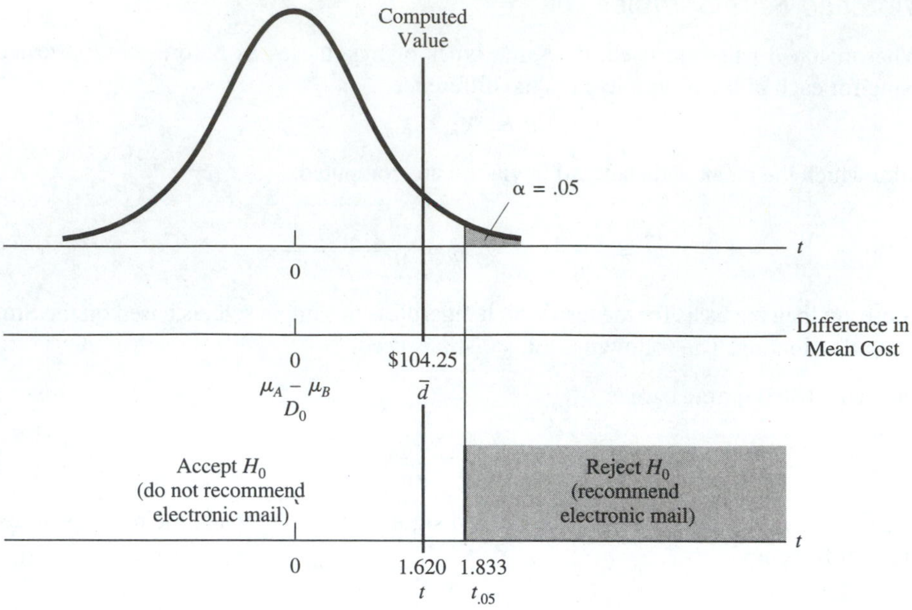

Large positive levels for t will refute H_0, and the test is upper-tailed. The decision rule is shown in Figure 11–21.

Step 4. *Collect the sample data and compute the value of the test statistic.* The sample results for the experiment are

$$\bar{d} = \$104.25 \qquad s_d = \$203.50$$

The computed value of the test statistic is

$$t = \frac{\$104.25}{\$203.50/\sqrt{10}} = 1.620$$

Step 5. *Make the decision.* The computed value of the test statistic falls inside the acceptance region. There is not a significant enough cost advantage to warrant recommending adoption of electronic mail, and the null hypothesis is accepted.

***Illustration**: Monte Carlo Simulation to Evaluate Testing Sequences*

Consider a second example based on Monte Carlo simulation. An industrial engineer is faced with a different type of evaluation, also involving testing. His concern is with the sequencing of final testing operations for robot units. Two alternatives are shown in Figure 11–22. The engineer believes that the greater variability in mechanical testing time might result in greater need for buffer stock under alternative A, increasing the overall unit-testing cost. But alternative B will cause difficulties with umbilical testing, increasing operator-time requirements. He wants to select the alternative that is, on the average, less costly.

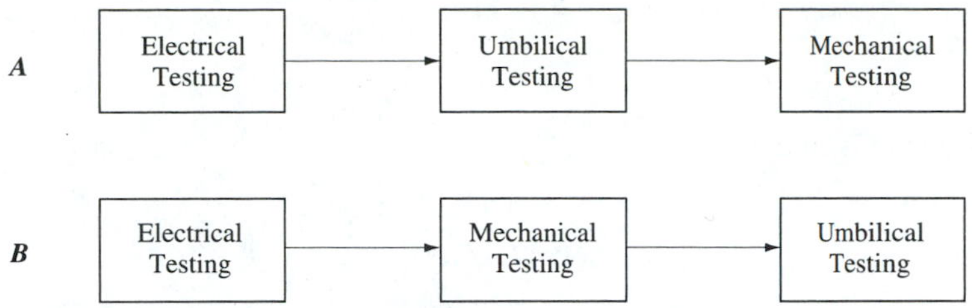

Figure 11-22 *Possible testing sequences for robot units.*

The robotics assembly operation has not been put into effect yet, so the engineer's evaluation must be based on a simulation. He has developed a model for finding unit costs under each alternative. Each involves a host of uncertain factors, and the engineer chooses to perform a Monte Carlo simulation of each alternative, using random numbers to establish the levels of the variables involved. The same set of random numbers will apply to both evaluations so that each trial robot unit tested under one simulated alternative will experience exactly the same conditions under the other sequence. In effect, the observations in the two simulations, or samples, may be paired.

The following steps apply for a simulation involving $n = 100$ robots tested under each sequence.

Step 1. *Formulate the hypotheses and specify the possible choices.* Letting μ_A and μ_B denote the respective mean costs, the engineer uses

$$H_0\colon \mu_A = \mu_B \qquad \text{or} \qquad \mu_A - \mu_B = 0 \qquad \text{(identical mean costs)}$$

$$H_1\colon \mu_A \neq \mu_B \qquad \text{or} \qquad \mu_A - \mu_B \neq 0 \qquad \text{(different mean costs)}$$

Accepting H_0 would require further testing. Rejecting H_0 because $\mu_A < \mu_B$ was concluded would involve selection of sequence A, while the opposite conclusion, $\mu_A > \mu_B$, would cause selection of B.

The two-sided H_0 designation reflects the engineer's neutral stance toward the candidate sequences. He views the two ways of incorrectly rejecting H_0 to be equally serious.

Step 2. *Select the test procedure and test statistic.* The Student t statistic computed from the matched-pair differences is used.

Step 3. *Establish the significance level and the acceptance and rejection regions for the decision rule*. The engineer chooses a 5% significance level. This must be split evenly between the two ways of incorrectly rejecting H_0. From Appendix Table G the critical value for $\alpha/2 = .025$ is $t_{.025} = 1.987$. The decision rule, shown in Figure 11–23, indicates that the testing sequence chosen must be consistent with the sample results obtained. If the null hypothesis is accepted, then no significant difference between the population means has been found. The engineer might in that case pick one sequence arbitrarily.

Step 4. *Collect the sample data and compute the value of the test statistic.* The following sample results were obtained for the matched-pairs differences

$$\bar{d} = \$15.43$$

$$s_d = \$62.47$$

so the t statistic is computed to be

$$t = \frac{\$15.43}{\$62.47/\sqrt{100}} = 2.47$$

Step 5. *Make the decision.* The computed value for t falls into the upper rejection region. The engineer must therefore reject H_0 and adopt sequence B.

Two-Sided Test Using Confidence Intervals The same conclusion could have been reached instead by first constructing a 95% confidence interval estimate for the difference in population means. The mean matched-pair difference $\bar{d} = \$15.43$ is a point estimate of the true difference in population means. This is the central value in the following confidence interval,

$$\mu_a - \mu_B = \bar{d} \pm t_{.025}\frac{s_d}{\sqrt{n}} = \$15.43 \pm 1.987\,\frac{\$62.47}{\sqrt{100}}$$
$$= \$15.43 \pm \$12.41$$

so that

$$\$3.02 \leq \mu_A - \mu_B \leq \$27.84$$

This interval could have been used in testing the original null hypothesis. Since the entire confidence interval lies above zero, again we see that H_0 should be rejected at the $\alpha = .05$ significance level.

Figure 11-23

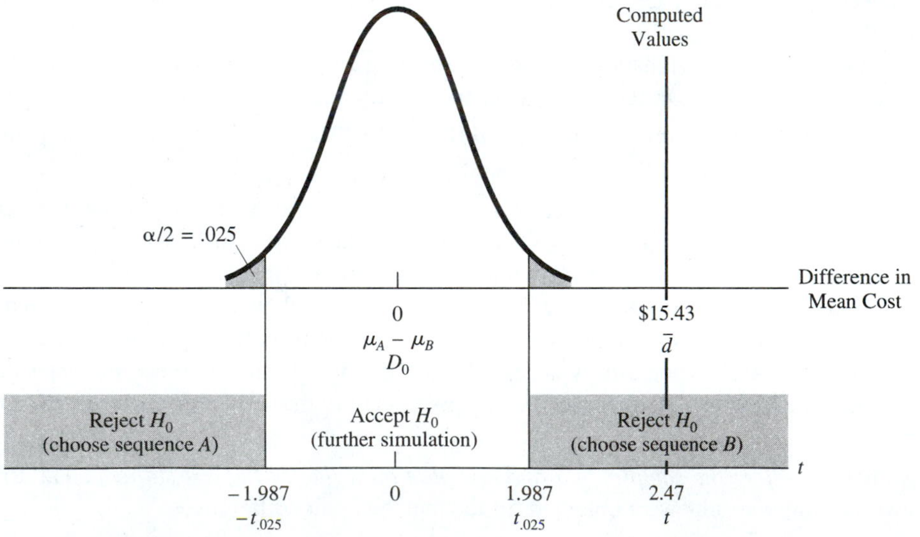

Testing with Bootstrapping Resamples

Restrictive assumptions regarding the underlying population must be made in order for the Student t distribution to apply. The key assumption is that, when sample sizes are small, the *populations* be normally distributed when drawing inferences regarding the difference in means. It is not possible with any certitude to verify these assumptions from the same sample data being used to compare the population means. Fortunately, statistical bootstrapping provides an alternative procedure for conducting a statistical test without forcing investigators to make any assumptions, verifiable or not, regarding the underlying populations.

***Illustration**: Comparing Algorithems for Drill Bit Prepositioning*

As an illustration, consider the sample data in Table 11–7 showing the total prepositioning distance traveled by the drill bit making holes in a printed circuit board. The data were obtained from simulations of two different methods for selecting the sequence for drilling holes in boards of comparable size and complexity. Method A involves a heuristic algorithm. Method B employs a genetic search algorithm.

We test the null hypothesis that the mean drill bit prepositioning distance is identical under the two methods, using $\alpha = .05$ significance level. If we conclude that the means are identical (accepting H_0), then neither method could be recommended without further testing. But if we conclude that one of the two methods, on the average, needs less traveling distance (rejecting H_0), that method could be recommended.

Figure 11–24 shows the results of a 1,000-trial resampling run with the data from Table 11–7. This was found by independently generating $n_A = 9$ resample observations from group A and another $n_B = 7$ from group B. The resample means were computed under each method for the two resamples, and the difference $\bar{X}_A - \bar{X}_B$ is computed each time. Those values were then used to construct the following 95% resampling interval estimate of the difference in population means:

$$5.3 \leq \mu_A - \mu_B \leq 51.8 \text{ inches} \qquad \text{(resampling)}$$

The interval lies entirely above zero, strong evidence that the two methods differ. Indeed, an examination of the histogram shows that less than 1 percent of the resamples gave $\bar{X}_A - \bar{X}_B$ differences that are negative, so that over 99% of the resamples gave mean distances under method B that were shorter than those under method A. The evidence is strong that the genetic search algorithm provides the shorter mean drill movement

Table 11–7
Sample results giving drill prepositioning distance traveled under two different routing methods

Method A Heuristic Algorithm	Method B Genetic Search
146 inches	57 inches
148	129
173	115
132	127
133	125
145	129
134	161
163	
133	

Figure 11–24
Resampling Stats run results for drill routing evaluation.

INSTRUCTIONS	EXPLANATION
COPY (146 148 173 132 133 145 134 163 133) A	Store data in A.
COPY (57 129 115 127 125 129 161) B	Store data in B.
MEAN A AM	Compute Mean of A; store in AM
MEAN B BM	Compute Mean of B; store in BM
STDEV A AS	Compute S. Dev. of A; store in AS
STDEV B BS	Compute S. Dev. of B; store in BS
PRINT AM AS	Print results for A
PRINT BM BS	Print results for B
REPEAT 1000	Generate 1,000 resamples:
SAMPLE 9 A A$	from A of size 9; store in A$;
SAMPLE 7 B B$	from B of size 7; store in B$;
MEAN A$ AM$	calculate A mean; store in AM$;
STDEV A$ AS$	calculate A s. dev.; store in AS$;
MEAN B$ BM$	calculate B mean; store in BM$;
STDEV B$ BS$	calculate B s. dev.; store in BS$;
SUBTRACT AM$ BM$ DELM	calculate difference in means; in DELM
SUBTRACT AS$ BS$ DELS	calculate difference in s. devs.; in DELS
SCORE DELM MDIFF	count DELM frequencies; store in MDIFF
SCORE DELS SDIFF	count DELS frequencies; store in SDIFF
END	.
PERCENTILE MDIFF (2.5 97.5) D	Print 95% resampling interval for MDIFF
PRINT D	Print interval.
HISTOGRAM MDIFF	Histogram of MDIFF.
PERCENTILE SDIFF (2.5 97.5) S	Print 95% resampling interval for SDIFF
PRINT S	Print interval.
HISTOGRAM SDIFF	Histogram of SDIFF.

RESULT

```
AM      =     145.22
AS      =     14.558
BM      =     120.43
BS      =     31.384

D       =     5.2937     51.762

  100+
F    +
r    +
e    +
q    +
u  75+
e    +                   * *
n    +                   *******
c    +                   *******
y    +                   ********
   50+                  ***********
*    +                 ************
     +                 ************
M    +               * ************ *
D    +               ****************
I  25+               *****************
F    +               ******************
F    +              *********************
     +             ***********************  *
     +          * ******************************
    0+-----------------------------------------------------
       |^^^^^^^^^|^^^^^^^^^|^^^^^^^^^|^^^^^^^^^|^^^^^^^^^|
      -20        0        20        40        60        80
```

Difference in Resample Means

distance, and the null hypothesis should be rejected and the genetic search method should be recommended.

Let's compare these results to the traditional confidence interval and t test. The applicable statistics obtained from Figure 11–24 are $\overline{X}_A = 145.22$, $\overline{X}_B = 120.43$, $s_A = 14.558$, and $s_B = 31.384$. Using these statistics, we compute the standard error (using the unequal variances case):

$$s_D = \sqrt{\frac{(14.558)^2}{9} + \frac{(31.384)^2}{7}} = 12.82$$

The degrees of freedom are

$$\text{df} = \frac{[(14.558)^2/9 + (31.384)^2/7]^2}{[(14.558)^2/9]^2/(9-1) + [(31.384)^2/7]^2/(7-1)} = 8.01 \qquad \text{or} \qquad 8$$

Appendix Table G provides the critical value $t_{.025} = 2.306$, so that the traditional 95% confidence interval is computed as

$$\mu_A - \mu_B = 145.22 - 120.43 \pm 2.306(12.82)$$
$$= 24.79 \pm 29.56$$

or

$$-4.77 \leq \mu_A - \mu_B \leq 54.35 \text{ inches} \qquad \text{(traditional)}$$

The above suggests that the null hypothesis of zero difference must be accepted—the opposite conclusion to that reached from the resampling run.

Which result should be used and what conclusion should we make? If there is good reason to believe that the underlying populations are indeed normally distributed, we should use the traditional confidence interval and accept the hypothesis of identical means, concluding that the heuristic algorithm and genetic search are not different. But if we do not have prior knowledge of the population distribution forms, the sample data in Table 11–7 are by themselves too skimpy to confirm or deny the normality assumption. In that case only the resampling results would not rest on possibly invalid assumptions—making them more appropriate, and we could comfortably reject the null hypothesis and recommend the genetic search procedure.

Problems

11–38 A chemical engineer is comparing two chemical processes. One involves a high-pressure purification stage. The second achieves purification at lower pressures but involves a catalyst. The levels of impurities (percentage of total volume) resulting from the two procedures are to be compared. The more expensive high-pressure process will be adopted only if it yields the lower mean impurity level. The chemical engineer wants only a 1% chance of adopting it when in fact that process is worse. His null hypothesis is that the mean impurity level from the high-pressure process is at least as great as that from the low-pressure procedure. Pilot batches have been run under the two methods, and the following data were obtained:

Low Pressure	High Pressure
$n_A = 50$	$n_B = 100$
$\bar{X}_A = 2.3$	$\bar{X}_B = 1.5$
$s_A = 1.0$	$s_B = 0.8$

Which process should the chemical engineer select?

11–39 A project engineer is comparing operating modes for a satellite tracking and identification radar. Her study is based on radar cross sections obtained from "chirp" pulses (short continuous transmissions of increasing frequency) and "burst" pulses (a series of very brief transmissions of constant frequency, punctuated by silent gaps). The cross sections are digitized and matched to templates stored in the memory of the system computer. The computer then identifies the satellite target. Although both procedures eventually provide correct identifications, the required time from target acquisition to final confirmation varies considerably and is uncertain. Since satellite position and atmospheric conditions are so crucial, and since only one acquisition radar and computer system was available for testing, independent sampling had to be used. Samples of 100 targets were evaluated under the two modes. The chirp mode's mean time was 35.3 seconds until identification, with a standard deviation of 8.1 seconds. Under the burst mode, the mean time was 33.1, with a standard deviation of 6.3.

At the 5% significance level, must the engineer accept or reject the null hypothesis that the chirp mode provides identification at least as fast as the burst mode?

11–40 Refer to the data for the maintenance policy example on page 329. The fix-only-when-broken policy will be adopted if it is found to be significantly cheaper. The null hypothesis is that this policy is at least as expensive as regular preventive maintenance. Using a 5% significance level, what action should be taken?

11–41 Consider the computer-aided-design example on page 329. Only if the sample evidence indicates that CAD is significantly faster will it be adopted. To make the two sets of times comparable, the log-in procedure at the drafting department must be eliminated for the traditional design observations. Subtract one day from each of those times. Then test the null hypothesis that CAD will on the average take at least as long as the traditional method. Use a 1% significance level. What action should be taken?

11–42 An engineering curriculum committee wishes to decide about high school mathematics requirements for entering freshmen. Although it is widely believed that high school calculus does not improve college calculus performance (making it a waste of time for fulfilling mathematics requirements), the committee has been tóld that high school calculus students outperform the others in the first *physics* course. A sampling investigation was performed involving 50 pairs of students, who were matched by quantitative SAT scores and first-term GPA. The control group (A) consisted of students who did not have high school calculus. The experimental group (B) contained those who did. The physics final examination score of the experimental student in each pair was subtracted from that of the control student, and the mean difference was found to be –2.4 points with a standard deviation of 4.6 points.

At the 1% significance level, should the committee accept or reject the null hypothesis that mean physics final examination scores are at least as high for those who did not have high school calculus as for those who did?

11–43 An automotive engineer is evaluating an experimental fuel-mixture control. Independent test runs have been made using random samples of cars where the new device has been installed and where standard carburetion was used. The following mileage data have been obtained:

Experimental	Control
$n_A = 15$	$n_B = 17$
$\bar{X}_A = 33.4$ mpg	$\bar{X}_B = 32.1$ mpg
$s_A = 1.3$	$s_B = 1.5$

Should the engineer accept or reject the null hypothesis that the standard carburetion provides at least as great mileage as that of the experimental method? Use $\alpha = .05$ and assume equal population variances.

11–44 A statistics instructor wishes to evaluate two methods for teaching statistics. One approach emphasizes computer applications, with all homework assignments requiring the use of "canned" programs in the central time-sharing library. One class of 20 students, forming the experimental group (A), was instructed under this format. The control group (B) was another statistics class of 25 students taught without any encouragement to use computers. The same final examination was given to the two groups. The instructor tested the null hypothesis that examination scores achieved by all students instructed under the two methods have identical means. The experimental group achieved a mean score of 78.5 with a standard deviation of 10.6; the controls achieved a mean of 81.2 and a standard deviation of 9.6. Using a 5% significance level, did the instructor accept or reject the null hypothesis? (Assume equal population variances.)

11–45 A chemical engineer is evaluating a new catalyst to determine if it speeds up a chemical procedure. A sample of 10 pilot batches was processed with the new catalyst (group A). Each test run was carefully matched with a second batch processed with the present catalyst (group B). Each pair of runs was conducted under nearly identical control settings, used the same raw-material stock, and involved almost the same time-phasing. The processing time for batch B was subtracted from its batch A partner, and the mean difference was found to be .53 hour with a standard deviation of .16 hour.

(a) Construct a 95% confidence interval estimate of the difference between mean processing times for all batches run with the two catalysts.

(b) Should the null hypothesis that the mean processing times are identical be accepted or rejected at the 5% significance level?

11–46 A test similar to the one in Problem 11–45 was performed independently in two different plants, one using the new catalyst (A) and the other using the present one (B). A total of 8 sample batches were processed using the new catalyst, while 15 runs were made with the present one. The following processing time results were obtained:

$$\bar{X}_A = 12.37 \text{ hours} \qquad \bar{X}_B = 12.93 \text{ hours}$$

$$s_A = 2.14 \qquad s_B = 1.86$$

(a) Construct a 95% confidence interval estimate of the difference between mean processing times. (Assume unequal population variances.)

(b) Should the null hypothesis that the mean processing times are identical be accepted or rejected at the 5% significance level?

11–47 Refer to the confidence interval you found in Problem 10–36 for the difference in engineers' mean incomes. Should the society accept or reject the null hypothesis that the mean incomes are the same? Use $\alpha = .05$.

11–48 Refer to the confidence interval you found in Problem 10–38 for the difference in MTBFs for two different transformers. Should the electrical engineer accept or reject the null hypothesis that the mean times between failures are the same? Use $\alpha = .01$.

11–49 Refer to the confidence interval you found in Problem 10–40 for the difference in mean completion times. Should the chief engineer accept or reject the null hypothesis that the mean times are the same for both electrical subcontractors? Use $\alpha = .05$.

11–50 Refer to the confidence interval you found in Problem 10–41 for the difference in mean annual engine maintenance cost. Should the foreman accept or reject the null hypothesis that the mean costs are identical for the two makes? Use $\alpha = .01$.

11–51 Refer to the paint drying-time data in Problem 10–42. Test the null hypothesis that paint A involves drying times that are on the average less than or equal to those of paint B. At the 1% significance level, must H_0 be accepted or rejected?

11–52 Refer to the student GPA data in Problem 10–43. Test the null hypothesis that regular engineering students achieve grade-point averages at least as high as those of cooperative students. At the 5% significance level, must H_0 be accepted or rejected?

11–53 *Computer exercise*. Refer to the armor-piercing velocity data on page 396. Using the original sample data to represent the underlying populations, generate for each type of armor plate 1,000 pairs of independent resample data sets of the same sizes as the original. Compute the resample means for each pair and find the difference. Using the differences in the two resample means, determine the 99% resampling interval for the difference in population means.

Does your evidence tend to support or refute the null hypothesis of identical mean piercing velocities for the two types of armor plating?

11–54 *Computer exercise*. Refer to the plankton illumination data on page 396. Using the original sample data to represent the underlying populations, generate for each plankton species 1,000 pairs of independent resample data sets of the same sizes as the original. Compute the resample means for each pair and find the difference. Using the differences in the two resample means, determine the 95% resampling interval for the difference in population means.

Does your evidence tend to support or refute the null hypothesis of identical mean illumination for the two plankton species?

Review Problems

11–55 For each of the following situations, (1) specify the test statistic, (2) give its critical value, and (3) indicate whether the test is lower-tailed, upper-tailed, or two-sided. In each case use $\alpha = .05$ and assume a large population unless otherwise indicated.

(a) $H_0: \mu \leq 50, \sigma = 10, n = 100$
(b) $H_0: \mu \geq 100, \sigma = ?, n = 100$
(c) $H_0: \mu = 75, \sigma = ?, n = 20$
(d) $H_0: \pi \leq .20, n = 100$
(e) $H_0: \pi \geq .40, n = 50, N = 200$
(f) $H_0: \mu \leq .50, \sigma = .15, n = 25, N = 50$
(g) $H_0: \mu \geq 1.46, \sigma = ?, n = 25$
(h) $H_0: \mu = 25, \sigma = 2, n = 200, N = 1{,}000$

11–56 The following data (minutes) apply for a random sample of processing times for a chemical reaction:

2.3	6.7	3.8	5.0	4.9	6.1
4.4	5.2	3.9	4.8	4.6	5.7
5.3	4.7	4.2	4.7	5.7	4.8

Can you conclude that the mean of all processing times exceeds 4 minutes? Use $\alpha = .05$.

11–57 The following data were obtained for the amount of time (seconds) taken by a proposed software system to compile a sample of short FORTRAN programs:

.6	1.0	1.8	4.8	4.2
4.4	2.4	3.5	3.9	3.9
4.7	3.0	2.3	2.9	1.0

At the 5% significance level, test the null hypothesis that the mean compilation time for all short FORTRAN programs run on the system is at least 4 seconds. Should it be accepted or rejected?

11–58 A pharmaceutical loader has been set to insert exactly 5 milligrams into each capsule of a particular drug. Periodically, a sample of 100 capsules is taken and the contents of each are measured. The standard deviation has been established at .2 milligram per capsule. Corrective action will be taken whenever the computed mean is significantly great or small.

(a) Determine a decision rule when a 1% chance is permitted for taking unnecessary action.

(b) Apply the above to find the probability of failing to take needed correction when the true mean is (1) 5.05 milligrams, (2) 4.97 milligrams, (3) 5.07 milligrams, (4) 4.98 milligrams.

11–59 A program director for a large computer maker is evaluating a circuit-designing software package. Not only is the program purported to ease the engineer's tasks, but its author also claims the finished products will be more economical because fewer logic elements will be needed. To test this theory, the software was borrowed and used to redesign 20 prototype circuits already completed in the traditional way. The resulting computer-generated designs all worked and involved a mean percentage reduction in components of 3.4% ($s = 4.2\%$).

(a) At the 1% significance level, can the project manager reject the null hypothesis that the software designs involve at least as many elements as do traditional ones?

(b) What is the smallest significance level at which the null hypothesis can be rejected?

11–60 Refer to the data in Problem 11–57.
Construct a 99% confidence interval estimate for the mean. Should the null hypothesis that $\mu = 3.50$ seconds be accepted or rejected at the 1% significance level?

11–61 Only 18 persons in a sample of 100 students indicated a desire for a full-scale summer program. Can it be concluded that the majority of students don't want such a program? Use $\alpha = .01$ seconds for incorrectly making that conclusion. Assume a large population and apply the normal approximation.

11–62 Specifications for a component require that the MTBF exceed 100 hours. The actual MTBF of any shipment is unknown, although it is assumed for each that the lifetime standard deviation is 10 hours. There should be only a 5% chance of concluding that a shipment meets specifications when it does not. A statistician takes a random sample of 50 items from a particular shipment of 300 and computes the MTBF to be 102 hours. What conclusion does he reach?

11–63 The failure rate of jet engine fan blades subjected to a particular test condition has been specified to be less than .05. An entire production batch found to violate that requirement will be scrapped; otherwise, they will be assembled into engines. Assuming just a 10% chance of erroneously assembling with a poor-quality batch of fan blades, determine the maximum number of failed blades in a sample of 100 that would still allow the batch to be used in assembly. (Use the binomial distribution directly, assuming the batch size is large.)

11–64 An engineer is interested in the heat-exchanging properties of liquids carrying the products of chemical reactions. The output of either stage A or stage B will be used to warm the water feeding one of the plant boilers. The following data represent independent random observations of the effluent from the two stages:

Temperature (°F)			
Stage A		Stage B	
185	233	197	248
206	250	235	234
217	206	245	224
251	215	211	231
190	224	225	216
		245	258

(a) Construct a 95% confidence interval estimate for the difference in mean temperatures of the effluents from the two stages. (Assume identical population variances.)

(b) Should the null hypothesis of identical mean temperatures be accepted or rejected at the 5% significance level?

11–65 The following assembly-time data have been obtained by an industrial engineer evaluating two methods. Each observation pair represents the number of seconds taken by a single operator to complete the required task using the respective method.

Operator	A Time (seconds)	B Time (seconds)	Operator	A Time (seconds)	B Time (seconds)
1	45	39	11	39	29
2	88	71	12	36	36
3	40	42	13	52	50
4	32	27	14	43	34
5	29	28	15	73	51
6	34	30	16	48	42
7	59	50	17	61	61
8	55	60	18	48	45
9	62	51	19	44	32
10	50	48	20	51	43

(a) Construct a 99% confidence interval estimate for the difference in mean assembly times under the two methods.

(b) Should the null hypothesis of identical means be accepted or rejected at the 1% significance level?

11–66 The following data have been collected by a student society comparing the grades achieved by full-time students to those of cooperative students:

Full-Time Students (A)				**Cooperative Students (B)**			
GPA	**SAT**	**GPA**	**SAT**	**GPA**	**SAT**	**GPA**	**SAT**
3.3	655	3.2	712	3.1	588	2.7	578
3.9	722	3.5	690	3.8	716	3.7	695
2.9	623	2.8	575	3.1	627	3.2	634
3.3	710	3.8	690	3.5	713	3.9	730

Answer the following, assuming that the samples were selected independently from populations with equal variances:

(a) Construct a 95% confidence interval estimate for the difference in population mean GPAs for full-time and cooperative students.

(b) Apply the t test with the null hypothesis that full-timers earn grades at least as high as cooperative students. At the 5% significance level, should that hypothesis be accepted or rejected?

CHAPTER TWELVE

Theory and Inferences in Regression Analysis

The regression techniques presented in Chapters 4 and 5 are based on the method of least squares. Those methods are by far the most common procedures because of the many desirable properties of the estimators they give. Those properties give rise to traditional methods for making a variety of inferences about the underlying population and relationships between the variables.

Like the inferential procedures encountered in the preceding chapters, there is a hidden cost in that certain assumptions must be assumed to apply. Those assumptions may be overly restrictive and sometimes false, possibly invalidating some conclusions. This chapter discusses how bootstrapping methods might be used as an alternative, thereby avoiding the difficulties with traditional regression inferences.

12–1 Assumptions and Properties of Linear Regression Analysis

It is very important that investigators be aware of the assumptions and properties of linear regression analysis. These provide the basis for extending the concepts and procedures of statistical inference to making predictions about the dependent variable and drawing conclusions regarding the nature of the regression line itself.

Assumptions of Linear Regression Analysis

To help explain the assumptions of linear regression, we will expand our data entry illustration. Consider each possible required time Y for all entries that are to be completed by materials-handling clerks for a specified number of inputs X. For any fixed X the values of Y constitute a population having some mean, which we denote by $\mu_{Y \cdot X}$. A separate population applies to each level for X.

Figure 12–1 shows how several Y populations may be characterized in the regression context. Notice that there are three dimensions. In addition to the axes for the independent and dependent variables, we include a third vertical axis for relative frequency. Each population of Ys may be represented by a frequency curve having center at some specified distance $\mu_{Y \cdot X}$ from the X-axis. We therefore refer to $\mu_{Y \cdot X}$ as the **conditional mean** of Y given X. There will be a different frequency curve for each X. Illustrated are

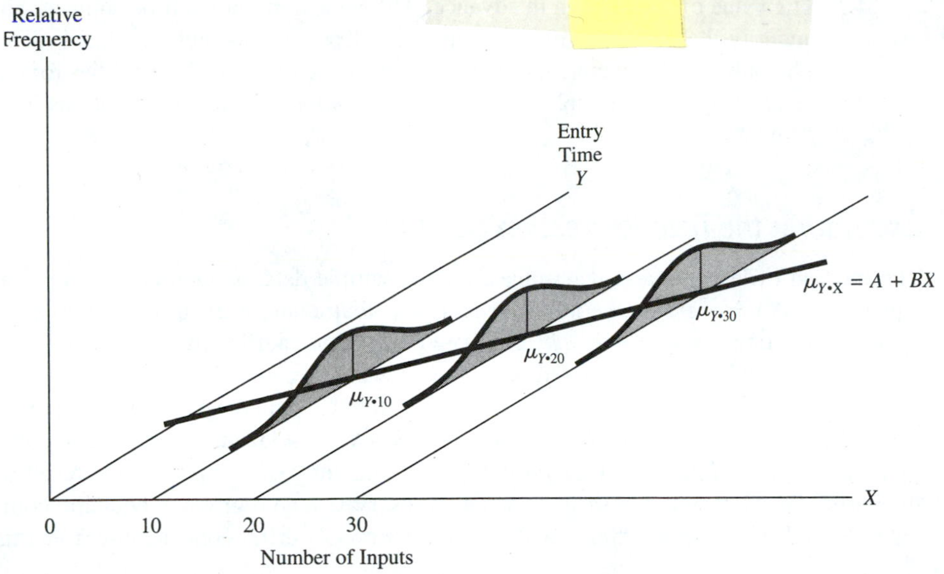

Figure 12–1 *Conditional population distributions with conditional means on the true regression line.*

the populations when the number of inputs is $X = 10$, $X = 20$, and $X = 30$. The respective conditional means are denoted by $\mu_{Y \cdot 10}$, $\mu_{Y \cdot 20}$, and $\mu_{Y \cdot 30}$. We can make the following four assumptions about the populations for Y:

1. The conditional means $\mu_{Y \cdot X}$ all lie on the same straight line, given by the equation

$$\mu_{Y \cdot X} = A + BX$$

which is the expression for the **true regression line**. This result follows from an assumption about individual observations. For the ith sample pair the level of the dependent variable may be expressed in terms of these parameters by

$$Y_i = A + BX_i + \epsilon_i$$

where ϵ_i is a **random error term**. For all i, the error term is assumed to be a random variable having expected value 0 and constant variance. It follows that

$$E(Y_i) = A + BX_i$$

so that for any given level X the corresponding observed Y has expected value equal to $\mu_{Y \cdot X}$.

2. All populations have the same standard deviation, denoted by $\sigma_{Y \cdot X}$, no matter what the value of X is. This follows from the assumption that $\text{Var}(\epsilon_i)$ is the same fixed value for all observations. Thus, treating $A + BX_i$ as a constant,

$$\text{Var}(Y_i) = \text{Var}(A + BX_i + \epsilon_i) = \text{Var}(\epsilon_i)$$

3. Successive sample observations are independent. This means that the value of successive Ys will not be affected by whatever other values are obtained. For this to be the case, the successive error terms are assumed to be independent random variables. It is further assumed that the error terms are normally distributed with mean 0 and standard deviation $\sigma_{Y \cdot X}$.

4. The value of X is known in advance. Although X might itself be an uncertain quantity, the regression model treats this independent variable as a fixed value. The only thing that matters is the uncertainty regarding Y. Probabilities for the levels of Y may be established using *conditional* probability distributions for a given X.

Estimating the True Regression Equation

The method of least squares determines, for the sample data, an estimated regression equation $\hat{Y}(X) = a + bX$. We now consider how that equation relates to the true regression equation $\mu_{Y \cdot X} = A + BX$. The **true regression coefficients** A and B are usually unknown and are ordinarily estimated from the sample counterparts, a and b.

Of course, the computed values for a and b will depend on the particular sample results obtained. The equation $\hat{Y}(X) = 30.04 + 2.854X$, found from the workstation data in Table 12–1, relating the time Y for completing the entry of X inputs, is the result of the particular 25 sample transactions actually observed. The regression equation computed for a different set of 25 transactions would probably differ considerably from this one.

Rationale for the Method of Least Squares

The estimated regression coefficients a and b found by the method of least squares are *unbiased estimators* of the true coefficients A and B. Thus, if the same regression procedure were applied to a series of random samples, then the average of the a's would tend toward A and the b's would cluster about B. As we have seen, unbiasedness is a very desirable property for an estimator. Another feature of estimators derived from least squares is that they are the *most efficient* of all possible unbiased estimators of the linear regression coefficients. Thus, a and b minimize the chance sampling error involved in establishing a relationship between X and Y. This makes predictions determined from $\hat{Y}(X) = a + bX$ the most reliable ones obtainable for a fixed sample size. The a and b computed by the least-squares procedure are also *consistent* estimators of A and B; this means that, on the average, they become progressively closer to their targets as the sample size is increased. (The standard error for both a and b decreases as n increases.)

Appropriateness of Model: Analysis of Residuals

A beginning point in evaluating the appropriateness of the theoretical regression model is an analysis of the residuals

$$e_i = Y_i - \hat{Y}(X_i)$$

We will illustrate using the workstation data in Table 12–1. The key intermediate calculations and regression results are given in Table 12–1. The residual values are computed and listed in the last column. Those values provide the residual scatter plot in Figure 12–2.

All traditional inferences, to be discussed in Section 12–3, are based on the properties just presented. If the underlying population and sampling process deviate from the

Table 12–1

Terminal Transaction Sample Data and Regression Results

Observation i	Transaction Time Y_i	Number of Entries X_i	Predicted Time from Regression Line $\hat{Y}(X_i) = 30.04 + 2.854X_i$	Residual $e_i = Y_i - \hat{Y}(X_i)$
1	66	2	35.748	30.252
2	77	19	84.266	–7.266
3	37	6	47.164	–10.164
4	106	23	95.682	10.318
5	55	10	58.580	–3.580
6	89	23	95.682	–6.682
7	52	9	55.726	–3.726
8	128	30	115.660	12.340
9	63	18	81.412	–18.412
10	104	25	101.390	2.610
11	76	19	84.266	–8.266
12	44	2	35.748	8.252
13	97	27	107.098	–10.098
14	109	28	109.952	–0.952
15	40	8	52.872	–12.872
16	124	29	112.806	11.194
17	98	29	112.806	–14.806
18	63	16	75.704	–12.704
19	131	33	124.222	6.778
20	41	3	38.602	2.398
21	111	34	127.076	–16.076
22	151	32	121.368	29.632
23	76	13	67.142	8.858
24	114	33	124.222	–10.222
25	143	35	.129.93	13.070
	$\sum Y = 2{,}195$	$\sum X = 506$		
	$\bar{X} = 20.24$		$\bar{Y} = 87.80$	
	$\sum X^2 = 13{,}130$		$\sum Y^2 = 220{,}445$	$\sum XY = 52{,}670$
	$\hat{Y} = 30.04 + 2.854X$		$s_{Y \cdot X} = 13.51$	

model described there, the regression analysis will be based on false assumptions. An analysis of the residuals can be helpful in uncovering model violations.

The workstation data residuals reflect the underlying linear relationship found earlier, with the individual e's falling within a narrow horizontal band. Such a plot has the ideal appearance for a regression analysis. Figure 12–3 shows three plots where key model assumptions are violated.

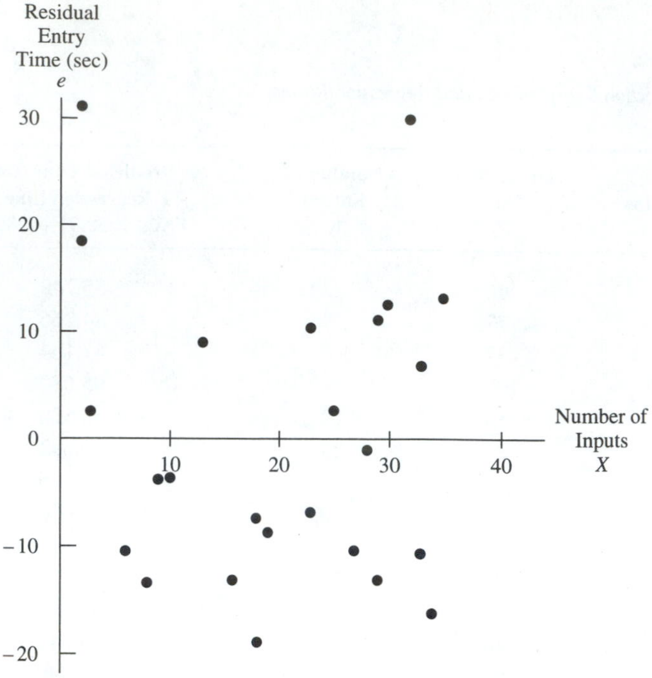

Figure 12–2
Residual scatter plot for workstation data entry.

The scatter plot in Figure 12–3(a) illustrates a *nonlinear* relationship, with the e's exhibiting a pronounced curve. As we saw in Chapter 5, one way to rectify this would be to transform Y or X.

Figure 12–3(b) shows a plot representing the regression analysis for a hypothetical experiment where the computer processing time Y for a network analysis is related to the number of nodes X. Notice that the band of e's gets wider as X increases. This indicates a *nonconstant variance* in the e's as the level for X becomes greater, and violates the theoretical assumption that errors have constant variance $\sigma^2_{Y \cdot X}$. That difficulty is sometimes rectified by transformation of the variables. For instance, when there is increasing variance in the e's, a regression with $Y' = \sqrt{Y}$ might provide a constant variance.

Figure 12–3(c) shows a residual plot of a hypothetical regression where Y is the time needed to adjust temperature settings on batches run in a chemical process involving final output of volume X. The horizontal axis is the *time sequence* of the observation (rather than X). Notice that the e's are increasing over time, indicating a *lack of independence*. This could be due to a variety of explained causes, such as operator fatigue. (That difficulty might be avoided by collecting all data from batches run on different days, at about the same time of day.)

Residual analysis extends to other types of model violations where a conventional scatter plot is less useful. One of these involves **outliers**, isolated observations that are so extreme that they don't appear to belong with the rest. If an identifiable cause exists (such as a power outage or an absent operator), the data point might be safely discarded.

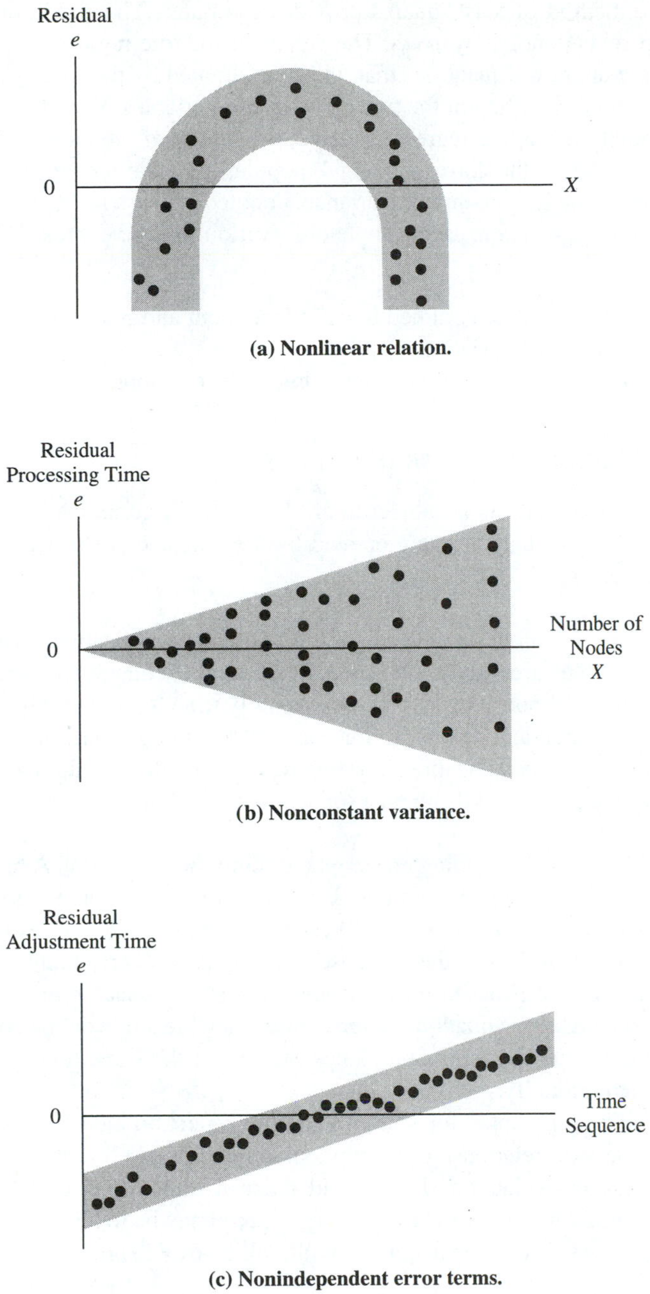

Figure 12–3 *Residual scatter plots reflecting violations of the underlying regression model.*

Assumptions of Multiple Regression

The theoretical assumptions of multiple regression analysis are straightforward extensions of the properties described for simple regression. As before, there is a true regression plane

$$\mu_{Y \cdot 12} = A + B_1 X_1 + B_2 X_2$$

for which the method of least squares provides estimates. The conditional mean of Y given X_1 and X_2 is denoted by $\mu_{Y\cdot 12}$. The values of the true regression coefficients, A, B_1, and B_2, are unknown quantities that will be estimated by their sample counterparts a, b_1, and b_2. For any setting of the two independent variables X_1 and X_2 there is a corresponding point on the true regression plane, the height of which is $\mu_{Y\cdot 12}$. This point serves as the center of the corresponding Y population. Like the simple linear model, each such population has a common variance, denoted by $\sigma^2_{Y\cdot 12}$.

As in simple regression, each sample observation may be expressed as follows:

$$Y_i = A + B_1X_{i1} + B_2X_{i2} + \epsilon_i$$

where the error terms ϵ_i are assumed to be independent and normally distributed, with the mean 0 and standard deviation $\sigma_{Y\cdot 12}$.

The above may be directly extended to higher dimensions.

Pitfalls in Multiple Regression Analysis

Multiple regression analysis is subject to all the potential complications of simple regression. But greater challenges are created by the existence of several variables and their interrelationships.

Multicollinearity Unique to multiple regression is a complication arising when two independent variables are highly correlated. For example, consider a regression evaluation for predicting a product's cost Y using assembly time X_1 and number of operations required X_2 as independent variables. Since assembly time may have a high correlation with number of operations, the three-variable regression relationship may provide sample data that are scattered tightly about a *single line* lying inside the estimated regression plane.

The problem is not in finding an estimated regression equation. A high correlation between X_1 and X_2 does not prevent that. But a second set of sample observations from the same underlying population would likely result in *very different* estimated regression coefficients, even though the linear scatter might be nearly identical for the two data sets. (Imagine two panes of glass suspended in space—each representing a different sample and regression equation—intersecting at a wide angle on the common line of scatter.) Because different regression planes can obtain the same line, such a situation exhibits **multicollinearity**.

Data exhibiting potential for multicollinearity reflect an inherent instability in the underlying regression relationship. An obvious remedy is to eliminate either X_1 or X_2 from the regression evaluation. That would make it impossible to draw conclusions about the removed variable. Another possible remedy may be to increase the number of sample observations so that the pattern of multicollinearity is broken.

Alternative Procedures: Weighted Least Squares

Although the least-squares method is by far the most common regression procedure, similar results may be obtained with different approaches. These alternatives might be used when key assumptions do not apply, particularly when the error terms are *not* normally distributed. Such a circumstance would be the case when outliers sometimes occur. Several alternative approaches utilize a type of *weighted* least squares method, giving greater weight to observations lying near an initial regression line found with standard methods and downweighting those farther away from that line. A second regression line is then found (using one of a variety of approaches) by applying the weights from the first iteration. Weighted least-squares procedures are complex and computationally intensive.

Problems

12–1 The true regression line for predicting the hours of drying time Y (under standard conditions) of a ceramic having a specified concentration X of an inert material (ppm) is

$$\mu_{Y \cdot X} = 3.0 - .001X$$

Ten simulated observations are generated. Individual Y values may be computed from the above by plugging in a level for X and adding the error term ϵ. Those error terms are normally distributed with mean zero and standard deviation $\sigma_{Y \cdot X} = .1$ hour. We represent a random normal deviate as z, so that error terms may be generated from

$$\epsilon = 0 + z\sigma_{Y \cdot X}$$

(a) Complete the following table for 10 levels of X and random normal deviates.

	Inert Material Concentration X	Theoretical Time $\mu_{Y \cdot X}$	Random Normal Deviate z	Error Term ϵ	Simulated Actual Time Y
(1)	50		.05		
(2)	50		−1.23		
(3)	100		−.87		
(4)	100		1.56		
(5)	200		.97		
(6)	250		.43		
(7)	300		−.98		
(8)	350		−.55		
(9)	400		.17		
(10)	500		1.08		

(b) Plot a scatter diagram using the above levels for Y and their matching value for X.

12–2 *Continuation.*
(a) Plot the *true* regression line on a graph.
(b) Using the above values for Y and X, determine the estimated regression equation.
(c) Plot on the same graph the *estimated* regression line.

12–3 *Continuation.*
(a) Compute $s_{Y \cdot X}$ using the regression results from Problem 12–2 (b).
(b) Why does $s_{Y \cdot X}$ differ from $\sigma_{Y \cdot X}$?

12–4 *Continuation.*

(a) Complete the following table.

(1) Concentration X	(2) Simulated Actual Time Y	(3) Theoretical Time $\mu_{Y \cdot X}$	(4) Estimated Time $\hat{Y}(X)$	(5) Error Term $\epsilon = Y - \mu_{Y \cdot X}$	(6) Residual $Y - \hat{Y}(X)$
(1) 50					
(2) 50					
(3) 100					
(4) 100					
(5) 200					
(6) 250					
(7) 300					
(8) 350					
(9) 400					
(10) 500					

(b) Compare the values in columns (3) and (4). Shouldn't these be identical?

(c) Compare the values in columns (5) and (6). Shouldn't these be identical?

12–5 Refer to the data in Problem 4–4 and your answers to that exercise. Compute for each observation the residual values. Plotting these, what do you conclude regarding consistency with the theoretical assumptions of the regression model?

12–6 Repeat Problem 12–5 using the data from Problem 4–7.

12–2 Assessing the Quality of the Regression

An important issue in regression analysis is assessing its overall quality. It is valuable to measure how well the regression explains the relationship between the dependent and independent variables. An index that quantifies the strength of that relationship is the *coefficient of determination.*

The Coefficient of Determination

The estimated regression equation $\hat{Y}(X)$ is found by minimizing the sum of the squared deviations about the regression line. For any particular observation there are *three* deviations, as indicated by the terms in the following expression:

$$(Y - \bar{Y}) = [\hat{Y}(X) - \bar{Y}] + [Y - \hat{Y}(X)]$$

The term on the left expresses the *total deviation* in the observed Y about the central value for the entire sample. Consider again the regression line in Figure 12–4 found earlier from the workstation data. For the 22nd sample observation we have $(Y - \bar{Y}) = 63.2$. This quantity may be expressed as the sum of two component deviations. The first of these indicates how the Y predicted from regression deviates from the

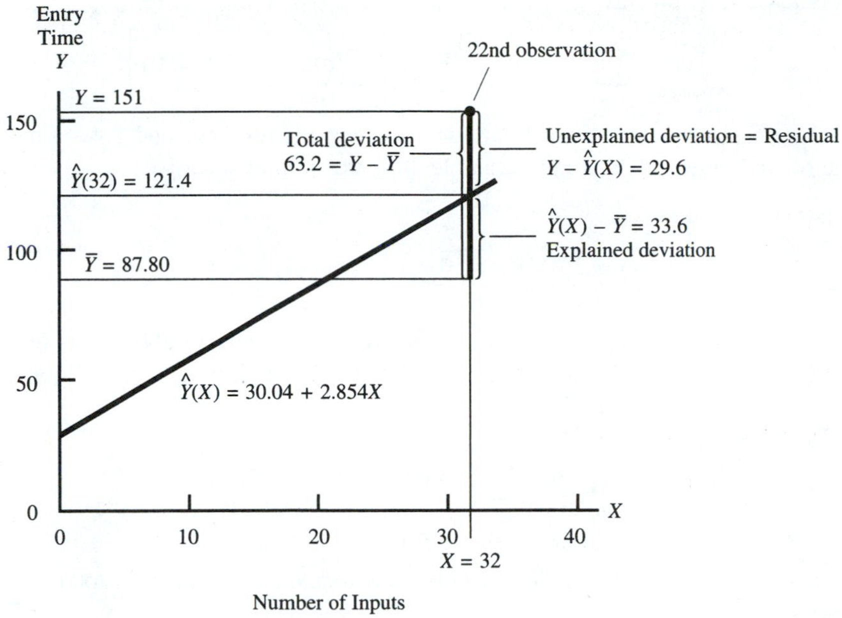

Figure 12–4 *Illustration of explained, unexplained, and total variation in regression analysis.*

center of the sample. From Figure 12–4 we find $[\hat{Y}(X) - \overline{Y}] = 33.6$, a difference that is explained by the regression line. That term is referred to as the **explained deviation**. The second term indicates how much the individual Y deviates from the line itself and is referred to as the **unexplained deviation**. For the 22nd transaction this quantity is $[Y - \hat{Y}(X)] = 29.6$.

Squaring these deviations and summing, we extend the deviation relationship to the entire collection of observations:

$$\begin{array}{ccccc} \text{Total variation} & = & \text{Explained variation} & + & \text{Unexplained variation} \\ \sum_{i=1}^{n}[Y_i - \overline{Y}]^2 & = & \sum_{i=1}^{n}[\hat{Y}(X_i) - \overline{Y}]^2 & + & \sum_{i=1}^{n}[Y_i - \hat{Y}(X_i)]^2 \\ SSTO & = & SSR & + & SSE \end{array}$$

The total variation expresses to what degree individual Ys deviate from their mean $\overline{Y}$ without regard to the regression relationship. This is summarized by the following.

TOTAL SUM OF SQUARES

$$SSTO = \sum_{i=1}^{n} [Y_i - \overline{Y}]^2 = \sum_{i=1}^{n} Y_i^2 - n\overline{Y}^2$$

As Figure 12–4 shows, the distance between Y and the regression line at X_i explains a portion of the deviation in the observed Y_i from its mean $\overline{Y}$. The explained variation summarizes the collective squared distances between the regression line $\hat{Y}(X)$ and the sample mean $\overline{Y}$. This is summarized by the following.

REGRESSION SUM OF SQUARES

$$SSR = \sum_{i=1}^{n} [\hat{Y}(X_i) - \bar{Y}]^2 = b\left[\sum_{i=1}^{n}(X_i - \bar{X})(Y_i - \bar{Y})\right]$$

The final component in total variation involves the residuals and gives a measure of the overall observed error in the sample, the error sum of squares.

ERROR SUM OF SQUARES

$$SSE = \sum_{i=1}^{n} [Y_i - \hat{Y}(X_i)]^2 = SSTO - SSR$$

The method of least squares chooses the regression coefficients a and b so that the above quantity is minimized. This term expresses the collective dispersion in Y about the regression line. The regression line leaves those deviations *unexplained.*

Rearranging terms, the identity

$$\begin{array}{ccccc} \text{Explained variation} & = & \text{Total variation} & - & \text{Unexplained variation} \\ SSR & = & SSTO & - & SSE \end{array}$$

is used to construct a useful index. Dividing the explained variation by total variation provides the

SAMPLE COEFFICIENT OF DETERMINATION

$$r^2 = \frac{\text{Explained variation}}{\text{Total variation}} = \frac{SSTO - SSE}{SSTO} = \frac{\sum[Y - \bar{Y}]^2 - \sum[Y - \hat{Y}(X)]^2}{\sum[Y - \bar{Y}]^2}$$

An equivalent expression is

$$r^2 = \frac{\text{Explained variation}}{\text{Total variation}} = 1 - \frac{\sum[Y - \hat{Y}(X)]^2}{\sum(Y - \bar{Y})^2}$$

The sample coefficient of determination expresses the proportion of the total variation in Y explained by the regression line.

It is convenient to express the sample coefficient of determination equivalently as

$$r^2 = 1 - \frac{s_{Y\cdot X}^2}{s_Y^2}\left(\frac{n-2}{n-1}\right)$$

For the workstation data this gives

$$r^2 = 1 - \frac{(13.51)^2}{(33.99)^2}\left(\frac{25-2}{25-1}\right) = .849$$

This indicates that 84.9% of the total variation in entry time Y may be explained by the regression line relating this variable to the number of inputs X. This leaves 15.1% of the variation unexplained by regression.

Relation to Correlation Coefficient The choice of r^2 to denote the sample coefficient of determination is no coincidence. The following relationship applies:

$$\text{Coefficient of determination} = (\text{Correlation coefficient})^2$$

Thus, the absolute value of the correlation coefficient is the square root of the coefficient of determination:

$$|r| = \sqrt{r^2}$$

The sign of the r will match the sign of the slope of the corresponding regression line. Thus, for the workstation illustration, we see that the sample correlation coefficient is

$$r = \sqrt{.849} = +.92$$

the value found in Chapter 4 for the same data.

This relationship provides another perspective on why the coefficient of determination gauges the strength of the X-Y relationship. In the special case when all the data points fall directly on the regression line, so there is a perfect correlation and $r = +1$ or $r = -1$, all of the variation in Y is explained and r^2 must then be equal to 1. At the other extreme, a horizontal regression line is obtained when X and Y exhibit zero correlation, so all of the variation in Y is unexplained and r^2 must then be equal to 0.

The Coefficient of Determination in Multiple Regression

The coefficient of determination applies in multiple regression as well, and is provided in the computer output of most regression packages. The regression runs made with the gasoline mileage data shown in Figures 4–10 and 4–11 give the following values for the coefficient of determination:

.9397 (using octane, speed, and load as independent predictors)

.9587 (adding stops as the fourth independent predictor)

Raising the number of independent variables in that case increases the amount of explained variation in Y. All other things being equal, a better set of predictions may be made from a regression equation that explains more of the variation in the dependent variable. When there are many candidate predictors, coefficients of determination can help in selecting the best set for future use. Stepwise procedures for doing this involve separate multiple regression evaluations for various combinations of variables, at each stage of which one variable is dropped or another is added to the current set of predictor variables.

Problems

12–7 Refer to the data in Problem 4–1 and your answers to that exercise. Calculate the sample coefficient of determination. What percentage of the variation in processing time Y is explained by the regression line using the number of data points X as the independent variable?

12–8 Refer to the data in Problem 4–2 and your answers to that exercise. Calculate the sample coefficient of determination. What percentage of the variation in pounds of final product Y is explained by the regression line using gallons of raw materials X as the independent variable?

12–9 Refer to the data in Problem 4–12 and your answer to that exercise. What percentage of the variation in reservoir storage Y could be explained by the estimated regression line in which season rainfall is the independent variable?

12–10 The coefficients of determination for various regression analyses with the engineers' earnings data from Problem 4–22 provide the following.

	Predictor					
Case	X_1	X_2	X_3	X_4	X_5	r^2
1	√					.362
2		√				.016
3			√			.016
4				√		.871
5					√	.006
6	√	√				.400
7	√		√			.427
8	√			√		.881
9	√				√	.370
10		√	√			.050
11		√		√		.880
12		√			√	.033
13			√	√		.898
14			√		√	.022
15				√	√	.909
16	√	√	√			.439
17	√	√		√		.886
18	√	√			√	.428
19	√		√	√		.899
20	√		√		√	.438
21	√			√	√	.917
22		√	√	√		.900
23		√	√		√	.072
24		√		√	√	.910
25			√	√	√	.934
26	√	√	√	√		.900
27	√	√	√		√	.463
28	√	√		√	√	.917
29	√		√	√	√	.935
30		√	√	√	√	.935
31	√	√	√	√	√	.935

(a) What do you recommend as the "best" set of independent variables for predicting income when the number of predictors is (1) one? (2) two? (3) three? and (4) four?

(b) Comment on the desirability of using a limited set of predictors versus a nearly complete set of the possible predictors.

12–3 Statistical Inferences Using the Regression Line

Inferences in regression analysis take two forms. One category involves predictions of the dependent variable. The second area concerns inferences regarding the regression coefficients A and B.

Prediction and Confidence Intervals in Regression Analysis

The main purpose of regression analysis is predicting the level of the dependent variable. Usually such a prediction will involve finding the level for the conditional mean $\mu_{Y \cdot X}$. For this purpose the estimated regression line provides the point estimate $\hat{Y}(X)$. To illustrate, we will continue with the workstation data provided in Table 12–1. We predict the mean time required for all entries involving 25 inputs. In that case $X = 25$, and the best point estimate of $\mu_{Y \cdot X}$ will be the fitted Y value from the regression line. We have

$$\begin{aligned} \hat{Y}(25) &= a + b(25) \\ &= 30.04 + 2.854(25) \\ &= 101.39 \end{aligned}$$

This value can also be used to estimate the time for a particular entry. To distinguish a mean value from an *individual value*, either of which is estimated for a given X by $\hat{Y}(X)$, we will use the special symbol Y_I.

As we have seen with estimates in general, the point estimate is deficient in acknowledging the presence of imprecision and uncertainty due to sampling error. In making predictions from the regression line, it is therefore desirable to extend the concept of the interval estimate. For this purpose, estimates of individual values of the dependent variable are usually expressed in the form of **prediction intervals**.

Confidence Intervals for the Conditional Mean

We establish prediction intervals just as we did the confidence intervals encountered in Chapter 10. For the mean these take the form

$$\mu = \bar{X} \pm t_{\alpha/2} \frac{s}{\sqrt{n}}$$

where $s/\sqrt{n}$ is the estimator of $\sigma_{\bar{X}}$.

In like fashion, a prediction interval for the conditional mean for Y may be established. We use

$$\mu_{Y \cdot X} = \hat{Y}(X) \pm t_{\alpha/2} \text{ estimated } \sigma_{\hat{Y}(X)}$$

The standard error of $\hat{Y}(X)$ is denoted by $\sigma_{\hat{Y}(X)}$ and represents the amount of variability in possible Ys for the specified X. In the workstation data-entry illustration, a somewhat different regression line from the one found earlier, such as $\hat{Y}(X) = 28 + 3.1X$ or $\hat{Y}(X) = 32 + 2.7X$, might be obtained after applying the least-squares procedure to the data obtained from another random sample collected in a similar fashion. Thus, a value for $\hat{Y}(25)$ that is slightly different from 101.39 might be computed from the regression line from another sample. For each setting of X, the potential set of $\hat{Y}(X)$ values will have a standard deviation of $\sigma_{\hat{Y}(X)}$.

Variability in $\hat{Y}(X)$ has two components. These are additive, so we have

$$\sigma^2_{\hat{Y}(X)} = \begin{array}{c}\text{Variability in the}\\ \text{mean of } Ys\end{array} + \begin{array}{c}\text{Variability caused by the}\\ \text{distance of } X \text{ from } \bar{X}\end{array}$$

The first source of variability is analogous to that for the sample mean, which, as we have seen, depends on the population standard deviation and the sample size. A second source is the distance separating X from $\bar{X}$. In Figure 12–5 the estimated regression lines have been plotted for different samples taken from the same data-entry population. Notice that although the Xs are of the same magnitude in each case and $\bar{X}$ is identical for each sample, the levels of a, b, and $\bar{Y}$ differ somewhat from sample to sample. The lines tend to diverge, with the total separation increasing as X becomes farther from $\bar{X}$. Thus, the values for $\hat{Y}(X)$ become more varied the farther X is from $\bar{X}$.

Although we won't derive it here, it may be established that for small sample sizes (generally those involving $n < 30$), the following expression provides the $100(1-\alpha)\%$

CONFIDENCE INTERVAL FOR THE CONDITIONAL MEAN USING SMALL SAMPLES

$$\mu_{Y\cdot X} = \hat{Y}(X) \pm t_{\alpha/2} s_{Y\cdot X} \sqrt{\frac{1}{n} + \frac{(X-\bar{X})^2}{\sum X^2 - (1/n)(\sum X)^2}}$$

The $1/n$ under the radical provides the first component of variability. The term involving $(X-\bar{X})^2$ establishes the magnitude of the second source of variability. Appendix

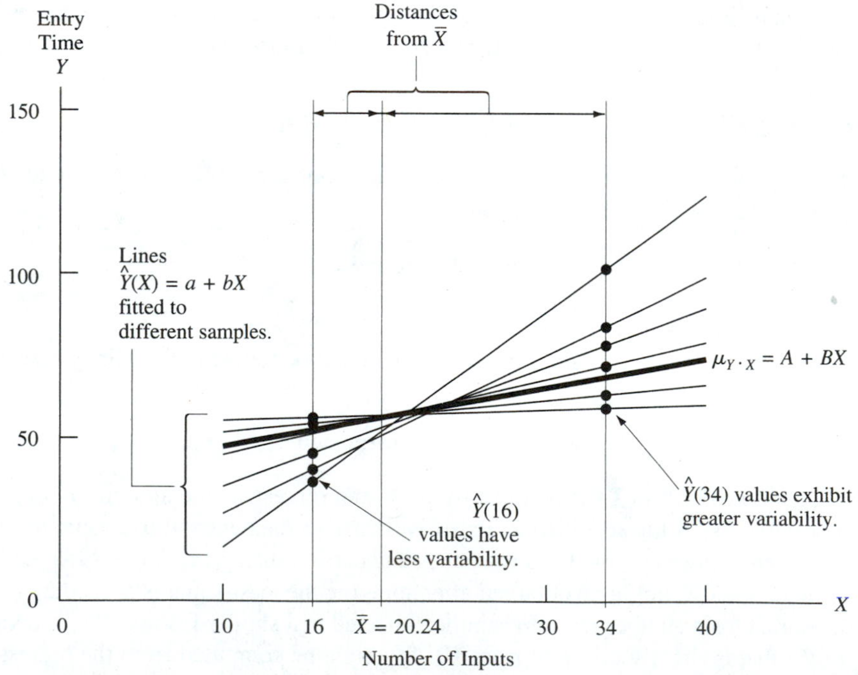

Figure 12–5 *Illustration of variability in predicted values for Y that is due to the particular estimated regression line obtained from the sample. Notice that this variability increases with the distance from the mean for the independent variable.*

Table G provides $t_{\alpha/2}$ for n–2 degrees of freedom. When transactions involve $X = 25$ inputs, the mean terminal time $\mu_{Y \cdot X}$ may be predicted with 95% confidence using the above expression. For 25–2 = 23 degrees of freedom, using $\alpha/2 = .025$, we have $t_{.025} = 2.069$, and we obtain

$$\mu_{Y \cdot 25} = \hat{Y}(25) \pm t_{.025} s_{Y \cdot X} \sqrt{\frac{1}{n} + \frac{(25 - \bar{X}^2)}{\sum X^2 - (1/n)\left(\sum X\right)^2}}$$

$$= 101.39 \pm 2.069(13.51)\sqrt{\frac{1}{25} + \frac{(25 - 20.24)^2}{13{,}130 - (1/25)\,(506)^2}}$$

$$= 101.39 \pm 6.10$$

or

$$95.29 \leq \mu_{Y \cdot 25} \leq 107.49$$

From this, we may conclude that the mean time to complete an entry involving 25 inputs is between 95.29 and 107.49 seconds. We are 95% confident this is so. (Which really means that if the regression procedure were repeated several times, and in each case $\hat{Y}(X)$ fitted to the data and an interval constructed similar to the above, then 95% of such intervals would contain the true value of $\mu_{Y \cdot X}$.

Analogous confidence intervals may be computed for other numbers X of inputs. Confidence limits may thereby be obtained over the entire range of X. This has been done for the workstation data. Figure 12–6 shows the resulting graphical display. Notice that the width of the confidence *band* increases with the distance X falls further above or below the mean.

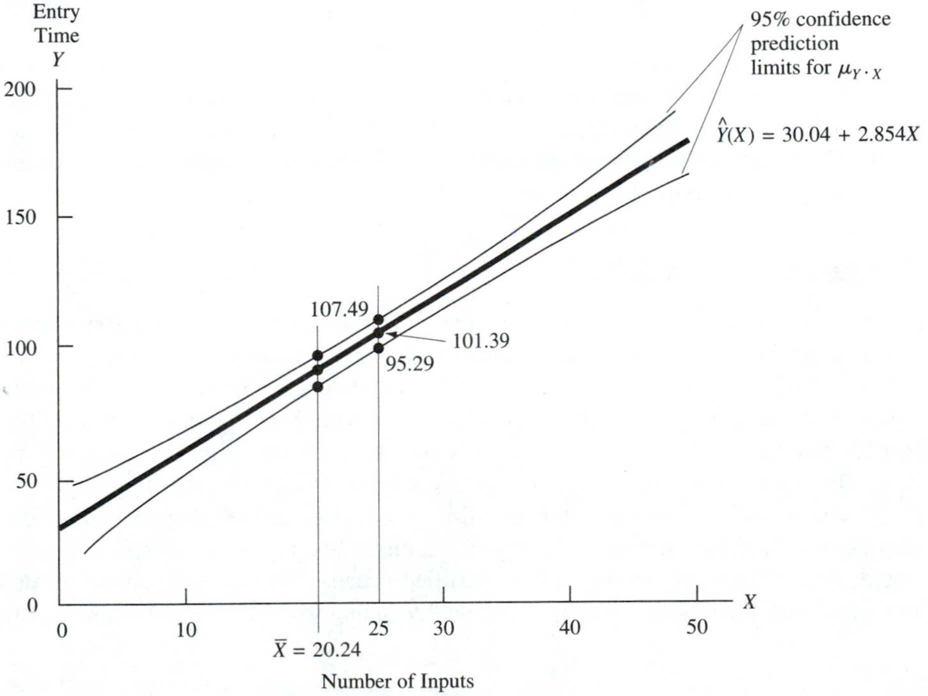

Figure 12–6 *Confidence limits for predictions of the mean workstation data-entry time.*

Prediction Intervals for an Individual Y Given X

Predicting an individual value of Y given X is similar to estimating the conditional mean. Consider a prediction of the next 25-input entry. The same point estimator applies as before, 101.39 seconds. As with the conditional mean, we account for sampling error by reporting our prediction in terms of a confidence interval. The following expression provides the $100(1-\alpha)\%$ prediction interval.

PREDICTION INTERVAL FOR AN INDIVIDUAL Y
USING SMALL SAMPLES

$$Y_I(X) = \hat{Y}(X) \pm t_{\alpha/2}s_{Y \cdot X}\sqrt{\frac{1}{n} + \frac{(X - \overline{X})^2}{\sum X^2 - (1/n)\left(\sum X\right)^2} + 1}$$

This expression is identical to the counterpart of $\mu_{Y \cdot X}$, except for the addition of $+1$ under the radical. That term reflects a third source of variability not encountered earlier. The component reflects the fact that *even if we knew the true regression line, individual Ys will still exhibit variability about their conditional mean.* Thus, given X, individual Ys will still have standard deviation $\sigma_{Y \cdot X}$, which is estimated by $s_{Y \cdot X}$.

We construct a 95% prediction interval for $Y_I(25)$, again using the workstation data. Substituting the appropriate values, we have

$$Y_I(25) = \hat{Y}(25) \pm t_{.025}s_{Y \cdot X}\sqrt{\frac{1}{n} + \frac{(25 - \overline{X}^2)}{\sum X^2 - (1/n)\left(\sum X\right)^2} + 1}$$

$$= 101.39 \pm 2.069(13.51)\sqrt{\frac{1}{25} + \frac{(25 - 20.24)^2}{13{,}130 - (1/25)\,(506)^2} + 1}$$

$$= 101.39 \pm 28.57$$

$$72.82 \leq Y_I(25) \leq 129.96$$

This interval is considerably wider than the one computed for the conditional mean. This should be no surprise, since it reflects the individual variability in Ys from entry to entry. This prediction interval expresses the single time for a *particular entry* involving 25 inputs, rather than the mean of them all. Its added width would be present even if the true regression line were known.

Dangers of Extrapolation

In making predictions from the regression line, it is important that investigators be cognizant of the dangers of extrapolation. This occurs whenever a value of Y is found for a level of X falling outside the observed range for that independent variable. In the workstation illustration, no sample observation involved more than 35 required inputs. Suppose that a forecast is desired for the mean time for very long entries involving $X = 100$ inputs. Would the estimated regression line established earlier be suitable?

There is no sure answer to this question. It could be argued that the data clerks might process such an entry considerably faster, on the average, than our regression line would predict. This conclusion could be justified because the repetitive nature of such long input runs permits an "economy of scale," reducing overall time requirements. But the reverse might just as well be argued: Longer entries will be more complex, and the data clerks will become so bogged down in detail that each entry will take longer than extrapolating the line would indicate; also, long input strings will be monotonous and therefore slower. Of course, either conclusion is pure conjecture—and that is precisely the point! The only way to know the response of Y to an unobserved X is to collect sample data that correspond. Extrapolation assigns values using a relationship that has been measured for different circumstances, so that any conclusions thereby reached are totally without supporting evidence.

Inferences Regarding the Regression Coefficients

The second class of inferences encountered in regression analysis involves the nature of the regression line. These generalizations apply to either the Y-intercept A or the slope B. The latter is often the greater concern in engineering, since it indicates the response Y to a one-unit change in X. Conclusions about A are usually of secondary importance. For example, a metallurgist might use regression analysis to develop a mathematical relationship between alloy concentrations and strength properties. The investigator's major concern may be determining the extra shearing force needed to permanently distort a rod for a unit change in alloy material, rather than predicting a particular force. Thus, B may be more important than $\mu_{Y \cdot X}$.

The Slope The level of B may be estimated from its sample counterpart, b. The standard error of b may be obtained from

$$s_b = \frac{s_{Y \cdot X}}{\sqrt{\sum X^2 - (1/n)(\sum X)^2}}$$

Using this expression, procedures have been established for making confidence interval estimates of B and testing its value.

The following expression is used to construct a $100(1-\alpha)\%$ confidence interval.

CONFIDENCE INTERVAL ESTIMATE OF THE TRUE SLOPE B

$$B = b \pm t_{\alpha/2} \frac{s_{Y \cdot X}}{\sqrt{\sum X^2 - (1/n)(\sum X)^2}}$$

where b is the computed slope of the estimated regression line and $t_{\alpha/2}$ involves $n-2$ degrees of freedom.

Again referring to the workstation data, we may construct a 95% confidence interval estimate of the slope B of the true regression line; this quantity represents the increase in entry time per additional input. Using $\alpha/2 = .025$ and $25 - 2 = 23$degrees of freedom, Appendix Table G provides $t_{.025} = 2.069$. We have

$$B = 2.854 \pm 2.069 \left(\frac{13.51}{\sqrt{13{,}130 - (1/25)\,(506)^2}} \right)$$
$$= 2.854 \pm .519$$

or

$$2.335 \leq B \leq 3.373$$

We therefore estimate with 95% confidence that there will be between 2.335 and 3.373 seconds of extra entry time for each additional data input.

It is also possible to test hypotheses regarding B. Ordinarily, such a test is used to establish whether *any* significant relationship exists between X and Y. A testing procedure will find whether or not the two variables are correlated, and if so, whether the slope of the true regression line is positive or negative. Such a test is useful when the only concern is the existence of a relationship or its direction.

To test the null hypothesis that $B = 0$, a two-sided procedure applies, and it is necessary only to construct the confidence interval estimate that matches the desired significance level. If the interval spans zero, H_0 must be accepted, and it must be rejected otherwise. A one-sided test will involve the following.

TEST STATISTIC FOR SLOPE *B* OF THE TRUE REGRESSION LINE

$$t = \frac{b}{\dfrac{s_{Y \cdot X}}{\sqrt{\sum X^2 - (1/n)(\sum X)^2}}}$$

The Intercept Inferences regarding the intercept are not as common as for the slope. The standard error of a is

$$s_a = s_{Y \cdot X}\sqrt{\frac{1}{n} + \frac{\bar{X}^2}{\sum(X - \bar{X})^2}}$$

From this, we get the following expression.

CONFIDENCE INTERVAL ESTIMATE OF THE TRUE INTERCEPT A

$$A = a \pm t_{\alpha/2} s_{Y \cdot X}\sqrt{\frac{1}{n} + \frac{\bar{X}^2}{\sum(X - \bar{X})^2}}$$

where $t_{\alpha/2}$ corresponds to $n - 2$ degrees of freedom.

Continuing with the workstation data, we find the 95% confidence interval estimate for the intercept of the true regression line. We first compute

$$\sum(X - \bar{X})^2 = \sum X^2 - n\bar{X}^2 = 13{,}130 - 25(20.24)^2 = 2{,}888.56$$

Then, for $25 - 2 = 23$ degrees of freedom, Appendix Table G provides $t_{.025} = 2.069$, and we compute:

$$A = 30.04 \pm 2.069(13.51)\sqrt{\frac{1}{25} + \frac{(20.24)^2}{2{,}888.56}} = 30.04 \pm 11.92$$

or

$$18.12 \leq A \leq 41.96$$

Using Bootstrapping to Make Inferences

The foregoing procedures all are based on the standard regression assumptions of Section 12–1. In practice, the shakiest of these is the normality of the error terms ϵ_i. Statistical bootstrapping provides alternative interval estimates that make *no* assumptions regarding the error terms.

Figure 12–7 shows the results obtained from a computer run using *Resampling Stats*. In a bootstrapping run the original data values constitute a surrogate population

```
READ FILE "B:TERMINAL.DAT" TIME ENTRIES
REGRESS NOPRINT TIME ENTRIES A
MAXSIZE YHAT 6000 YYHAT 6000 RESID 6000 RRESID 6000
COPY (25) X25
CLEAR YH25
REPEAT 100
     GENERATE 25 1,25 B
     TAKE TIME B TIME$
     TAKE ENTRIES B ENTRIES$
     REGRESS NOPRINT TIME$ ENTRIES$ A$
     SCORE A$ Z1 Z2
     MULTIPLY Z1 X25 YH25
     ADD Z2 YH25 YH25
     END
PRINT A
PERCENTILE Z1 (2.5 97.5) C1
PRINT C1
HISTOGRAM Z1
PERCENTILE Z2 (2.5 97.5) C2
PRINT C2
HISTOGRAM Z2
PERCENTILE YH25 (2.5 97.5) C3
PRINT C3
HISTOGRAM YH25
PERCENTILE RESID (2.5 97.5) C4
PRINT C4
```

```
A          =       2.8537      30.04
C3         =       95.762     107.34

  15+
    +
F   +
r   +
e   +
q 10+
    +                                *
*   +                                *
    +                  *   *   **    *
Y   +                  *   *  ***   **
H  5+                  *  *** ****  **
2   +                  ** *** ***** **  *
5   +               *  ****** ********  *
    +               *  ******************
    +          * ***************************
   0+-------------------------------------------
      |^^^^^^^^^|^^^^^^^^^|^^^^^^^^^|^^^^^^^^^|
     90        95        100       105       110
```

Figure 12–7 *Printout of bootstrapping run made with workstation data using* Resampling Stats.

that is sampled repeatedly with replacement. Altogether, 100 resamples and regressions are performed. For each resample of size $n = 25$, the least-squares regression coefficients are recomputed. In the run shown in Figure 12–7 a tally is made of those coefficients so that frequency distributions may be constructed for the a's and b's. In that run the value of $\hat{Y}(25)$ is computed for each successive resample. The histogram for these is shown at the bottom of Figure 12–7.

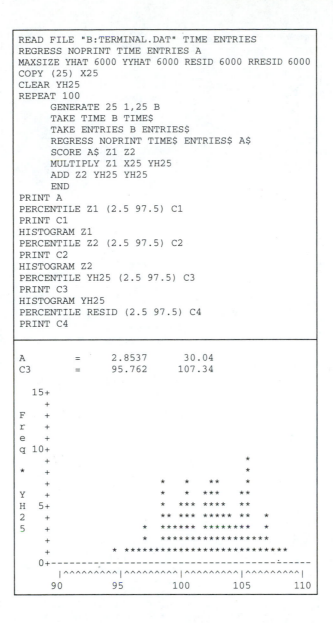

Figure 12–8
Histograms for regression parameters from bootstrapping run with workstation data.

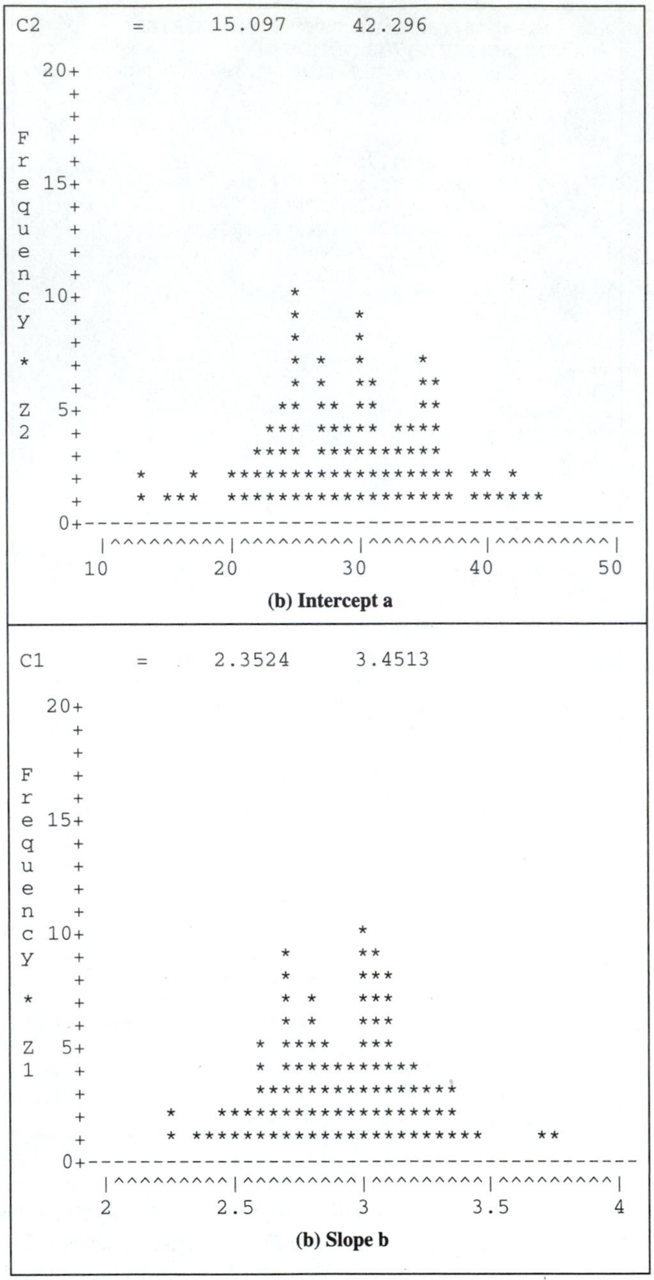

Notice the pronounced bimodality in the values for $\hat{Y}(25)$. A bell-shaped histogram is what we would expect under the conditions assumed by the theoretical regression model.

The computer run provides the 2.5 and 97.5 percentiles for $\hat{Y}(25)$, giving the following 95% resampling interval estimate of $\mu_{Y \cdot X}$:

$$95.762 \leq \mu_{Y \cdot X} \leq 107.34$$

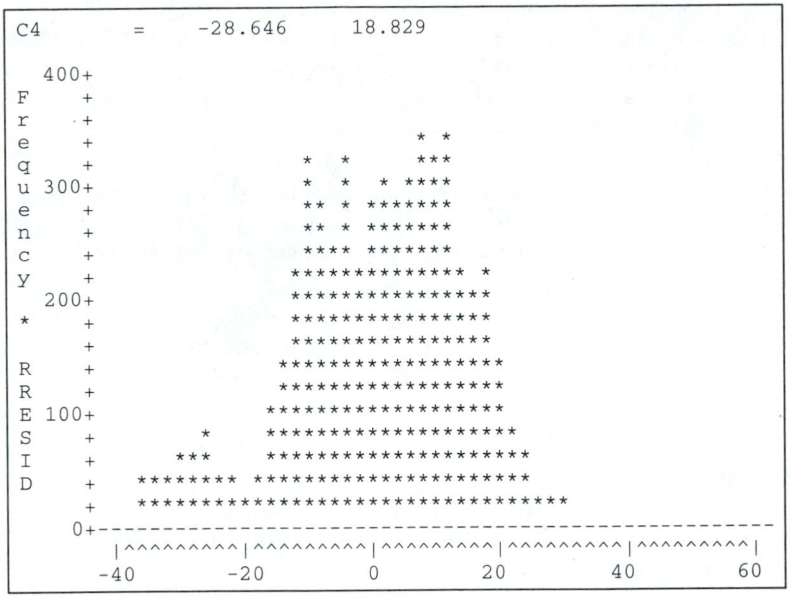

Figure 12–9 *Combined histogram of residuals from all bootstrapping run regressions.*

The above interval agrees quite well with the earlier one found using the traditional method.* A similar set of intervals for a and b are found in the same computer run and are shown in Figure 12–8. Again, these agree closely with the traditional counterparts computed earlier.

The bootstrapping run computes the residual values for each data point within each resample regression, so that altogether $100 \times 25 = 2{,}500$ residuals were computed. The histogram for these is plotted in Figure 12–9. Notice that there is a pronounced bimodality and that the histogram deviates very much from the bell curve that should apply under the theoretical assumptions. The evidence is consistent with a nonnormal error distribution.

Problems

12–11 Refer to the data in Problem 4–1 and your answers to that exercise.

(a) Construct a 95% confidence interval estimate of the conditional mean computer processing time when the number of given data points is 250.

(b) Construct a 95% prediction interval estimate of the computer processing time for an individual job having 320 data points.

*As an alternative, a *centered 95%* reampling interval may be computed using the value of 101.39 computed for $\hat{Y}(25)$ from the original regression equation:

$$101.39 - (107.34 - 101.39) \le \mu_{Y \cdot X} \le 101.39 + (101.39 - 95.762)$$

or

$$95.44 \le \mu_{Y \cdot X} \le 106.918$$

The above corrects for backwardness. Similar alternative intervals may be found for the regression coefficients. Simulations show that the uncentered intervals are nearly as good in estimating the underlying parameter. (For a detailed discussion, see Lawrence G. Hamilton, *Regression with Graphics: A Second Course in Applied Statistics*, Brooks/Cole, 1992, pages 320–24.)

12–12 Refer to the data in Problem 4–2 and your answers to that exercise.

(a) Construct a 99% confidence interval estimate of the conditional mean final product weight when the raw-materials volume is 20 gallons.

(b) Construct a 99% prediction interval estimate of the final product weight in an individual batch having 25 gallons of raw material.

12–13 Refer to Problem 4–1 and your answers to that exercise.

(a) Construct a 90% confidence interval estimate of the slope of the true regression line.

(b) At the 10% significance level, must you accept or reject the null hypothesis that processing time and the number of data points are uncorrelated variables?

12–14 Refer to Problem 4–2 and your answers to that exercise.

(a) Construct a 95% confidence interval estimate of the slope of the true regression line.

(b) At the 5% significance level, must you accept or reject the null hypothesis that final product weight and raw-materials volume are uncorrelated variables?

12–15 Using your intermediate calculations and answers to Problem 4–3, construct a 95% confidence interval estimate for Poisson's ratio for the experimental alloy represented in that exercise.

12–16 Using your intermediate calculations and answers to Problem 4–4, answer the following:

(a) Construct a 95% confidence interval estimate of the mean cost of batches involving 1,000 items.

(b) Construct a 95% confidence interval estimate for the variable cost of items produced.

12–17 Refer to the student data in Problem 4–5.

(a) Construct a 95% confidence interval estimate of the "unconditional" mean examination score for the students in the sampled population.

(b) Using the sample mean GPA as the given level for X, construct a 95% confidence interval estimate for the conditional mean for examination score (Y).

12–18 Refer to Problem 4–6 and your intermediate calculations and answers to that exercise. Construct a 95% confidence interval estimate of the modulus of elasticity for the steel represented by the test data.

12–19 Researchers performed an experiment to evaluate an electromagnetic bearing for the vibrational control mechanism for supercritical shafts. The following data were obtained in tests where the shaft was displaced:

Programmed Damping Rate X	**Damping Coefficient** Y
0	0.013
199	0.123
272	0.178
397	0.279
496	0.357
595	0.407
744	0.501

Source: Bradfield et al., *Proceedings of the Institution of Mechanical Engineers*, Vol. 201, No. 3, March 1987, pp. 201–212.

(a) Find the estimated regression equation.
(b) For a programmed damping rate of 400, construct a 95% confidence interval estimate of the *mean* level of the damping coefficient.
(c) Construct a 95% prediction interval estimate for the level of the damping coefficient for a particular *individual* trial displacement when the programmed damping rate is 500.

12–20 *Computer Exercise*. Refer to the data in Problem 4–1 and your answers to Problems 12–11 and 12–13. Conduct 100 resampling regressions, each with the same sample size as the original.
(a) Obtain the 95% resampling interval estimate for the conditional mean processing time when $X = 250$. Does this coincide closely with the confidence interval computed with the traditional procedure?
(b) Obtain the 90% resampling interval estimate for the slope B of the true regression line. Does this coincide closely with the confidence interval computed with the traditional procedure?

12–21 *Computer Exercise*. Refer to the data in Problem 4–2 and your answers to Problems 12–12 and 12–14. Conduct 100 resampling regressions, each with the same sample size as the original.
(a) Obtain the 99% resampling interval estimate for the conditional mean product weight when $X = 20$ gallons. Does this coincide closely with the confidence interval computed with the traditional procedure?
(b) Obtain the 95% resampling interval estimate for the slope B of the true regression line. Does this coincide closely with the confidence interval computed with the traditional procedure?

12–4 Inferences in Multiple Regression Analysis

As in the simple regression analysis, a variety of inferences may be drawn in multiple regression. The existence of multiple independent variables makes for a richer mix of procedures, including confidence intervals for the conditional mean, prediction intervals for individual Ys, tests for nonzero regression coefficients, and estimates of regression coefficients.

Confidence and Prediction Intervals

Although they are conceptually similar to the intervals found for simple regression applications in Section 12–3, confidence and prediction intervals in multiple regression evaluations are much more challenging to compute. That is almost always done with computer assistance.

To illustrate, consider the data in Table 12–2, providing the observed results from compilations of a sample of source programs. Compilation time serves as the dependent variable Y, which may be predicted using three independent variables: number of lines of code X_1, number of variables X_2, and number of subroutines X_3.

Table 12–2
Sample Data for Compilations of Source Programs

Program	Compilation Time (seconds) Y	Number of Lines X_1	Number of Variables X_2	Number of Subroutines X_3
1	132	301	7	5
2	421	12,106	17	6
3	74	723	11	2
4	75	6,123	8	0
5	41	887	9	2
6	201	9,235	0	1
7	148	5,521	9	5
8	60	4,689	2	0
9	308	1,119	25	4
10	338	1,136	36	3
11	238	1,317	2	5
12	275	485	0	7
13	203	880	14	3
14	157	1,482	15	3
15	99	624	5	2
16	27	692	1	3
17	30	586	3	0
18	62	2,046	2	1
19	377	2,111	8	2
20	227	3,047	6	5

A computer run was made with the compilation data using Minitab. The following estimated regression equation was obtained from the computer printout in Figure 12–10.

$$\hat{Y} = 3.12 + .01187X_1 + 5.803X_2 + 29.354X_3$$

We may use the above to predict the compilation time when the number of program lines is $X_1 = 1{,}000$, the number of variables is $X_2 = 10$, and the number of subroutines is $X_3 = 10$. The corresponding point estimate for the mean compilation time is computed by the program:

$$\hat{Y} = 3.12 + .01187(1{,}000) + 5.803(10) + 29.354(10) = 366.6 \text{ seconds}$$

The above is the center of the 95% confidence interval for the conditional mean compilation time $\mu_{Y\cdot 123}$ at those levels for the independent predictors:

$$219.6 \leq \mu_{Y\cdot 123} \leq 513.5$$

Centered at the same value is the following 95% prediction interval for an individual compilation time for the same levels:

$$136.9 \leq Y_I \leq 596.3$$

The second interval is considerably wider than the one for the conditional mean. That is because individual Y values will vary more than the means.

Figure 12–10 *Portion of Minitab printout for multiple regression run using compilation data.*

```
MTB > Regress c1 3 c2-c4;
SUBC>   Predict 1000 10 10.
SUBC>   Predict c2 c3 c4.
```

```
The regression equation is

Time = 3.1 + 0.0119 Lines + 5.80 No.Vars + 29.4 No.Subs

Predictor        Coef       Stdev   t-ratio        p
Constant         3.12       39.75      0.08    0.938
Lines        0.011870    0.005920      2.01    0.062
No.Vars         5.803       2.141      2.71    0.015
No.Subs        29.354       9.377      3.13    0.006

s = 83.26        R-sq = 60.3%      R-sq(adj) = 52.9%

Unusual Observations
Obs.   Lines        Time        Fit  Stdev.Fit   Residual    St.Resid
  2    12106       421.0      421.6       66.1       -0.6     -0.01 X
 19     2111       377.0      133.3       20.9      243.7      3.02R

R denotes an obs. with a large st. resid.
X denotes an obs. whose X value gives it large influence.
```

```
     Fit  Stdev.Fit          95% C.I.              95% P.I.
   366.6       69.3    (  219.6,   513.5)   (  136.9,   596.3) X
```

Above intervals are for Time when Lines=1,000, No.Vars=10, and No.Subs=10.

```
     Fit  Stdev.Fit          95% C.I.              95% P.I.
   194.1       31.4    (  127.6,   260.6)   (    5.4,   382.7)
   421.6       66.1    (  281.5,   561.7)   (  196.2,   647.0) X
   134.2       24.4    (   82.5,   186.0)   (  -49.7,   318.2)
     .
     .
     .
   220.9       28.3    (  161.0,   280.8)   (   34.4,   407.3)
```

Above intervals are for Time predicted for each actual observation.

```
X  denotes a row with X values away from the center
```

Inferences Regarding Regression Coefficients

The estimated intercept a and partial regression coefficients $b_1, b_2, \ldots$ are statistical estimators of the true regression coefficients $A, B_1, B_2, \ldots$ The following null hypothesis

$$H_0: B_1 = B_2 = \cdots = B_{m-1} = 0$$

is tested. The regression results are statistically significant when this null hypothesis is rejected. The procedure for making this determination is based on the *analysis-of-variance* approach (described in more detail in Chapter 13), which focuses on sums of squares.

To test the null hypothesis that all partial regression coefficients are zero, we must compare the explained and unexplained variations in Y. This is accomplished by a statistical test using a composite test statistic. This test statistic is based on the *means* of the squared deviations for the explained and unexplained variations. The first of these is the following.

REGRESSION MEAN SQUARE

$$MSR = \frac{SSR}{m-1}$$

where m is the total number of variables in the analysis. The denominator $m - 1$ is referred to as the number of degrees of freedom for *SSR*. The second mean is the following.

ERROR MEAN SQUARE

$$MSE = \frac{SSE}{n - m} = S^2_{Y \cdot 12 \cdots}$$

where $n - m$ is the number of degrees of freedom for the unexplained variation.

Recall that the sample variance is found by averaging squared deviations about the central value. *MSR* and *MSE* are both computed that way, and thus both mean squares are actually sample variances. Since these form the basis for the test, the procedure is called analysis of *variance* (even though the null hypothesis itself says nothing about variances).

The following expression is the regression test statistic.

REGRESSION TEST STATISTIC

$$F = \frac{\text{Variance explained by regression}}{\text{Unexplained variance}} = \frac{MSR}{MSE}$$

This quantity has the F distribution (introduced in Chapter 9). Large values for F tend to refute the null hypothesis. Critical values may be read from Appendix Table J. There are two sets of entries: the lightface ones apply when the upper-tail area is $\alpha = .05$, and the boldface values correspond to $\alpha = .01$. Two parameters must be specified. These are the divisors used in computing the respective mean squares; for this test $m - 1$ applies for the numerator and $n - m$ for the denominator.

Figure 12–11 shows that portion of the *Minitab* run that gives the analysis of variance results for the multiple regression with the compilation data. There we see that $SSR = 168{,}548$ and $SSE = 110{,}919$.

Dividing by the respective degrees of freedom, the mean squares are

$$MSR = \frac{168{,}548}{4-1} = 56{,}183 \qquad MSE = \frac{110{,}919}{20-4} = 6{,}932$$

These indicate that the variation in compilation time Y that is explained by that regression is quite high in relation to that left unexplained. This results in the computed value for the test statistic of

Figure 12–11 *Portion of* Minitab *printout for multiple regression run using compilation data.*

```
Analysis of Variance

SOURCE        DF        SS          MS         F        p
Regression     3    168548       56183      8.10    0.002
Error         16    110919        6932
Total         19    279467

SOURCE        DF    SEQ SS
Lines          1     27056
No. Vars       1     73556
No. Subs       1     67936
```

$$F = \frac{56{,}183}{6{,}932} = 8.10$$

Using degrees of freedom of $m - 1 = 3$ for the numerator and $n - m = 16$ for the denominator, Appendix Table J provides the critical value $F_{.01} = 5.29$. Since the computed value is greater, the null hypothesis of zero-valued partial regression coefficients must be *rejected* at the $\alpha = .01$ significance level. We must conclude that B_1, B_2, and B_3 are not all equal to zero. Stated equivalently, the regression results are significant at the $\alpha = .01$ level. (Since $F = 8.10$ is larger than the tabled critical value, this conclusion would apply at a lower significance level than .01. Figure 12–11 gives the rounded probability as .002.)

Problems

12–22 Refer to the gasoline mileage experimental data in Table 4–7. Consider mileage forecasts made from runs using 90-octane gasoline at 50 mph with a 500-pound load.

(a) Determine the 95% confidence interval estimate for the conditional mean gasoline mileage.

(b) Determine the 95% prediction interval estimate for the gasoline mileage from a particular test run.

12–23 Refer to the data in Problem 4–14 and your answers to that exercise. Construct a 95% confidence interval for the mean river flow when rainfall is 20 inches and 150,000 cfps are released from upstream dams.

12–24 Refer to the data in Problem 4–15 and your answer to that exercise. Construct a 95% confidence interval for the mean river flow when rainfall is 20 inches and average daily temperature is 60° F.

12–25 Refer to the data in Problem 4–16 and your answer to that exercise.

(a) Construct a 95% confidence interval for the mean final product yield when settling time is 10 hours and 100 pounds of solid catalyst are used.

(b) Repeat (a), predicting instead the final product yield of a single batch.

12–26 Refer to the data in Problem 4–26 and your answer to that exercise.

(a) Construct a 99% confidence interval for the mean processing time for all jobs requiring 50 k bytes of RAM and 10 k bytes of input.

(b) Repeat (a), predicting instead the processing time for a single job.

12–27 Refer to Problem 4–18 and your answer to that exercise.

(a) Construct a 95% confidence interval for the mean river flow when rainfall is 20 inches, 150,000 cfps are released from upstream dams, and the average daily temperature is 60° F.

(b) Compare your interval from (a) to the counterparts found in Problems 12–23 and 12–24. What do you notice? Explain.

12–28 Consider the following data.

Y	X_1	X_2
10	5	4
15	6	5
24	10	8
30	9	12
18	4	10

Using given levels $X_1 = 8$ and $X_2 = 9$, construct 95% interval estimates for (a) the conditional mean and (b) an individual prediction.

12–29 Refer to Problem 4–18 and your answer to that exercise. Are the regression results significant?

12–30 Refer to Problem 4–22 and your answer to that exercise.

(a) Construct 95% confidence intervals for the mean income of all engineers falling into each of the following groups:

	Professional Experience	Engineering Education	Positions Held	Number of Employees	Patents Held
(1)	5 yr	5 yr	2	0	2
(2)	10	4	5	10	0
(3)	10	6	2	0	10
(4)	15	4	3	25	1

(b) Using the same groups as above, construct 95% prediction intervals for the income of a single engineer having those characteristics.

(c) Are the regression results significant?

12–31 *Computer Exercise.* Refer to the compilation data in Table 12–2. Conduct 100 resampling regressions, each with the same sample size as the original. Obtain the 95% resampling interval estimate for the conditional mean compilation time for programs having 1,000 lines, 10 variables, and 10 subroutines. Does this coincide closely with the confidence interval on page 436?

12–32 *Computer Exercise.* Refer to the data in Table 4–7 and your answers to Problem 12–22. Conduct 100 resampling regressions, each with the same sample size as the original. Obtain the 95% resampling interval estimate for the conditional mean gasoline mileage for trips made with 90-octane gasoline at 50 mph carrying a load of 500 pounds. Does this coincide closely with the confidence interval found in Problem 12–22 (a)?

12–33 The following disabling injury data apply to California workers:

	Injuries per Million Hours			Injuries per Million Hours	
Year	*Burns*	*Amputations*	*Year*	*Burns*	*Amputations*
1953	0.6165	0.2330	1970	0.4726	0.1151
1954	0.5537	0.2045	1971	0.4426	0.0961
1955	0.6091	0.1934	1972	0.4522	0.0905
1956	0.5551	0.1871	1973	0.5171	0.0853
1957	0.5184	0.1740	1974	0.5084	0.0734
1958	0.5290	0.1680	1975	0.4652	0.0494
1959	0.5684	0.1529	1976	0.5249	0.0462

	Injuries per Million Hours			Injuries per Million Hours	
Year	*Burns*	*Amputations*	*Year*	*Burns*	*Amputations*
1960	0.5123	0.1315	1977	0.5840	0.0584
1961	0.5032	0.1382	1978	0.5939	0.0394
1962	0.4915	0.1273	1979	0.5914	0.0340
1963	0.5005	0.1321	1980	0.5492	0.0366
1964	0.5226	0.1498	1981	0.5009	0.0486
1965	0.5503	0.1553	1982	0.4608	0.0408
1966	0.5098	0.1783	1983	0.4527	0.0427
1967	0.5178	0.1483	1984	0.4691	0.0452
1968	0.4999	0.1397	1985	0.4483	0.0474
1969	0.5037	0.1252			

Source: Robinson, *American Journal of Public Health*, Vol. 78, No. 3, March 1986, p. 279.

(a) Treating year as the independent variable and burns as the dependent variable, plot a scatter diagram.
(b) Making any suitable transformations of variables, find the estimated regression equation.
(c) *Extrapolating*, predict the burn rate for 1990. Discuss why you cannot attach any statistical confidence to such a number.

12–34 Repeat Problem 12–33 using amputations as the dependent variable.

12–35 Refer to the data in Problem 4–24.
(a) Determine the predicted reaction time under the following retort temperatures:
(1) 220°F, (2) 245°F, (3) 250°F.
(b) Construct a 95% confidence interval for the mean reaction time under each of the temperatures in (a).
(c) Construct a 95% prediction interval for an individual batch reaction time under each of the temperatures in (a).

12–36 Refer to the data in Problem 4–24.
(a) Compute the coefficient of determination. From this, determine the value for the correlation coefficient. What proportion of the variation in Y is explained by the estimated regression line?
(b) Construct a 95% confidence interval estimate for the slope of the true regression line. At the 5% significance level, must the null hypothesis of zero slope be accepted or rejected?
(c) In light of your answer to (b), should the null hypothesis of zero correlation between reaction time and retort temperature be accepted or rejected? Use $\alpha = .05$.

12–37 Refer to the data in Problem 4–25.
(a) Determine the predicted mileages at the following speeds: (1) 50 mph, (2) 55 mph, (3) 60 mph.
(b) Construct a 95% confidence interval for the mean mileage at each of the speeds in (a).
(c) Construct a 95% prediction interval for an individual track-run mileage at each of the speeds in (a).

12–38 Refer to the data in Problem 4–25.

(a) Compute the coefficient of determination. From this determine the value for the correlation coefficient. What proportion of the variation in Y is explained by the estimated regression line?

(b) Construct a 99% confidence interval estimate for the slope of the true regression line. At the 1% significance level, must the null hypothesis of zero slope be accepted or rejected?

(c) In light of your answer to (b), should the null hypothesis of zero correlation between mileage and speed be accepted or rejected? Use $\alpha = .01$.

12–39 Refer to Problem 5–20 and your answer to that exercise.

(a) Using the appropriate results, construct a 95% confidence interval for the mean processing time for jobs involving 10 k bytes of input when (1) no peripheral memory is needed and (2) peripheral memory is needed.

(b) Construct 95% confidence intervals for the same jobs using your interactive estimated multiple regression results from Problem 5–20.

12–40 Refer to Problem 5–28 and your answer to that exercise. Predict the cost of a 50-thousand-square-foot warehouse. Construct a 95% confidence interval for the mean cost for all such warehouses.

12–41 Refer to Problem 5–29 and your answer to that exercise. Predict the cost of a 50-thousand-square-foot warehouse when the mean winter/spring temperature is 40 degrees. Construct a 95% confidence interval for the mean cost for all such warehouses.

12-42 Refer to Problem 5-30 and your answer to that exercise. Predict the cost of a 50-thousand-square-foot warehouse when the mean temperature is 40° for winter/spring and 70° for summer/fall. Construct a 95% confidence interval for the mean cost for all such warehouses.

12–43 Refer to Problem 5–31 and your answer to that exercise. Predict the cost of a 50-thousand-square-foot warehouse when the mean temperature is 40° for winter/spring and 70° for summer/fall and the energy cost index is 100. Construct a 95% confidence interval for the mean cost for all such warehouses.

CHAPTER THIRTEEN

Analysis of Variance

This chapter describes procedures for evaluating several quantitative populations in terms of their respective means. To begin, we introduce a test that determines whether or not those parameters take on a common value. In Chapter 11 we encountered such a test in which just two populations are considered. Our present concern is with any number of populations greater than two, and our central focus is placed on a statistical application called **analysis of variance**. Although the terminology can be confusing, this procedure is actually concerned with levels of *means*.

Such a tool can be very useful in engineering evaluations, where there is need for a variety of multiple comparisons. For instance, software engineers must select the best of several candidate routines by evaluating them in terms of their mean job-completion times. The processing times required under a particular routine establish a target population that may be sampled by running test jobs. The collective results from all such samples can be used as the basis for a final choice. Likewise, a chemical engineer can use sample batches run under various temperature settings to establish the optimal level for future production runs to synthesize a compound.

Careful planning is especially important when several samples are used in making inferences about their counterpart populations. Because a large number of expensive or hard-to-get observations must usually be made, the efficiency of the statistical procedure is especially important. The analysis-of-variance procedure itself encompasses *experimental design* considerations that guide the investigator in setting up the details of a statistical experiment. This chapter provides some insight into the design of such experiments.

13–1 Framework for a Single-Factor Analysis

The procedure for a single-factor analysis compares several populations. Each of these groups represents a different **treatment**. This term originated with pioneering medical and biological applications. The methods developed for such investigations now have universal application in most experimental fields, and, regardless of application, any factor that the investigator wishes to evaluate is referred to as a treatment. Thus, a software engineer might refer to each candidate algorithm as a separate "treatment" to be administered to the data being processed. Similarly, a chemical engineer may view any particular retort temperature setting as a different treatment of the input compound. In all cases sample data are generated by applying the respective treatments to randomly chosen units. Analysis of the data so obtained can indicate if the treatments differ.

To help our presentation of the analysis-of-variance procedure, we will use a detailed illustration that extends the bacterial leaching investigation introduced in Chapter 11.

Testing for Equality of Means

The data in Table 13–1 have been obtained for three different strains of *thiobacillus*, each representing a different bacterial leaching treatment. A mining engineer wishes to evaluate the strains in terms of their respective performance in the recovery of copper from low-grade ores.

Illustration: *Comparing Bacteria Strains for Leaching Copper Ore*

Each observation represents the copper yield (lb/ton) obtained from a several-ton batch of ore removed from a random location in the deposit area. The yields obtained may be treated as random samples from the respective target populations of all batches that might be processed by the indicated bacteria strain.

Two variables are involved in this statistical evaluation. Type of bacteria is referred to as the **treatment variable** or factor. Although usually qualitative, treatment variables might also represent discrete quantities or numerical categories. The treatment plays a role analogous to the independent variable in regression analysis. Copper yield is the **response variable**, which must be quantitative. It has an interpretation similar to the dependent variable in a regression analysis.

Table 13–1

Experimental Layout with Sample Copper Yields (lb/ton) from Leaching with Three Strains of Bacteria

	***Thiobacillus* Treatment**		
Observation i	Strain 1	Strain 2	Strain 3
1	31	33	25
2	35	41	27
3	33	37	24
4	29	32	26
5	35	34	37
6	41	43	39
7	53	45	33
8	—	43	36
9	—	41	22
10	—	27	—
Totals	257	376	269
Means	$\bar{X}_{.1} = 36.71$	$\bar{X}_{.2} = 37.60$	$\bar{X}_{.3} = 29.89$

$$\bar{\bar{X}}_{..} = \frac{257 + 376 + 269}{7 + 10 + 9} = 34.69$$

The mining engineer must determine if the alternative strains of bacteria differ in their efficacy at leaching recoverable copper. He will begin by testing the null hypothesis that the population mean yield per ton of ore is the same for each treatment. Denoting the *j*th treatment population mean by μ_j, we have the following.

NULL HYPOTHESIS

$$H_0: \mu_1 = \mu_2 = \mu_3$$

This implies a complementary alternative: In at least one pair, the μs differ.

Figure 13–1 should help illustrate the underlying concepts. When the sample data are pooled, they look like observations from a single population with high variability, as in (a). When viewed separately, however, the same copper-yield data look like three separate populations with smaller variances, as shown in (b). Diagram (c) illustrates the single copper-yield frequency curve assumed by the null hypothesis.

Although we have become accustomed to using means to gauge differences between groups of data, we can also use the respective variabilities for this purpose. When several groups, each of which clusters at different levels, are *pooled*, they will generally exhibit greater dispersion than when viewed singly. This happens to be the case with the sample data from the bacterial leaching investigation. Notice that the frequency curve in Figure 13–1 (a) is flatter and has longer tails than do any of those in (b). The greater spread in the top curve can be attributed mainly to differences in sample treatment means. If the individual distributions have similar means and variances, the pooled sample will be indistinguishable from them.

In organizing evaluations, it is convenient to arrange sample data in the columnar format of Table 13–1. Such an arrangement is referred to as the **experimental layout**. In the usual layout, observed data are arranged in a matrix, with each observed value of the response variable denoted by X_{ij}. The subscript *i* represents the observation, or row, and the subscript *j* refers to the treatment, or column. Thus, $X_{53} = 37$ lb/ton indicates the copper yield from the 5th observation (row 5) with treatment by strain 3 (column 3).

The Single-Factor Model

The models underlying analysis of variance treat each sample observation in terms of components. Here we describe the single-factor model.

SINGLE-FACTOR MODEL

$$X_{ij} = \mu_j + \epsilon_{ij}$$

This expresses each sample observation in terms of two components. X_{ij} denotes the *i*th observation under treatment *j*. Altogether, there are *c* treatments (one for each column) and $j = 1, 2, \ldots, c$. There are n_j observations made under the *j*th treatment, so $i = 1, 2, \ldots, n_j$.

In the bacterial leaching illustration, there are $c = 3$ treatments and $n_1 = 7$, $n_2 = 10$, and $n_3 = 9$. It is convenient to work with the combined sample.

COMBINED SAMPLE SIZE

$$n_T = \sum_{j=1}^{c} n_j$$

Figure 13–2
Underlying concepts of analysis of variance.

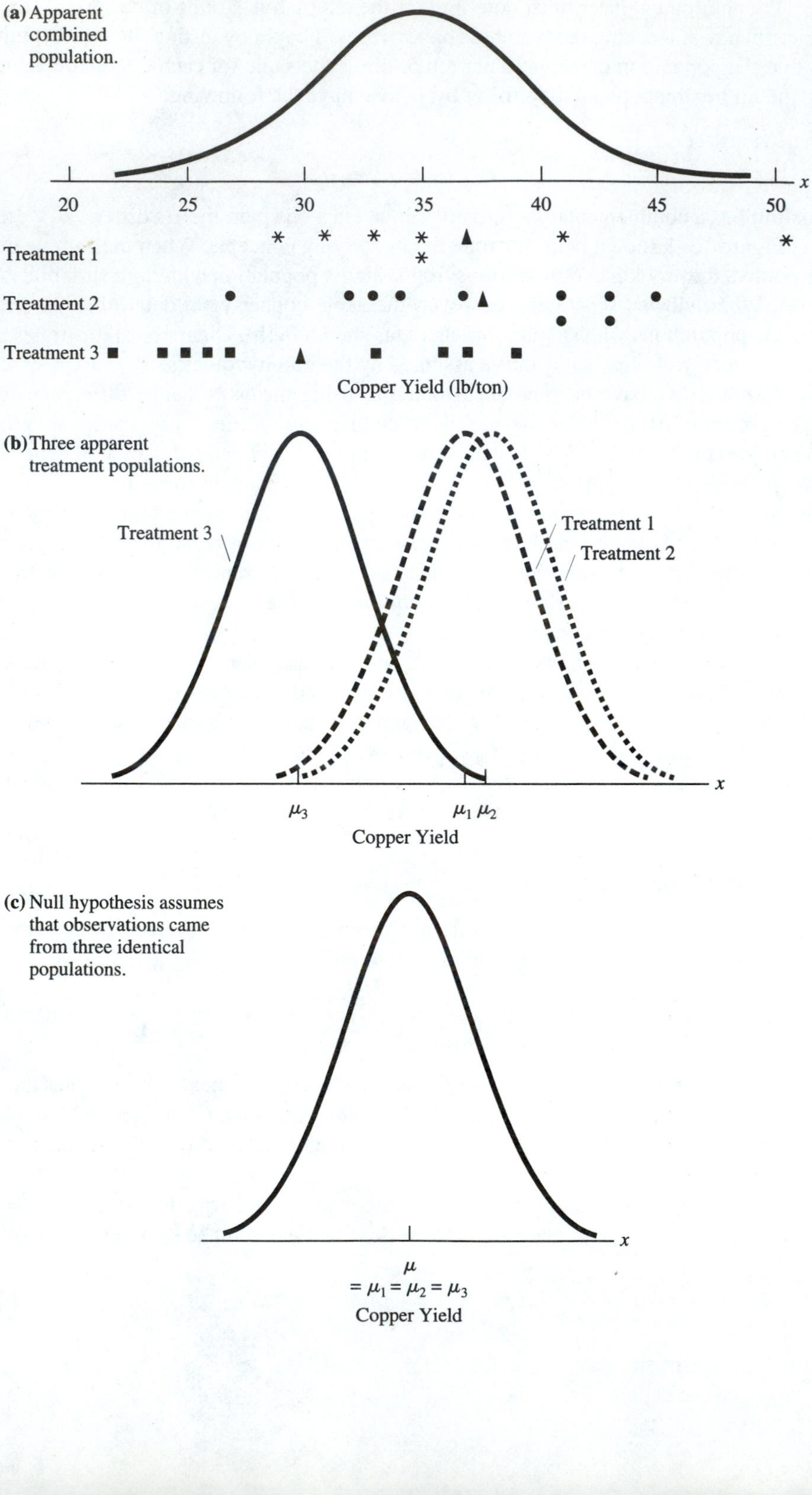

For the present illustration,

$$n_T = n_1 + n_2 + n_3 = 7 + 10 + 9 = 26$$

The **treatment population mean** μ_j is a constant of unknown value. The **error term** for each observation is denoted as ϵ_{ij}. The error terms are assumed to be independent, normally distributed random variables with mean 0 and standard deviation σ. Under this model, each sample observation is expected to be equal to the respective treatment population mean and to have a common variance σ^2, regardless of treatment j.

In a general sense, each treatment may be viewed as the level of a factor. The jth treatment population mean μ_j may also be referred to as the **mean for factor-level *j***. Each population mean may then be expressed as the sum of two components:

$$\mu_j = \mu_{\cdot} + B_j$$

The first term is the **overall population mean** $\mu_{\cdot}$. The dot in the subscript indicates that the overall mean is an average. The model assumes that the overall population mean is the average of the individual treatment population means:

$$\mu_{\cdot} = \frac{\sum_{j=1}^{c} \mu_j}{c}$$

The term B_j is defined as follows.

EFFECT FOR FACTOR-LEVEL j

$$B_j = \mu_j - \mu_{\cdot}$$

The null hypothesis of equal factor-level (treatment population) means may be stated equivalently as

$$H_0\colon B_1 = B_2 = \cdots = B_c$$

Sample Means Treating each column as a separate sample, we calculate the following.

SAMPLE MEAN FOR FACTOR-LEVEL (TREATMENT) j

$$\bar{X}_{\cdot j} = \frac{\sum_{i=1}^{n_j} X_{ij}}{n_j}$$

This indicates that the observations are averaged within each column. The subscript dot in the first position notation indicates that the averaging is done over the rows (is).

In the bacterial leaching investigation, the sample mean for the first treatment is calculated using the yields in column 1 of Table 13–1:

$$\bar{X}_{\cdot 1} = \frac{31 + 35 + 33 + 29 + 35 + 41 + 53}{7} = 36.71 \text{ lb/ton}$$

The other treatment sample means are $\bar{X}_{\cdot 2} = 37.60$ and $\bar{X}_{\cdot 3} = 29.89$.

Since the expected value of each treatment sample mean is

$$E(\bar{X}_{\cdot j}) = \mu_j$$

the above quantities are unbiased estimators of the respective treatment population means. The $\bar{X}_{.j}$'s are also least-squares and maximum-likelihood estimators of the respective μ_j's.

The mean of the pooled sample data is also needed in the evaluation. The following expression is used.

GRAND MEAN FOR POOLED SAMPLE RESULTS

$$\bar{\bar{X}}_{..} = \frac{\sum_{j=1}^{c}\sum_{i=1}^{n_j} X_{ij}}{n_T}$$

We use two overbars to denote the grand mean ("called X-double-bar"). The two dots in the subscript indicate that the averaging is done over both rows and columns. The grand mean for the bacterial leaching illustration is $\bar{\bar{X}}_{..} = 34.69$ lb/ton.

Problems

13–1 Express $\bar{\bar{X}}_{..}$ equivalently as a weighted average of the $\bar{X}_{.j}$'s.

13–2 Show that the B_j's must sum to zero.

13–3 Assuming for the bacterial leaching illustration that $\mu_1 = 36$, $\mu_2 = 38$, and $\mu_3 = 29$, complete the following:

(a) Compute $\mu_{..}$.

(b) Compute for each treatment the value for B_j.

(c) Construct a table giving for each observation the associated error term.

13–2 Single-Factor Analysis of Variance

The analysis-of-variance procedure can help translate the sample data into action. First, the null hypothesis of identical treatment population means is accepted or rejected. Then, if significant differences are found in the μ_j's, a variety of comparisons may be made. Additional conceptual framework must be erected before we are ready to perform the basic test.

Deviations about Sample Means

There are three types of deviations, as shown in Figure 13–2 for the bacterial leaching illustration. The total deviations for each individual observation are shown in (a) as bars connecting the data point to the grand mean. To illustrate, if we subtract the grand mean from the first observation under treatment by bacteria strain 1, we obtain the

$$\text{Total deviation: } X_{11} - \bar{\bar{X}}_{..} = 31 - 34.69 = -3.69$$

which is negative since X_{11} lies below $\bar{\bar{X}}_{..}$.

Figure 13–2(b) plots the **treatment deviations** obtained for each factor level. These are found by subtracting the grand mean from the corresponding treatment (column) sample mean. For bacteria strain 1, the following difference provides the

$$\text{Treatment deviation: } \bar{X}_{.1} - \bar{\bar{X}}_{..} = 36.71 - 34.69 = 2.02$$

$(\overline{X}_{\cdot 11} - \overline{\overline{X}}_{\cdot\cdot})$

Strain 1

Strain 2

Strain 3

X

25   30   35   40   45   50

Copper Yield (lb/ton)

$\overline{\overline{X}}_{\cdot\cdot}$

(a) Total deviations about grand mean.

$(\overline{X}_{\cdot 1} - \overline{\overline{X}}_{\cdot\cdot})$

Strain 1

Strain 2

Strain 3

X

25   30   35   40   45   50

Copper Yield (lb/ton)

$\overline{X}_{\cdot 3}$ $\overline{\overline{X}}_{\cdot\cdot}$ $\overline{X}_{\cdot 1}\overline{X}_{\cdot 2}$

(b) Treatment deviations of factor-level means about grand mean.

$(\overline{X}_{\cdot 11} - \overline{\overline{X}}_{\cdot\cdot})$

Strain 1

Strain 2

Strain 3

X

25   30   35   40   45   50

Copper Yield (lb/ton)

$\overline{X}_{\cdot 3}$ $\overline{\overline{X}}_{\cdot\cdot}$ $\overline{X}_{\cdot 1}\overline{X}_{\cdot 2}$

(c) Error deviations about factor-level (treatment) means.

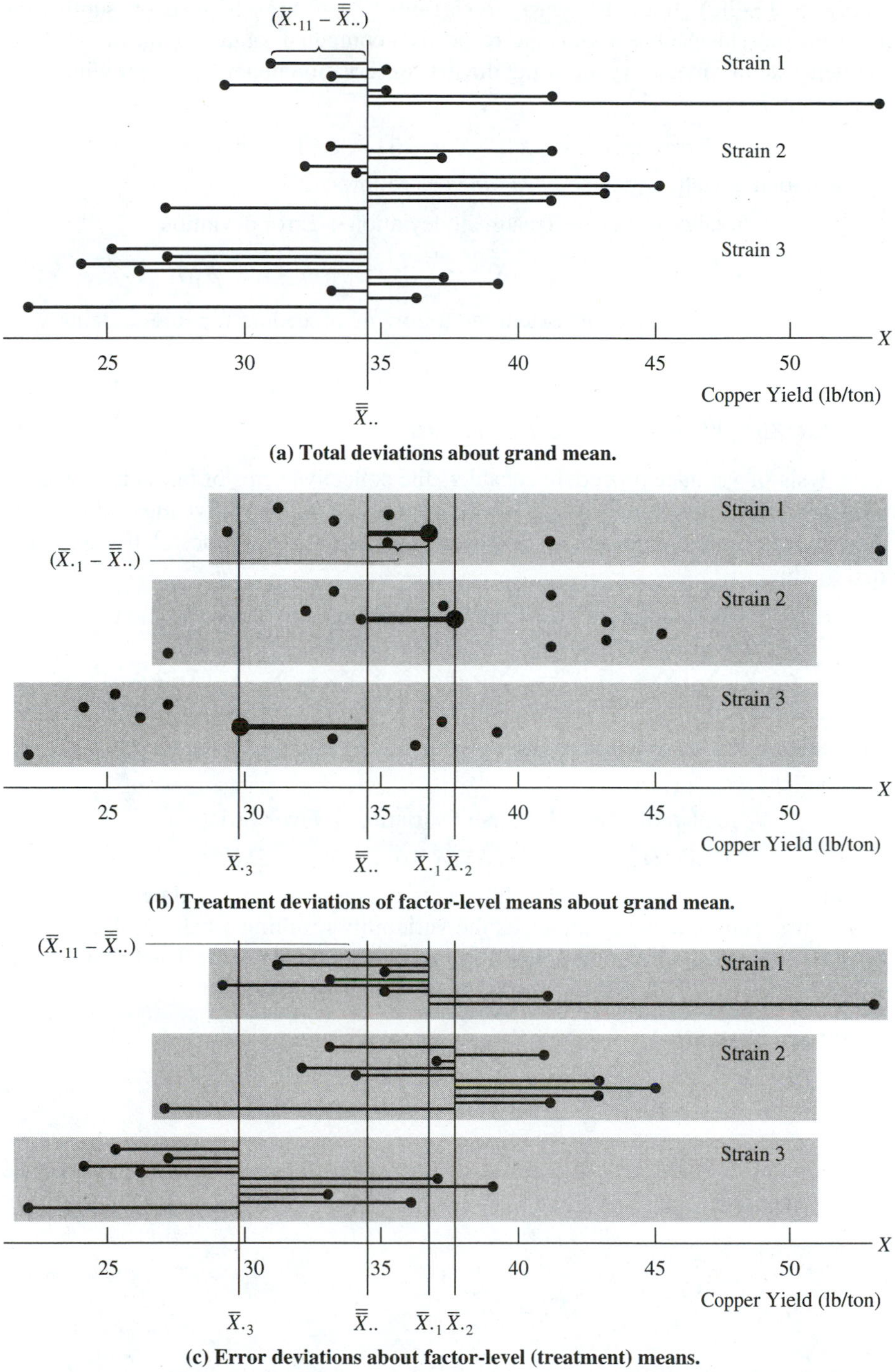

Figure 13–4
Deviation plots for bacterial leaching illustration.

Figure 13–2(c) shows the **error deviations**. These are obtained by subtracting from each individual observation the respective computed value for the factor-level (treatment) sample mean. Again, using the first observation under treatment with strain 1, we obtain the

$$\text{Error deviation: } X_{11} - \bar{X}_{.1} = 31 - 36.71 = -5.71$$

Each total deviation may be *partitioned* as follows:

$$\begin{array}{ccccc} \text{Total deviation} & = & \text{Treatment deviation} & + & \text{Error deviation} \\ (X_{11} - \bar{\bar{X}}_{..}) & = & (\bar{X}_{.1} - \bar{\bar{X}}_{..}) & + & (X_{11} - \bar{X}_{.1}) \end{array}$$

This expression justifies to some extent the following procedure for summarizing variation in sample data.

Summarizing Variation in Sample Data

The analysis-of-variance procedure considers the collective sums of individual *squared* deviations. These express the amount of *variation* about the respective means that is exhibited by the sample values. The next expression summarizes the relationships obtained by this:

$$\begin{array}{ccccc} \text{Total variation} & = & \text{Treatment variation} & + & \text{Error variation} \\ \sum_{j=1}^{c}\sum_{i=1}^{n_j}(X_{ij} - \bar{\bar{X}}_{..})^2 & = & \sum_{j=1}^{c} n_j(\bar{X}_{.j} - \bar{\bar{X}}_{..})^2 & + & \sum_{j=1}^{c}\sum_{i=1}^{n_j}(X_{ij} - \bar{X}_{.j})^2 \end{array}$$

Using the *sum of squares* notation encountered in regression analysis, we can write this relationship as

$$\begin{array}{ccccc} \text{Total variation} & = & \text{Treatment variation} & + & \text{Error variation} \\ SSTO & = & SSTR & + & SSE \end{array}$$

Each sum of squares may be computed from the expressions described below.

The first sum of squares expresses the variability resulting when all observations are treated as a combined sample coming from a common population. The following expression is used to calculate this quantity, the total sum of squares.

TOTAL SUM OF SQUARES

$$SSTO = \sum_{j=1}^{c}\sum_{i=1}^{n_j}(X_{ij} - \bar{\bar{X}}_{..})^2 = SSTR + SSE$$

The variability *between* sample groups or treatment columns is summarized by the following equation.

TREATMENTS SUM OF SQUARES

$$SSTR = \sum_{j=1}^{c} n_j(\bar{X}_{.j} - \bar{\bar{X}}_{..})^2$$

There is one term in this summation for each of the c treatments (columns). In order that all observations be represented once, each squared deviation is multiplied by the number of observations n_j for that column. Since any differences between groups might be explained by different responses to treatment, rather than to chance alone, *SSTR* is sometimes referred to as **explained variation**.

The variability *within* sample groups or treatment columns is summarized by the squared deviations of individual observations from their respective sample means. It is determined by the following expression.

ERROR SUM OF SQUARES

$$SSE = \sum_{j=1}^{c}\sum_{i=1}^{n_j}(X_{ij} - \overline{X}_{\cdot j})^2 = SSTO - SSTR$$

Since this quantity summarizes the differences between individual sample values that are due to chance and have no identifiable cause, *SSE* is sometimes referred to as **unexplained variation**.

The ANOVA Table

Table 13–2 shows a tabular arrangement of the primary results from an analysis of variance. The summary is aptly called the ANOVA table and organizes the calculations in a useful format. Because it is a time-consuming and error-prone chore to compute the sums of squares by hand, that task is best done with computer assistance.
Figure 13–3 shows the ANOVA table generated for the bacterial leaching data using *Minitab*.

The following sums of squares are listed in the SS column. (Minitab uses the term "factor" as a synonym for treatments.)

$$\begin{aligned} SSTR &= 320.8 \\ SSE &= \underline{1,034.7} \\ SSTO &= 1,355.5 \end{aligned}$$

Table 13–2
ANOVA Table for Single-Factor Analysis

Variation	Degrees of Freedom	Sum of Squares	Mean Square	*F*
Explained by Treatments (between columns)	$c-1$	$SSTR = \sum_{j=1}^{c} n_j(\overline{X}_{\cdot j} - \overline{\overline{X}}_{\cdot\cdot})^2$	$MSTR = \dfrac{SSTR}{c-1}$	$\dfrac{MSTR}{MSE}$
Error or Unexplained (within columns)	$n_T - c$	$SSE = \sum_{j=1}^{c}\sum_{i=1}^{n_j}(X_{ij} - \overline{X}_{\cdot j})^2$	$MSE = \dfrac{SSE}{n_T - c}$	
Total	$n_T - 1$	$SSTO = \sum_{j=1}^{c}\sum_{i=1}^{n_j}(X_{ij} - \overline{\overline{X}}_{\cdot\cdot})^2$		

Figure 13–3
ANOVA table for bacterial leaching data computed by Minitab.

```
MTB > AOVOneway c1-c3.

ANALYSIS OF VARIANCE
SOURCE     DF        SS        MS        F        p
FACTOR      2     320.8     160.4     3.57    0.045
ERROR      23    1034.7      45.0
TOTAL      25    1355.5
```

A Basis for Comparison: Mean Squares

Total variation has two components: explained variation (*SSTR*) and unexplained variation (*SSE*). These may be used to determine if the explained variation is great enough to significantly refute the null hypothesis of identical treatment population means. That question will be settled by the test statistic (to be described shortly), which reflects the relative magnitudes of those two variations.

In any sampling experiment, unexplained variation within a sample is a natural result of chance sampling error. Under H_0, however, the amount of variation between sample treatment groups should be small, and the respective sample means should be close in value. If the responses to the several treatments are identical, then explained and unexplained variations should be about the same magnitude.

The sums of squares cannot be directly compared because *SSTR* and *SSE* are each computed by totaling a different number of squared deviations. The former involves c terms, and the latter n_T terms. To make them comparable, each sum of squares must first be converted to an average. These sample variances, referred to as **mean squares**, are estimators of the respective population variances. To make these estimators unbiased, proper divisors are chosen, which are defined as follows.

TREATMENTS MEAN SQUARE,
ERROR MEAN SQUARE

$$MSTR = \frac{SSTR}{c-1} \qquad MSE = \frac{SSE}{n_T - c}$$

The theoretical assumptions underlying analysis of variance lead to the property that both *MSTR* and *MSE* have a common expected value, σ^2. These are a standard feature of the ANOVA table and are automatically computed by the popular statistical software packages. The following values apply to the bacterial leaching data.

$$MSTR = 160.4 \quad [= 320.8/(3-1)]$$
$$MSE = 45.0 \quad [= 1{,}034.7/(26-3)]$$

The treatments mean square is almost four times as large as the error mean square. If H_0 is true and the treatment populations have identical means, these quantities ought to be close in value. Such a large discrepancy would be unlikely if the response of copper yield to treatment were identical for the three strains of bacteria.

Before we can determine how unlikely this result is, we must develop additional concepts.

The F Statistic

Each source has a number of **degrees of freedom**, the divisors used in calculating the mean squares. The test statistic, which will now be described, is the quantity in the column labeled F.

The treatments mean square $MSTR$ may be referred to as the variance explained by treatments. Likewise, the error mean square MSE can be called the unexplained variance. The *ratio* of these two quantities will be used to test the null hypothesis of equal population treatment means.

TEST STATISTIC FOR ANALYSIS OF VARIANCE

$$F = \frac{\text{Variance explained by treatments}}{\text{Unexplained variance}} = \frac{MSTR}{MSE}$$

Applying this to the bacterial leaching investigation, we have

$$F = \frac{160.40}{45.00} = 3.57$$

Under the null hypothesis the values of F are expected to be close to 1 since $MSTR$ and MSE are each unbiased estimators of the common population variance σ^2 and thus have a common expected value. This statistic has a sampling distribution referred to as the F distribution. It has two parameters, each of which expresses a number of degrees of freedom.

Degrees of Freedom The divisor $c-1$ used in computing $MSTR$ is the number of degrees of freedom associated with using that statistic to estimate σ^2. Recall that this particular sum of squares involves calculating the grand mean $\bar{\bar{X}}_{..}$, which itself can be expressed in terms of the individual treatment sample means, the $\bar{X}_{.j}$'s. For a fixed level of $\bar{\bar{X}}_{..}$, all but one of the c $\bar{X}_{.j}$'s are free to vary.

Analogously, the divisor $n_T - c$ used in calculating MSE is the number of degrees of freedom associated with using MSE to estimate σ^2. This is easy to see since each $\bar{X}_{.j}$ represents the mean of a column involving n_j quantities. Thus, with $\bar{X}_{.j}$ at a fixed level, only $n_j - 1$ of the X_{ij}'s are free to assume any value. Since there are c columns, the number of free variables is

$$\sum_{j=1}^{c} n_j - 1 = n_T - c$$

A pair of degrees of freedom is associated with the F statistic. These sum to $n_T - 1$, the combined sample size minus 1. In the bacterial leaching investigation we have

Degrees of Freedom	
Numerator, $c-1$	$3-1 = 2$
Denominator, $n_T - c$	$26-3 = 23$
Total, $n_T - 1$	$26-1 = 25$

The theoretical sampling distribution for this test statistic is the F distribution, introduced in Chapter 9. Appendix Table J lists the critical values F_α for either $\alpha = .05$ or $\alpha = .01$.

The Hypothesis-Testing Steps

Using analysis of variance, the mining engineer in the bacterial leaching investigation performed the following steps:

Step 1. *Formulate the hypotheses and specify the possible choices.* The null hypothesis is that the treatments by the three strains of bacteria yield identical mean copper-yield responses

$$H_0\colon \mu_1 = \mu_2 = \mu_3$$

The mining engineer will either accept H_0 and conclude that the treatment means are equal or reject H_0 and find them not all equal.

Step 2. *Select the test procedure and test statistic.* The F statistic is used.

Step 3. *Establish the significance level and the acceptance and rejection regions for the decision rule.* The mining engineer selected a type I error probability of $\alpha = .05$ for incorrectly concluding that the bacterial leaching treatment means are not equal. For this investigation the degrees of freedom are 2 for the numerator and 23 for the denominator. Appendix Table J gives $F_{.05} = 3.42$. The acceptance and rejection regions for this test, which is *upper-tailed*, are shown in Figure 13–4.

Step 4. *Collect the sample data and compute the value of the statistic.* Included in this step are the preliminary calculations and construction of the ANOVA table. Earlier, the computed value of $F = 3.57$ was found.

Step 5. *Make the decision.* The computed test statistic falls in the rejection region. The engineer must therefore reject H_0 and conclude that the treatment means are not all equal. (The opposite conclusion would have been reached with $\alpha = .01$, in which case the critical value would have been $F_{.01} = 5.66$.)

Figure 13–5

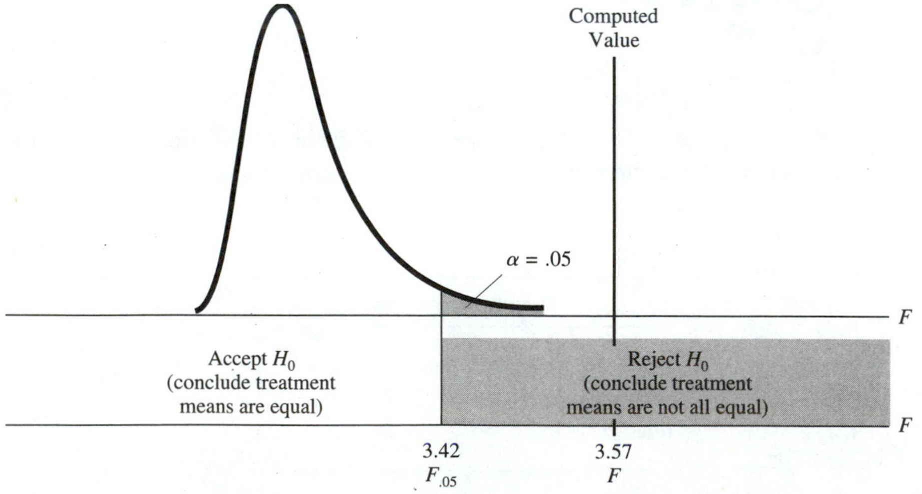

Deciding What To Do: Comparing Treatment Means

By rejecting the null hypothesis of equal population mean copper yield from ore treated under each bacteria strain, the mining engineer knows that at least one strain can be eliminated from further consideration. To identify the inferior strain or strains, additional comparisons must be made. This step will involve a pair-by-pair comparison. The basic method for doing this has already been encountered in Chapter 11. However, these earlier procedures must be amended when population pairs are compared after performing an analysis of variance. Section 13–3 describes those multiple comparisons.

More Discriminating Testing Procedures

The procedure just described uses only one treatment variable to explain differences between populations. It is therefore classified as a **one-way** analysis of variance. However, additional factors may explain those differences even more precisely. Section 13–5 describes **two-factor** analysis-of-variance procedures. There, a second factor—grade of ore deposit—is used in addition to bacteria strain to explain copper yields. Analysis-of-variance techniques exist even for **three-factor** and higher experimental layouts.

The following example shows that the single-factor analysis-of-variance procedure might uncover no significant differences in population means. This may be the conclusion even when it seems obvious that the opposite should be concluded.

Example: *Evaluating Methods for Solving Quadratic Assignment Problems*

Two researchers have developed an algorithm that uses Monte Carlo simulation to pick intermediate solutions in solving quadratic assignment problems. Although it is not necessarily faster than traditional heuristic methods, it is believed that the proposed procedure will provide better, less costly solutions.

The researchers tested their model under a complex experimental design that provides for a range of settings for four parameters. Table 13–3 shows the results from one test. Each observation represents a set of computer runs made with problems of the same size, with bigger problems used for successive rows.

Figure 13–5 provides a computer printout of the ANOVA table for these results, obtained using Minitab. Notice that the unexplained variation (error) dwarfs that explained by treatments. The computed value for F is therefore too small to warrant rejection of the null hypothesis of identical means, and in accordance with the single-factor model, we should *accept* the null hypothesis of identical population means, concluding that average best costs do not differ under the four procedures.

Figure 13–3
ANOVA table for testing methods for solving quadratic assignment problems, computed by Minitab.

```
MTB > AOVOneway c1-c4.

ANALYSIS OF VARIANCE
SOURCE      DF        SS         MS         F         p
FACTOR       3      2138        713      0.00     1.000
ERROR       28  32461878    1159353
TOTAL       31  32464016
```

Table 13–3
Sample Results for Evaluation of Methods for Solving Quadratic Assignment Problems

	Average Best Cost with Procedure (treatment)			
Observation i	(1) Simulated Annealing	(2) CRAFT	(3) CRAFT Biased Sampling	(4) Revised Hillier
1	\$25	\$28.2	\$27.6	\$25
2	43	44.2	43.6	43
3	74	79.6	74.8	73.2
4	107	110	108.2	107.4
5	291	296.2	294.8	298.4
6	578.2	606	581	582.2
7	1,308.0	1,339	1,321	1,324.6
8	3,099.8	3,197.8	3,124	3,114.2
	$\bar{X}_{.1} = \$690.75$	$\bar{X}_{.2} = \$712.63$	$\bar{X}_{.3} = \$696.88$	$\bar{X}_{.4} = \$696.00$
		$\bar{\bar{X}}_{..} = \$699.06$		

Source: Wilhelm, Mickey R. and Thomas L. Ward, "Solving Quadratic Assignment Problems by 'Simulated Annealing.'" *IEE Transactions*. March 1987, pp. 107–118.

The sample data in Table 13–3 show that the column values all increase with subsequent observations, and that within each row the values are nearly the same. That is the case because each observation row represents a changing set of circumstances that have a similar effect on the response under all treatments. We will see in Section 13–6 how that fact may be incorporated to help reduce the unexplained variation by utilizing a second factor, or *blocking variable*.

The Type II Error

As with all hypothesis-testing procedures, whenever H_0 is accepted there is the chance that this is an incorrect action. Although generally judged to be the less serious error, this potential for type II error cannot be ignored. There can be many ways for H_0 not to be true, and the F distribution is not very helpful in quantifying type II errors in terms of β probabilities. As with earlier testing procedures, the chance of such errors can be kept within acceptable bounds by using large sample sizes.

Violations of the Underlying Model

The F test is predicated on the assumptions regarding error terms. As with regression analysis, violations of the model may be serious and could invalidate the conclusions. Two violations can be especially serious.

First, the error terms, and hence the X_{ij}'s and $\bar{X}_{.j}$'s, are assumed to be normally distributed. As with procedures based on the Student t distribution, this condition is usually of little consequence when the population frequency curves are not highly skewed.

A second violation arises when the variance in error terms is nonconstant. To remedy this, we could transform the response variable, perhaps using $X' = \sqrt{X}$ or $X' = \log(X)$. But as a practical matter, the nonconstant variance is most serious in the more complex models involving random effects (not described in this book). When factor effects are fixed (as in this book), the standard procedures still perform well and are therefore *robust* against this violation.

Problems

13–4 The superintendent at a chemical processing facility wishes to compare temperature settings at various stages of processing. Her response variable is the percentage impurity of the output at each stage, each of which is to be tested independently. In each test the null hypothesis is that the mean percentage of impurities is the same under all four temperature settings. A total of 10 batches has been run under each setting for each stage. For the following results, find the applicable critical value and indicate whether H_0 should be accepted or rejected when a 1% significance level applies:
(a) Stage 13, $F = 4.28$ (b) Stage 4, $F = 4.62$ (c) Stage 11, $F = 5.43$

13–5 A chemical engineer wishes to assess the effect of different pressure settings on the mean yield of a final product from a given level of raw ingredients. The following sample data (in grams/liter) have been obtained:

	Pressure			
Sample Batch	**(1) Low**	**(2) Moderate**	**(3) Strong**	**(4) High**
1	28	30	31	29
2	26	29	29	27
3	29	30	33	30
4	30	30	33	31
5	28	28	29	27
6	31	32	33	32
7	26	29	28	27
8	32	32	32	32
9	25	28	27	27
10	29	30	32	30

(a) Find the critical value of the test statistic and identify the acceptance and rejection regions. Use $\alpha = .01$.
(b) Construct the ANOVA table.
(c) Should the null hypothesis of identical mean yields under the various pressure settings be accepted or rejected?

13–6 Figure 13–6 shows the results of a computer run using SPSS. The sampling experiment considers the mean purification time (hours) of four filtering systems.
What can you conclude regarding mean filtering times of the respective systems?

Figure 13–6
SPSS analysis-of-variance printout.

```
   VARIABLE   TIME               PURIFICATION TIME IN HOURS
BY VARIABLE   TYPE               FILTERING SYSTEM

                                        ANALYSIS OF VARIANCE

           SOURCE          D.F.    SUM OF SQUARES     MEAN SQUARES      F RATIO

    BETWEEN GROUPS            3          150.0808          50.0269        4.349

    WITHIN GROUPS           129         1491.0981          11.5589

    TOTAL                   132         1641.1789
```

13–7 The following data have been obtained by a systems analyst:

	Disk Drive		
Test Run	**(1) Flip-Flop**	**(2) Hard-Core**	**(3) Hy-Discus**
1	32%	17%	−5%
2	−11	23	8
3	14	15	12
4	9	7	10
5	16	13	—
6	8	—	—

The response expresses the percentage reduction in processing time over that taken when the identical job was run using existing disk hardware. The analyst wishes to test the null hypothesis that the three test drives yield identical mean responses.

(a) Find the critical value of the test statistic and identify the acceptance and rejection regions. Use $\alpha = .05$.

(b) Construct the ANOVA table.

(c) Should the null hypothesis of identical mean percentage reductions in processing times with different disk drives be accepted or rejected?

13–8 An industrial engineering student helped an entomological research team evaluate various location strategies for gypsy moth scent-lure traps. The following data were obtained:

	Trap Location Strategy				
Observation	**(1) Scattered**	**(2) Concentrated**	**(3) Host Plant**	**(4) Aerial**	**(5) Ground**
1	90%	99%	95%	98%	87%
2	92	97	96	98	93
3	94	98	97	99	90
4	93	98	97	99	91
5	—	99	96	—	89

The response variable is the estimated percentage of the native male population trapped.

(a) Find the critical value of the test statistic and identify the acceptance and rejection regions. Use $\alpha = .01$.

(b) Construct the ANOVA table.

(c) Should the null hypothesis of identical mean percentage native males trapped under the various strategies be accepted or rejected?

13–9 A curriculum committee at an engineering school is evaluating basic laboratory policies for freshmen. They requested an experiment where laboratory sections were taught under three different syllabi. The following mean scores on a standard examination were achieved by each sample class in this experiment:

	Laboratory Syllabus		
Class	(1) Theoretical	(2) Empirical	(3) Mixed Mode
1	78	77	83
2	85	86	91
3	64	71	75
4	77	75	78
5	81	80	82
6	75	77	80

What should the committee conclude regarding the mean scores of all laboratory classes taught under the respective syllabi? Use $\alpha = .05$.

13–10 A petroleum engineer for a large oil company is evaluating secondary recovery techniques for oil wells. The following sample data provide the daily increase (barrels) in pumped crude obtained from sample wells:

	Secondary Treatment			
Well	(1) Explosive Fracture	(2) Water Injection	(3) Steam Injection	(4) Controlled Pumping
1	−5	11	15	8
2	11	14	10	10
3	13	22	25	18
4	−5	0	2	1
5	25	35	40	28
6	21	24	28	27
7	11	15	19	17
8	33	30	28	32
9	105	224	328	276

(a) Can the engineer conclude that some secondary treatments might be better than others? Use $\alpha = .01$.
(b) What factors other than secondary treatment type might explain variation in increased daily production?

13–3 Comparative Analysis of Treatments

Continuing with the bacterial leaching investigation of the previous section, we now consider what the investigator does next after rejecting the null hypothesis of identical treatment means. The mining engineer must find the best bacteria strain or at least narrow his search to fewer alternatives. How do we translate the analysis of variance into such tangible action? This involves either estimating individual population means or comparing pairs of population means in terms of their estimated differences.

Since several pairs of means can be compared, we encounter a new complication arising from the *collective* nature of the inferences made. Every confidence interval is a separate statement that affects the credibility of the other estimates, and all must be accounted for through construction of interval estimates somewhat wider than when only single estimates are made.

Single Inferences for a Mean or a Pairwise Difference

To set the stage, we first adapt the procedures of the previous chapters for making *single inferences.*

Confidence Interval for a Treatment Population Mean Recall that for a single sample and population, the unknown μ may be estimated by a $100(1-\alpha)\%$ confidence interval of the following form.

$$\mu = \overline{X} \pm t_{\alpha/2}\frac{s}{\sqrt{n}}$$

Slight modification must be made when several samples and populations are involved. Recall that a common variance σ^2 is assumed for all treatment populations (even when H_0 of identical means has been rejected). Rather than isolating the observations in a single sample group and using the sample variance s^2 to estimate σ^2, a more refined estimator is provided by the unexplained variance *MSE*, which reflects information from *pooled* samples. The following expression is used.

SINGLE CONFIDENCE INTERVAL ESTIMATE OF μ_j

$$\mu_j = \overline{X}_{.j} \pm t_{\alpha/2}\sqrt{\frac{MSE}{n_j}}$$

The sample mean $\overline{X}_{.j}$ for the *j*th treatment is used as the estimator of μ_j. The sample size is the number of observations n_j for that group. The number of degrees of freedom for $t_{\alpha/2}$ is $n_T - c$, the same quantity used as the divisor in computing *MSE*.

The above may be applied to the results from the bacterial leaching investigation. The 95% confidence interval estimate of the mean copper yield from *thiobacillus* strain

1 is constructed using $t_{.025} = 2.069$, taken from Appendix Table G for $n_T - c = 26 - 3 = 23$ degrees of freedom. Substituting $\overline{X}_{.1} = 36.71$ and $MSE = 44.99$, we have

$$\begin{aligned}\mu_1 &= \overline{X}_{.1} \pm t_{.025}\sqrt{\frac{MSE}{n_1}} \\ &= 36.71 \pm 2.069\sqrt{\frac{44.99}{7}} \\ &= 36.71 \pm 5.25 \text{ lb/ton}\end{aligned}$$

or

$$31.46 \leq \mu_1 \leq 41.96 \text{ lb/ton}$$

Many popular statistical software packages will automatically compute confidence intervals for the treatment population means. Figure 13–7 provides the report generated by Minitab with the bacterial leaching data. The pooled standard deviation is $\sqrt{MSE} = \sqrt{44.99} = 6.707$.

Estimating a Pairwise Difference in Means In Chapter 11 we saw how to compare two population means using a confidence interval for their difference,

$$\mu_A - \mu_B = (\overline{X}_A - \overline{X}_B) \pm t_{\alpha/2} s_D$$

where s_D is the standard error for the difference $D = \overline{X}_A - \overline{X}_B$ and is the estimator of

$$\sigma_D = \sigma\sqrt{\frac{1}{n_A} + \frac{1}{n_B}}$$

where σ is the common population standard deviation. Using MSE as an unbiased estimator of σ^2 and using number subscripts instead of letters, the following equivalent expression is used in establishing the $100(1-\alpha)\%$ single confidence interval.

SINGLE CONFIDENCE INTERVAL FOR THE DIFFERENCE $\mu_j - \mu_k$

$$\mu_j - \mu_k = (\overline{X}_{.j} - \overline{X}_{.k}) \pm t_{\alpha/2}\sqrt{MSE\left(\frac{1}{n_j} + \frac{1}{n_k}\right)}$$

where, again, the number of degrees of freedom is $n_T - c$.

Continuing with the bacterial leaching investigation and using 95% confidence, we have the following estimated difference in mean copper yield between strains 2 and 3:

$$\begin{aligned}\mu_2 - \mu_3 &= (\overline{X}_{.2} - \overline{X}_{.3}) \pm t_{.025}\sqrt{MSE\left(\frac{1}{n_2} + \frac{1}{n_3}\right)} \\ &= (37.60 - 29.89) \pm 2.069\sqrt{44.99\left(\frac{1}{10} + \frac{1}{9}\right)} \\ &= 7.71 \pm 6.38\end{aligned}$$

or

$$1.33 \leq \mu_2 - \mu_3 \leq 14.09 \qquad \text{(significant)}$$

Figure 13–7
Individual confidence intervals for bacterial leaching illustration, reported by Minitab.

```
                                          INDIVIDUAL 95 PCT CI'S FOR MEAN
                                          BASED ON POOLED STDEV
LEVEL       N       MEAN      STDEV   ----------+---------+---------+------
    1       7     36.714      8.118                (---------*----------)
    2      10     37.600      5.910                   (--------*--------)
    3       9     29.889      6.373    (--------*--------)
                                      ----------+---------+---------+------
POOLED STDEV =     6.707                       30.0      35.0      40.0
```

This interval suggests that the true difference in treatment population mean yield responses between strains 2 and 3 lies somewhere between 1.33 and 14.09 lb/ton. Since this confidence interval lies totally above zero, we may conclude that μ_2 and μ_3 differ significantly at the $\alpha = .05$ level.

As with the individual treatment population means, some statistical software packages will compute confidence intervals for pairwise mean differences. Figure 13–8 shows the Minitab report listing these. The two further confidence intervals extracted from there are reported below in the usual format.

$$-7.725 \leq \mu_1 - \mu_2 \leq 5.953 \qquad \text{(not significant)}$$

$$-0.168 \leq \mu_1 - \mu_3 \leq 13.819 \qquad \text{(not significant)}$$

We see that μ_1 does not differ significantly from either μ_2 or μ_3.

Unfortunately, when several confidence intervals are combined, overall confidence and level of significance must be sacrificed.

Multiple Comparisons and Collective Inferences

The procedures just described have a glaring deficiency: *Each inference applies only individually*. Only a single estimate can be made with $100(1-\alpha)\%$ confidence, and only one test can be performed at the α significance level. Those methods cannot be used in drawing a *family* of inferences involving several μ's and differences in means.

This inadequacy can be alleviated by one of two remedies: (1) Construct *wider* confidence intervals yielding less precise estimates or (2) *reduce* the confidence levels.

Figure 13–8
Confidence intervals for pairwised differences with bacterial leaching data, reported by Minitab.

```
MTB > oneway c11 c12;
SUBC> fisher .05.

Fisher's pairwise comparisons

    Family error rate = 0.119
Individual error rate = 0.0500

Critical value = 2.069

Intervals for (column level mean) - (row level mean)

                 1          2

      2     -7.725
             5.953

      3     -0.168      1.335
            13.819      14.08
```

Table 13–4
Results of Metallurgical Evaluation

	Alloy (treatment)			
Observation i	(1) Carbonized	(2) Titanium	(3) Composite	(4) Secret
1	13,550	13,440	16,430	17,130
2	14,240	13,750	15,880	17,770
3	11,970	12,880	16,930	16,550
4	12,560	14,110	17,330	16,230
5	13,220	12,330	15,350	18,030
	$\bar{X}_{\cdot 1} = 13,108$	$\bar{X}_{\cdot 2} = 13,302$	$\bar{X}_{\cdot 3} = 16,384$	$\bar{X}_{\cdot 4} = 17,142$
		$\bar{\bar{X}}_{\cdot\cdot} = 14,984$		

Figure 13–8 shows for the three bacterial-leaching pairwise differences that the collective significance level (family error rate) is

$$\alpha = .119 \qquad \text{(collective)}$$

This means also that the collective confidence level is

$$100(1-\alpha) = 100(1-.119)\% = 88.1\% \qquad \text{(collective)}$$

The same issue arises when the confidence intervals for the individual treatment population means are combined, so that the collective confidence level must be smaller than what would apply to each estimate singly. Furthermore, together the two families of inferences—one for μs and a second for pairwise differences—would require further reductions in collective confidence (increases in collective significance level).

Another example should be helpful in illustrating collective inferences.

Illustration: Metallurgical Evaluation of Strength of Materials

Consider the results in Table 13–4 obtained for a metallurgical evaluation of the strength of materials. The sampling experiment involves $r = 5$ observations where test rods were stressed until permanent distortion. Rods of the same size were made of four different steel alloys. With each alloy as a separate treatment, $c = 4$. The response variable is the maximum stress level (psi) achieved. The ANOVA table in Table 13–5 shows that

Table 13–5
ANOVA Table for Metallurgical Evaluation

Variation	Degrees of Freedom	Sum of Squares	Mean Square	F
Explained by Treatments: Alloy	$c-1 = 3$	$SSTR = 64,827,320$	$MSTR = 21,609,106.67$	34.70
Error or Unexplained	$n_T - c = (r-1)c = 16$	$SSE = 9,963,760$	$MSE = 622,735.00$	
Total	$n_T - 1 = rc - 1 = 19$	$SSTO = 74,791,080$		

Figure 13–9
Plot of results from metallurgical evaluation.

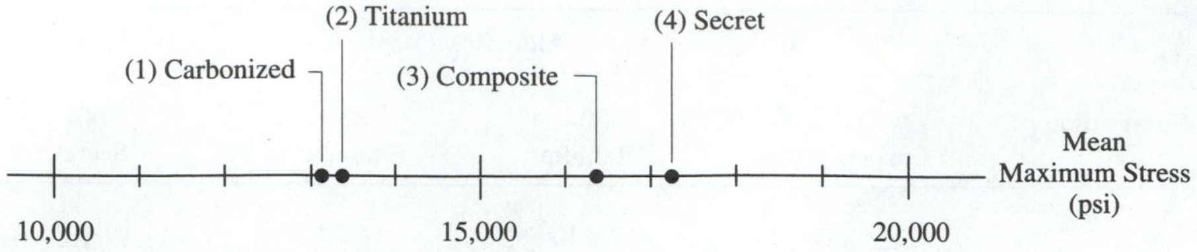

$F = 34.70$. Since the computed value exceeds $F_{.01} = 5.29$, at least one pair of the treatments provides significantly different population means.

Figure 13–9 shows a plot of the sample data, which indicates that alloys (1) and (2) both have substantially lower response than either of alloys (3) and (4). The first pair provides very close sample means, as does the latter.

Figure 13–10 reports the interval estimates for the pairwise differences in treatment population means, each computed at an individual 99% confidence level. The collective confidence level is $100(1-.0446) = 95.54\%$. These estimates may be easier to follow when arranged in the standard format.

Carbonized – titanium:	$-1,652 \le \mu_1 - \mu_2 \le 1,264$	
Carbonized – composite:	$-4,734 \le \mu_1 - \mu_3 \le -1,818$	(significant)
Carbonized – secret:	$-5,492 \le \mu_1 - \mu_4 \le -2,576$	(significant)

Figure 13–10
Confidence intervals for pairwise differences in mean alloy stress, computed by Minitab.

```
MTB > oneway c11 c12;
SUBC> fisher .01.

Fisher's pairwise comparisons

    Family error rate = 0.0446
Individual error rate = 0.0100

Critical value = 2.921

Intervals for (column level mean) - (row level mean)

               1          2          3

    2      -1652
            1264

    3      -4734      -4540
           -1818      -1624

    4      -5492      -5298      -2216
           -2576      -2382        700
```

Titanium – composite: $-4{,}540 \le \mu_2 - \mu_3 \le -1{,}624$ (significant)

Titanium – secret: $-5{,}298 \le \mu_2 - \mu_4 \le -2{,}382$ (significant)

Composite – secret: $-2{,}216 \le \mu_3 - \mu_4 \le 700$

These intervals show significant difference between the composite alloy and either the carbonized or titanium alloys and also between the secret alloy and those two. Neither the titanium vs. carbonized nor the secret vs. composite pairings results in significant differences.

Problems

13–11 The following ANOVA table and sample means were obtained by a builders' trade group. This organization wishes to determine if the mean amount of time taken by local governments to approve housing projects differs regionally. The regions were (1) East, (2) South, (3) Midwest, and (4) West. A sample of 10 projects was collected from each.

Variation	Degrees of Freedom	Sum of Squares	Mean Square	*F*
Treatments	3	5.30	1.77	4.12
Error	36	15.48	.43	
Total	39	20.78		

$\bar{X}_{.1} = 2.3$ yr $\quad \bar{X}_{.2} = 1.8$ yr $\quad \bar{X}_{.3} = 1.9$ yr $\quad \bar{X}_{.4} = 1.6$ yr $\quad \bar{\bar{X}}_{..} = 1.9$ yr

(a) Construct a 95% confidence interval for the mean of treatment population (1).

(b) Construct a 95% confidence interval for the difference $\mu_1 - \mu_2$.

(c) Does your answer to (b) indicate that the mean approval times differ significantly, at the 5% level, between East and South?

13–12 Referring to Problem 13–7:

(a) Construct a 95% confidence interval for the mean percentage reduction in processing time using flip-flop disk drives.

(b) Construct a 95% confidence interval for the difference in mean percentage reduction using the flip-flop disk drive versus the hard-core disk drive.

13–13 Referring to Problem 13–8, determine whether the scattered and concentrated trap location strategies result in significantly different mean percentages of trapped native male gypsy moths. Use $\alpha = .05$.

13–14 The computer printout in Figure 13–11 was obtained, using SPSS, by an industrial engineer studying queue "discouragement." The percentage of arriving customers who left before receiving service was determined for five sample time periods, each under three different queue disciplines.

(a) Construct interval estimates for the pairwise differences for all combinations of treatment population means, using individual error rates of .01.

(b) For which pairs, if any, do the mean percentages of discouraged customers differ significantly?

Figure 13–11
SPSS computer printouts for investigation of queue discouragement.

GROUP	COUNT	MEAN	STANDARD DEVIATION
1 FIFO	5	25.0000	8.7500
2 SIRO	5	35.0000	8.4167
3 PRIORITY	5	39.0000	7.7235
TOTAL	15	33.0000	5.4772

VARIABLE LEAVE PER. PERCENTAGE LEAVING QUEUE
BY VARIABLE DISCIPLINE QUEUE DISCIPLINE

ANALYSIS OF VARIANCE

SOURCE	D.F.	SUM OF SQUARES	MEAN SQUARES	F RATIO
BETWEEN GROUPS	2	520.0000	260.0000	8.667
WITHIN GROUPS	12	360.0000	30.0000	
TOTAL	14	880.0000		

13–15 Refer to Problem 13–5.

(a) Using individual error rates of .01, construct interval estimates for all pairwise differences in treatment population means.

(b) For which pairs, if any, do the mean yields differ significantly?

13–16 Refer to Problem 13–8.

(a) Using individual error rates of .01, construct interval estimates for all the pairwise differences in treatment population means.

(b) For which pairs, if any, do mean trapping percentages significantly differ?

13–4 Designing the Experiment

The analysis-of-variance procedure may be extended to a variety of experimental evaluations. Each may be classified in several major categories, or **experimental designs**. Selecting an appropriate experimental design is crucial when planning multivariate analyses.

Example: *Evaluation Sequences for Etching and Layering Microcircuit Chips*

An industrial engineer is concerned with finding the best sequence for etching and layering microcircuit chips. Although even for a simple chip the number of possible combinations would be astronomical, she has identified four basic sequences—E1, E2, E3, and E4—for etching chips and five—L1, L2, L3, L4, and L5—for layering them. She proposes a series of test batch runs for each combination of both factors: E1-L1, E1-L2, . . . , E2-L1, E2-L2, . . ., and so on.

This experiment involves *two* factors: etching sequence and layering sequence. Altogether, there are $4 \times 5 = 20$ distinct treatments. As her response variable, the engineer proposes using the percentage of acceptable chips per batch.

Although this chapter is concerned largely with a single-factor experimental design, more elaborate structures do exist. These incorporate two or more factors, and there is a separate treatment for each combination in levels of these factors.

Another type of two-factor design involves treatments from only one of the factors. The second factor is a **blocking variable** that serves primarily to reduce the amount of unexplained variation in the response variable.

Example: *Methods for Training Assemblers*

A plant superintendent evaluates three training methods for assemblers: (1) on-the-job only, (2) classroom—then on-line, and (3) on-line—then classroom. A random sample of trainees is selected to undergo each method, or treatment. Because of trainer idiosyncrasies, each trainer indoctrinates one sample group under each treatment. Performance rating scores are obtained as the response variable.

Trainer is the blocking variable. Differences in trainer effectiveness are thereby effectively eliminated as a source of unexplained variation in scores. Trainer is not a treatment per se since it is the *method* of training that is being evaluated.

In Section 13–6 an expanded bacterial leaching experiment will again be evaluated using *site* as the blocking variable. Such an experiment is called a **randomized block design**. The sample ore batches from each site are randomly assigned bacterial treatments, nullifying differences between batches from different ore-bearing sites. Thus, no site systematically influences the measured bacterial efficacy. The assignment of ore batches to bacteria type is done randomly, a procedure called **randomization**.

In all of the illustrations described in this chapter the categories for each variable have been set in advance. This is an ideal arrangement, where the investigator has full control of the study—often referred to as a **fixed-effects experiment**. Often, however, the collection of data is not totally under the investigator's control, and levels for the treatment or blocking variables are really themselves random. Such an investigation is called a **random-effects experiment**. An extreme example arises in evaluating aircraft landing accidents, where crash distance from the airport (the response variable) might serve as the basis for comparing these events in terms of primary cause (the treatment variable). Using aircraft type as the blocking variable, an air safety investigator must take whatever data are available and obviously cannot create the conditions under which an observation is made.

The **interaction** between two variables might also be considered. Procedures for such evaluations are described in Section 13–5. There each combination of treatment and blocking levels are sampled more than once—a process referred to as **replication**. The bacterial leaching experiment described earlier involved replication, with each treatment administered to from seven to ten batches.

When an observation is made of each combination of factor levels, the experiment has a **complete factorial design**. Since it may be impossible to observe all combinations of categories, some investigations leave out certain groupings. These involve **fractional factorial designs**. Such a design might arise in a chemical processing investigation, similar to that of the preceding section, where the apparatus is not strong enough to withstand high temperatures and high pressures simultaneously; thus, certain cells in the rectangular layout might be impossible to observe.

Problems

13–17 Establish design guidelines for a sampling experiment in each of the following situations. For each, suggest (1) appropriate factor(s) and possible levels that define the treatments, (2) a meaningful response variable, and (3) a possible blocking variable.

(a) A field-support manager is concerned with on-site time spent by customer engineers on calls to home-office management.

(b) A personnel director evaluates the placement of ads seeking engineering job applicants.

(c) A research engineer for a tire manufacturer evaluates tires in terms of conditions of use.

(d) A thermodynamics instructor compares how various strategies of assigning homework influence student achievement.

13–18 Give an example of a two-factor experiment where interactions might be an important consideration.

13–19 Refer to Problem 13–5. Suggest a blocking variable that the chemical engineer might use to reduce the level of unexplained variation.

13–20 Refer to Problem 13–10. Suggest a blocking variable that the petroleum engineer might use to reduce the level of unexplained variation.

13–5 Two-Factor Analysis of Variance

The preceding sections cover experiments where there is a *single* treatment or factor. Those procedures extend to experiments involving two factors, so that each combination of both factors defines a separate treatment. For example, a chemical process might be evaluated under different levels for both temperature and pressure. Such an experiment could also provide for *interactions* between the levels of the two factors. For instance, chemical yield may vary not only as the temperature or the pressure change, but also with various combinations of temperature and pressure.

The analysis-of-variance procedures of this section involve two factors, denoted as A and B. Each factor involves several levels, so that the experimental layout takes the form of a matrix as shown in Table 13–6. There are r levels (rows) for factor A and c levels (columns) for factor B. A separate treatment (cell) applies for each combination of factor levels.

Table 13–6 *Experimental Layout for Analysis of Variance with Two Factors*

Levels for Factor A	Treatment Population Means Factor B Levels (1)	(2)	(3)	(4)	(5)	Factor A–Level Population Means
(1)	μ_{11}	μ_{12}	μ_{13}	μ_{14}	μ_{15}	$\mu_{1\cdot}$
(2)	μ_{21}	μ_{22}	μ_{23}	μ_{24}	μ_{25}	$\mu_{2\cdot}$
(3)	μ_{31}	μ_{32}	μ_{33}	μ_{34}	μ_{35}	$\mu_{3\cdot}$
(4)	μ_{41}	μ_{42}	μ_{43}	μ_{44}	μ_{45}	$\mu_{4\cdot}$
Factor B–Level Population Means	$\mu_{\cdot 1}$	$\mu_{\cdot 2}$	$\mu_{\cdot 3}$	$\mu_{\cdot 4}$	$\mu_{\cdot 5}$	$\mu_{\cdot\cdot}$

Problems

The Populations and Means

Under this layout, there are $r \times c$ treatment populations, each having the following mean.

TREATMENT (CELL) POPULATION MEAN

$$\mu_{ij} = \text{Population mean with factor } A \text{ at level } i \text{ and factor } B \text{ at level } j$$

There is a separate population for each factor level as well. The respective means are represented by the margins of Table 13–6, where the listed value is as follows.

FACTOR-LEVEL (ROW) POPULATION MEAN

$$\mu_{i\cdot} = \text{Factor } A \text{ population mean at level } i$$

FACTOR-LEVEL (COLUMN) POPULATION MEAN

$$\mu_{\cdot j} = \text{Factor } B \text{ population mean at level } j$$

Included in the two-factor layout is the **overall population mean** $\mu_{\cdot\cdot}$, which applies regardless of the particular factor level. The factor level (marginal) means $\mu_{i\cdot}$ and $\mu_{\cdot j}$ are equal to the mean of the treatment (cell) means μ_{ij} in their respective rows and columns:

$$\mu_{i\cdot} = \frac{\sum_{j=1}^{c} \mu_{ij}}{c} \qquad \text{and} \qquad \mu_{\cdot j} = \frac{\sum_{i=1}^{r} \mu_{ij}}{r}$$

The overall mean $\mu_{\cdot\cdot}$ is equal to the mean of the factor-level (marginal) means for both the rows and the columns:

$$\mu_{\cdot\cdot} = \frac{\sum_{i=1}^{r} \mu_{i\cdot}}{r} \qquad \text{and} \qquad \mu_{\cdot\cdot} = \frac{\sum_{j=1}^{c} \mu_{\cdot j}}{c}$$

The overall population mean is also equal to the mean of all treatment (cell) means:

$$\mu_{\cdot\cdot} = \frac{\sum_{i=1}^{r}\sum_{j=1}^{c} \mu_{ij}}{rc}$$

The Underlying Model

The additive model for two-factor analysis of variance expresses the cell means as the sum of four components:

$$\mu_{ij} = \mu_{\cdot\cdot} + A_i + B_j + (AB)_{ij}$$

The first component is the overall mean. The following components are the effects parameters. There are three types of parameters.

TWO-FACTOR EFFECT PARAMETERS

A_i = main effect for factor A and level i

B_j = main effect for factor B and level j

$(AB)_{ij}$ = interaction effect when factor A is at level i and factor B is at level j

The effects parameters A_i, B_j, and $(AB)_{ij}$ are related to the population means as follows:

$$A_i = \mu_{i.} - \mu_{..}$$
$$B_j = \mu_{.j} - \mu_{..}$$
$$(AB)_{ij} = \mu_{ij} - \mu_{i.} - \mu_{.j} + \mu_{..}$$

These parameters are in the same units as the response variable. The following relationships apply:

$$\sum_i A_i = \sum_j B_j = \sum_i (AB)_{ij} = \sum_j (AB)_{ij} = 0$$

The effects parameters are helpful in explaining how the factor levels influence the response variable and any interactions between factors. The following examples illustrate the various cases.

Figure 13–12 shows the hypothetical means that might exist for an experiment that has paint drying time as the response variable. Factor A is temperature, with three levels, and factor B is humidity, also with three levels. The plot of the means shows that factor B has *nonzero* effects, since the mean drying times differ for each level of humidity. The horizontal lines indicate that factor A has *zero* effects, reflecting that mean drying times for this paint do not differ with temperature.

Figure 13–12
Paint drying experiment.

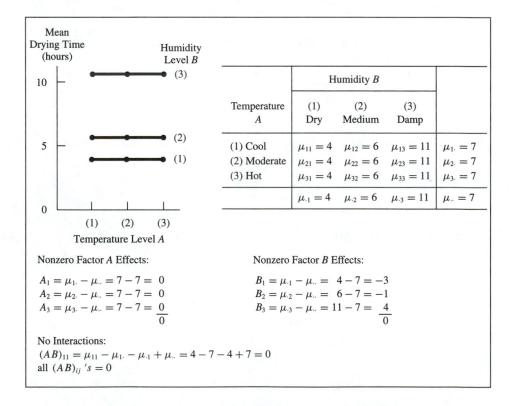

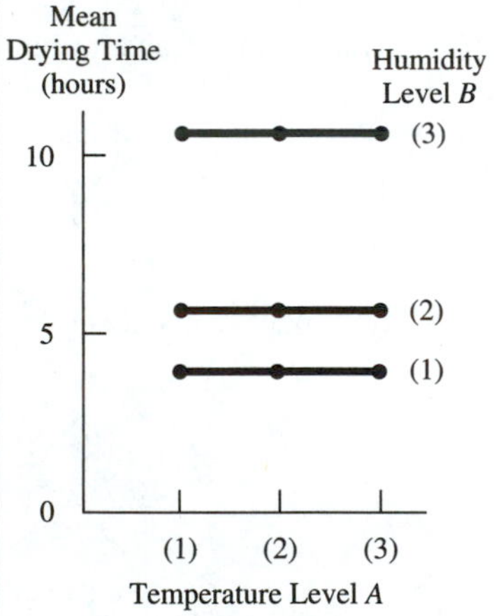

Temperature A	Humidity B (1) Dry	(2) Medium	(3) Damp	
(1) Cool	$\mu_{11} = 4$	$\mu_{12} = 6$	$\mu_{13} = 11$	$\mu_{1.} = 7$
(2) Moderate	$\mu_{21} = 4$	$\mu_{22} = 6$	$\mu_{23} = 11$	$\mu_{2.} = 7$
(3) Hot	$\mu_{31} = 4$	$\mu_{32} = 6$	$\mu_{33} = 11$	$\mu_{3.} = 7$
	$\mu_{.1} = 4$	$\mu_{.2} = 6$	$\mu_{.3} = 11$	$\mu_{..} = 7$

Nonzero Factor A Effects:

$$A_1 = \mu_{1.} - \mu_{..} = 7 - 7 = 0$$
$$A_2 = \mu_{2.} - \mu_{..} = 7 - 7 = 0$$
$$A_3 = \mu_{3.} - \mu_{..} = 7 - 7 = \underline{0}$$
$$0$$

Nonzero Factor B Effects:

$$B_1 = \mu_{.1} - \mu_{..} = 4 - 7 = -3$$
$$B_2 = \mu_{.2} - \mu_{..} = 6 - 7 = -1$$
$$B_3 = \mu_{.3} - \mu_{..} = 11 - 7 = \underline{4}$$
$$0$$

No Interactions:
$(AB)_{11} = \mu_{11} - \mu_{1.} - \mu_{.1} + \mu_{..} = 4 - 7 - 4 + 7 = 0$
all $(AB)_{ij}$'s $= 0$

Figure 13–13 illustrates a similar example involving an industrial engineering evaluation of how assembly rates are affected by line configuration A (two levels) and locomotion B (three levels). The plots show increasing assembly rates for factor A from (1) the series configuration to (2) the parallel arrangement and also for factor B from (1) carts to (2) roller surfaces and then to (3) conveyor belts. Thus, both factors have nonzero effects. The parallel plot lines indicate that the effects of one factor are equal under any level of the other factor. This indicates that there are *no interactions* among the factors.

The presence of interactions is shown in Figure 13–14, where a hypothetical experiment assesses how the performance of a digital computer, in millions of floating-point operations per second (FLOPS), relates to program complexity A (four levels) and microprocessor architecture B. The latter factor has two levels: (1) regular and (2) reduced instruction set computations (RISC) designs. Both factors tend to provide higher mean responses for successive levels. Moreover, the performance gap between levels of factor B, (1) regular and (2) RISC architectures, narrows as the level for factor A (program complexity) increases. Thus, there are underlying *interactions* between the factors. The presence of the interactions is reflected by the nonparallel plots.

Of course, for all these examples, the levels for the population means and the effects parameters will ordinarily be unknown. Inferences regarding these unknown quantities will be made using sample data. We next consider testing hypotheses regarding their levels.

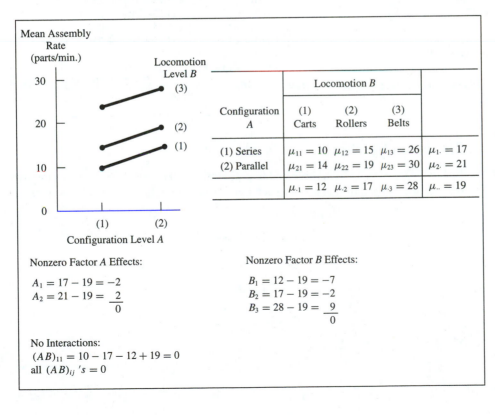

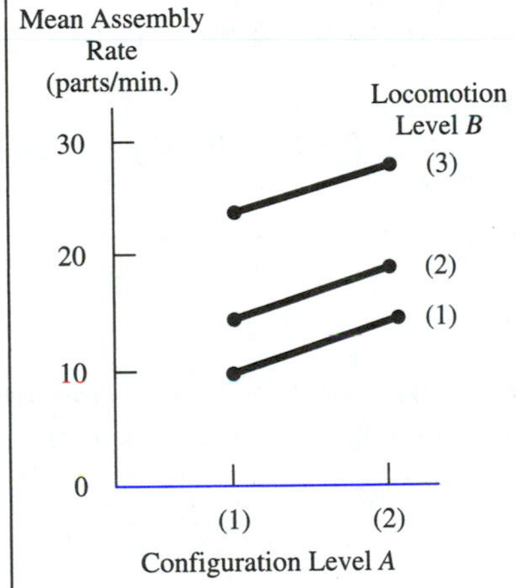

Configuration A	Locomotion B (1) Carts	(2) Rollers	(3) Belts	
(1) Series	$\mu_{11} = 10$	$\mu_{12} = 15$	$\mu_{13} = 26$	$\mu_{1\cdot} = 17$
(2) Parallel	$\mu_{21} = 14$	$\mu_{22} = 19$	$\mu_{23} = 30$	$\mu_{2\cdot} = 21$
	$\mu_{\cdot 1} = 12$	$\mu_{\cdot 2} = 17$	$\mu_{\cdot 3} = 28$	$\mu_{\cdot\cdot} = 19$

Nonzero Factor A Effects:

$$A_1 = 17 - 19 = -2$$
$$A_2 = 21 - 19 = \underline{\ 2\ }$$
$$0$$

Nonzero Factor B Effects:

$$B_1 = 12 - 19 = -7$$
$$B_2 = 17 - 19 = -2$$
$$B_3 = 28 - 19 = \underline{\ 9\ }$$
$$0$$

No Interactions:
$(AB)_{11} = 10 - 17 - 12 + 19 = 0$
all $(AB)_{ij}$'s $= 0$

Figure 13–13 Assembly layout experiment.

Figure 13–14
Computer performance evaluation.

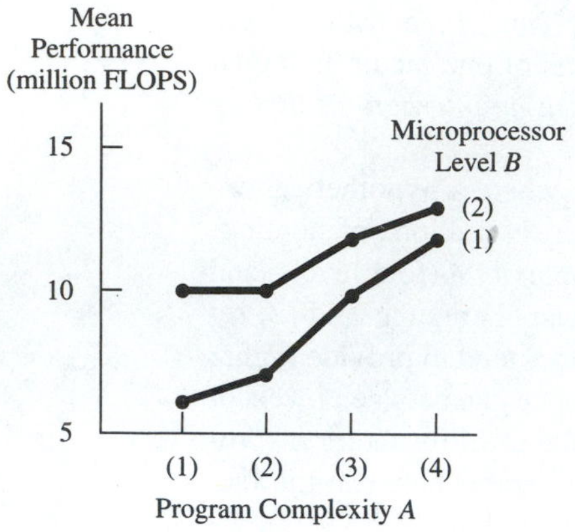

Program Complexity A	Microprocessor Architecture B (1) Regular	(2) RISC	
(1) Low	$\mu_{11} = 6$	$\mu_{12} = 10$	$\mu_{1\cdot} = 8.0$
(2) Moderate	$\mu_{21} = 7$	$\mu_{22} = 10$	$\mu_{2\cdot} = 8.5$
(3) High	$\mu_{31} = 10$	$\mu_{32} = 12$	$\mu_{3\cdot} = 11.0$
(4) Very High	$\mu_{41} = 12$	$\mu_{42} = 13$	$\mu_{4\cdot} = 12.5$
	$\mu_{\cdot 1} = 8.75$	$\mu_{\cdot 2} = 11.25$	$\mu_{\cdot\cdot} = 10.0$

Nonzero Factor A Effects:

$$\begin{aligned} A_1 &= 8.0 - 10.0 = -2.0 \\ A_2 &= 8.5 - 10.0 = -1.5 \\ A_3 &= 11.0 - 10.0 = 1.0 \\ A_4 &= 12.5 - 10.0 = \underline{2.5} \\ & \qquad\qquad\quad 0.0 \end{aligned}$$

Nonzero Factor B Effects:

$$\begin{aligned} B_1 &= 8.75 - 10.0 = -1.25 \\ B_2 &= 11.25 - 10.0 = \underline{1.25} \\ & \qquad\qquad\quad 0.00 \end{aligned}$$

Interactions:

$$\begin{aligned} (AB)_{11} &= 6 - 8.0 - 8.75 + 10 = -.75 \\ (AB)_{12} &= 10 - 8.0 - 11.25 + 10 = .75 \\ (AB)_{21} &= 7 - 8.5 - 8.75 + 10 = -.25 \\ (AB)_{22} &= 10 - 8.5 - 11.25 + 10 = .25 \\ (AB)_{31} &= 10 - 11.0 - 8.75 + 10 = .25 \\ (AB)_{32} &= 12 - 11.0 - 11.25 + 10 = -.25 \\ (AB)_{41} &= 12 - 12.5 - 8.75 + 10 = .75 \\ (AB)_{42} &= 13 - 12.5 - 11.25 + 10 = \underline{-.75} \\ & \qquad\qquad\qquad\qquad\quad 0.00 \end{aligned}$$

The Null Hypotheses

In a two-factor evaluation, *three* null hypotheses may be tested. The first two involve the factor-level population means, each hypothesis being that there is no main effect:

$$H_0 : \mu_{1\cdot} = \mu_{2\cdot} = \cdots = \mu_{r\cdot} \qquad \text{(factor } A)$$

$$[\text{all} A_i\text{'s} = 0]$$

$$H_0 : \mu_{\cdot 1} = \mu_{\cdot 2} = \cdots = \mu_{\cdot c} \qquad \text{(factor } B)$$

$$[\text{all} B_j\text{'s} = 0]$$

The third hypothesis is that there are no interaction effects:

$$H_0 : \mu_{ij} = \mu_{..} + A_i + B_j \quad \text{for all cells } (i, j) \qquad \text{(interactions)}$$

$$[\text{all } (AB)_{ij}\text{'s} = 0]$$

Sample Data in Two-Factor Experiments

The procedure requires that there be exactly n observations per treatment cell and that the same sample size apply to each combination of factor levels. Denoting by X_{ijk} the kth sample observation when factor A is at level i and B is at level j, the additive model provides

$$X_{ijk} = \mu_{ij} + \epsilon_{ijk}$$

or

$$X_{ijk} = \mu_{..} + A_i + B_j + (AB)_{ij} + \epsilon_{ijk}$$

The last quantity ϵ_{ijk} expresses the residual value or **error term**. The analytical framework presented here assumes that the ϵ's are independent and normally distributed, with mean zero and common standard deviation σ.

The interaction effects cannot be assessed unless $n > 1$. In that case sampling is done with **replication**.

Illustration: *Effectiveness of Timing and Positioning of Engineering Employment Newspaper Ads*

The employment director for a large aerospace company wishes to assess the effectiveness of newspaper display advertising, with day of week as factor A ($r = 7$ levels—one for each day), and section as factor B ($c = 3$ levels—news, sports, and business). He conducted a sampling experiment wherein $n = 3$ ads were run under each day-section combination. As his response variable the director chose the number of resumes received. The sample data from this experiment are shown in Table 13–7.

The sample results in Table 13–7 are arranged in the same basic layout introduced earlier for the population means. The counterpart *sample* means are computed as follows.

CELL SAMPLE MEAN FOR TREATMENT COMBINATION (i, j)

$$\bar{X}_{ij\cdot} = \frac{\sum_{k=1}^{n} X_{ijk}}{n}$$

ROW SAMPLE MEAN FOR FACTOR A AT LEVEL i

$$\bar{X}_{i\cdot\cdot} = \frac{\sum_{j=1}^{c} \bar{X}_{ij\cdot}}{c}$$

COLUMN SAMPLE MEAN FOR FACTOR B AT LEVEL j

$$\bar{X}_{\cdot j\cdot} = \frac{\sum_{i=1}^{r} \bar{X}_{ij\cdot}}{r}$$

Table 13–7
Sample Data for Experiment on Timing and Positioning of Engineering Employment Newspaper Ads

Day of Week A	Newspaper Section B (1) News	(2) Sports	(3) Business	Factor A Sample Mean
(1) Sunday	$X_{111}(17)$ $X_{112}(18)$ $X_{113}(20)$	$X_{121}(21)$ $X_{122}(18)$ $X_{123}(15)$	$X_{131}(6)$ $X_{132}(6)$ $X_{133}(8)$	
Cell Mean	$\overline{X}_{11\cdot} = 18.33$	$\overline{X}_{12\cdot} = 18.00$	$\overline{X}_{13\cdot} = 6.67$	$\overline{X}_{1\cdot\cdot} = 14.33$
(2) Monday	$X_{211}(10)$ $X_{212}(9)$ $X_{213}(7)$	$X_{221}(4)$ $X_{222}(2)$ $X_{223}(6)$	$X_{231}(9)$ $X_{232}(11)$ $X_{233}(13)$	
Cell Mean	$\overline{X}_{21\cdot} = 8.67$	$\overline{X}_{22\cdot} = 4.00$	$\overline{X}_{23\cdot} = 11.00$	$\overline{X}_{2\cdot\cdot} = 7.89$
(3) Tuesday	$X_{311}(9)$ $X_{312}(9)$ $X_{313}(10)$	$X_{321}(5)$ $X_{322}(4)$ $X_{323}(6)$	$X_{331}(7)$ $X_{332}(8)$ $X_{333}(10)$	
Cell Mean	$\overline{X}_{31\cdot} = 9.33$	$\overline{X}_{32\cdot} = 5.00$	$\overline{X}_{33\cdot} = 8.33$	$\overline{X}_{3\cdot\cdot} = 7.56$
(4) Wednesday	$X_{411}(8)$ $X_{412}(7)$ $X_{413}(7)$	$X_{421}(5)$ $X_{422}(9)$ $X_{423}(8)$	$X_{431}(7)$ $X_{432}(8)$ $X_{433}(8)$	
Cell Mean	$\overline{X}_{41\cdot} = 7.33$	$\overline{X}_{42\cdot} = 7.33$	$\overline{X}_{43\cdot} = 7.67$	$\overline{X}_{4\cdot\cdot} = 7.44$
(5) Thursday	$X_{511}(4)$ $X_{512}(4)$ $X_{513}(3)$	$X_{521}(7)$ $X_{522}(5)$ $X_{523}(5)$	$X_{531}(9)$ $X_{532}(7)$ $X_{533}(5)$	
Cell Mean	$\overline{X}_{51\cdot} = 3.67$	$\overline{X}_{52\cdot} = 5.67$	$\overline{X}_{53\cdot} = 7.00$	$\overline{X}_{5\cdot\cdot} = 5.44$
(6) Friday	$X_{611}(12)$ $X_{612}(12)$ $X_{613}(11)$	$X_{621}(11)$ $X_{622}(11)$ $X_{623}(12)$	$X_{631}(10)$ $X_{632}(9)$ $X_{633}(8)$	
Cell Mean	$\overline{X}_{61\cdot} = 11.67$	$\overline{X}_{62\cdot} = 11.33$	$\overline{X}_{63\cdot} = 9.00$	$\overline{X}_{6\cdot\cdot} = 10.67$
(7) Saturday	$X_{711}(12)$ $X_{712}(17)$ $X_{713}(17)$	$X_{721}(14)$ $X_{722}(13)$ $X_{723}(14)$	$X_{731}(13)$ $X_{732}(11)$ $X_{733}(9)$	
Cell Mean	$\overline{X}_{71\cdot} = 15.33$	$\overline{X}_{72\cdot} = 13.67$	$\overline{X}_{73\cdot} = 11.00$	$\overline{X}_{7\cdot\cdot} = 13.33$
Factor B Sample Mean	$\overline{X}_{\cdot1\cdot} = 10.62$	$\overline{X}_{\cdot2\cdot} = 9.29$	$\overline{X}_{\cdot3\cdot} = 8.67$	$\overline{\overline{X}}_{\cdot\cdot\cdot} = 9.52$

GRAND OVERALL SAMPLE MEAN

$$\bar{\bar{X}}_{...} = \frac{\sum_{i=1}^{r}\sum_{j=1}^{c}\bar{X}_{ij.}}{rc} = \frac{\sum_{i=1}^{r}\bar{X}_{i..}}{r} = \frac{\sum_{j=1}^{c}\bar{X}_{.j.}}{c}$$

In accordance with the underlying model, these sample means ($\bar{X}$s) are unbiased least-squares estimators of the respective unknown treatment population means (μs).

Analytical Framework

As with the single-factor procedures, the analytical framework may be explained in terms of the following three types of deviations:

$$\text{Total deviation} = \text{Treatments deviation} + \text{Error deviation}$$

$$(X_{ijk} - \bar{\bar{X}}_{...}) = (\bar{X}_{ij.} - \bar{\bar{X}}_{...}) + (X_{ijk} - \bar{X}_{ij.})$$

This formula *partitions* the total deviation about the grand mean into the treatments and error deviation components. The two-factor analysis further partitions the treatments deviation as follows:

$$\begin{matrix}\text{Treatments} \\ \text{deviation}\end{matrix} = \begin{matrix}A\text{: Main effect} \\ \text{deviation}\end{matrix} + \begin{matrix}B\text{: Main effect} \\ \text{deviation}\end{matrix} + \begin{matrix}AB\text{: Interaction effect} \\ \text{deviation}\end{matrix}$$

$$(\bar{X}_{ij.} - \bar{\bar{X}}_{...}) = (\bar{X}_{i..} - \bar{\bar{X}}_{...}) + (\bar{X}_{.j.} - \bar{\bar{X}}_{...}) + (\bar{X}_{ij.} - \bar{X}_{i..} - \bar{X}_{\cdot j \cdot} + \bar{\bar{X}}_{...})$$

As with the single-factor experiments, the collective sums of individual squared deviations are used to express the amount of variation by groups. The first expression is generalized in terms of sums of squares as

$$\text{Total variation} = \text{Treatments variation} + \text{Error variation}$$

$$SSTO = SSTR + SSE$$

The treatments sum of squares is partitioned as follows:

$$\begin{matrix}\text{Treatments} \\ \text{variation}\end{matrix} = \begin{matrix}A\text{: Main effect} \\ \text{variation}\end{matrix} + \begin{matrix}B\text{: Main effect} \\ \text{variation}\end{matrix} + \begin{matrix}AB\text{: Interaction effect} \\ \text{variation}\end{matrix}$$

$$SSTR = SSA + SSB + SSAB$$

The respective sums of squares form the basis for evaluation. This is done through the ANOVA table.

The Two-Factor ANOVA Table

The general organization of the two-factor ANOVA table is shown in Table 13–8. The expressions for computing the respective sums of squares are also shown. Recall that the mean squares are obtained by dividing these by the respective degrees of freedom. The latter are obtained from the dimensions—reduced by 1—of the experimental layout matrix.

Notice that there are three test statistics, or values, for F—one for testing each of the above null hypotheses.

Values for the ANOVA table are almost always obtained with computer assistance. Figure 13–15 shows the computer results for the advertising experiment.

Table 13–8
Expressions for Elements in Two-Factor Analysis and General Form of the ANOVA Table

Variation	Degrees of Freedom	Sum of Squares	Mean Square	F
Explained by Factor A (between rows)	$r-1$	$SSA = nc\sum_{i=1}^{r}(\overline{X}_{i..}-\overline{\overline{X}}_{...})^2$	$MSA = \dfrac{SSA}{r-1}$	$\dfrac{MSA}{MSE}$
Explained by Factor B (between columns)	$c-1$	$SSB = nr\sum_{j=1}^{c}(\overline{X}_{.j.}-\overline{\overline{X}}_{...})^2$	$MSB = \dfrac{SSB}{c-1}$	$\dfrac{MSB}{MSE}$
Explained by Interactions (between cells)	$(r-1)(c-1)$	$SSAB = n\sum_{i=1}^{r}\sum_{j=1}^{c}(\overline{X}_{ij.}-\overline{X}_{i..}-\overline{X}_{.j.}+\overline{\overline{X}}_{...})^2$	$MSAB = \dfrac{SSAB}{(r-1)(c-1)}$	$\dfrac{MSAB}{MSE}$
Error or Unexplained (residual)	$rc(n-1)$	$SSE = \sum_{i=1}^{r}\sum_{j=1}^{c}\sum_{k=1}^{n}(X_{ijk}-\overline{X}_{ij.})^2$	$MSE = \dfrac{SSE}{rc(n-1)}$	
Total	$nrc-1$	$SSTO = \sum_{i=1}^{r}\sum_{j=1}^{c}\sum_{k=1}^{n}(X_{ijk}-\overline{\overline{X}}_{...})^2$		

Testing the null hypothesis that the mean numbers of resumes received are the same for all levels of factor A (day of week), the computed value of the test statistic is

$$F = \frac{MSA}{MSE} = 40.52$$

Reading Appendix Table J at 6 degrees of freedom for the numerator and 42 degrees of freedom for the denominator, and using $\alpha = .01$, we find that the critical value is $F_{.01} = 3.26$. Since the computed value exceeds this amount, the null hypothesis of no main factor A effects from day-of-week advertising placement must be *rejected*.

Figure 13–15 *ANOVA table for advertising experiment, computed with Minitab.*

```
MTB > Twoway c3 c1 c2;
SUBC>   Means c1 c2.

ANALYSIS OF VARIANCE  Resumes

Source         DF        SS        MS       F      P
Day             6   598.159    99.693   40.52  0.000
Section         2    41.810    20.905    8.50  0.001
Day*Section    12   388.413    32.368   13.16  0.000
Error          42   103.333     2.460
Total          62  1131.714
```

Also at the $\alpha = .01$ significance level, the computed value of $F = 8.50$ leads to *rejection* of the null hypothesis of identical mean resumes received for all levels of factor B, newspaper section. (The critical value using 2 and 42 degrees of freedom is $F_{.01} = 5.15$.) We must conclude that at least one of the factor B main effects is nonzero.

We may also test the null hypothesis that there are *no interactions*. Again using $\alpha = .01$, the computed value of $F = 13.16$, we see that the null hypothesis must also be *rejected.* (The critical value using 12 and 42 degrees of freedom is $F_{.01} = 2.64$.) We must conclude that a nonzero interaction effect exists for at least one treatment combination (cell).

Deciding What To Do

The preceding data indicate that the mean number of resumes received differs both by day of week and by newspaper section. Furthermore, the presence of interactions means that on some days, ads in a particular section might provide a higher mean response than on other days.

Multiple Comparison Procedures Depending on which of the null hypotheses is rejected the next phase of the analysis involves comparing specific factor-level means, or individual treatment (cell) means. Depending on whether or not the sample data indicate the presence of interactions, one of the following approaches will be taken:

1. *Significant interactions are found.* Search the sample data for significant differences among treatment population (cell) means. The sample results may suggest further interesting comparisons.
2. *No interactions are significant.* Compare all factor-level means in each set for which the F test results in rejection of the null hypothesis of identical means. Isolate those pairs of means that differ significantly.

The placement director in the present illustration would take approach 1.

Although it is possible to construct confidence-interval estimates for the pairwise differences between cell means, there is spotty software availability for this option, and the prospect of hand computations places these beyond the scope of this text. The estimated standard error for those estimates is

$$\hat{s}_D = \sqrt{\frac{2MSE}{n}} = \sqrt{\frac{2(2.460)}{3}} = 1.28$$

Any difference between cell means less than about three standard errors should be treated as insignificant.

It should be helpful to examine a graph such as the one in Figure 13–16, where the mean responses for each treatment cell are plotted, using a separate axis for factor A (day of week) and different line weights for factor B (section). There we see that the greatest cell mean differences occur between news or sports and business on Sundays. A similar finding, less pronounced, applies for Saturday, while on Monday or Thursday the business ads provide greater mean resume returns than sports.

We might conclude that weekend news or sports would be the most effective ad placement, with news or sports much better on Sundays. The best weekday seems to be

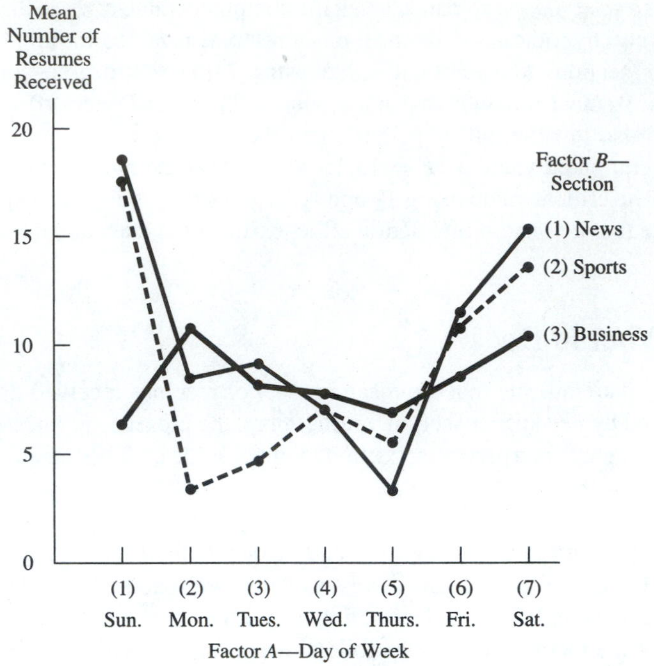

Figure 13–16 *Sample treatment (cell) means for engineering-placement advertisement evaluation.*

Friday, when similar sectional ranking applies—although business gives a higher response on Mondays than any other non-Friday placement.

The following example illustrates an experiment in which no significant interactions are found.

Table 13–9
Sample Results for Chemical Processing Parameter Evaluation

Factor *A* Temperature	Factor *B* Pressure (1) 150 psi	(2) 200 psi	(3) 250 psi	(4) 300 psi	Factor *A* Mean
(1) 300° F	28.5	30.7	30.5	34.6	
	31.4	33.4	29.7	32.3	$\bar{X}_{1..} = 31.39$
(2) 325° F	27.6	31.5	31.1	31.4	
	28.9	29.8	30.9	30.7	$\bar{X}_{2..} = 30.24$
(3) 350° F	27.9	28.3	30.4	30.3	
	28.2	28.0	30.0	29.6	$\bar{X}_{3..} = 29.09$
(4) 375° F	27.5	27.3	29.0	29.3	
	26.6	29.1	29.2	31.0	$\bar{X}_{4..} = 28.62$
(5) 400° F	26.4	26.8	27.7	28.1	
	27.2	26.0	26.3	27.8	$\bar{X}_{5..} = 27.04$
Factor *B* Mean	$\bar{X}_{.1.} = 28.02$	$\bar{X}_{.2.} = 29.09$	$\bar{X}_{.3.} = 29.48$	$\bar{X}_{.4.} = 30.51$	$\bar{\bar{X}}_{...} = 29.28$

Example: *Chemical Processing-Parameter Evaluation*

The data in Table 13–9 represent the sample results of an investigation to establish operating parameters for processing a chemical. Two factors are considered. Factor A is temperature setting (degrees Fahrenheit, $r = 5$ levels) and factor B is pressure (psi, $c = 4$ levels). The response variable is yield (grams per liter of raw material). Three null hypotheses may be tested:

1. H_0: The mean yield (response) in grams per liter of raw material is the same under all temperature settings.
2. H_0: The mean yield (response) in grams per liter of raw material is the same under all pressure settings.
3. H_0: There are no interactions between temperature and pressure.

Figure 13–17
ANOVA table for chemical processing parameter evaluation, using Minitab.

Analysis of Variance for Yield

Source	DF	SS	MS	F	P
Temp.	4	86.8250	21.7063	22.92	0.000
Pressure	3	31.7650	10.5883	11.18	0.000
Temp.*Pressure	12	16.6850	1.3904	1.47	0.216
Error	20	18.9400	0.9470		
Total	39	154.2150			

Figure 13–17 shows the ANOVA table for this experiment. A value of $F = 1.47$ was found for the interactions, for which the prob value is an insignificant level of .216. The null hypothesis of no interactions must be *accepted.*

Since there are significant differences amongst the temperature treatment means, pairwise comparisons may be made of those. Figure 13–18 provides confidence interval limits for all pairings, several of which differ significantly. A similar conclusion applies to the pressure treatment means, whose pairwise confidence-interval limits are also listed there.

Figure 13–18
Confidence intervals for pairwise differences in mean yields for chemical processing parameter evaluation, using Minitab.

TEMPERATURE

Family error rate = 0.0705
Individual error rate = 0.0100

	300	325	350	375
325	-0.740 3.040			
350	0.410 4.190	-0.740 3.040		
375	0.873 4.652	-0.277 3.502	-1.427 2.352	
400	2.460 6.240	1.310 5.090	0.160 3.940	-0.302 3.477

PRESSURE

Family error rate = 0.0471
Individual error rate = 0.0100

	150	200	250
200	-3.313 1.173		
250	-3.703 0.783	-2.633 1.853	
300	-4.733 -0.247	-3.663 0.823	-3.273 1.213

Problems

13–21 Two researchers evaluating an algorithm for solving a dynamic lot-sizing problem formulated the original problem with a mathematical model allowing up to two sources and two price breaks. Their solution procedure is based on the branch-and-bound method. Table 13–10 shows the results obtained from several computer runs, each made using two similarly configured problems. The response variable used to assess the algorithm's efficiency is the average amount of CPU time.

(a) Construct the ANOVA table.

(b) At the $\alpha = .05$ level of significance, indicate whether the following null hypotheses should be accepted or rejected:

(1) H_0: There are no interactions between problem duration and complexity.

(2) H_0: Means for all problem durations are identical.

(3) H_0: Means for all levels of complexity are identical.

Table 13–10 *Sample Data (average CPU time in seconds) for Computer Evaluation of Algorithm Tested on Problems with Various Durations and Levels of Complexity*

	Complexity *B*			
Duration *A*	**(1) Simple**	**(2) Low**	**(3) Medium**	**(4) High**
(1) Short	.09	.09	.20	.30
	.13	.16	.39	.30
(2) Middle	.11	.15	.39	2.48
	.20	.32	2.46	7.12
(3) Long	.31	.31	.64	.75
	.49	.33	1.40	1.99

Source: Erengue, S. Selcuk, and Suleyman Tufekci. "A Branch-and-Bound Algorithm for a Single-Item Multi-Source Dynamic Lot-Sizing Problem with Capacity Constraints," *IIEE Transactions,* March 1987. pp. 73–80.

13–22 A project engineer for a large construction firm conducted a sampling study to determine if the mean job completion delay of electrical jobs differs between subcontractors and also by type of job. The following sample data (days late) were obtained for jobs completed by four major subcontractors.

(a) Construct the ANOVA table.

(b) At the $\alpha = .05$ level of significance, indicate whether the following null hypotheses should be accepted or rejected:

(1) H_0: There are no interactions between job type and subcontractor.

(2) H_0: Means for all job types are identical.

(3) H_0: Means for all subcontractors are identical.

	Subcontractor			
Job Type	**(1) Big M**	**(2) H-V Eng.**	**(3) CC&B**	**(4) Jones**
(1) Remodeling	8.5	−3.0	7.5	13.0
	11.0	−7.5	9.0	11.5
	10.5	−4.5	7.5	20.5

(2) New Building	22.6	−2.0	34.0	45.0
	27.0	−8.5	32.0	39.5
	11.0	−4.5	24.0	35.5
(3) Groundwork	3.0	2.0	4.0	5.5
	−2.0	−3.0	1.5	2.5
	−1.0	−2.0	.5	4.0

13–23 An engineering admissions committee wishes to establish better guidelines for high schools regarding preparation for entering freshmen. A random sample of high school seniors has been selected, and their scores on a freshman placement test have been determined. The following average test scores apply to students grouped by science courses taken and highest mathematics level.

	Science Preparation			
Mathematics Placement	**(1) Chemistry Only**	**(2) Physics Only**	**(3) Chemistry and Physics**	**(4) Advanced Placement**
(1) Trigonometry	57	60	73	74
	63	68	79	85
	66	67	65	67
	54	61	67	62
(2) Analysis	59	70	82	75
	64	62	77	79
	63	66	79	70
	62	62	66	74
(3) Calculus	67	69	83	76
	70	72	79	81
	72	75	77	88
	65	60	73	75

(a) Construct the ANOVA table.
(b) At the $\alpha = .05$ level of significance, indicate whether the following null hypotheses should be accepted or rejected:
  (1) H_0: There are no interactions between mathematics and science preparations.
  (2) H_0: Means for all mathematics preparations are identical.
  (3) H_0: Means for all science preparations are identical.

13–24 The project engineer in Problem 13–22 also tested the two factors with average percentage cost overrun as the response variable. Only one observation per treatment combination was available. The following data apply:

	Subcontractor			
Job Type	**(1) Big M**	**(2) H-V Eng.**	**(3) CC&B**	**(4) Jones**
(1) Remodeling	15%	2%	13%	8%
(2) New Building	13	5	22	15
(3) Groundwork	5	−2	4	2

(a) Construct the ANOVA table.

(b) At the $\alpha = .05$ level of significance, indicate whether the following null hypotheses should be accepted or rejected:

(1) H_0: Means for all job types are identical.

(2) H_0: Means for all subcontractors are identical.

13–25 An engineering society wishes to assess the effects of member's specialty (A, with six levels) and employment sector (B, with four levels) on professional income. The following data represent observations of one randomly chosen person from each category combination.

$$SSA = 640{,}000 \qquad SSB = 710{,}000 \qquad SSTO = 1{,}500{,}000$$

(a) Construct a single-factor ANOVA table using specialty as the only treatment. At the $\alpha = .01$ significance level, can you conclude that the treatment means differ?

(b) Construct a single-factor ANOVA table using employment sector as the only treatment. At the $\alpha = .01$ significance level, can you conclude that the treatment means differ?

(c) Construct a two-factor ANOVA table ($n = 1$) using both specialty and employment sector as treatments. What can you conclude regarding the respective null hypotheses of identical mean incomes for specialities and for employment sectors at the $\alpha = .01$ level?

(d) Do any of your conclusions change in the two-factor evaluation from what they are in the respective single-factor evaluation?

13–26 The researchers in Problem 13–21 obtained a second response variable, average length of candidate list, in evaluating their algorithm for solving a dynamic lot-sizing problem. Figure 13–19 shows a partial SAS printout of the results. What conclusion should you reach?

13–27 Two researchers experimented with parameter settings for their "simulated annealing" algorithm for solving quadratic assignment problems. A portion of their results are shown in Table 13–11, where the response variable is the percentage by which their algorithm's solution falls above the best-known solution for problems having the same parameters.

From these sample data, test at the 1% significance level null hypotheses of identical responses for all levels of each of two parameters—interchange complexity (factor A) and epoch interval (factor B)—plus the null hypothesis of no interactions from those levels. What conclusions do you reach?

Analysis of Variance Procedure

Dependent Variable: TIME

Source	DF	Sum of Squares	Mean Square	F Value
Model	11	14.32500000	1.30200000	22.33
Error	12	0.70000000	0.05833333	
Corrected Total	23	15.02500000		

R-Square	C.V.	Root MSE	TIME Mean
0.953411	778.39377	0.241523	1.88000000

Source	DF	Anova SS	Mean Square	F Value
DURATION	2	2.19000000	1.09500000	32.53
COMPLEX	3	7.29800000	2.43266667	18.77
DURATION*COMPLEX	6	4.38700000	0.73116667	13.82

Level of DURATION	N	-------------TIME------------- Mean
LONG	8	1.70000000
MIDDLE	8	2.30000000
SHORT	8	1.63000000

Level of COMPLEX	N	-------------TIME------------- Mean
HIGH	6	2.70000000
LOW	6	1.50000000
MEDIUM	6	2.03000000
SIMPLE	6	1.27000000

Level of DURATION	Level of COMPLEX	N	-------------TIME------------- Mean
LONG	HIGH	2	2.30000000
LONG	LOW	2	1.10000000
LONG	MEDIUM	2	2.10000000
LONG	SIMPLE	2	1.30000000
MIDDLE	HIGH	2	4.10000000
MIDDLE	LOW	2	1.80000000
MIDDLE	MEDIUM	2	2.20000000
MIDDLE	SIMPLE	2	1.10000000
SHORT	HIGH	2	1.70000000
SHORT	LOW	2	1.60000000
SHORT	MEDIUM	2	1.80000000
SHORT	SIMPLE	2	1.40000000

Figure 13–19 *SAS printout of analysis of variance for computer algorithm evaluation using list length as the response variable.*

13–28 Refer to Problem 13–23 and your answers to that exercise. Assuming that no significant interactions are found, find 95% individual interval estimates for each pairwise difference in factor-level means. What may you conclude?

13–29 Refer to Problem 13–24 and your answers to that exercise. Since $n = 1$, testing for interactions is impossible. Using individual confidence levels of 95%, find interval estimates for each pairwise difference in factor-level means. What may you conclude?

13–30 Refer to the chemical processing example on page 479. Assuming that no significant interactions are found, find 95% individual interval estimates for each pairwise difference in factor-level means. What may you conclude?

13–31 Refer to Problem 13–22 and your answers to that exercise.

(a) Using job type as the horizontal axis and mean days late as the vertical axis, plot on a common graph the sample cell means for each contractor.

(b) For which cell pairs do the means significantly differ?

(c) Give an overall summary statement.

Table 13–11 *Sample Data (percentage above best cost) for Simulated Annealing Solution of Quadratic Assignment Problems (with 30 departments and .25 error constant)*

Interchange Complexity A	Epoch Interval B			
	(1) $e = 5$	(2) $e = 15$	(3) $e = 25$	(4) $e = 50$
(1) $N' = 10$	4.05	4.35	4.59	6.67
	3.97	4.55	4.78	6.44
	4.04	4.46	4.92	6.40
	4.02	4.49	5.02	6.37
	3.92	4.55	4.99	6.52
(2) $N' = 50$	2.81	2.01	1.82	2.33
	2.74	2.13	1.87	2.17
	2.74	2.19	1.99	2.18
	2.68	2.16	1.96	2.08
	2.78	2.26	2.11	2.24
(3) $N' = 100$	2.62	1.91	1.55	1.99
	2.65	1.87	1.59	2.18
	2.74	1.78	1.70	1.95
	2.70	1.75	1.64	1.73
	2.79	1.84	1.57	2.15

Source: Wilheim, Mickey R., and Thomas L. Ward,"Solving Quadratic Assignment Problems by 'Simulated Annealing,'" *IEE Transactions,* March 1987, pp. 107–118.

Note: Only the means are actual values. The cell entries are for purposes of illustration only.

13–32 Refer to Problem 13–26 and your answers to that exercise.
(a) Plot the cell mean response against levels for factor A, with one set of data points for each level of factor B.
(b) For which cell pairs do the means differ significantly?
(c) Give an overall summary statement.

13–33 Refer to Problem 13–27 and your answers to that exercise. Repeat (a)–(c) as in the preceding problem.

13–6 The Randomized Block Design

An important class of two-factor studies uses a randomized block design. Here the second factor is used primarily to reduce the amount of unexplained variation in response, thereby increasing the efficacy of the analysis of variance over what would be achieved with a single factor (treatment variable).

Illustration: *Bacterial Leaching Investigation*

To introduce the procedure, we extend the original bacterial leaching investigation, so that *deposit site* of the ore is the second factor in explaining copper-yield variation. The mining engineer used ore taken from ten different major sites within the deposit complex. These were selected for their distinctive material composition, and the ores pro-

Table 13–12
Layout for Bacterial Leaching Investigation and Expanded Sample Results under the Randomized Block Design

Block i	***Thiobaccilus* Treatment j**			**Mean Yield (lb/ton)**
	Strain 1	**Strain 2**	**Strain 3**	
Site 1	31	33	25	$\bar{X}_{1\cdot} = 29.67$
Site 2	35	41	27	$\bar{X}_{2\cdot} = 34.33$
Site 3	33	37	24	$\bar{X}_{3\cdot} = 31.33$
Site 4	29	32	26	$\bar{X}_{4\cdot} = 29.00$
Site 5	35	34	29	$\bar{X}_{5\cdot} = 32.67$
Site 6	41	43	37	$\bar{X}_{6\cdot} = 40.33$
Site 7	53	45	39	$\bar{X}_{7\cdot} = 45.67$
Site 8	41 new	43	33	$\bar{X}_{8\cdot} = 39.00$
Site 9	40 new	41	36	$\bar{X}_{9\cdot} = 39.00$
Site 10	31 new	27	22 new	$\bar{X}_{10\cdot} = 26.67$
Mean Yield (lb/ton)	$\bar{X}_{\cdot 1} = 36.90$	$\bar{X}_{\cdot 2} = 37.60$	$\bar{X}_{\cdot 3} = 29.80$	$\bar{\bar{X}}_{\cdot\cdot} = 34.77$

vided represent the wide range of types found at the mine. Each of the sites is referred to as a **block**. A random sample was then taken from each site or block. Table 13–12 shows the original bacterial leaching data arranged by block and treatment.

The earlier bacterial leaching data involved an unequal number of observations in each treatment column since not all blocks (sites) were represented in the earlier experiment. To meet the design requirements, new sample observations have been added to the original data under strains 1 and 3.*

A sampling procedure involving this type of layout is called a **randomized block design**. Although the terminology evokes neighboring rectangles, much like city blocks, there is no reason why the deposit sites must be contiguous or even the same size and shape. A block can be any homogeneous grouping of sample units. For example, such a design might be used to evaluate different languages (FORTRAN, BASIC, Pascal) in computer instruction, where test students might be blocked by available computer mode (batch, time-share, minicomputer), a contributing factor to programming proficiency in any given language. Although of secondary concern, a block might influence response by a similar order of magnitude as that of a particular treatment, and this factor is sometimes referred to as a **blocking variable**.

*Of course, neither the single-factor nor the randomized blocking procedures should ordinarily be used on the same sample data. The same data set is used here only for purposes of illustration.

Analytical Framework and Theoretical Model

The underlying model for a two-factor randomized block design treats each observation as the following sum:

$$X_{ij} = \mu_{..} + B_i + T_j + \epsilon_{ij}$$

where $\mu_{..}$ denotes the overall population mean, B_i is the blocking effect for block i, T_j is the treatment effect, and ϵ_{ij} denotes the error term. The ϵ's are assumed to be independent, normally distributed random variables with mean zero and standard deviation σ. There are r blocks (levels for the blocking variable), one for each row. In addition, there is a separate column for each treatment (level of the treatments variable), and there are c of these. There must be exactly one observation under each treatment and for each block, and every treatment is represented exactly once by each block. Thus, $n = 1$ observation per cell.

The two-factor notation is simplified, with X_{ij} (with no overbar) used in place of $\overline{X}_{ij\cdot}$, and with no third dot needed in the subscripts in symbols representing the various sample means. The analytical framework may be explained in terms of the following partitioning of the deviations:

$$\begin{array}{ccccccc}\text{Total deviation} & = & \text{Treatments deviation} & + & \text{Blocks deviation} & + & \text{Error deviation} \\ (X_{ij} - \overline{\overline{X}}_{..}) & = & (\overline{X}_{.j} - \overline{\overline{X}}_{..}) & + & (\overline{X}_{i.} - \overline{\overline{X}}_{..}) & + & (X_{ij} - \overline{X}_{i.} - \overline{X}_{.j} + \overline{\overline{X}}_{..})\end{array}$$

Table 13–13

Randomized Block Design: Expressions and General Form of the ANOVA Table

Variation	Degrees of Freedom	Sum of Squares	Mean Square	F
Explained by Treatments (between columns)	$c - 1$	$SSTR = r\sum_{j=1}^{c}(\overline{X}_{.j} - \overline{\overline{X}}_{..})^2$	$MSTR = \dfrac{SSTR}{c-1}$	$\dfrac{MSTR}{MSE}$
Explained by Blocks (between rows)	$r - 1$	$SSBL = c\sum_{i=1}^{r}(\overline{X}_{i.} - \overline{\overline{X}}_{..})^2$	$MSBL = \dfrac{SSBL}{r-1}$	*
Error or Unexplained (residual)	$(r-1)(c-1)$	$SSE = \sum_{i=1}^{r}\sum_{j=1}^{c}(X_{ij} - \overline{X}_{i.} - \overline{X}_{.j} + \overline{\overline{X}}_{..})^2$	$MSE = \dfrac{SSE}{(r-1)(c-1)}$	
Total	$rc - 1$	$SSTO = \sum_{i=1}^{r}\sum_{j=1}^{c}(X_{ij} - \overline{\overline{X}}_{..})^2$		

*In the randomized block design F is not ordinarily computed for blocks.

As with the earlier procedures, the collective sums of individual squared deviations are used to express the amount of variation by groups:

$$\text{Total variation} = \text{Treatments variation} + \text{Blocks variation} + \text{Error variation}$$
$$SSTO = SSTR + SSBL + SSE$$

The respective sums of squares form the basis for evaluation. This is done through the ANOVA table, which has the generalized form shown in Table 13–13, where the expressions are given for the sums of squares, mean squares, and number of degrees of freedom.

The null hypothesis of equal treatment population means is as follows:

$$H_0\text{: } \mu_{\cdot 1} = \mu_{\cdot 2} = \cdots = \mu_{\cdot c}$$

This is tested in the usual manner, by computing the F statistic,

$$F = \frac{\text{Variation explained by treatments}}{\text{Unexplained variation}} = \frac{MSTR}{MSE}$$

Returning to the bacterial leaching illustration, we obtain the ANOVA table in Table 13–14. The computed value for the test statistic is $F = 28.83$.

Suppose that the engineer chooses a smaller significance level, $\alpha = .01$, than used before. The acceptance and rejection regions shown in Figure 13–20 now apply. The critical value $F_{.01} = 6.01$ was read from Appendix Table J using $c-1 = 2$ degrees of freedom for the numerator and $(r-1)(c-1) = 18$ degrees of freedom for the denominator.

Table 13–14
Two-Way ANOVA Table for Bacterial Leaching Investigation

Variation	Degrees of Freedom	Sum of Squares	Mean Square	F
Explained by Treatments (between columns)	$c-1=2$	$SSTR = 372.47$	$MSTR = \frac{372.47}{2} = 186.23$	$\frac{MSTR}{MSE} = \frac{186.23}{6.46} = 28.83$
Explained by Blocks between rows)	$r-1=9$	$SSBL = 980.70$	$MSBL = \frac{980.70}{9} = 108.97$	
Error or Unexplained (within columns)	$(r-1)(c-1) = 18$	$SSE = 116.20$	$MSE = \frac{116.20}{18} = 6.46$	
Total	$rc-1 = 29$	$SSTO = 1{,}469.37$		

Figure 13–20

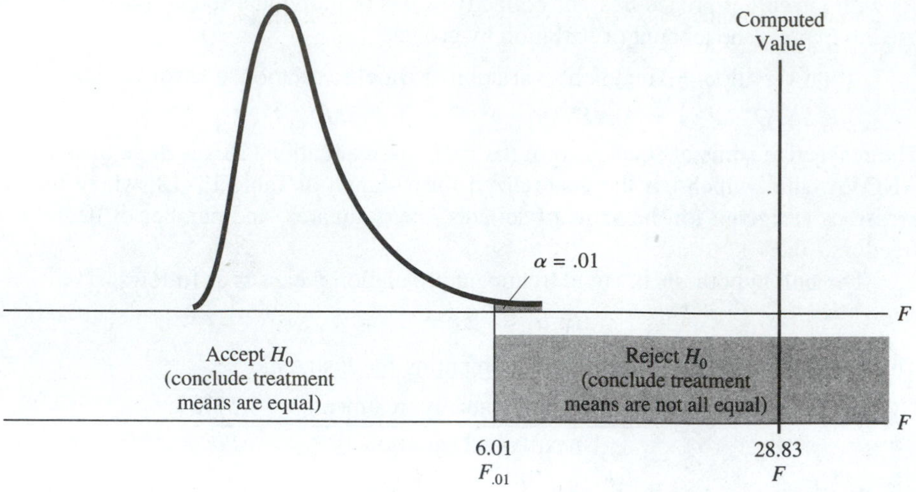

The computed value $F = 28.83$ falls substantially inside the rejection region, providing strong evidence that the mean copper yields are not all equal for different bacteria strains. The two-factor test provides a lower significance level than the single-factor procedure for nearly the same data. Although several degrees of freedom were lost by including the blocking variable, these were more than made up by the ensuing reduction in previously unexplained variation.

Table 13–15

Sample Results for Evaluation of Methods for Solving Quadratic Assignment Problems

	Average Best Cost with Procedure (treatment)				
Problem (block)	**(1) Simulated Annealing**	**(2) CRAFT**	**(3) CRAFT-Biased Sampling**	**(4) Revised Hillier**	**Mean Cos**
1	$25	$28.2	$27.6	$25	$\bar{X}_{1.} =$ $26.45
2	43	44.2	43.6	43	$\bar{X}_{2.} =$ 43.45
3	74	79.6	74.8	73.2	$\bar{X}_{3.} =$ 75.40
4	107	110	108.2	107.4	$\bar{X}_{4.} =$ 108.15
5	291	296.2	294.8	298.4	$\bar{X}_{5.} =$ 295.10
6	578.2	606	581	582.2	$\bar{X}_{6.} =$ 586.85
7	1,308.0	1,339	1,321	1,324.6	$\bar{X}_{7.} =$ 1,323.15
8	3,099.8	3,197.8	3,124	3,114.2	$\bar{X}_{8.} =$ 3,133.95
Mean Cost	$690.75 $= \bar{X}_{.1}$	$712.625 $= \bar{X}_{.2}$	$696.875 $= \bar{X}_{.3}$	$696.00 $= \bar{X}_{.4}$	$\bar{\bar{X}}_{..} =$ $699.0625

Source: Wilheim, Mickey R., and Thomas L. Ward. "Solving Quadratic Assignment Problems by 'Simulated Annealing,'" *IEE Transactions,* March 1987, pp. 107–118.

Example: *Evaluating Methods for Solving Quadratic Assignments Problems (continued)*

Table 13–15 shows the sample results from an experiment testing four procedures for solving quadratic assignment problems. The earlier single-factor analysis of those data resulted in a computed value of $F = .0006$; thus, that earlier statistical procedure leads to the conclusion that the mean response (average best cost) is identical under all methods.

The increased power of the randomized block design is illustrated now as we use the fact that successive sample observations involve problems of progressively larger size. That is, we now treat *problem size* as the blocking variable.

Table 13–16 is the ANOVA table. The computed value of $F = 3.219$ under the randomized block design has increased over 5,000-fold from the single-factor value found earlier. The results are now significant enough at the 5% level to *reject* the null hypothesis of identical mean costs.

Table 13–16

ANOVA Table for Evaluation of Methods for Solving Quadratic Assignment Problems

Variation	Degrees of Freedom	Sum of Squares	Mean square	F
Explained by Treatments: Method (between columns)	$c-1=3$	$SSTR = 2{,}138$	$MSTR = \frac{SSTR}{3} = 712.5$	$\frac{MSTR}{MSE} = 3.219$
Explained by Blocks: Problem Size (between rows)	$r-1=7$	$SSBL = 32{,}457{,}230$	$MSBL = \frac{SSBL}{7} = 4{,}636{,}747$	
Error or Unexplained (residual)	$(r-1)(c-1)=21$	$SSE = 4{,}648$	$MSE = \frac{SSE}{21} = 221.3$	
Total	$rc-1=31$	$SSTO = 32{,}464{,}015$		

Problems

13–34 Suppose that the sample data in each row of Problem 13–5 represent observed yields using raw ingredients from a different supplier. Using supplier as a blocking variable, complete the following:

(a) Construct the ANOVA table.

(b) Compare the value of *SSE* to the one obtained in Problem 13–5. Does the blocking variable appear promising in reducing unexplained variation in yield?

(c) At the $\alpha = .01$ significance level, should the null hypothesis of identical mean yield under each pressure setting be accepted or rejected?

13–35 A quality-assurance engineer evaluates three different test stands to assess the characteristics of printed circuit boards. In conducting the test, he uses a different one of four circuit types for successive sample boards tested on the three stands. Prior to testing, each sample unit has been carefully prepared, with bugs and anomalies added. The engineer's response variable is the percentage of known defects detected. The following data apply:

	Percentage of Known Defects Detected		
Circuit	**(1) Stand *W***	**(2) Stand *Y***	**(3) Stand *Z***
1	92%	87%	78%
2	76	78	61
3	99	95	89
4	98	92	88

At the 5% significance level, what should the engineer conclude regarding the mean percentages of known defects detected?

13–36 Refer to the information in Problem 13–9. Suppose that each row represents a different instructor, with the three scores representing a separate class taught by that instructor using a different syllabus. Treating instructor as the blocking variable, complete the following:

(a) Construct the ANOVA table.

(b) Compare the value of *SSE* to the one obtained in Problem 13–9. Does the blocking variable appear promising in reducing unexplained variation in score?

(c) At the $\alpha = .01$ significance level, should the null hypothesis of identical mean examination score under each laboratory syllabus be accepted or rejected?

Figure 13–21 *SAS printout for investigation of facility layout*

```
                         FACILITY LAYOUT

                  Analysis of Variance Procedure

Dependent Variable: CROSS TRAFFIC

                                 Sum of              Mean
 Source                  DF      Squares            Square      F Value

Model                     5  234555.00000    46911.00000          5.76

Error                     6   48885.00000     8147.50000

Corrected Total          11  283440.00000

               R-Square          C.V.        Root MSE   CROSS TRAFFIC Mean

               0.827530        12.0031      0.2635000           752.000000


Source                   DF      Anova SS    Mean Square       F Value

BLOCK                     3    178444.000     59481.3333          7.30
TRTMENT                   2     56111.000     28055.5000          3.44
```

13–37 Refer to the information in Problem 13–10. Suppose that each row represents a different oil field, with the four increased daily yields representing a different well within the same field. Using oil field as the blocking variable, complete the following:
(a) Construct the ANOVA table.
(b) Compare the value of *SSE* to the one obtained in Problem 13–10. Does the blocking variable appear promising in reducing unexplained variation in yield increase?
(c) At the $\alpha = .01$ significance level, should the null hypothesis of identical mean daily increase in oil production under each method of secondary recovery be accepted or rejected?

13–38 The partial *SAS* computer printout in Figure 13–21 was obtained by an industrial engineer studying the level of cross traffic (response, in item-feet) for three facility layouts (treatments). As her blocking variable she used product, four different types. At the 5% significance level, can the engineer conclude that the mean cross-traffic levels differ?

Review Problems

13–39 An engineer for a bridge authority wishes to evaluate several types of paint in terms of the time to onset of potentially damaging levels of oxidation. The following data (in months) have been obtained:

	Type of Paint			
Test Stand	**(1)**	**(2)**	**(3)**	**(4)**
1	5	6	5	7
2	6	8	7	6
3	5	9	7	5
4	5	9	7	5
5	6	8	7	6
6	5	9	8	7
7	7	8	—	6
8	—	9	—	7

(a) Construct the ANOVA table.
(b) Should the engineer conclude that the mean times are different for the paint types? Use $\alpha = .01$.

13–40 Refer to Problem 13–39 and your answer to that exercise.
(a) Using individual confidence levels of 99%, construct interval estimates for all six pairwise differences in mean paint lifetimes.
(b) Which paint, if any, provides a significantly longer lifetime for each pair?
(c) Which paint provides a significantly longer lifetime than all others?

13–41 A petroleum engineer is evaluating drilling bit shapes to be used in penetrating a deep stratum covering an oil field. On randomly selected days, each of the four shapes was used in boring ten portions of a particular shaft. The following data represent the observed drilling rate, or increased shaft depth (in feet) per 8-hour drilling day:

	Drill Bit Shape			
Observations	**(1)**	**(2)**	**(3)**	**(4)**
1	3	5	9	5
2	10	8	5	12
3	8	11	13	15
4	7	8	6	10
5	12	9	8	11
6	14	13	10	12
7	8	8	4	13
8	7	6	5	14
9	3	4	2	11
10	10	11	7	9

(a) Construct the ANOVA table.

(b) At the 1% significance level, can the engineer conclude that the mean drilling rate differs between bit shapes?

13–42 Refer to Problem 13–41 and your answer to that exercise.

(a) Construct individual 99% confidence interval estimates for the difference in mean daily depths between all six pairs of bits.

(b) Which bit, if any, bores significantly deeper for each pair?

(c) Which bit bores significantly deeper than all others?

13–43 A software engineer conducted an experiment where four C language compilers were evaluated. In his experiment he first compiled a randomly selected program with the present compiler; he than recompiled the same program with the experimental routine. For each set of sample runs, the percentage reduction in compilation time was computed. The following results were obtained:

	Percentage Reduction in Compilation Time			
Observation i	**(1) Super C (U.S./cust.)**	**(2) C Squared (U.S./gen.)**	**(3) C Plus (imp./cust.)**	**(4) Triple C (imp./gen.)**
1	38.6	19.7	1.3	8.6
2	25.5	15.3	5.2	13.5
3	49.3	26.2	−6.7	12.6
4	42.2	21.9	2.0	7.3
5	36.0	22.2	0.5	11.5

Two of the compilers were written in the United States. Each compiler is either customizable or general-purpose.

(a) In this evaluation, a different set of five source programs were used with each version of C. Can you suggest how the engineer might have increased the efficiency of his test by using a slightly different experimental procedure?

(b) The researcher wants to estimate (1) all individual population means and (2) all pairwise differences in means. How many estimates will be made altogether?

(c) Construct individual 99% confidence interval estimates for (1) − (2) above.

(d) Which compiler provides the greatest percentage reduction in compilation time?

13–44 Two researchers obtained the data in Table 13–17 for their experiment with the dynamic lot-size algorithm. Here, a different response variable, number of subproblems, applies.

(a) Construct the ANOVA table.

(b) At the $\alpha = .05$ level of significance, indicate whether the following null hypotheses should be accepted or rejected:

(1) H_0: There are no interactions between problem duration and complexity.

(2) H_0: Means for all problem durations are identical.

(3) H_0: Means for all levels of complexity are identical.

Table 13–17 *Sample Data (average number of subproblems) for Computer Evaluation of Algorithm Tested on Problems with Various Durations and Levels of Complexity*

	Complexity *B*			
Duration *A*	**(1) Simple**	**(2) Low**	**(3) Medium**	**(4) High**
(1) Short	4.2 6.6	3.8 8.2	7.8 14.6	11.0 11.0
(2) Middle	3.0 6.2	4.6 9.8	9.4 62.0	64.6 190.2
(3) Long	6.6 12.0	7.0 7.0	9.0 21.4	11.8 34.2

Source: Erengue, S. Selcuk, and Suleyman Tufekci, "A Branch-and-Bound Algorithm for a Single-Item Multi-Source Dynamic Lot-Sizing Problem with Capacity Constraints," *IIEE Transactions,* March 1987, pp. 73–80.

13–45 Refer to Problem 13–44 and your answers to that problem.

(a) Plot the cell mean response against levels for factor A, with one set of data points for each level of factor B.

(b) For which cell pairs do the means appear to significantly differ? Do you find anything unusual? How do you explain this?

(c) Construct individual 99% confidence interval estimates for the differences in factor-B level means. Indicate which ones are significantly different.

13–46 An engineering consultant to a painting contractor is evaluating three brands of exterior paint in terms of drying time. Using drying condition as the blocking variable, he obtained the following data (in hours):

	Brand of Paint		
Drying Condition	**(1)** ***A***	**(2)** ***B***	**(3)** ***C***
(1) Direct Sun	11.3	12.4	10.9
(2) Shade	14.2	14.3	13.7
(3) Humid	14.9	15.2	14.7
(4) Dry	12.2	11.3	12.5

At the 5% significance level, what should the engineer conclude regarding the mean drying times of the paints?

13–47 A petroleum refinery supervisor wishes to evaluate alternative procedures for removal of sulfur compounds from distillates in an intermediate stage of processing. The following sample data have been obtained for pilot runs under each procedure. The values represent the percentage reduction in sulfur content of the raw distillate processed at that stage. Each test run was made with crude from one of five major origins, which serve as levels for the blocking variable.

	Alternative Sulfur Removal Procedure			
Crude Source	**(1) Evaporator**	**(2) Precipitator**	**(3) Slurry Reactor**	**(4) Tertiary Processing**
(1) Alaska	80%	91%	88%	94%
(2) Arabia	30	28	10	40
(3) Louisiana	72	65	47	65
(4) Texas	65	57	62	72
(5) Venezuela	48	53	58	71

At the 5% significance level, what should the engineer conclude regarding the mean reductions in sulfur content?

CHAPTER FOURTEEN

Experimental Design

This chapter describes statistical procedures falling into the experimental design family of topics. These methods are particularly valuable to engineers since they focus on exactly the same issues faced by the designer of a product or system. A chemical engineer may use experimental design procedures to help determine the best conditions—temperatures, pressures, catalysts—for processing raw feedstocks. Those statistical procedures might likewise help an automotive engineer select the right balance of sizes and strengths for car suspension elements or speed a metallurgist's search for the best combination of components for a jet engine fan blade.

Like regression techniques, experimental design considers several variables simultaneously. But whereas regression attempts to find an accurate mathematical relationship between the variables, experimental design is only tangentially concerned with expressing that relationship. The focus is instead on establishing levels for the various variables that seem to give the best response: target chemical yield, robust driving performance, and sustained strength at high temperatures. The statistical investigation applies only to the particular engineering decision at hand—achieving the best product or system design.

We may envision the engineer's role in arriving at the physical design as an iterative process involving some trial and error, with improvements made along the way and with occasional blind alleys. Statistical experimental design can have the same dynamics, so that one experiment may not only answer earlier questions but suggest as well new ones that may be answered only after further experimentation.

14–1 Issues in Experimental Design

Experimental design considers simultaneously several variables, called **factors**, and attempts to establish how various levels of these relate to one or more **response variables** whose **level** is to be optimized. Table 14–1 shows some examples. Notice that the factors themselves may assume either numerical values or discrete categories. More than one response variable may be considered in an investigation, although a separate statistical analysis is generally required for each, so that new response variables can be introduced in successive stages of an evaluation.

Table 14–1
Examples of Situations Employing Experimental Design

Situation	Response Variable	Factor Variables	Levels
a. Selecting a tire design	Gasoline mileage (mpg) Wear (% treadlife)	Tread pattern Tread depth Track width Tire pressure	Webbed, cross-hatched .5, .75 cm 4, 6 cm 30, 40 psi
b. Choosing paint	Drying time (hrs) Durability (years)	Brand Location Drying conditions Number of coats	A, B, C Sun, shade Dry, humid 2, 3
c. Selecting chemical processing parameters	Yield (g/l) Impurity level (ppm)	Temperature Tank pressure Settling time	300, 400 °F 150, 200, 250, 300 psi 30, 45, 60 min
d. Choosing sulfur removal procedure	Percentage reduction in sulfur	Crude source Procedure Original sulfur level	Alaska, Arabia, Gulf Evaporator, precipitator, slurry reactor, tertiary 1500 ppm, 2000 ppm

We will discuss the underlying concepts and present the techniques using the experiences of an automotive design engineering director charged with the responsibility of establishing specifications for tires to be used as standard equipment on the model XG2 passenger car. She plans to make a series of experiments to establish optimal levels for several parameters and to find the best tread pattern configuration.

The first tire experiment will involve tread width W, tread gap size G, and tire inflation pressure P. Her response variable will be remaining tire lifetime Y after 10,000 miles of driving on a test track.

The underlying relationships among the three tire experiment factors and between themselves and the remaining lifetime response are key to reaching a decision regarding choice of levels. Figure 14–1 suggests one function relating tread width W and remaining tire lifetime Y. Although the exact shape of the curve is not known, it is reasonable to assume that the underlying function would be continuous and reflect a lower value for Y when W is very small or very great, peaking somewhere over the contemplated range of tread widths.

Inclusion of a second factor complicates the relationships. Figure 14–2(a) shows a possible relation between tread width W, gap size G, and remaining tire lifetime Y. Such three-dimensional displays are more easily represented with the contour plot in Figure 14–2(b), which shows a separate contour curve for various levels of response. The geometry of the relationships is further complicated when all three variables are evaluated by incorporating tire inflation pressure P.

The thrust of experimental design is selecting *levels* for each of the factors—not establishing the mathematical form tying together those variables with each other and with the response. Exact shapes for the various curves are not needed. Although the regression techniques of Chapters 4 and 5 might be used in conjunction with test-run data to determine with some accuracy the curve shape, the director has no interest in the par-

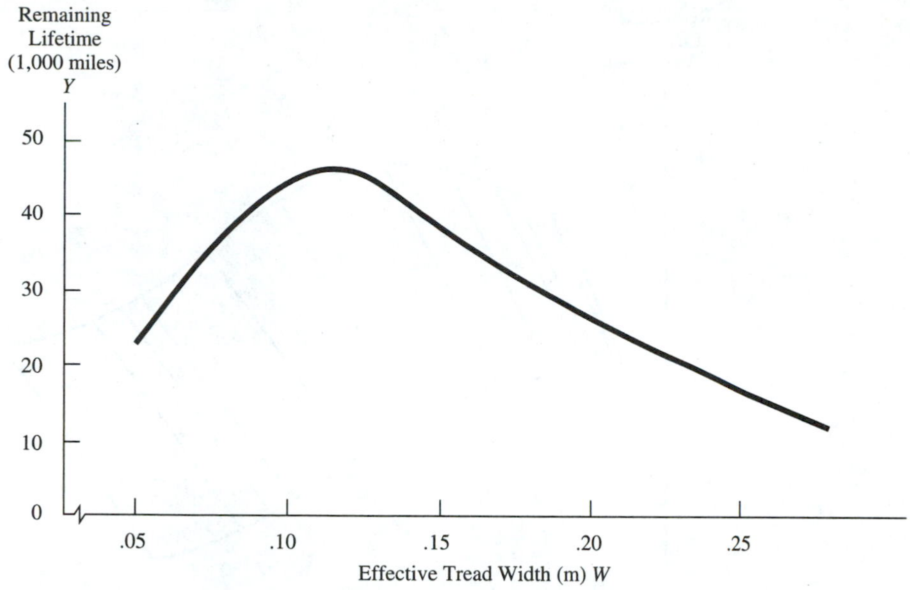

Figure 14–1
Possible true relation between tire lifetime and tread width.

ticular functional form. She wants to find only which width to recommend as the standard. To do this, she needs only assume that some form of iso-contours exists. These will be approximated by *straight lines*. The rationale for this is that over a limited range, a straight line may provide a very good approximation to almost any continuous function. And, straight lines only require two points for a complete description that can lead to meaningful conclusions.

The Factorial Design

Experimental designs involve an evaluation involving a series of tests, called **runs**, each conducted under prescribed test conditions. Test conditions are distinguished by the particular levels chosen for the several factors in the evaluation. The key goal is to discover what similarities and simplifications are consistent with the unknown underlying relationships. An evaluation employing a factorial design efficiently accomplishes this.

> A **factorial design** is one in which there is a test condition for each combination of levels for the factors considered in the experiment.

Definition

A factorial design allows for the testing of every level of each factor under all possible combinations with the other factor levels.

In the tire evaluation the director might consider three tread widths (.15, .20, and .25 m), two gap sizes (.002 and .003 m), and four tire inflation pressures (20, 25, 30, and 35 psi). A factorial design would then involve

$$3 \times 2 \times 4 = 24$$

test conditions. There would be one or more separate test runs, each involving four tires installed on a test vehicle, then driven 10,000 miles, for every combination of factor levels.

Figure 14–2
Possible true two-variable relationship for tire design.

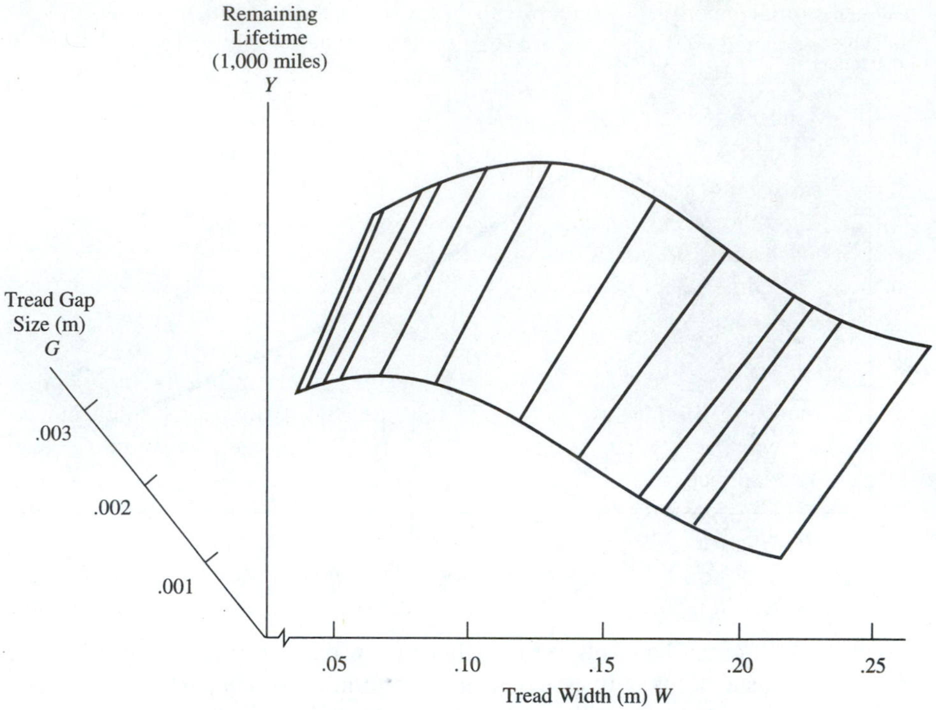

(a) Complete representation.

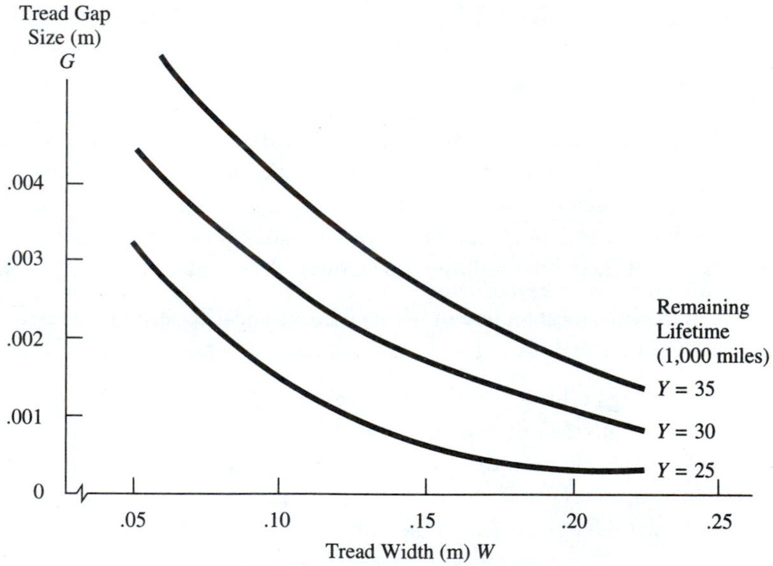

(b) Contour representation.

The factorial experiment would be substantially more efficient than a series of separate tests conducted one-at-a-time for each factor. The latter would still require that the remaining factors be fixed at some level, in which case a total of

$$\text{Number of factors} \times \text{Number of level combinations} = 3 \times 24 = 72$$

sets of runs would be needed because the same number of test conditions would still need to be evaluated once for each factor. A factorial design avoids such redundancy in testing and allows joint conclusions to be reached regarding the entire family of factors.

Joint conclusions are valid from one-at-a-time experiments only when the effects on response (lifetime) from each factor (width, etc.) are *additive*. Independent experiments cannot detect nonadditive behavior. For example, a wider tire might be found to increase remaining lifetime by 2,000 miles per .05 m, while a 5 psi increase in tire inflation pressure increases it by 3,000 miles. But together, greater width and higher pressure might not raise lifetime by 2,000 + 3,000 miles, but perhaps just 3,500 miles, because of the way the two factors *interact*. A factorial design can provide a useful expression of that interaction.

Section 14–2 considers in detail the special case of two-level factorial experiments. By allowing for only *two levels* of each factor, the analytical structure is simplified, and only two points are needed to fathom the approximate linear relationship between variables. In the tire evaluation only two levels each are used for tread width, gap size, and inflation pressure. That experimental design yields

$$2 \times 2 \times 2 = 8$$

test conditions. More generally, in an n-factor evaluation the two-level factorial design will have 2^n test conditions.

Any loss in resolution by having only two levels for each factor is more than offset by the efficiency and simplicity of the procedure. Keep in mind that further experiments may follow, and the results of the earlier evaluation can suggest new levels to consider for one or more factors.

Section 14–3 briefly describes other experimental designs and allied approaches for attacking the same problems.

Problems

14–1 For each of the following situations list possible responses, variables, and levels.
- (a) Workstation keyboard layout
- (b) Simulated toll bridge traffic lanes and fee collection booths
- (c) Bicycle trips
- (d) Automobile engine performance

14–2 Determine the number of test conditions for a factorial design suggested by each possible experiment listed in Table 14–1. In doing this, the listed factor levels will define the test conditions. Assume in each case only one response variable.

14–3 Give one example of an experiment for a design evaluation.
- (a) List possible response variables, important factors, and meaningful levels for each.
- (b) How many test conditions will there be when a factorial design is used involving (1) all levels identified in (a) and (2) just two levels for each factor.

14–2 The Two-Level Factorial Design for Experiments

The number of factor-level combinations determines the number of conditions that must be explicitly accounted for in order to evaluate a design. For instance, suppose that a communications relay is to be tested in terms of input stream bauds (four levels), power supply (two levels), main circuit (two levels), and housing shield thickness (five levels). The possible number of test conditions would be

$$4 \times 2 \times 2 \times 5 = 80$$

In this section we present a method based upon just two levels for each factor. With little loss in experimental efficacy, the experiment may be modified to require just two input stream densities (say 10,000 and 50,000 bauds) and two shield thicknesses (such as .0015 and .0020 m). That would reduce the number of test conditions to

$$2^4 = 16$$

possibilities. Should the experimental results indicate that stream density or shield thickness have a significant effect on the response variable, further testing might be done using new levels for one or both of those factors.

Table 14–2 summarizes the results of the engineering director's tire-evaluation experiment involving several test cars. The first experimental factor is effective tire tread width W; the tires used were either .150 meter or .175 meter wide. The second factor is tread gap size G, the tires in a run having gaps either .002 or .003 meter wide. The final factor is tire pressure P, with all tires in each run set at 25 psi or 30 psi. Altogether the evaluation involves $2^3 = 8$ experimental test conditions, one for each combination of the factor levels. Each test car was driven 10,000 miles on a test track with a new set of tires under just one of those test conditions. Two cars were driven independently under each of the 8 conditions. The test car tires were rotated, pressure checked, and wheels balanced and aligned every 1,000 miles. Each pair of test cars were identically maintained and driven by the same drivers on a rotation basis; any difference between the results achieved could therefore be attributed to chance causes.

Table 14–2
Test Conditions and Responses for Tire Experiment

Test Condition j	Effective Tread Width (m) W	Tread Gap Size (m) G	Tire Pressure (psi) P	Average Remaining Lifetime (1,000 mi) Y_j
1	.150(−)	.002(−)	25(−)	29
2	.175(+)	.002(−)	25(−)	33
3	.150(−)	.003(+)	25(−)	23
4	.175(+)	.003(+)	25(−)	26
5	.150(−)	.002(−)	30(+)	18
6	.175(+)	.002(−)	30(+)	42
7	.150(−)	.003(+)	30(+)	18
8	.175(+)	.003(+)	30(+)	36

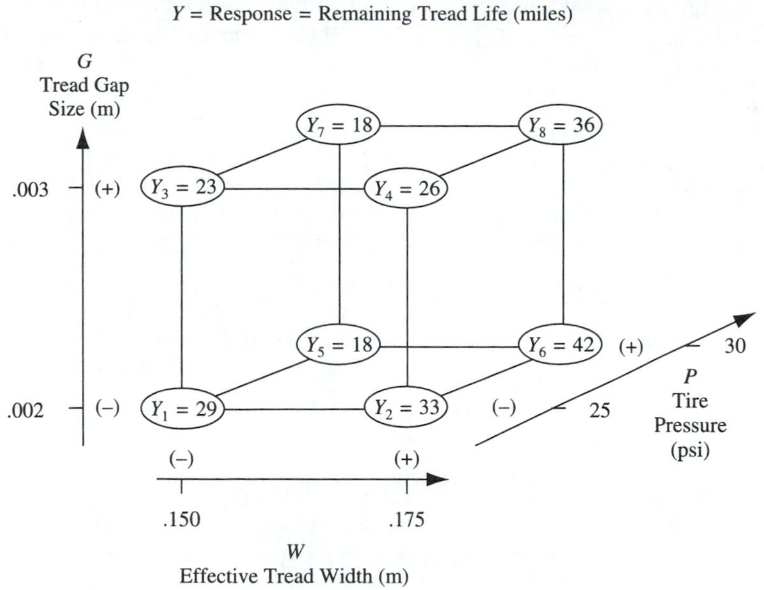

Figure 14–3 *Graphical representation of results with tire experiment.*

Levels are sought for the factors that will optimize overall response, average miles of remaining tread life Y. The responses obtained are listed in Table 14–2 for each test condition.

The test conditions in a two-level factorial experiment are traditionally numbered in the order listed. This corresponds to the geometrical representation in Figure 14–3, where each factor has an axis scaled for just the two levels prescribed for the experiment. The smaller level for each variable is coded with a $(-)$, the other with a $(+)$. Should a variable be qualitative, such as tread design, the levels for that variable would be the attributes (e.g., webbed versus sinuous patterns), and the $(-)$ and $(+)$ designations would then be arbitrary. This symbolic coding is especially helpful in describing the steps in the evaluation.

It is particularly convenient to schematically represent each experimental condition to a matching vertex on a cube. The responses obtained when the respective factors are at the indicated levels are given in the ovals at corners. For the tire experiment those values are the average tread life remaining Y_j, in thousands of miles, in the tires from the two test runs made under the jth set of experimental conditions.

Main Effects

The procedure involves estimating for each factor the main effect, for which we make the following:

Definition

> The **main effect** for a single factor expresses the average change in response associated with changing the factor from its $(-)$ level to its $(+)$ level.

The main effect is computed by averaging all of the responses achieved when the factor is at the $(+)$ level, so that all combinations of the other factor levels are represented exactly once, and computing the analogous average for that same factor at the $(-)$ level. The difference in these two averages is the effect.

For the tire experiment the effect for tread width is computed below. To avoid confusion, we employ the symbol E_W to distinguish the effect of the tread width from the factor W itself.

Tread Width Effect:

$$E_W = \left(\frac{Y_2 + Y_4 + Y_6 + Y_8}{4}\right) - \left(\frac{Y_1 + Y_3 + Y_5 + Y_7}{4}\right)$$
$$= \frac{33 + 26 + 42 + 36}{4} - \frac{29 + 23 + 18 + 18}{4}$$
$$= 34.25 - 22.00 = 12.25$$

The above estimates that the effect of increasing effective tread width from .150 m to .175 m is an average increase in remaining tire lifetime of 12.25 thousand miles.

Analogously, the effect for tread gap size is computed.

Tread Gap Size Effect:

$$E_G = \left(\frac{Y_3 + Y_4 + Y_7 + Y_8}{4}\right) - \left(\frac{Y_1 + Y_2 + Y_5 + Y_6}{4}\right)$$
$$= \frac{23 + 26 + 18 + 36}{4} - \frac{29 + 33 + 18 + 42}{4}$$
$$= 25.75 - 30.5 = -4.75$$

The above estimates that the effect of increasing tread gap size from .002 m to .003 m is a reduction in remaining tire lifetime of 4.75 thousand miles.

Finally, the effect for tire pressure is computed.

Tire Pressure Effect:

$$E_P = \left(\frac{Y_5 + Y_6 + Y_7 + Y_8}{4}\right) - \left(\frac{Y_1 + Y_2 + Y_3 + Y_4}{4}\right)$$
$$= \frac{18 + 42 + 18 + 36}{4} - \frac{29 + 33 + 23 + 26}{4}$$
$$= 28.5 - 27.75 = .75$$

The above estimates that the effect of increasing tire pressure from 25 psi to 30 psi is an increase in remaining tire lifetime of .75 thousand miles.

Graphical Representation The main effects for the tire-experiment factors are represented in Figure 14–4. The two levels of each factor are represented as opposite faces of the summary cube. The four corners on one face give the responses obtained when that factor is at the $(-)$ level, the other four corners having the response when that factor is at the $(+)$ level. The main effect is found by subtracting the average of the $(-)$ responses from the average of the $(+)$ responses.

Additivity and Non-Additivity of Effects Although the true form of the response surface is unknown, this procedure assumes that it is linear over the range of levels set in the experiment for each variable. The premise being that over a limited range a straight line should provide an adequate approximation to most nonlinear functions.

We have computed $E_W = 12.25$ and $E_G = -4.75$. For the moment, let's presume that these main effects are true and not subject to sampling error. Then, increasing tire

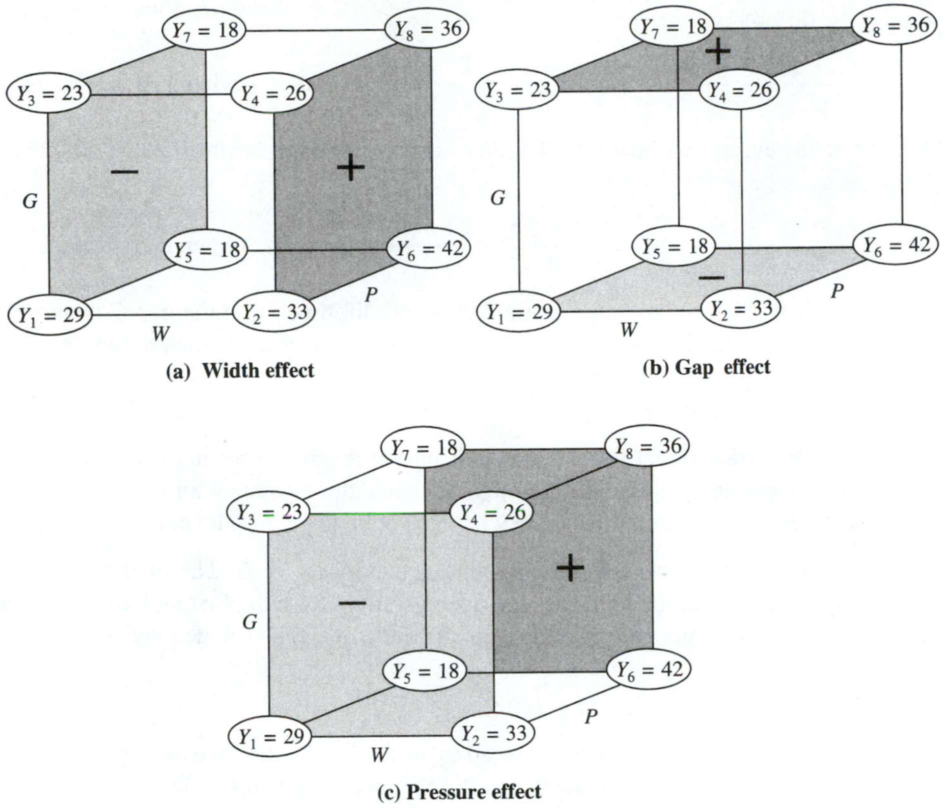

Figure 14–4 *Graphical representation of tire experiment main effect.*

tread width from the (−) level of .150 m to the (+) level of .175 m adds 12.25 thousand miles to the average tire life. Furthermore, an increase in tread gap size from .002 m (−) to .003 m (+) reduces the remaining tire life by 4.75 thousand miles. If the two effects were additive, then the combined effect would be the sum of the two main effects:

$$E_W + E_G = 12.25 + (-4.75) = 7.5$$

and then increasing both factors to their (+) levels should then increase tire lifetimes by 7.5 thousand miles over what it would be when the two variables are both at the (−) levels.

There are two reasons why this may not be so. First, the main effects themselves have been computed from experimental data, so that because of sampling error they are only imprecise estimates of the true effects. We will account for sampling error later through construction of confidence intervals and determination of statistical significance. Second, as we shall see next, the effects may not be additive, in which case, by themselves the main effects do not express dependencies among the factors. Those are detected by the interaction effects.

Interaction Effects

If we separately compute the width effect when the tread gap is narrow and when it is wide, we get different values from the $E_W = 12.25$ computed earlier. To demonstrate, we first hold gap size at .002 m (−) and ignore the other test conditions, computing one average response when W is at the (+) level and another when W is at the (−) level, and taking the difference,

$$\frac{Y_2 + Y_6}{2} - \frac{Y_1 + Y_5}{2} = \frac{33 + 42}{2} - \frac{29 + 18}{2} = 14.0 \qquad \text{(when gap is .002 m)}$$

We obtain the average width effect of 14.0. Analogously, when the tread gap is at .003 m (+), the average width effect is

$$\frac{Y_4 + Y_8}{2} - \frac{Y_3 + Y_7}{2} = \frac{26 + 36}{2} - \frac{23 + 18}{2} = 10.5 \qquad \text{(when gap is .003 m)}$$

Tread width W therefore has a greater effect on tire lifetime when the gap G is narrow than when it is wide. The width and gap size variables interact in such a way that the separate effects are not additive.

We make the following:

Definition

The **interaction effect** for two factors is half of the difference in the average effects separately computed for one factor when the second factor is first at its lower (−) level and then again when it is at its upper (+) level.

We have computed the average effects for W at two levels for G. The final measure is then computed by subtracting the second average effect from the first and dividing that difference by 2. The interaction effect for tread width and gap size, denoted as $W \times G$, is thus

$$W \times G = (10.5 - 14.0)/2 = -1.75$$

(The same result would have been achieved by instead computing the average tread gap size G effect with narrow and wide treads W and computing half the difference in those two values.)

Before we can properly interpret this interaction effect, we must acknowledge the presence of sampling error. But even if we were certain about the true value for $W \times G$, we should not combine that quantity arithmetically with E_W and E_G, since the linearity in response is only presumed. The issue here is not one of establishing the function $Y = f(W, G, P)$. Rather, the engineering director wants only to select levels for the factors (i.e., set standard equipment specifications for tires).

Generally, the interaction effect for two variables is found by first computing the average response determined for the test conditions in which both variables have the same sign code (both at − or both at +). A second average response is then obtained under the remaining test conditions (under which the two variables have opposite sign codes). The interaction is the difference between those averages.

Continuing with the tire-experiment data, we compute

Tread Width and Gap Size Interaction:

$$\begin{aligned} W \times G &= \left(\frac{Y_1 + Y_4 + Y_5 + Y_8}{4}\right) - \left(\frac{Y_2 + Y_3 + Y_6 + Y_7}{4}\right) \\ &= \frac{29 + 26 + 18 + 36}{4} - \frac{33 + 23 + 42 + 18}{4} \\ &= 27.25 - 29 = -1.75 \end{aligned}$$

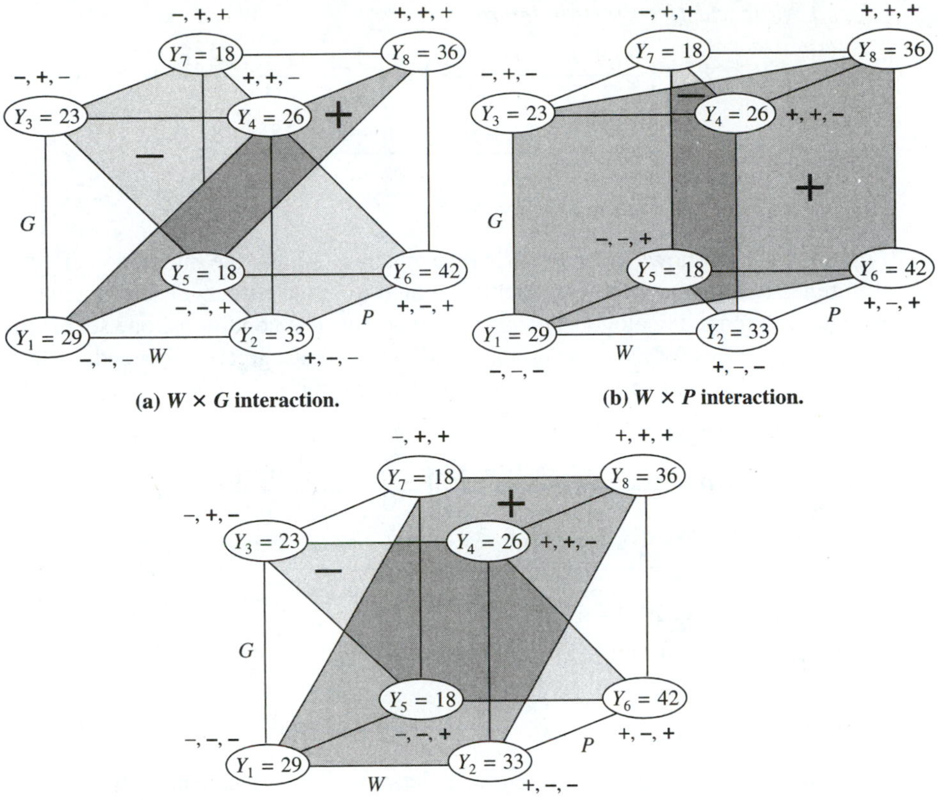

(a) $W \times G$ interaction.

(b) $W \times P$ interaction.

(c) $G \times P$ interaction.

Figure 14–5 *Geometric repesentation of two-variable interaction effects.*

Graphical Representation A geometrical portrayal of each interaction is provided by the cubes in Figure 14–5. Interaction $W \times G$ is presented in cube (a), where each corner of the cube is labeled in accordance with the sign-code vector whose elements coincide with the individual factor-level sign codes. Corner 1 has the sign-code vector (−, −, −) corresponding to the codes (−) for each variable represented by the levels for test condition 1, and there W and G have concurring sign codes. Corner 2 has the vector (+, −, −) since test condition 2 involves W at the higher level (+), with G and P at their lower levels (−), and the sign codes for W and G disagree. Those two corners lie on opposite planes, in keeping with our previous graphic representation of effects. For each cube, the "+" plane has corners with concurring sign codes, and the second "−" plane is described by the corners with disagreeing signs. In making these designations, the sign code for the variable whose effect is not measured gets ignored. In cube (a) factor P is ignored; its sign code appears in lighter type as the third element in each corner vector.

The interaction effect is computed by separately averaging the response values for the two planes, found by subtracting the average response for the corners of the "−" plane from the average for the corners for the "+" plane.

There are two other two-variable interactions for the three-variable tire experiment, computed analogously.

Tread Width and Tire Pressure Interaction:

$$W \times P = \left(\frac{Y_1 + Y_3 + Y_6 + Y_8}{4}\right) - \left(\frac{Y_2 + Y_4 + Y_5 + Y_7}{4}\right)$$
$$= \frac{29 + 23 + 42 + 36}{4} - \frac{33 + 26 + 18 + 18}{4}$$
$$= 32.5 - 23.75 = 8.75$$

Figure 14–5(b) shows the geometric interpretation. Again, the interaction effect is the difference in response averages, each computed from the corners describing two opposite planes formed analogously from the respective test conditions, one set having concurring sign codes for the two variables and the other disagreeing sign codes. (In cube (b) G is the ignored variable, its sign codes appearing in light type.)

Tread Gap Size and Tire Pressure Interaction:

$$G \times P = \left(\frac{Y_1 + Y_2 + Y_7 + Y_8}{4}\right) - \left(\frac{Y_3 + Y_4 + Y_5 + Y_6}{4}\right)$$
$$= \frac{29 + 33 + 18 + 36}{4} - \frac{23 + 26 + 18 + 42}{4}$$
$$= 29.0 - 27.25 = 1.75$$

The geometric interpretation is shown in Figure 14–5(c).

Short Cut Calculations

The sign codes prove helpful in shortcut calculations of the effects, as shown in Table 14–3. There each test condition is represented as a row in the main body of the table, with the sign codes appearing in the columns for the respective factor levels. Applying the signs to the mean responses in the Y_j column, the main effects may be easily calculated without having to look at a formula or graph. The respective effects are computed by dividing the respective Y-sum by the value listed in the divisor row.

The main effects are thereby quickly calculated:

$$E_W = (-29 + 33 - 23 + 26 - 18 + 42 - 18 + 36)/4 = 49/4 = 12.25$$
$$E_G = (-29 - 33 + 23 + 26 - 18 - 42 + 18 + 36)/4 = -19/4 = -4.75$$
$$E_P = (-29 - 33 - 23 - 26 + 18 + 42 + 18 + 36)/4 = 3/4 = .75$$

The same process applies to the interaction effects. The product of the respective sign codes for the applicable variables gives the interaction sign code, following arithmetic rules, with $(-) \times (-) = (+)$, $(-) \times (+) = (-)$, and so on. The sign codes for interaction column $W \times G$ are the products of the sign codes from column W and column G. The sign codes in the other interaction columns are computed analogously. The interaction effects are computed like the above from Table 14–3:

$$W \times G = (+29 - 33 - 23 + 26 + 18 - 42 - 18 + 36)/4 = -7/4 = -1.75$$
$$W \times P = (+29 - 33 + 23 - 26 - 18 + 42 - 18 + 36)/4 = 35/4 = 8.75$$
$$G \times P = (+29 + 33 - 23 - 26 - 18 - 42 + 18 + 36)/4 = 7/4 = 1.75$$

These are the same values computed earlier.

Table 14–3
Calculation of Tire-Experiment Effects and for Estimating Standard Deviation

Remaining Tread (1,000 miles) Car 1 Y_{1j}	Car 2 Y_{2j}	Test Condition j	Mean Response Y_j	Factor Levels −:.150 +:.175 W	−:.002 +:.003 G	−:25 +:30 P	Interactions $W \times G$	$W \times P$	$G \times P$	$W \times G \times P$	Standard Deviation Estimation $Y_{1j} - Y_{2j}$	$(Y_{1j} - Y_{2j})^2$
27	31	1	29	−	−	−	+	+	+	−	−4	16
34	32	2	33	+	−	−	−	−	+	+	2	4
24	22	3	23	−	+	−	−	+	−	+	2	4
24	28	4	26	+	+	−	+	−	−	−	−4	16
20	16	5	18	−	−	+	+	−	−	+	4	16
42	42	6	42	+	−	+	−	+	−	−	0	0
17	19	7	18	−	+	+	−	−	+	−	−2	4
37	35	8	36	+	+	+	+	+	+	+	2	4
Σ			225	49	−19	3	−7	35	7	−5		64
Divisor			8	4	4	4	4	4	4	4		
Effect			28.125	12.25	−4.75	.75	−1.75	8.75	1.75	−1.25		

The Three-Variable Interaction Effect

Finally, the three-variable interaction $W \times G \times P$ may be computed. The sign code for test condition 1 is the product of the sign codes from columns W, G, and P:

$$(-) \times (-) \times (-) = (-)$$

For condition 2 the product is

$$(+) \times (-) \times (-) = (+)$$

and so on. The sign codes thereby established are listed in the $W \times G \times P$ column. Using those as the guide to add or subtract the respective Y value, we compute

$$W \times G \times P = (-29 + 33 + 23 - 26 + 18 - 42 - 18 + 36)/4 = -5/4 = -1.25$$

The graphical representation of the three-factor interaction effect is clumsy to give. It involves two opposite tetrahedrons (four-sided pyramids), one involving corners with positive sign-code products, the second with negative products. The average response from each tetrahedron is calculated, with the three-variable interaction equal to the difference obtained by subtracting the negative-code-product average from the opposite average.

Evaluating the Results

We must translate the computed effects into action. The process involves (1) sifting the statistical summaries to identify the significant effects, and (2) choosing a course of action or designing further experiments. The first action is based on confidence intervals constructed for each effect.

Constructing Confidence Intervals The effects are formed from means and differences in means. Just as unknown population means are imprecisely and uncertainly measured by their sample counterparts, so too are the true levels of the effects. These may be estimated only in a statistical sense, subject to sampling error.

The estimation procedure is based on the underlying run-to-run variability, which is assumed to be of the same magnitude regardless of the particular test conditions set for any run. For instance, in the tire-evaluation experiment, two cars made 10,000-mile test runs with tires at 25 psi and having narrow tire treads with small gaps. The tires from those cars are expected to yield remaining tire tread lifetimes that differ by the same amount as the tires from two matching test runs made under any other of the seven test conditions. This underlying response variability is summarized by the **overall response variance** σ_Y^2.

The response variance may be estimated from the sample results for any particular test condition for which there has been **replication**, an experimental design having separate runs at duplicate variable levels. The theoretical arguments are straightforward. Denoting by Y_{ij} the response obtained for the ith replicated run made under test condition j, and denoting by n_j the number of duplicate tests for that test condition, the following sample mean and variance apply:

$$Y_j = \frac{\sum_{i=1}^{n_j} Y_{ij}}{n_j} \qquad \text{and} \qquad s_j^2 = \frac{\sum_{i=1}^{n_j} \left(Y_{ij} - Y_j\right)^2}{n_j - 1}$$

(For simplicity, and in keeping with the earlier notation, the overbar has been dropped for the mean response, Y_j.) In the special case where $n_j = 2$, the sample response variance reduces to half the squared difference in the pair of single-run responses:

$$s_j^2 = \frac{\left(Y_{1j} - Y_{2j}\right)^2}{2}$$

For each test condition a similar sample variance applies, each of which serves as an estimate of the overall response variance σ_Y^2. The average of these variances provides the

POOLED ESTIMATE OF THE OVERALL RESPONSE VARIANCE IN 2-LEVEL REPLICATED FACTORIAL DESIGN:

$$\hat{\sigma}_Y^2 = \frac{\sum_{j=1}^{N_T} \left(Y_{1j} - Y_{2j}\right)^2}{2N_T}$$

where N_T denotes the number of test conditions in the experiment. The number of degrees of freedom for this estimate is N_T.

To illustrate, we continue with the tire experiment summarized in Table 14–3. The first two columns provide the remaining tire lifetimes from each pair of cars tested under the $N_T = 8$ listed test conditions. The last two columns list the differences and squared difference terms, the sum of which is 64. Thus,

$$\hat{\sigma}_Y^2 = \frac{64}{2(8)} = 4 \qquad \text{or} \qquad \hat{\sigma}_Y = 2$$

Each effect is the difference in two separately computed means, each involving the average response under half of the test conditions. Assuming a replicated experiment, there are altogether N_T individual responses for each of those means, so that

$$\begin{aligned}\text{Var (Effect)} &= \text{Var(1st } Y \text{ mean} - \text{2nd } Y \text{ mean)} \\ &= 2\text{Var}(Y \text{ mean}) = 2\left(\frac{\sigma_Y^2}{N_T}\right) = \frac{2\sigma_Y^2}{N_T}\end{aligned}$$

The following expression is used to estimate the

STANDARD ERROR OF AN EFFECT IN A REPLICATED FACTORIAL EXPERIMENT:

$$s_{\text{effect}} = \sqrt{\frac{2}{N_T}}\hat{\sigma}_Y$$

The above may be used in calculating a confidence interval estimate for the true level of an effect. The following expression applies for constructing the

100(1−α)% CONFIDENCE INTERVAL ESTIMATE OF AN EFFECT IN A FACTORIAL EXPERIMENT

$$\text{True effect} = \text{Computed effect} \pm t_{\alpha/2} s_{\text{effect}}$$

where the number of degrees of freedom is N_T.

To illustrate, we return to the tire experiment. The standard error is

$$s_{\text{effect}} = \sqrt{\frac{2}{8}}(2) = 1.0$$

The following 99% confidence interval is computed for the main tire width effect, using $t_{.005} = 3.355$ read from Appendix Table G for 8 degrees of freedom:

$$W = 12.25 \pm 3.355(1.0) = 12.25 \pm 3.355$$

Thus, we are 99% confident that the true effect of increasing the tire tread width from .150 m to .175 m is an increase of somewhere between 8.90 and 15.61 thousand miles in remaining tread life. Furthermore, since the stated confidence interval lies totally above zero, we may conclude that the effect is *statistically significant* (and we can reject the null hypothesis that tire width has zero effect).

But no final conclusion can be reached regarding tread width until we examine how it interacts with the other variables.

Table 14–4 shows the 99% confidence intervals for all seven effects measured in the tire experiment. Those effects having intervals overlapping zero are not statistically significant from what would be expected from noise (zero effect). We find that the only statistically significant main effects are tire tread width W and tread gap size G. The latter estimate tells us that increasing tread gap size from .002 m to .003 m will have an effect on remaining tread life somewhere between −8.10 and −1.40 thousand miles. By

Table 14–4
Separate Confidence Intervals for Effects in Tire Experiment

Effect	99% Confidence Interval	
Main		
Tread width W	12.25 ± 3.355	(significant)
Tread gap size G	-4.75 ± 3.355	(significant)
Tire pressure P	$.75 \pm 3.355$	
Interactions		
$W \times G$	-1.75 ± 3.355	
$W \times P$	8.75 ± 3.355	(significant)
$G \times P$	1.75 ± 3.355	
$W \times G \times P$	-1.25 ± 3.355	

itself, tire pressure P has no significant main effect. Nevertheless, we cannot ignore this variable without first evaluating its interactions.

Table 14–4 indicates that three of the four interactions do not differ significantly from zero. Only $W \times P$ is statistically significant, and we must conclude that tire pressure does have a significant influence on tire life, but that is manifested only through an interaction with tread width. By itself that interaction is estimated to have an effect on tire lifetime of somewhere between 5.40 and 12.11 thousand miles.

But it is not meaningful to interpret an interaction effect by itself, and its existence clouds our ability to give separate interpretations to the main effects for the factors involved. When there is an interaction between two factors, they are best interpreted together. That may be done by plotting the response variable on one of the factors separately for the levels of the second factor. This is done in Figure 14–6, which shows how average remaining tire lifetime Y is related to tread width W in terms of two sepa-

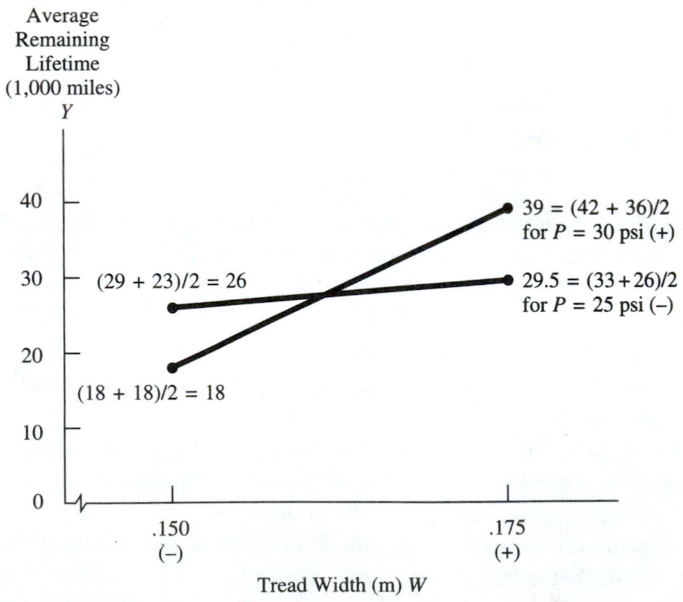

Figure 14–6
Average response for tread width at two pressure levels.

rate lines, one for each level of tire pressure. The Y values for the two points defining each line are the average response obtained from the runs made at the respective W and P levels.

Both lines indicate that the wider tire, on the average, produces better wear, but that the wider tire provides much better average response when $P = 30$ psi than at $P = 25$ psi. The narrower tire width is actually better than the wide one at the low level for P.

These results favor selection of the wider tread, but that tire life is very sensitive to inflation pressure.

What engineering decision can be made from these results? The issue of tread gap size has been resolved—use the smaller gaps. And, in further testing, the wider tires may be assumed to yield longer life, but only at the higher inflation pressure.

Collective Confidence Level The tire experiment involves seven different confidence intervals, all computed from the same sample data. Although each interval represents a 99% confidence level, as a *group* the collective level of confidence is smaller. When several interval estimates are made from the same data, then we must use the following reduced

COLLECTIVE CONFIDENCE LEVEL:

$$100(1 - N_I \alpha)\%$$

where $N_I =$ the number of interval estimates made at the individual $100(1 - \alpha)\%$ level.

In the tire experiment $N_I = 7$ effects were estimated, each at the 99% individual level of confidence, so that $\alpha = .01$. As a group, those intervals provide only a $[1 - 7(.01)]\% = 93\%$ *collective* level of confidence. The only way to increase the overall confidence level is to raise the individual levels. For example, at 99.5% individual confidence, so that $\alpha = .005$, we use instead $t_{.0025} = 3.833$ to construct the intervals. The resulting limits are somewhat wider than presented in Table 14–4, but the overall collective confidence would then be $[1 - 7(.005)]\% = 96.5\%$. In this example the same overall conclusions would have been reached even at the higher confidence levels.

Higher-Dimensional Factorial Experiments

The two-level factorial experiment may be done in more dimensions, although a staggered experiment may be more efficient. Consider again the director's tire evaluation, which continued into a second stage involving a completely new experiment and objective. Table 14–5 summarizes the design for this follow-on study in which tread pattern,

Table 14–5
Design for Second Tire Experiment

Response	Variables	Levels
Gasoline mileage (mpg) Y	Tread pattern T	Webbed (−), cross-hatched (+)
	Groove depth D	.0050 m (−), .0075 m (+)
	Tread width W	.150 m (−), .175 m (+)
	Average track velocity V	45 mph (−), 60 mph (+)

tread depth, tread width, and average track velocity are considered. Although tire wearability is still important, this study employs a *different* response variable, gasoline mileage (mpg), which is to be maximized. One variable, tire tread width W, is carried forward from the first tire experiment. Altogether, four variables apply, three of them new to the second experiment: tread pattern T, groove depth D, and average track velocity V.

Altogether there are $2^4 = 16$ test conditions. Figure 14–7 shows the geometric representation of the experiment. As before, three variables W, D, and T are represented by three axes and the test conditions by the corners of a cube. Two cubes are used to represent the fourth variable, average track velocity V.

Table 14–6 shows the experimental results and the effects calculations. As before, there is one main effect for each variable. There is a potential interaction for each

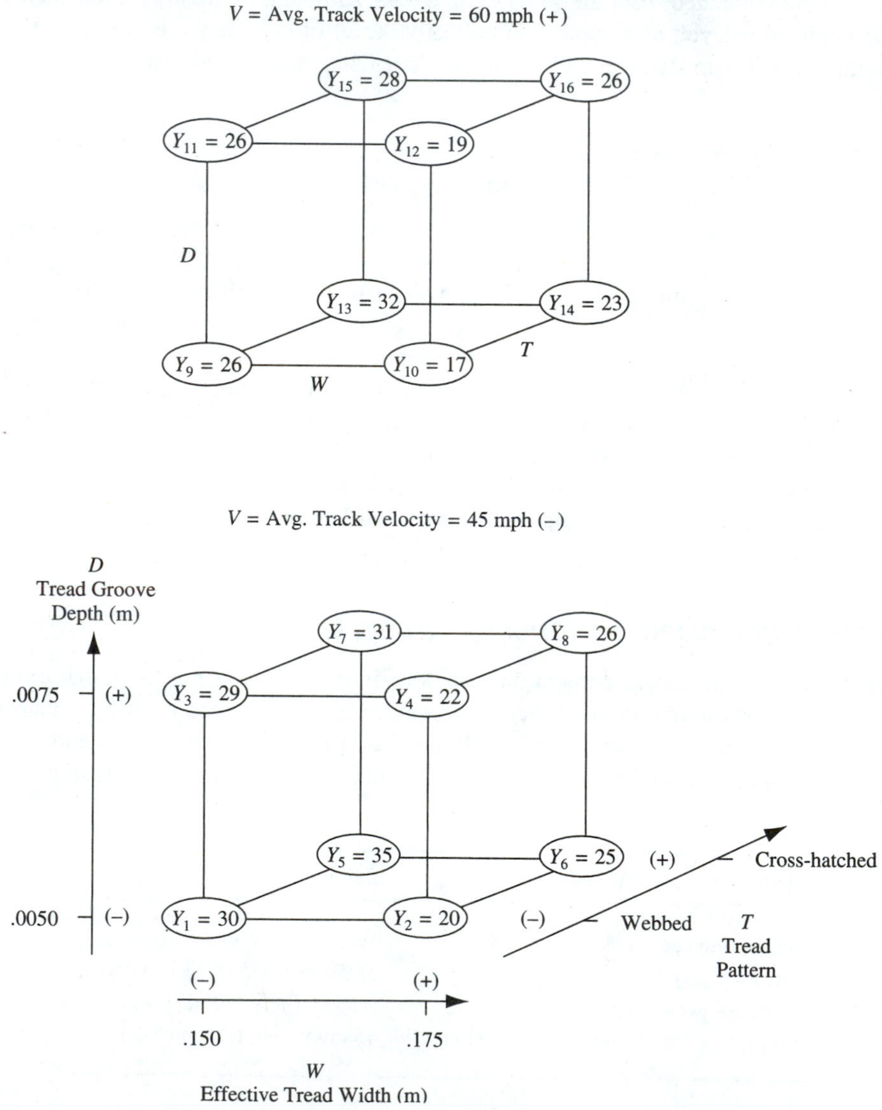

Figure 14–7
Geometrical representation of the second, four-variable tire experiment.

combination of variables: six for the variable pairs, four for the triples, and one for all four variables. As with the first tire experiment, the statistically significant effects must be identified before any meaningful conclusions can be reached. The procedure for doing this differs from the first experiment, since there was no replication. (It is common in higher-dimensional experiments to forgo replication.)

Constructing Confidence Intervals for Unreplicated Experiments When there is no replication, and thus only one run per test condition, the response variance cannot be estimated from squared differences in response from individual runs. Instead, the standard error of the effects is estimated directly from the calculated higher-order interaction effects themselves. That rests on the assumption that the three- and four-variable interactions have a negligible true effect, so that the computed levels for those effects represent only measurements of experimental error. Thus, the variance in effects is estimated from computed results by taking the average of the squared higher-order effects. For the second tire experiment, we have

$$s_{\text{effect}}^2 = \frac{(W \times D \times T)^2 + (W \times D \times V)^2 + (W \times T \times V)^2 + (D \times T \times V)^2 + (W \times D \times T \times V)^2}{5}$$

$$= \frac{(.88)^2 + (.13)^2 + (.38)^2 + (.13)^2 + (.38)^2}{5} = .2194$$

and

$$s_{\text{effect}} = \sqrt{.2194} = .47$$

There are 5 degrees of freedom in the above calculations, one for each higher-order effect being averaged.

Using the data from Table 14–6, the following 99% confidence interval estimate is computed for the effect of tire tread width W, using $t_{.005} = 4.032$, read from Appendix Table G for 5 degrees of freedom:

$$W = -7.38 \pm 4.032(.47) = -7.38 \pm 1.90 \quad \text{mpg}$$

Table 14–7 shows the confidence intervals for all remaining main effects and two-variable interaction effects. The collective level of confidence for the ten estimates is $100[1 - 10(.01)] = 90\%$. All main effects except for groove depth D have a significant effect on gasoline mileage, although that variable's interaction effect with the tread width W is significant too. This means that W and D should be analyzed together. The only significant interaction effect is $W \times D$, which allows us to reach conclusions singly from the main effects for tread pattern T and track velocity V. The cross-hatched tread pattern (+) has a gasoline mileage advantage over the webbed design (−) that is estimated to fall somewhere between 2.73 and 6.53 mpg. Increasing track velocity from 45 mph (−) to 60 mph (+) reduces gasoline mileage by some amount estimated to lie between .73 and 4.53 mpg.

Figure 14–8 plots the average gasoline mileage against tread width W. There we see that increasing W from .150 m (−) to .175 m (+) has resulted in a drop in average gasoline mileage for both groove depths—falling from 30.75 mpg to 21.25 mpg when $D = .0050$ m (−) and from 28.5 mpg to 23.25 mpg when $D = .0075$ m (+). In both cases the wider tire gives poorer gasoline mileage, but the dropoff is less severe for the deeper tire tread.

Table 14–6
Calculation of Second Tire Experiment Effects

Test condition j	Mean Response (mpg) Y_j	−: .150 +: .175 W	.0050 .0075 D	Web. C.H. T	45 60 V	$W \times D$	$W \times T$	$W \times V$
		Factor Levels				**Two-Factor Interactions**		
1	30	−	−	−	−	+	+	+
2	20	+	−	−	−	−	−	−
3	29	−	+	−	−	−	+	+
4	22	+	+	−	−	+	−	−
5	35	−	−	+	−	+	−	+
6	25	+	−	+	−	−	+	−
7	31	−	+	+	−	−	−	+
8	26	+	+	+	−	+	+	−
9	26	−	−	−	+	+	+	−
10	17	+	−	−	+	−	−	+
11	26	−	+	−	+	−	+	−
12	19	+	+	−	+	+	−	+
13	32	−	−	+	+	+	−	−
14	23	+	−	+	+	−	+	+
15	28	−	+	+	+	−	−	−
16	26	+	+	+	+	+	+	+
Σ		−59	−1	37	−21	17	7	5
Divisor		8	8	8	8	8	8	8
Effect		−7.38	−.13	4.63	−2.63	2.13	.88	.63

The two experiments indicate that the wider tread improves wearability but reduces fuel efficiency, although less so for deeper tread grooves. The cross-hatched pattern is the most fuel-efficient design, although the influence of tire tread pattern on wearability has not been examined. Tread gap size should be kept small to promote tire wearability, but its effect on fuel consumption was not examined. The fuel consumption advantage of lower speed is no surprise, while tire pressure's effect on that response has not been examined. Higher tire pressure has been found to increase the wearability of

Two-Factor Interactions			**Multiple Interactions**				
$D \times T$	$D \times V$	$T \times V$	$W \times D \times T$	$W \times D \times V$	$W \times T \times V$	$D \times T \times V$	$W \times D \times T \times V$
+	+	+	−	−	−	−	+
+	+	+	+	+	+	−	−
−	−	+	+	+	−	+	−
−	−	+	−	−	+	+	+
−	+	−	+	−	+	+	−
−	+	−	−	+	−	+	+
+	−	−	−	+	+	−	+
+	−	−	+	−	−	−	−
+	−	−	−	+	+	+	−
+	−	−	+	−	−	+	+
−	+	−	+	−	+	−	+
−	+	−	−	+	−	−	−
−	−	+	+	+	−	−	+
−	−	+	−	−	+	−	−
+	+	+	−	−	−	+	−
+	+	+	+	+	+	+	+
$\frac{-7}{8}$	$\frac{3}{8}$	$\frac{5}{8}$	$\frac{7}{8}$	$\frac{1}{8}$	$\frac{3}{8}$	$\frac{1}{8}$	$\frac{3}{8}$
−.88	.38	.63	.88	.13	.38	.13	.38

the wider tire, but reduce it for the narrower tire. All other things being equal, it has been a well-established fact that gasoline mileage is better at higher tire pressures.

Although further testing might resolve lingering issues, there was no time for that. The engineering director therefore set the recommended specifications. Deciding that the poor gasoline mileage of the wider tire was far more important than the wearability advantage, she recommended tires with the cross-hatched tread pattern. For those, she chose the narrower tread width, the smaller tread gap, and the shallower grooves. Since

Table 14–7
Separate Confidence Intervals for Effects in Second Tire Experiment

Effect	99% Confidence Interval	
Main		
Tread width W	-7.50 ± 1.90	(significant)
Groove depth D	$-.13 \pm 1.90$	
Tread pattern T	4.63 ± 1.90	(significant)
Track velocity V	-2.63 ± 1.90	(significant)
Interactions		
$W \times D$	2.13 ± 1.90	(significant)
$W \times T$	$.88 \pm 1.90$	
$W \times V$	$.63 \pm 1.90$	
$D \times T$	$-.88 \pm 1.90$	
$D \times V$	$.38 \pm 1.90$	
$T \times V$	$.63 \pm 1.90$	

the first experiment established that tire inflation pressure of 25 psi gave greater wearability for tires of these dimensions, that level was suggested in the specifications, although owners would also be told that better gasoline mileage might be expected from higher tire pressure.

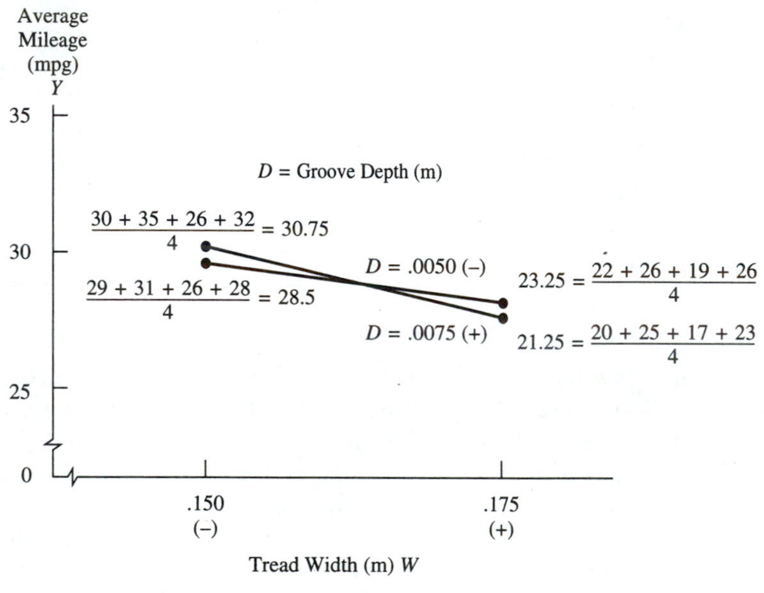

Figure 14–8
Average response for tread width at two groove depths.

Problems

14–4 Suppose that an automotive engineer obtained the following gasoline mileage data in establishing operating standards for a new car model. The car will have two engines, 110 hp and 150 hp; two suspension systems, standard and tight; two passenger loadings, 200 pounds and 500 pounds. Two test cars were driven under each experimental condition.

Test Condition j	Engine Type (hp) E	Suspension S	Passenger Loading (pounds) L	Gasoline Mileage (mpg) Y_{1j}	Y_{2j}
1	110 (−)	standard (−)	200 (−)	32	36
2	150 (+)	standard (−)	200 (−)	39	37
3	110 (−)	tight (+)	200 (−)	26	30
4	150 (+)	tight (+)	200 (−)	29	33
5	110 (−)	standard (−)	500 (+)	24	22
6	150 (+)	standard (−)	500 (+)	36	32
7	110 (−)	tight (+)	500 (+)	28	24
8	150 (+)	tight (+)	500 (+)	29	29

(a) Calculate the main and interaction effects.
(b) Construct 99% individual confidence intervals for the true levels of each effect.
(c) Provide a statement summarizing the results.

14–5 A chemical engineer for an adhesives company has tested two glue types under two formulations and thickness of application. The following data were obtained.

Test Condition j	Glue Type G	Proportion of Inert Material I	Application Layer L	Separation Force (lbs) Y_{1j}	Y_{2j}
1	standard (−)	.50 (−)	thin (−)	62	58
2	epoxy (+)	.50 (−)	thin (−)	81	89
3	standard (−)	.75 (+)	thin (−)	55	63
4	epoxy (+)	.75 (+)	thin (−)	101	89
5	standard (−)	.50 (−)	thick (+)	81	77
6	epoxy (+)	.50 (−)	thick (+)	132	142
7	standard (−)	.75 (+)	thick (+)	90	96
8	epoxy (+)	.75 (+)	thick (+)	155	165

(a) Calculate the main and interaction effects.
(b) Construct 95% individual confidence intervals for the true levels of each effect.
(c) Provide a statement summarizing the results.

14–6 The following data were obtained from a two-level factorial experiment involving two runs for each test condition. The standard deviation in response is $\hat{\sigma}_Y = .6$.
(a) Construct individual 90% confidence intervals for each true effect. Then indicate which effects are significantly nonzero. What is the collective level of confidence for the family of estimates?

Effect	Response
(1) A	5.3
(2) B	−1.5
(3) C	2.4
(4) $A \times B$	1.7
(5) $A \times C$	.5
(6) $B \times C$	−3.4
(7) $A \times B \times C$	.5

(b) Repeat the above using instead individual 95% confidence intervals.
(c) Is the number of significant effects found in (b) greater or smaller than found in (a)? What should you generally expect?

14–7 A chemical engineer is using the following experimental batch results to select chemical processing parameters. These data were obtained from two experimental batches processed under each test condition.

Average Yield (g/l) Y	Temperature (°F) T	Tank Pressure (psi) P	Settling Time (min) S
40	300	150	30
38	400	150	30
41	400	200	60
41	400	150	60
36	300	200	30
45	300	150	60
34	400	200	30
43	300	200	60

(a) Using temperature as the first factor column and tank pressure as the second, organize the data into a table with test conditions that follow the standard form for sign codes.
(b) Calculate the main and interaction effects.
(c) The standard deviation in response has been estimated to be 1.0 g/l. Construct 99% individual confidence intervals for the true levels of each effect.
(d) Provide a statement summarizing the results.

14–8 The following data were obtained from a two-level factorial experiment involving one run for each test condition.
(a) There is no replication. Use the three-factor and four-factor interaction effects to estimate the standard error in effect.
(b) Construct individual 95% confidence intervals for each true main and two-factor interaction effect. Then indicate which effects are significantly nonzero. What is the collective level of confidence for the family of estimates?
(c) Repeat the above using instead individual 99% confidence intervals.
(d) Is the number of significant effects found in (c) greater or smaller than found in (b)? What should you generally expect?

Effect	Response
(1) J	104
(2) K	53
(3) L	−7
(4) M	−25
(5) $J \times K$	6
(6) $J \times L$	9
(7) $J \times M$	−11
(8) $K \times L$	−13
(9) $K \times M$	21
(10) $L \times M$	15
(11) $J \times K \times L$	5
(12) $J \times K \times M$	−7
(13) $J \times L \times M$	−3
(14) $K \times L \times M$	11
(15) $J \times K \times L \times M$	2

14–9 An industrial engineer is evaluating settings for a conveyor system. The following experimental results were obtained from test runs made while the equipment was run with excessive loads. There was no replication.

Test Condition j	Roller Tolerance R	Maintenance Cycle (hrs) C	Avg. Line Speed (ft/sec) S	Avg.Load (lbs) L	Hours to Failure Y_j
1	.125″(−)	1 (−)	1 (−)	50 (−)	70
2	.250″(+)	1 (−)	1 (−)	50 (−)	51
3	.125″(−)	2 (+)	1 (−)	50 (−)	89
4	.250″(+)	2 (+)	1 (−)	50 (−)	72
5	.125″(−)	1 (−)	2 (+)	50 (−)	53
6	.250″(+)	1 (−)	2 (+)	50 (−)	34
7	.125″(−)	2 (+)	2 (+)	50 (−)	73
8	.250″(+)	2 (+)	2 (+)	50 (−)	48
9	.125″(−)	1 (−)	1 (−)	100 (+)	50
10	.250″(+)	1 (−)	1 (−)	100 (+)	42
11	.125″(−)	2 (+)	1 (−)	100 (+)	75
12	.250″(+)	2 (+)	1 (−)	100 (+)	54
13	.125″(−)	1 (−)	2 (+)	100 (+)	50
14	.250″(+)	1 (−)	2 (+)	100 (+)	27
15	.125″(−)	2 (+)	2 (+)	100 (+)	61
16	.250″(+)	2 (+)	2 (+)	100 (+)	36

(a) Calculate the main and interaction effects.
(b) Construct 99% individual confidence intervals for the true levels of each main and two-factor interaction effect.
(c) Provide a statement summarizing the results.

14–10 The following experimental results were obtained from environmental testing of paint. There was no replication.

Test Condition j	Brand B	Location L	Drying Conditions D	Number of Coats C	Drying Time (hours) Y_j
1	X	shade	dry	2	10
2	Y	shade	dry	2	9
3	X	sun	dry	2	11
4	Y	sun	dry	2	9
5	X	shade	humid	2	12
6	Y	shade	humid	2	12
7	X	sun	humid	2	13
8	Y	sun	humid	2	13
9	X	shade	dry	3	11
10	Y	shade	dry	3	8
11	X	sun	dry	3	12
12	Y	sun	dry	3	9
13	X	shade	humid	3	12
14	Y	shade	humid	3	9
15	X	sun	humid	3	13
16	Y	sun	humid	3	12

(a) Calculate the main and interaction effects.
(b) Construct 99% individual confidence intervals for the true levels of each main and two-factor interaction effect.
(c) Provide a statement summarizing the results.

14–3 Other Approaches to Experimental Design

There is room in this book to barely scratch the surface in presenting experimental design methodology. In this section two alternative approaches to factorial designs are described. The section concludes with some comments on extensions of the factorial design.

Analysis of Variance

The alternative analysis-of-variance procedure introduced in Chapter 13 may be used in multifactor decision making. The analysis of variance tests the null hypothesis that the mean response will be identical for all levels for the factor under investigation. The whole process—similar in objective to that for comparing two means encountered in Chapter 11—allows three or more levels (treatment populations) each to be considered simultaneously for one or more factors.

Unlike the experimental design procedure presented in Section 14–2, the validity of an analysis of variance rests on a foundation of explicit theoretical assumptions. Of crucial importance is the linearity and additivity of effects. Conceptual approximations tend to cloud an analysis of variance, however. Experimental design methodology allows for approximations and is generally not hamstrung by simplifying assumptions. Analysis of variance certainly has important engineering applications, even though that approach may lack the force of the experimental design procedures of this chapter. The methods presented earlier in this chapter have a universal and direct application to the basic engineering function—designing products and systems.

Optimization of Quality Loss Function: Taguchi Method

As noted in Chapter 3, traditional statistical process control has been criticized because it places the quality emphasis at the very end of the product cycle rather than at the beginning. Modern thinking moves much of the quality-control emphasis to the engineering phase, where attention is given to designing the final product or system so that it will have robustness when confronted with detrimental forces. It was this focus on "total quality" that led Genichi Taguchi to foster a new approach to be used while a new product or system is still in its design phase. The statistical procedure for doing this is often referred to as the Taguchi method.

Taguchi advocates using a quality loss function to help direct the product design effort. In doing this, quality is considered in a much broader sense than the traditional "defective versus satisfactory." This Taguchi method and statistical design of experiments have common elements and similarities. The goals are essentially the same, with Taguchi's quality loss function operating analogously to a response variable that is to be minimized. The nature of the experiments is very similar as well. Both approaches use data taken from experiments in which various levels of design parameters are systematically tried out.

Like the factorial design, the Taguchi method considers any number of factors simultaneously, typically with three or more levels evaluated for each. As we have seen, a classical factorial experiment for such an evaluation could therefore involve millions, even billions, of possible test conditions. Obviously, testing so many cases would be impractical, so the factorial designs can involve just a small handful of factors. The experimental design approach outlined in Section 14–2 handles the bigger problems by breaking the investigation into successive stages, each encompassing just a small number of factors.

The Taguchi method is much less constrained by problem size. It can incorporate all relevant factors simultaneously, without necessarily breaking the evaluation into stages. Its test conditions do not follow a factorial design of the type suggested earlier in the chapter, but rather they are defined by means of an *orthogonal array*.

Table 14–8 shows the orthogonal array suggested for the problem of choosing one of three levels for each of 13 different factors. These variables are the parameters that govern a car's steering performance, such as spring stiffness and shock absorber stiffness.

Table 14–8
Orthogonal Array for Taguchi Experiment with Car Steering Parameters

Test Condition	Factor and Levels (1, 2, or 3)												
	A	*B*	*C*	*D*	*E*	*F*	*G*	*H*	*I*	*J*	*K*	*L*	*M*
1	1	1	1	1	1	1	1	1	1	1	1	1	1
2	1	1	1	1	2	2	2	2	2	2	2	2	2
3	1	1	1	1	3	3	3	3	3	3	3	3	3
4	1	2	2	2	1	1	1	2	2	2	3	3	3
5	1	2	2	2	2	2	2	3	3	3	1	1	1
6	1	2	2	2	3	3	3	1	1	1	2	2	2
7	1	3	3	3	1	1	1	3	3	3	2	2	2
8	1	3	3	3	2	2	2	1	1	1	3	3	3
9	1	3	3	3	3	3	3	2	2	2	1	1	1
10	2	1	2	3	1	2	3	1	2	3	1	2	3
11	2	1	2	3	2	3	1	2	3	1	2	3	1
12	2	1	2	3	3	1	2	3	1	2	3	1	2
13	2	2	3	1	1	2	3	2	3	1	3	1	2
14	2	2	3	1	2	3	1	3	1	2	1	2	3
15	2	2	3	1	3	1	2	1	2	3	2	3	1
16	2	3	1	2	1	2	3	3	1	2	2	3	1
17	2	3	1	2	2	3	1	1	2	3	3	1	2
18	2	3	1	2	3	1	2	1	2	3	2	3	1
19	3	1	3	2	1	3	2	1	3	2	1	3	2
20	3	1	3	2	2	1	3	2	1	3	2	1	3
21	3	1	3	2	3	2	1	3	2	1	3	2	1
22	3	2	1	3	1	3	2	2	1	3	3	2	1
23	3	2	1	3	2	1	3	3	2	1	1	3	2
24	3	2	1	3	3	2	1	1	3	2	2	1	3
25	3	3	2	1	1	3	2	3	2	1	2	1	3
26	3	3	2	1	2	1	3	1	3	2	3	2	1
27	3	3	2	1	3	2	1	2	1	3	1	3	2

Source: Genichi Taguchi and Don Glausing, "Robust Quality," *Harvard Business Review*, January–February, 1990.

Unlike the factorial design, not all combinations of factor levels are examined, but every factor level is represented in an equal proportion of test conditions. This drastically reduces the number of test cases that need to be considered. (A factorial design for the car steering choices would require $3^{13} = 1{,}594{,}323$ test conditions.) The Taguchi method thereby obtains great testing efficiency, although at the price of being incapable of measuring interactions between factors.

Another difference in that approach is how the experimental data are interpreted. As we have seen, the experimental design approach finds those effects that are statistically significant. The Taguchi method instead determines for each factor a signal-to-noise ratio, identifying as important those levels having high signal-to-noise ratios. The Taguchi method does not rely on any particular probability distributions—or on any theoretical assumptions upon which they implicitly rest—in identifying "significant" factor levels.

The Taguchi method is not without drawbacks. Orthogonal layouts can be complex, especially when there is an uneven number of levels or more than three. The definition of a meaningful quality loss function itself is not a trivial matter, and the signal-to-noise ratio calculation can be elaborate. Nevertheless, this approach is an attractive alternative to the two-level factorial design; its compelling advantage is the low number of test cases needed. Unfortunately, there is not sufficient space in this book to present a detailed description of the Taguchi method.

Fractional Factorial and Other Experimental Designs

We have presented in detail only the two-level factorial experimental design. A major weakness of this approach is the large number of test conditions needed when there is more than a handful of factors. With five factors, a factorial design needs 32 test cases, which may be too many to accommodate. One possibility is to reduce the number of factors. But by using refinements it may still be possible to stay within budget allotments or time constraints without arbitrarily reducing the number of factors. This might be achieved by employing a fractional design. By ignoring unimportant interactions, it is possible to reduce the number of test conditions. A half-fraction design would consider only the main effects and two-factor effects. More details about other experimental design issues may be found in the References in the Appendix to this book.

Problems

14–11 The following orthogonal design applies to an experiment.

Test Condition	Factor and Levels (1, 2, or 3)			
	A	*B*	*C*	*D*
1	1	1	1	1
2	1	2	2	2
3	1	3	3	3
4	2	1	2	3
5	2	2	3	1
6	2	3	1	2
7	3	1	3	2
8	3	2	1	3
9	3	3	2	1

How many test conditions would be required for a complete factorial design?

Review Problems

14–12 A two-level factorial design will be used to evaluate an engineering project having five factors: *A* (low, high), *B* (10,20), *C* (negative, positive), *D* (*X*,*Y*), and *E* (100, 200). Construct a table listing the possible test conditions.

14–13 A chemical engineer is setting chemical chamber parameters for a chemical process. The following experimental results were obtained.

Test Condition j	Catalyst C	Pressure (psi) P	Temperature (°C) T	Percentage Yield Y_{1j}	Y_{2j}
1	A	50	300	43	49
2	B	50	300	38	34
3	A	100	300	55	41
4	B	100	300	36	40
5	A	50	400	54	52
6	B	50	400	41	45
7	A	100	400	47	53
8	B	100	400	41	39

You have been retained as a consultant to evaluate these results. What do you recommend in order to maximize percentage yield?

14–14 A mining engineer wishes to finalize a design for ore cars. The following experimental results were obtained from hauling test cars filled with extra-heavy loads. There was no replication.

Test Condition j	Wheel Diameter (inches) W	Axle Length (inches) A	Wheel Base (feet) B	Load (tons) L	Time to Failure (hrs) Y_j
1	8	36	5	2	17
2	10	36	5	2	15
3	8	40	5	2	18
4	10	40	5	2	14
5	8	36	6	2	17
6	10	36	6	2	21
7	8	40	6	2	22
8	10	40	6	2	19
9	8	36	5	3	7
10	10	36	5	3	11
11	8	40	5	3	9
12	10	40	5	3	12
13	8	36	6	3	10
14	10	36	6	3	8
15	8	40	6	3	8
16	10	40	6	3	11

You have been retained as a consultant to evaluate these results. What do you recommend in order to maximize time between failures?

APPENDIX A

Tables

Table A
Cumulative Values for the Binomial Probability Distribution

$$B(r; n, \pi) = \Pr[R \le r] = \sum_{k=0}^{r} b(k; n, \pi)$$

	r	$\pi = .01$	$\pi = .05$	$\pi = .10$	$\pi = .20$	$\pi = .30$	$\pi = .40$	$\pi = .50$
$n = 1$	0	0.9900	0.9500	0.0900	0.8000	0.7000	0.6000	0.5000
	1	1.0000	1.0000	1.0000	1.0000	1.0000	1.0000	1.0000
$n = 2$	0	0.9801	0.9025	0.8100	0.6400	0.4900	0.3600	0.2500
	1	0.9999	0.9975	0.9900	0.9600	0.9100	0.8400	0.7500
	2	1.0000	1.0000	1.0000	1.0000	1.0000	1.0000	1.0000
$n = 3$	0	0.9703	0.8574	0.7290	0.5120	0.3430	0.2160	0.1250
	1	0.9997	0.9927	0.9720	0.8960	0.7840	0.6480	0.5000
	2	1.0000	0.9999	0.9990	0.9920	0.9730	0.9360	0.8750
	3	1.0000	1.0000	1.0000	1.0000	1.0000	1.0000	1.0000
$n = 4$	0	0.9606	0.8145	0.6561	0.4096	0.2401	0.1296	0.0625
	1	0.9994	0.9860	0.9477	0.8192	0.6517	0.4752	0.3125
	2	1.0000	0.9995	0.9963	0.9728	0.9163	0.8208	0.6875
	3	1.0000	1.0000	0.9999	0.9984	0.9919	0.9744	0.9375
	4	1.0000	1.0000	1.0000	1.0000	1.0000	1.0000	1.0000
$n = 5$	0	0.9510	0.7738	0.5905	0.3277	0.1681	0.0778	0.0313
	1	0.9990	0.9774	0.9185	0.7373	0.5282	0.3370	0.1875
	2	1.0000	0.9988	0.9914	0.9421	0.8369	0.6826	0.5000
	3	1.0000	1.0000	0.9995	0.9933	0.9692	0.9130	0.8125
	4	1.0000	1.0000	1.0000	0.9997	0.9976	0.9898	0.9688
	5				1.0000	1.0000	1.0000	1.0000
$n = 10$	0	0.9044	0.5987	0.3487	0.1074	0.0282	0.0060	0.0010
	1	0.9957	0.9139	0.7361	0.3758	0.1493	0.0464	0.0107
	2	0.9999	0.9885	0.9298	0.6778	0.3828	0.1673	0.0547
	3	1.0000	0.9990	0.9872	0.8791	0.6496	0.3823	0.1719
	4	1.0000	0.9999	0.9984	0.9672	0.8497	0.6331	0.3770
	5	1.0000	1.0000	0.9999	0.9936	0.9526	0.8338	0.6230
	6	1.0000	1.0000	1.0000	0.9991	0.9894	0.9452	0.8281
	7				0.9999	0.9999	0.9877	0.9453
	8				1.0000	1.0000	0.9983	0.9893
	9						0.9999	0.9990
	10						1.0000	1.0000

Table A (continued)

	r	π = .01	π = .05	π = .10	π = .20	π = .30	π = .40	π = .50
$n = 20$	0	0.8179	0.3585	0.1216	0.0115	0.0008	0.0000	0.0000
	1	0.9831	0.7358	0.3917	0.0692	0.0076	0.0005	0.0000
	2	0.9990	0.9245	0.6769	0.2061	0.0355	0.0036	0.0002
	3	1.0000	0.9841	0.8670	0.4114	0.1071	0.0160	0.0013
	4	1.0000	0.9974	0.9568	0.6296	0.2375	0.0510	0.0059
	5	1.0000	0.9997	0.9887	0.8042	0.4164	0.1256	0.0207
	6	1.0000	1.0000	0.9976	0.9133	0.6080	0.2500	0.0577
	7	1.0000	1.0000	0.9996	0.9679	0.7723	0.4159	0.1316
	8	1.0000	1.0000	0.9999	0.9900	0.8867	0.5956	0.2517
	9	1.0000	1.0000	1.0000	0.9974	0.9520	0.7553	0.4119
	10				0.9994	0.9829	0.8725	0.5881
	11				0.9999	0.9949	0.9435	0.7483
	12				1.0000	0.9987	0.9790	0.8684
	13					0.9997	0.9935	0.9423
	14					1.0000	0.9984	0.9793
	15						0.9997	0.9941
	16						1.0000	0.9987
	17							0.9998
	18							1.0000
$n = 50$	0	0.6050	0.0769	0.0052	0.0000	0.0000	0.0000	0.0000
	1	0.9106	0.2794	0.0338	0.0002	0.0000	0.0000	0.0000
	2	0.9862	0.5405	0.1117	0.0013	0.0000	0.0000	0.0000
	3	0.9984	0.7604	0.2503	0.0057	0.0000	0.0000	0.0000
	4	0.9999	0.8964	0.4312	0.0185	0.0002	0.0000	0.0000
	5	1.0000	0.9622	0.6161	0.0480	0.0007	0.0000	0.0000
	6	1.0000	0.9882	0.7702	0.1034	0.0025	0.0000	0.0000
	7	1.0000	0.9968	0.8779	0.1094	0.0073	0.0001	0.0000
	8	1.0000	0.9992	0.9421	0.3073	0.0183	0.0002	0.0000
	9	1.0000	0.9998	0.9755	0.4437	0.0402	0.0008	0.0000
	10	1.0000	1.0000	0.9906	0.5836	0.0789	0.0022	0.0000
	11	1.0000	1.0000	0.9968	0.7107	0.1390	0.0057	0.0000
	12	1.0000	1.0000	0.9990	0.8139	0.2229	0.0133	0.0002
	13	1.0000	1.0000	0.9997	0.8894	0.3279	0.0280	0.0005
	14	1.0000	1.0000	0.9999	0.9393	0.4468	0.0540	0.0013
	15	1.0000	1.0000	1.0000	0.9692	0.5692	0.0955	0.0033

Table A (continued)

	r	$\pi = .01$	$\pi = .05$	$\pi = .10$	$\pi = .20$	$\pi = .30$	$\pi = .40$	$\pi = .50$
$n = 50$	16				0.9856	0.6839	0.1561	0.0077
	17				0.9937	0.7822	0.2369	0.0164
	18				0.9975	0.8594	0.3356	0.0325
	19				0.9991	0.9152	0.4465	0.0595
	20				0.9997	0.9522	0.5610	0.1013
	21				0.9999	0.9749	0.6701	0.1611
	22				1.0000	0.9877	0.7660	0.2399
	23					0.9944	0.8438	0.3359
	24					0.9976	0.9022	0.4439
	25					0.9991	0.9427	0.5561
	26					0.9997	0.9686	0.6641
	27					0.9999	0.9840	0.7601
	28					1.0000	0.9924	0.8389
	29						0.9966	0.8987
	30						0.9986	0.9405
	31						0.9995	0.9675
	32						0.9998	0.9836
	33						0.9999	0.9923
	34						1.0000	0.9967
	35							0.9987
	36							0.9995
	37							0.9998
	38							1.0000

Table A (continued)

	r	π = .01	π = .05	π = .10	π = .20	π = .30	π = .40	π = .50
n = 100	0	0.3660	0.0059	0.0000	0.0000	0.0000	0.0000	0.0000
	1	0.7358	0.0371	0.0003	0.0000	0.0000	0.0000	0.0000
	2	0.9206	0.1183	0.0019	0.0000	0.0000	0.0000	0.0000
	3	0.9816	0.2578	0.0078	0.0000	0.0000	0.0000	0.0000
	4	0.9966	0.4360	0.0237	0.0000	0.0000	0.0000	0.0000
	5	0.9995	0.6160	0.0576	0.0000	0.0000	0.0000	0.0000
	6	0.9999	0.7660	0.1172	0.0001	0.0000	0.0000	0.0000
	7	1.0000	0.8720	0.2061	0.0003	0.0000	0.0000	0.0000
	8	1.0000	0.9369	0.3209	0.0009	0.0000	0.0000	0.0000
	9	1.0000	0.9718	0.4513	0.0023	0.0000	0.0000	0.0000
	10	1.0000	0.9885	0.5832	0.0057	0.0000	0.0000	0.0000
	11	1.0000	0.9957	0.7030	0.0126	0.0000	0.0000	0.0000
	12	1.0000	0.9985	0.8018	0.0253	0.0000	0.0000	0.0000
	13	1.0000	0.9995	0.8761	0.0469	0.0001	0.0000	0.0000
	14	1.0000	0.9999	0.9274	0.0804	0.0002	0.0000	0.0000
	15	1.0000	1.0000	0.9601	0.1285	0.0004	0.0000	0.0000
	16	1.0000	1.0000	0.9794	0.1923	0.0010	0.0000	0.0000
	17	1.0000	1.0000	0.9900	0.2712	0.0022	0.0000	0.0000
	18	1.0000	1.0000	0.9954	0.3621	0.0045	0.0000	0.0000
	19	1.0000	1.0000	0.9980	0.4602	0.0089	0.0000	0.0000
	20	1.0000	1.0000	0.9992	0.5595	0.0165	0.0000	0.0000
	21	1.0000	1.0000	0.9997	0.6540	0.0288	0.0000	0.0000
	22	1.0000	1.0000	0.9999	0.7389	0.0479	0.0001	0.0000
	23	1.0000	1.0000	1.0000	0.8109	0.0755	0.0003	0.0000
	24				0.8686	0.1136	0.0006	0.0000
	25				0.9125	0.1631	0.0012	0.0000
	26				0.9442	0.2244	0.0024	0.0000
	27				0.9658	0.2964	0.0046	0.0000
	28				0.9800	0.3768	0.0084	0.0000
	29				0.9888	0.4623	0.0148	0.0000
	30				0.9939	0.5491	0.0248	0.0000
	31				0.9969	0.6331	0.0398	0.0001
	32				0.9984	0.7107	0.0615	0.0002
	33				0.9993	0.7793	0.0913	0.0004
	34				0.9997	0.8371	0.1303	0.0009
	35				0.9999	0.8839	0.1795	0.0018

Table A (continued)

	r	π = .01	π = .05	π = .10	π = .20	π = .30	π = .40	π = .50
n = 100	36				0.9999	0.9201	0.2386	0.0033
	37				1.0000	0.9470	0.3068	0.0060
	38					0.9660	0.3822	0.0105
	39					0.9790	0.4621	0.0176
	40					0.9875	0.5433	0.0284
	41					0.9928	0.6225	0.0443
	42					0.9960	0.6967	0.0666
	43					0.9979	0.7635	0.0967
	44					0.9989	0.8211	0.1356
	45					0.9995	0.8689	0.1841
	46					0.9997	0.9070	0.2421
	47					0.9999	0.9362	0.3086
	48					0.9999	0.9577	0.3822
	49					1.0000	0.9729	0.4602
	50						0.9832	0.5398
	51						0.9900	0.6178
	52						0.9942	0.6914
	53						0.9968	0.7579
	54						0.9983	0.8159
	55						0.9991	0.8644
	56						0.9996	0.9033
	57						0.9998	0.9334
	58						0.9999	0.9557
	59						1.0000	0.9716
	60							0.9824
	61							0.9895
	62							0.9940
	63							0.9967
	64							0.9982
	65							0.9991
	66							0.9996
	67							0.9998
	68							0.9999
	69							1.0000

Source: Lawrence L. Lapin, *Quantitative Methods for Business Decisions*, 4th ed. Copyright © 1987 Harcourt Brace Jovanovich, Inc., New York. Reproduced by permission of the publisher.

Table B
Exponential Functions

y	e^y	e^{-y}	y	e^y	e^{-y}
0.00	1.0000	1.000000	4.00	54.598	.018316
0.10	1.1052	.904837	4.10	60.340	.016573
0.20	1.2214	.818731	4.20	66.686	.014996
0.30	1.3499	.740818	4.30	73.700	.013569
0.40	1.4918	.670320	4.40	81.451	.012277
0.50	1.6487	.606531	4.50	90.017	.011109
0.60	1.8221	.548812	4.60	99.484	.010052
0.70	2.0138	.496585	4.70	109.95	.009095
0.80	2.2255	.449329	4.80	121.51	.008230
0.90	2.4596	.406570	4.90	134.29	.007447
1.00	2.7183	.367879	5.00	148.41	.006738
1.10	3.0042	.332871	5.10	164.02	.006097
1.20	3.3201	.301194	5.20	181.27	.005517
1.30	3.6693	.272532	5.30	200.34	.004992
1.40	4.0552	.246597	5.40	221.41	.004517
1.50	4.4817	.223130	5.50	244.69	.004087
1.60	4.9530	.201897	5.60	270.43	.003698
1.70	5.4739	.182684	5.70	298.87	.003346
1.80	6.0496	.165299	5.80	330.30	.003028
1.90	6.6859	.149569	5.90	365.04	.002739
2.00	7.3891	.135335	6.00	403.43	.002479
2.10	8.1662	.122456	6.10	445.86	.002243
2.20	9.0250	.110803	6.20	492.75	.002029
2.30	9.9742	.100259	6.30	544.57	.001836
2.40	11.023	.090718	6.40	601.85	.001662
2.50	12.182	.082085	6.50	665.14	.001503
2.60	13.464	.074274	6.60	735.10	.001360
2.70	14.880	.067206	6.70	812.41	.001231
2.80	16.445	.060810	6.80	897.85	.001114
2.90	18.174	.055023	6.90	992.27	.001008
3.00	20.086	.049787	7.00	1096.6	.000912
3.10	22.198	.045049	7.10	1212.0	.000825
3.20	24.533	.040762	7.20	1339.4	.000747
3.30	27.113	.036883	7.30	1480.3	.000676
3.40	29.964	.033373	7.40	1636.0	.000611
3.50	33.115	.030197	7.50	1808.0	.000553
3.60	36.598	.027324	7.60	1998.2	.000501
3.70	40.447	.024724	7.70	2208.3	.000453
3.80	44.701	.022371	7.80	2440.6	.000410
3.90	49.402	.020242	7.90	2697.3	.000371

Table B
(continued)

y	e^y	e^{-y}	y	e^y	e^{-y}
8.00	2981.0	.000336	9.00	8103.1	.000123
8.10	3294.5	.000304	9.10	8955.3	.000112
8.20	3641.0	.000275	9.20	9897.1	.000101
8.30	4023.9	.000249	9.30	10938	.0000914
8.40	4447.1	.000225	9.40	12088	.0000827
8.50	4914.8	.000204	9.50	13360	.0000749
8.60	5431.7	.000184	9.60	14765	.0000677
8.70	6002.9	.000167	9.70	16318	.0000613
8.80	6634.2	.000151	9.80	18034	.0000555
8.90	7332.0	.000136	9.90	19930	.0000502
9.00	8103.1	.000123	10.00	22026	.0000454

Table C *Cumulative Probability Values for the Poisson Distribution*

$$P(x; \lambda, t) = \Pr[X \leq x] = \sum_{k=0}^{x} p(k; \lambda, t)$$

	λt									
x	1.0	2.0	3.0	4.0	5.0	6.0	7.0	8.0	9.0	10.0
0	0.3679	0.1353	0.0498	0.0183	0.0067	0.0025	0.0009	0.0003	0.0001	0.0000
1	0.7358	0.4060	0.1991	0.0916	0.0404	0.0174	0.0073	0.0030	0.0012	0.0005
2	0.9197	0.6767	0.4232	0.2381	0.1247	0.0620	0.0296	0.0138	0.0062	0.0028
3	0.9810	0.8571	0.6472	0.4335	0.2650	0.1512	0.0818	0.0424	0.0212	0.0103
4	0.9963	0.9473	0.8153	0.6288	0.4405	0.2851	0.1730	0.0996	0.0550	0.0293
5	0.9994	0.9834	0.9161	0.7851	0.6160	0.4457	0.3007	0.1912	0.1157	0.0671
6	0.9999	0.9955	0.9665	0.8893	0.7622	0.6063	0.4497	0.3134	0.2068	0.1301
7	1.0000	0.9989	0.9881	0.9489	0.8666	0.7440	0.5987	0.4530	0.3239	0.2202
8		0.9998	0.9962	0.9786	0.9319	0.8472	0.7291	0.5926	0.4557	0.3328
9		1.0000	0.9989	0.9919	0.9682	0.9161	0.8305	0.7166	0.5874	0.4579
10			0.9997	0.9972	0.9863	0.9574	0.9015	0.8159	0.7060	0.5830
11			0.9999	0.9991	0.9945	0.9799	0.9466	0.8881	0.8030	0.6968
12			1.0000	0.9997	0.9980	0.9912	0.9730	0.9362	0.8758	0.7916
13				0.9999	0.9993	0.9964	0.9872	0.9658	0.9262	0.8645
14				1.0000	0.9998	0.9986	0.9943	0.9827	0.9585	0.9165
15					0.9999	0.9995	0.9976	0.9918	0.9780	0.9513
16					1.0000	0.9998	0.9990	0.9963	0.9889	0.9730
17						0.9999	0.9996	0.9984	0.9947	0.9857
18						1.0000	0.9999	0.9993	0.9976	0.9928
19							0.9999	0.9997	0.9989	0.9965
20							1.0000	0.9999	0.9996	0.9984
21								1.0000	0.9998	0.9993
22									0.9999	0.9997
23									1.0000	0.9999
24										0.9999
25										1.0000

Source: Lawrence L. Lapin, *Quantitative Methods for Business Decisions*, 4th ed. Copyright © 1987 Harcourt Brace Jovanovich, Inc., New York. Reproduced by permission of the publisher.

Table C (continued)

	λt									
x	11.0	12.0	13.0	14.0	15.0	16.0	17.0	18.0	19.0	20.0
0	0.0000	0.0000	0.0000	0.0000	0.0000	0.0000	0.0	0.0	0.0	0.0
1	0.0002	0.0001	0.0000	0.0000	0.0000	0.0000	0.0000	0.0000	0.0000	0.0
2	0.0012	0.0005	0.0002	0.0001	0.0000	0.0000	0.0000	0.0000	0.0000	0.0000
3	0.0049	0.0023	0.0011	0.0005	0.0002	0.0001	0.0000	0.0000	0.0000	0.0000
4	0.0151	0.0076	0.0037	0.0018	0.0009	0.0004	0.0002	0.0001	0.0000	0.0000
5	0.0375	0.0203	0.0107	0.0055	0.0028	0.0014	0.0007	0.0003	0.0002	0.0001
6	0.0786	0.0458	0.0259	0.0142	0.0076	0.0040	0.0021	0.0010	0.0005	0.0003
7	0.1432	0.0895	0.0540	0.0316	0.0180	0.0100	0.0054	0.0029	0.0015	0.0008
8	0.2320	0.1550	0.0998	0.0621	0.0374	0.0220	0.0126	0.0071	0.0039	0.0021
9	0.3405	0.2424	0.1658	0.1094	0.0699	0.0433	0.0261	0.0154	0.0089	0.0050
10	0.4599	0.3472	0.2517	0.1757	0.1185	0.0774	0.0491	0.0304	0.0183	0.0108
11	0.5793	0.4616	0.3532	0.2600	0.1847	0.1270	0.0847	0.0549	0.0347	0.0214
12	0.6887	0.5760	0.4631	0.3585	0.2676	0.1931	0.1350	0.0917	0.0606	0.0390
13	0.7813	0.6815	0.5730	0.4644	0.3632	0.2745	0.2009	0.1426	0.00984	0.0061
14	0.8540	0.7720	0.6751	0.5704	0.4656	0.3675	0.2808	0.2081	0.1497	0.1049
15	0.9074	0.8444	0.7636	0.6694	0.5681	0.4667	0.3714	0.2866	0.2148	0.1565
16	0.9441	0.8987	0.8355	0.7559	0.6641	0.5660	0.4677	0.3750	0.2920	0.2211
17	0.9678	0.9370	0.8905	0.8272	0.7489	0.6593	0.5640	0.4686	0.3784	0.2970
18	0.9823	0.9626	0.9302	0.8826	0.8195	0.7423	0.6549	0.5622	0.4695	0.3814
19	0.9907	0.9787	0.9573	0.9235	0.8752	0.8122	0.7363	0.6509	0.5606	0.4703
20	0.9953	0.9884	0.9750	0.9521	0.9170	0.8682	0.8055	0.7307	0.6472	0.5591
21	0.9977	0.9939	0.9859	0.9711	0.9469	0.9108	0.8615	0.7991	0.7255	0.6437
22	0.9989	0.9969	0.9924	0.9833	0.9672	0.9418	0.9047	0.8551	0.7931	0.7206
23	0.9995	0.9985	0.9960	0.9907	0.9805	0.9633	0.9367	0.8989	0.8490	0.7875
24	0.9998	0.9993	0.9980	0.9950	0.9888	0.9777	0.9593	0.9317	0.8933	0.8432
25	0.9999	0.9997	0.9990	0.9974	0.9938	0.9869	0.9747	0.9554	0.9269	0.8878
26	1.0000	0.9999	0.9995	0.9987	0.9967	0.9925	0.9848	0.9718	0.9514	0.9221
27		0.9999	0.9998	0.9994	0.9983	0.9959	0.9912	0.9827	0.9687	0.9475
28		1.0000	0.9999	0.9997	0.9991	0.9978	0.9950	0.9897	0.9805	0.9657
29			1.0000	0.9999	0.9996	0.9989	0.9973	0.9940	0.9881	0.9782

Table C (continued)

	λt									
x	11.0	12.0	13.0	14.0	15.0	16.0	17.0	18.0	19.0	20.0
30				0.9999	0.9998	0.9994	0.9985	0.9967	0.9930	0.9865
31				1.0000	0.9999	0.9997	0.9992	0.9982	0.9960	0.9919
32					0.9999	0.9999	0.9996	0.9990	0.9978	0.9953
33					1.0000	0.9999	0.9998	0.9995	0.9988	0.9973
34						1.0000	0.9999	0.9997	0.9994	0.9985
35							0.9999	0.9999	0.9997	0.9992
36							1.0000	0.9999	0.9998	0.9996
37								1.0000	0.9999	0.9998
38									1.0000	0.9999
39										0.9999
40										1.0000

Table D
Cumulative Probability Distribution Function for the Standard Normal Distribution

$$\Phi(z) = \Pr[Z \leq z] = \int_{-\infty}^{z} \frac{1}{\sqrt{2\pi}} e^{\frac{-y^2}{2}} dy$$

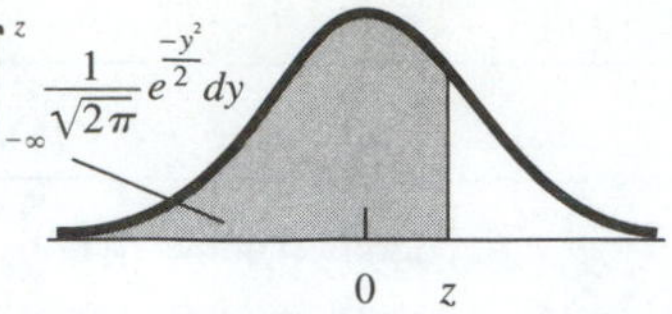

Normal Deviate z	.00	.01	.02	.03	.04	.05	.06	.07	.08	.09
−4.0	.00003									
−3.2	.0007	.0007	.0006	.0006	.0006	.0006	.0006	.0005	.0005	.0005
−3.1	.0010	.0009	.0009	.0009	.0008	.0008	.0008	.0008	.0007	.0007
−3.0	.0014	.0013	.0013	.0012	.0012	.0011	.0011	.0011	.0010	.0010
−2.9	.0019	.0018	.0018	.0017	.0016	.0016	.0015	.0015	.0014	.0014
−2.8	.0026	.0025	.0024	.0023	.0023	.0022	.0021	.0021	.0020	.0019
−2.7	.0035	.0034	.0033	.0032	.0031	.0030	.0029	.0028	.0027	.0026
−2.6	.0047	.0045	.0044	.0043	.0041	.0040	.0039	.0038	.0037	.0036
−2.5	.0062	.0060	.0059	.0057	.0055	.0054	.0052	.0051	.0049	.0048
−2.4	.0082	.0080	.0078	.0075	.0073	.0071	.0069	.0068	.0066	.0064
−2.3	.0107	.0104	.0102	.0099	.0096	.0094	.0091	.0089	.0087	.0084
−2.2	.0139	.0136	.0132	.0129	.0125	.0122	.0119	.0116	.0113	.0110
−2.1	.0179	.0174	.0170	.0166	.0162	.0158	.0154	.0150	.0146	.0143
−2.0	.0228	.0222	.0217	.0212	.0207	.0202	.0197	.0192	.0188	.0183
−1.9	.0287	.0281	.0274	.0268	.0262	.0256	.0250	.0244	.0239	.0233
−1.8	.0359	.0351	.0344	.0336	.0329	.0322	.0314	.0307	.0301	.0294
−1.7	.0446	.0436	.0427	.0418	.0409	.0401	.0392	.0384	.0375	.0367
−1.6	.0548	.0537	.0526	.0516	.0505	.0495	.0485	.0475	.0465	.0455
−1.5	.0668	.0655	.0643	.0630	.0618	.0606	.0594	.0582	.0571	.0559
−1.4	.0808	.0793	.0778	.0764	.0749	.0735	.0721	.0708	.0694	.0681
−1.3	.0968	.0951	.0934	.0918	.0901	.0885	.0869	.0853	.0838	.0823
−1.2	.1151	.1131	.1112	.1093	.1075	.1056	.1038	.1020	.1003	.0985
−1.1	.1357	.1335	.1314	.1292	.1271	.1251	.1230	.1210	.1190	.1170
−1.0	.1587	.1562	.1539	.1515	.1492	.1469	.1446	.1423	.1401	.1379
−0.9	.1841	.1814	.1788	.1762	.1736	.1711	.1685	.1660	.1635	.1611
−0.8	.2119	.2090	.2061	.2033	.2005	.1977	.1949	.1922	.1894	.1867
−0.7	.2420	.2388	.2358	.2327	.2296	.2266	.2236	.2206	.2177	.2148
−0.6	.2743	.2709	.2676	.2643	.2611	.2578	.2546	.2514	.2482	.2451
−0.5	.3085	.3050	.3015	.2981	.2946	.2912	.2877	.2843	.2810	.2776
−0.4	.3446	.3409	.3372	.3336	.3300	.3264	.3228	.3192	.3156	.3121
−0.3	.3821	.3783	.3745	.3707	.3669	.3632	.3594	.3557	.3520	.3483
−0.2	.4207	.4168	.4129	.4090	.4052	.4013	.3974	.3936	.3897	.3859
−0.1	.4602	.4562	.4522	.4483	.4443	.4404	.4364	.4325	.4286	.4247
−0.0	.5000	.4960	.4920	.4880	.4840	.4801	.4761	.4721	.4681	.4641

Table D (continued)

Normal Deviate z	.00	.01	.02	.03	.04	.05	.06	.07	.08	.09
0.0	.5000	.5040	.5080	.5120	.5160	.5199	.5239	.5279	.5319	.5359
0.1	.5398	.5438	.5478	.5517	.5557	.5596	.5636	.5675	.5714	.5753
0.2	.5793	.5832	.5871	.5910	.5948	.5987	.6026	.6064	.6103	.6141
0.3	.6179	.6217	.6255	.6293	.6331	.6368	.6406	.6443	.6480	.6517
0.4	.6554	.6591	.6628	.6664	.6700	.6736	.6772	.6808	.6844	.6879
0.5	.6915	.6950	.6985	.7019	.7054	.7088	.7123	.7157	.7190	.7224
0.6	.7257	.7291	.7324	.7357	.7389	.7422	.7454	.7486	.7518	.7549
0.7	.7580	.7612	.7642	.7673	.7704	.7734	.7764	.7794	.7823	.7852
0.8	.7881	.7910	.7939	.7967	.7995	.8023	.8051	.8078	.8106	.8133
0.9	.8159	.8186	.8212	.8238	.8264	.8289	.8315	.8340	.8365	.8389
1.0	.8413	.8438	.8461	.8485	.8508	.8531	.8554	.8577	.8599	.8621
1.1	.8643	.8665	.8686	.8708	.8729	.8749	.8770	.8790	.8810	.8830
1.2	.8849	.8869	.8888	.8907	.8925	.8944	.8962	.8980	.8997	.9015
1.3	.9032	.9049	.9066	.9082	.9099	.9115	.9131	.9147	.9162	.9177
1.4	.9192	.9207	.9222	.9236	.9251	.9265	.9279	.9292	.9306	.9319
1.5	.9332	.9345	.9357	.9370	.9382	.9394	.9406	.9418	.9429	.9441
1.6	.9452	.9463	.9474	.9484	.9495	.9505	.9515	.9525	.9535	.9545
1.7	.9554	.9564	.9573	.9582	.9591	.9599	.9608	.9616	.9625	.9633
1.8	.9641	.9649	.9656	.9664	.9671	.9678	.9686	.9693	.9699	.9706
1.9	.9713	.9719	.9726	.9732	.9738	.9744	.9750	.9756	.9761	.9767
2.0	.9772	.9778	.9783	.9788	.9793	.9798	.9803	.9808	.9812	.9817
2.1	.9821	.9826	.9830	.9834	.9838	.9842	.9846	.9850	.9854	.9857
2.2	.9861	.9864	.9868	.9871	.9875	.9878	.9881	.9884	.9887	.9890
2.3	.9893	.9896	.9898	.9901	.9904	.9906	.9909	.9911	.9913	.9916
2.4	.9918	.9920	.9922	.9925	.9927	.9929	.9931	.9932	.9934	.9936
2.5	.9938	.9940	.9941	.9943	.9945	.9946	.9948	.9949	.9951	.9952
2.6	.9953	.9955	.9956	.9957	.9959	.9960	.9961	.9962	.9963	.9964
2.7	.9965	.9966	.9967	.9968	.9969	.9970	.9971	.9972	.9973	.9974
2.8	.9974	.9975	.9976	.9977	.9977	.9978	.9979	.9979	.9980	.9981
2.9	.9981	.9982	.9982	.9983	.9984	.9984	.9985	.9985	.9986	.9986
3.0	.9986	.9987	.9987	.9988	.9988	.9989	.9989	.9989	.9990	.9990
3.1	.9990	.9991	.9991	.9991	.9992	.9992	.9992	.9992	.9993	.9993
3.2	.9993	.9993	.9994	.9994	.9994	.9994	.9994	.9995	.9995	.9995
3.3	.99952									
3.4	.99966									
3.5	.99977									
3.6	.99984									
3.7	.99989									
3.8	.99993									
3.9	.99995									
4.0	.99997									
4.5	1.00000									

Table E
Critical Normal Deviate Values

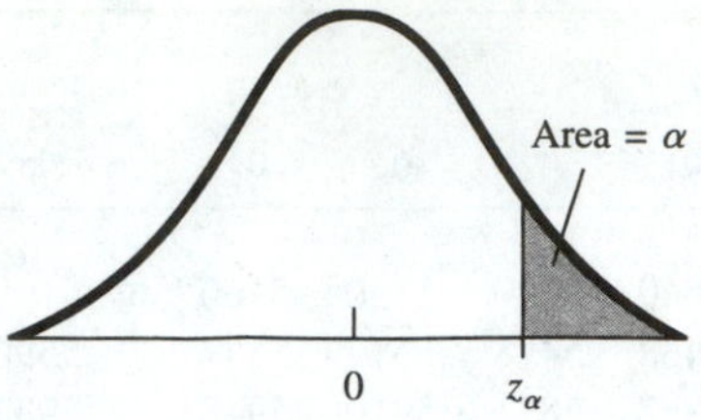

Upper-Tail Area α	Normal Deviate z_α
.10	1.28
.05	1.64
.025	1.96
.01	2.33
.005	2.57
.001	3.08
.0005	3.30

Table F
Random Numbers

12651	61646	11769	75109	86996	97669	25757	32535	07122	76763
81769	74436	02630	72310	45049	18029	07469	42341	98173	79260
36737	98863	77240	76251	00654	64688	09343	70278	67331	98729
82861	54371	76610	94934	72748	44124	05610	53750	95938	01485
21325	15732	24127	37431	09723	63529	73977	95218	96074	42138
74146	47887	62463	23045	41490	07954	22597	60012	98866	90959
90759	64410	54179	66075	61051	75385	51378	08360	95946	95547
55683	98078	02238	91540	21219	17720	87817	41705	95785	12563
79686	17969	76061	83748	55920	83612	41540	86492	06447	60568
70333	00201	86201	69716	78185	62154	77930	67663	29529	75116
14042	53536	07779	04157	41172	36473	42123	43929	50533	33437
59911	08256	06596	48416	69770	68797	56080	14223	59199	30162
62368	62623	62742	14891	39247	52242	98832	69533	91174	57979
57529	97751	54976	48957	74599	08759	78494	52785	68526	64618
15469	90574	78033	66885	13936	42117	71831	22961	94225	31816
18625	23674	53850	32827	81647	80820	00420	63555	74489	80141
74626	68394	88562	70745	23701	45630	65891	58220	35442	60414
11119	16519	27384	90199	79210	76965	99546	30323	31664	22845
41101	17336	48951	53674	17880	45260	08575	49321	36191	17095
32123	91576	84221	78902	82010	30847	62329	63898	23268	74283
26091	68409	69704	82267	14751	13151	93115	01437	56945	89661
67680	79790	48462	59278	44185	29616	76531	19589	83139	28454
15184	19260	14073	07026	25264	08388	27182	22557	61501	67481
58010	45039	57181	10238	36874	28546	37444	80824	63981	39942
56425	53996	86245	32623	78858	08143	60377	42925	42815	11159
82630	84066	13592	60642	17904	99718	63432	88642	37858	25431
14927	40909	23900	48761	44860	92467	31742	87142	03607	32059
23740	22505	07489	85986	74420	21744	97711	36648	35620	97949
32990	97446	03711	63824	07953	85965	87089	11687	92414	67257
05310	24058	91946	78437	34365	82469	12430	84754	19354	72745
21839	39937	27534	88913	49055	19218	47712	67677	51889	70926
08833	42549	93981	94051	28382	83725	72643	64233	97252	17133
58336	11139	47479	00931	91560	95372	97642	33856	54825	55680
62032	91144	75478	47431	52726	30289	42411	91886	51818	78292
45171	30557	53116	04118	58301	24375	65609	85810	18620	49198
91611	62656	60128	35609	63698	78356	50682	22505	01692	36291
55472	63819	86314	49174	93528	73604	78614	78849	23096	72825
18573	09729	74091	53994	10970	86557	65661	41854	26037	53296
60866	02955	90288	82136	83644	94455	06560	78029	98768	71296
45043	55608	82767	60890	74646	79485	13619	98868	40857	19415
17831	09737	79473	75945	28394	79334	70577	38048	03607	06932
40137	03981	07585	18128	11178	32601	27994	05641	22600	86064
77776	31343	14576	97706	16039	47517	43300	59080	80392	63189
69605	44104	40103	95635	05635	81673	68657	09559	23510	95875
19916	52934	26499	09821	87331	80993	61299	36979	73599	35055
02606	58552	07678	56619	65325	30705	99582	53390	46357	13244
65183	73160	87131	35530	47946	09854	18080	02321	05809	04898
10740	98914	44916	11322	89717	88189	30143	52687	19420	60061
98642	89822	71691	51573	83666	61642	46683	33761	47542	23551
60139	25601	93663	25547	02654	94829	48672	28736	84994	13071

Source: Reprinted from page 103 of *A Million Random Digits with 100,000 Normal Deviates*, by the Rand Corporation. New York: The Free Press, 1955.

Table G
Student t Distribution

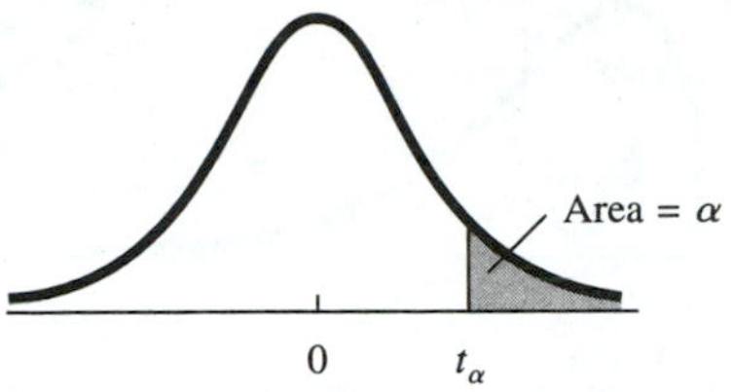

The following table provides the values of t_α that correspond to a given upper-tail area α and a specified number of degrees of freedom.

Degrees of Freedom	Upper-Tail Area α									
	.4	.25	.1	.05	.025	.01	.005	.0025	.001	.0005
1	.325	1.000	3.078	6.314	12.706	31.821	63.657	127.32	318.31	636.62
2	.289	.816	1.886	2.920	4.303	6.965	9.925	14.089	22.327	31.598
3	.277	.765	1.638	2.353	3.182	4.541	5.841	7.453	10.214	12.924
4	.271	.741	1.533	2.132	2.776	3.747	4.604	5.598	7.173	8.610
5	.267	.727	1.476	2.015	2.571	3.365	4.032	4.773	5.893	6.869
6	.265	.718	1.440	1.943	2.447	3.143	3.707	4.317	5.208	5.959
7	.263	.711	1.415	1.895	2.365	2.998	3.499	4.029	4.785	5.408
8	.262	.706	1.397	1.860	2.306	2.896	3.355	3.833	4.501	5.041
9	.261	.703	1.383	1.833	2.262	2.821	3.250	3.690	4.297	4.781
10	.260	.700	1.372	1.812	2.228	2.764	3.169	3.581	4.144	4.587
11	.260	.697	1.363	1.796	2.201	2.718	3.106	3.497	4.025	4.437
12	.259	.695	1.356	1.782	2.179	2.681	3.055	3.428	3.930	4.318
13	.259	.694	1.350	1.771	2.160	2.650	3.012	3.372	3.852	4.221
14	.258	.692	1.345	1.761	2.145	2.624	2.977	3.326	3.787	4.140
15	.258	.691	1.341	1.753	2.131	2.602	2.947	3.286	3.733	4.073
16	.258	.690	1.337	1.746	2.120	2.583	2.921	3.252	3.686	4.015
17	.257	.689	1.333	1.740	2.110	2.567	2.898	3.222	3.646	3.965
18	.257	.688	1.330	1.734	2.101	2.552	2.878	3.197	3.610	3.922
19	.257	.688	1.328	1.729	2.093	2.539	2.861	3.174	3.579	3.883
20	.257	.687	1.325	1.725	2.086	2.528	2.845	3.153	3.552	3.850
21	.257	.686	1.323	1.721	2.080	2.518	2.831	3.135	3.527	3.819
22	.256	.686	1.321	1.717	2.074	2.508	2.819	3.119	3.505	3.792
23	.256	.685	1.319	1.714	2.069	2.500	2.807	3.104	3.485	3.767
24	.256	.685	1.318	1.711	2.064	2.492	2.797	3.091	3.467	3.745
25	.256	.684	1.316	1.708	2.060	2.485	2.787	3.078	3.450	3.725
26	.256	.684	1.315	1.706	2.056	2.479	2.779	3.067	3.435	3.707
27	.256	.684	1.314	1.703	2.052	2.473	2.771	3.057	3.421	3.690
28	.256	.683	1.313	1.701	2.048	2.467	2.763	3.047	3.408	3.674
29	.256	.683	1.311	1.699	2.045	2.462	2.756	3.038	3.396	3.659
30	.256	.683	1 310	1.697	2.042	2.457	2.750	3.030	3.385	3.646
40	.255	.681	1.303	1.684	2.021	2.423	2.704	2.971	3.307	3.551
60	.254	.679	1.296	1.671	2.000	2.390	2.660	2.915	3.232	3.460
120	.254	.677	1.289	1.658	1.980	2.358	2.617	2.860	3.160	3.373
∞	.253	.674	1.282	1.645	1.960	2.326	2.576	2.807	3.090	3.291

Source: E. S. Pearson and H. O. Hartley, *Biometrika Tables for Statisticians.* Vol. I. London: Cambridge University Press, 1966. Partly derived from Table III of Fisher and Yates, *Statistical Tables for Biological, Agricultural, and Medical Research*, published by Longman Group Ltd., London (previously published by Oliver & Boyd, Edinbugh, 1963). Reproduced with permission of the authors and publishers.

Table H
Chi-Square Distribution

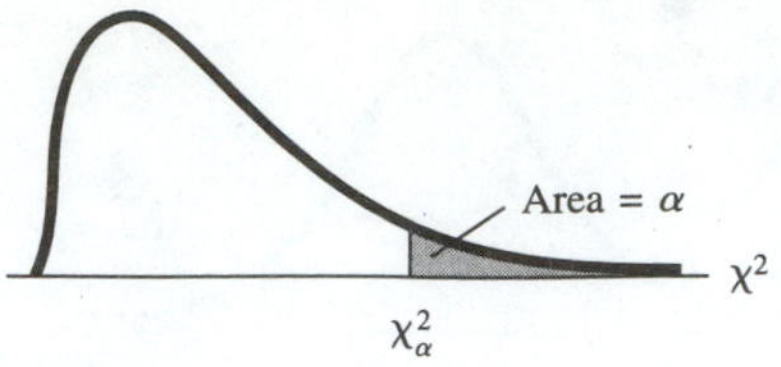

The following table provides the values of χ^2_α that correspond to a given upper-tail area α and a specified number of degrees of freedom.

Degrees of Freedom	Upper-tail Area α						
	.99	.98	.95	.90	.80	.70	.50
1	.000157	.000628	.00393	.0158	.0642	.148	.455
2	.0201	.0404	.103	.211	.446	.713	1.386
3	.115	.185	.352	.584	1.005	1.424	2.366
4	.297	.429	.711	1.064	1.649	2.195	3.357
5	.554	.752	1.145	1.610	2.343	3.000	4.351
6	.872	1.134	1.635	2.204	3.070	3.828	5.348
7	1.239	1.564	2.167	2.833	3.822	4.671	6.346
8	1.646	2.032	2.733	3.490	4.594	5.527	7.344
9	2.088	2.532	3.325	4.168	5.380	6.393	8.343
10	2.558	3.059	3.940	4.865	6.179	7.267	9.342
11	3.053	3.609	4.575	5.578	6.989	8.148	10.341
12	3.571	4.178	5.226	6.304	7.807	9.034	11.340
13	4.107	4.765	5.892	7.042	8.634	9.926	12.340
14	4.660	5.368	6.571	7.790	9.467	10.821	13.339
15	5.229	5.985	7.261	8.547	10.307	11.721	14.339
16	5.812	6.614	7.962	9.312	11.152	12.624	15.338
17	6.408	7.255	8.672	10.085	12.002	13.531	16.338
18	7.015	7.906	9.390	10.865	12.857	14.440	17.338
19	7.633	8.567	10.117	11.651	13.716	15.352	18.338
20	8.260	9.237	10.851	12.443	14.578	16.266	19.337
21	8.897	9.915	11.591	13.240	15.445	17.182	20.337
22	9.542	10.600	12.338	14.041	16.314	18.101	21.337
23	10.196	11.293	13.091	14.848	17.187	19.021	22.337
24	10.856	11.992	13.848	15.659	18.062	19.943	23.337
25	11.524	12.697	14.611	16.473	18.940	20.867	24.337
26	12.198	13.409	15.379	17.292	19.820	21.792	25.336
27	12.879	14.125	16.151	18.114	20.703	22.719	26.336
28	13.565	14.847	16.928	18.939	21.588	23.647	27.336
29	14.256	15.574	17.708	19.768	22.475	24.577	28.336
30	14.953	16.306	18.493	20.599	23.364	25.508	29.336

Table H (continued)

Degrees of Freedom	Upper-Tail Area α						
	.30	.20	.10	.05	.02	.01	.001
1	1.074	1.642	2.706	3.841	5.412	6.635	10.827
2	2.408	3.219	4.605	5.991	7.824	9.210	13.815
3	3.665	4.642	6.251	7.815	9.837	11.345	16.268
4	4.878	5.989	7.779	9.488	11.668	13.277	18.465
5	6.064	7.289	9.236	11.070	13.388	15.086	20.517
6	7.231	8.558	10.645	12.592	15.033	16.812	22.457
7	8.383	9.803	12.017	14.067	16.622	18.475	24.322
8	9.524	11.030	13.362	15.507	18.168	20.090	26.125
9	10.656	12.242	14.684	16.919	19.679	21.666	27.877
10	11.781	13.442	15.987	18.307	21.161	23.209	29.588
11	12.899	14.631	17.275	19.675	22.618	24.725	31.264
12	14.011	15.812	18.549	21.026	24.054	26.217	32.909
13	15.119	16.985	19.812	22.362	25.472	27.688	34.528
14	16.222	18.151	21.064	23.685	26.873	29.141	36.123
15	17.322	19.311	22.307	24.996	28.259	30.578	37.697
16	18.418	20.465	23.542	26.296	29.633	32.000	39.252
17	19.511	21.615	24.769	27.587	30.995	33.409	40.790
18	20.601	22.760	25.989	28.869	32.346	34.805	42.312
19	21.689	23.900	27.204	30.144	33.687	36.191	43.820
20	22.775	25.038	28.412	31.410	35.020	37.566	45.315
21	23.858	26.171	29.615	32.671	36.343	38.932	46.797
22	24.939	27.301	30.813	33.924	37.659	40.289	48.268
23	26.018	28.429	32.007	35.172	38.968	41.638	49.728
24	27.096	29.553	33.196	36.415	40.270	42.980	51.179
25	28.172	30.675	34.382	37.652	41.566	44.314	52.620
26	29.246	31.795	35.563	38.885	42.856	45.642	54.052
27	30.319	32.912	36.741	40.113	44.140	46.963	55.476
28	31.391	34.027	37.916	41.337	45.419	48.278	56.893
29	32.461	35.139	39.087	42.557	46.693	49.588	58.302
30	33.530	36.250	40.256	43.773	47.962	50.892	59.703

Source: From Table IV of Fisher and Yates, *Statistical Tables for Biological, Agricultural and Medical Research*, published by Longman Group Ltd., London (previously published by Oliver & Boyd, Edinburgh, 1963). Reproduced with permission of the authors and publishers.

Table I
Conversion Table for Correlation Coefficient and Fisher's Z′

ρ	Z′	ρ	Z′	ρ	Z′	ρ	Z′
.00	.0000	.25	.2554	.50	.5493	.75	.973
.01	.0100	.26	.2661	51	.5627	.76	.996
.02	.0200	.27	.2769	.52	.5763	.77	1.020
.03	.0300	.28	.2877	.53	.5901	.78	1.045
.04	.0400	.29	.2986	.54	.6042	.79	1.071
.05	.0500	.30	.3095	.55	6184	.80	1.099
.06	.0601	.31	.3205	.56	.6328	.81	1.127
.07	.0701	.32	.3316	.57	.6475	.82	1.157
.08	.0802	.33	.3428	.58	.6625	.83	1.188
.09	.0902	.34	.3541	.59	.6777	.84	1.221
.10	.1003	.35	.3654	.60	.6931	.85	1.256
.11	.1104	.36	.3769	.61	.7089	.86	1.293
.12	.1206	.37	.3884	.62	.7250	.87	1.333
.13	.1307	.38	.4001	.63	.7414	.88	1.376
.14	.1409	.39	.4118	.64	.7582	.89	1.422
.15	.1511	.40	.4236	.65	.7753	.90	1.472
.16	.1614	.41	.4356	.66	.7928	.91	1.528
.17	.1717	.42	.4477	.67	.8107	.92	1.589
.18	.1820	.43	.4599	.68	.8291	.93	1.658
.19	.1923	.44	.4722	.69	.8480	.94	1.738
.20	.2027	.45	.4847	.70	.8673	.95	1.832
.21	.2132	.46	.4973	.71	.8872	.96	1.946
.22	.2237	.47	.5101	.72	.9076	.97	2.092
.23	.2342	.48	.5230	.73	.9287	.98	2.298
.24	.2448	.49	.5361	.74	.9505	.99	2.647

Source: Abridged from Table 14 in E. S. Pearson and H. O. Hartley, *Biometrika Tables for Statisticians,* Vol. 1. Cambridge, England: Cambridge University Press, on behalf of The Biometrika Society, 1966. By permission of the authors and publishers.

Table J
Critical Values for F Distribution

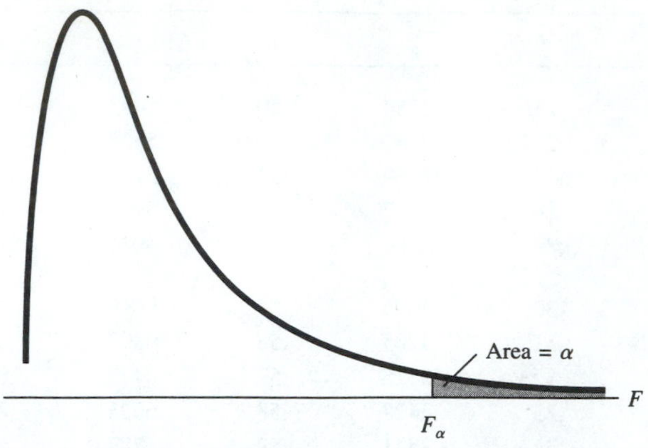

The following table provides the values of F_α that correspond to a given upper-tail area α and a specified degrees of freedom pair. The values of F_{05} are in lightface type, while those for F_{01} are given in boldface type.

Degrees of Freedom for Denominator	Degrees of Freedom for Numerator											
	1	2	3	4	5	6	7	8	9	10	11	12
1	161	200	216	225	230	234	237	239	241	242	243	244
	4,052	**4,999**	**5,403**	**5,625**	**5,764**	**5,859**	**5,928**	**5,981**	**6,022**	**6,056**	**6,082**	**6,106**
2	18.51	19.00	19.16	19.25	19.30	19.33	19.36	19.37	19.38	19.39	19.40	19.41
	98.49	**99.01**	**99.17**	**99.25**	**99.30**	**99.33**	**99.34**	**99.36**	**99.38**	**99.40**	**99.41**	**99.42**
3	10.13	9.55	9.28	9.12	9.01	8.94	8.88	8.84	8.81	8.78	8.76	8.74
	34.12	**30.81**	**29.46**	**28.71**	**28.24**	**27.91**	**27.67**	**27.49**	**27.34**	**27.23**	**27.13**	**27.05**
4	7.71	6.94	6.59	6.39	6.26	6.16	6.09	6.04	6.00	5.96	5.93	5.91
	21.20	**18.00**	**16.69**	**15.98**	**15.52**	**15.21**	**14.98**	**14.80**	**14.66**	**14.54**	**14.45**	**14.37**
5	6.61	5.79	5.41	5.19	5.05	4.95	4.88	4.82	4.78	4.74	4.70	4.68
	16.26	**13.27**	**12.06**	**11.39**	**10.97**	**10.67**	**10.45**	**10.27**	**10.15**	**10.05**	**9.96**	**9.89**
6	5.99	5.14	4.76	4.53	4.39	4.28	4.21	4.15	4.10	4.06	4.03	4.00
	13.74	**10.92**	**9.78**	**9.15**	**8.75**	**8.47**	**8.26**	**8.10**	**7.98**	**7.87**	**7.79**	**7.72**
7	5.59	4.74	4.35	4.12	3.97	3.87	3.79	3.73	3.68	3.63	3.60	3.57
	12.25	**9.55**	**8.45**	**7.85**	**7.46**	**7.19**	**7.00**	**6.84**	**6.71**	**6.62**	**6.54**	**6.47**
8	5.32	4.46	4.07	3.84	3.69	3.58	3.50	3.44	3.39	3.34	3.31	3.28
	11.26	**8.65**	**7.59**	**7.01**	**6.63**	**6.37**	**6.19**	**6.03**	**5.91**	**5.82**	**5.74**	**5.67**
9	5.12	4.26	3.86	3.63	3.48	3.37	3.29	3.23	3.18	3.13	3.10	3.07
	10.56	**8.02**	**6.99**	**6.42**	**6.06**	**5.80**	**5.62**	**5.47**	**5.35**	**5.26**	**5.18**	**5.11**
10	4.96	4.10	3.71	3.48	3.33	3.22	3.14	3.07	3.02	2.97	2.94	2.91
	10.04	**7.56**	**6.55**	**5.99**	**5.64**	**5.39**	**5.21**	**5.06**	**4.95**	**4.85**	**4.78**	**4.71**
11	4.84	3.98	3.59	3.36	3.20	3.09	3.01	2.95	2.90	2.86	2.82	2.79
	9.65	**7.20**	**6.22**	**5.67**	**5.32**	**5.07**	**4.88**	**4.74**	**4.63**	**4.54**	**4.46**	**4.40**

Table J (continued)

Degrees of Freedom for Denominator	Degrees of Freedom for Numerator: 14	16	20	24	30	40	50	75	100	200	500	∞
1	245	246	248	249	250	251	252	253	253	254	254	254
	6,142	**6,169**	**6,208**	**6,234**	**6,258**	**6,286**	**6,302**	**6,323**	**6,334**	**6,352**	**6,361**	**6,366**
2	19.42	19.43	19.44	19.45	19.46	19.47	19.47	19.48	19.49	19.49	19.50	19.50
	99.43	**99.44**	**99.45**	**99.46**	**99.47**	**99.48**	**99.48**	**99.49**	**99.49**	**99.49**	**99.50**	**99.50**
3	8.71	8.69	8.66	8.64	8.62	8.60	8.58	8.57	8.56	8.54	8.54	8.53
	26.92	**26.83**	**26.69**	**26.60**	**26.50**	**26.41**	**26.30**	**26.27**	**26.23**	**26.18**	**26.14**	**26.12**
4	5.87	5.84	5.80	5.77	5.74	5.71	5.70	5.68	5.66	5.65	5.64	5.63
	14.24	**14.15**	**14.02**	**13.93**	**13.83**	**13.74**	**13.69**	**13.61**	**13.57**	**13.52**	**13.48**	**13.46**
5	4.64	4.60	4.56	4.53	4.50	4.46	4.44	4.42	4.40	4.38	4.37	4.36
	9.77	**9.68**	**9.55**	**9.47**	**9.38**	**9.29**	**9.24**	**9.17**	**9.13**	**9.07**	**9.04**	**9.02**
6	3.96	3.92	3.87	3.84	3.81	3.77	3.75	3.72	3.71	3.69	3.68	3.67
	7.60	**7.52**	**7.39**	**7.31**	**7.23**	**7.14**	**7.09**	**7.02**	**6.99**	**6.94**	**6.90**	**6.88**
7	3.52	3.49	3.44	3.41	3.38	3.34	3.32	3.29	3.28	3.25	3.24	3.23
	6.35	**6.27**	**6.15**	**6.07**	**5.98**	**5.90**	**5.85**	**5.78**	**5.75**	**5.70**	**5.67**	**5.65**
8	3.23	3.20	3.15	3.12	3.08	3.05	3.03	3.00	2.98	2.96	2.94	2.93
	5.56	**5.48**	**5.36**	**5.28**	**5.20**	**5.11**	**5.06**	**5.00**	**4.96**	**4.91**	**4.88**	**4.86**
9	3.02	2.98	2.93	2.90	2.86	2.82	2.80	2.77	2.76	2.73	2.72	2.71
	5.00	**4.92**	**4.80**	**4.73**	**4.64**	**4.56**	**4.51**	**4.45**	**4.41**	**4.36**	**4.33**	**4.31**
10	2.86	2.82	2.77	2.74	2.70	2.67	2.64	2.61	2.59	2.56	2.55	2.54
	4.60	4.52	4.41	4.33	4.25	4.17	4.12	4.05	4.01	3.96	3.93	3.91
11	**2.74**	**2.70**	**2.65**	**2.61**	**2.57**	**2.53**	**2.50**	**2.47**	**2.45**	**2.42**	**2.41**	**2.40**
	4.29	4.21	4.10	4.02	3.94	3.86	3.80	3.74	3.70	3.66	3.62	3.60

Table J (continued)

Degrees of Freedom for Denominator	Degrees of Freedom for Numerator											
	1	2	3	4	5	6	7	8	9	10	11	12
12	4.75	3.89	3.49	3.26	3.11	3.00	2.92	2.85	2.80	2.76	2.72	2.69
	9.33	**6.93**	**5.95**	**5.41**	**5.06**	**4.82**	**4.65**	**4.50**	**4.39**	**4.30**	**4.22**	**4.16**
13	4.67	3.80	3.41	3.18	3.02	2.92	2.84	2.77	2.72	2.67	2.63	2.60
	9.07	**6.70**	**5.74**	**5.20**	**4.86**	**4.62**	**4.44**	**4.30**	**4.19**	**4.10**	**4.02**	**3.96**
14	4.60	3.74	3.34	3.11	2.96	2.85	2.77	2.70	2.65	2.60	2.56	2.53
	8.86	**6.51**	**5.56**	**5.03**	**4.69**	**4.46**	**4.28**	**4.14**	**4.03**	**3.94**	**3.86**	**3.80**
15	4.54	3.68	3.29	3.06	2.90	2.79	2.70	2.64	2.59	2.55	2.51	2.48
	8.68	**6.36**	**5.42**	**4.89**	**4.56**	**4.32**	**4.14**	**4.00**	**3.89**	**3.80**	**3.73**	**3.67**
16	4.49	3.63	3.24	3.01	2.85	2.74	2.66	2.59	2.54	2.49	2.45	2.42
	8.53	**6.23**	**5.29**	**4.77**	**4.44**	**4.20**	**4.03**	**3.89**	**3.78**	**3.69**	**3.61**	**3.55**
17	4.45	3.59	3.20	2.96	2.81	2.70	2.62	2.55	2.50	2.45	2.41	2.38
	8.40	**6.11**	**5.18**	**4.67**	**4.34**	**4.10**	**3.93**	**3.79**	**3.68**	**3.59**	**3.52**	**3.45**
18	4.41	3.55	3.16	2.93	2.77	2.66	2.58	2.51	2.46	2.41	2.37	2.34
	8.28	**6.01**	**5.09**	**4.58**	**4.25**	**4.01**	**3.85**	**3.71**	**3.60**	**3.51**	**3.44**	**3.37**
19	4.38	3.52	3.13	2.90	2.74	2.63	2.55	2.48	2.43	2.38	2.34	2.31
	8.18	**5.93**	**5.01**	**4.50**	**4.17**	**3.94**	**3.77**	**3.63**	**3.52**	**3.43**	**3.36**	**3.30**
20	4.35	3.49	3.10	2.87	2.71	2.60	2.52	2.45	2.40	2.35	2.31	2.28
	8.10	**5.85**	**4.94**	**4.43**	**4.10**	**3.87**	**3.71**	**3.56**	**3.45**	**3.37**	**3.30**	**3.23**
21	4.32	3.47	3.07	2.84	2.68	2.57	2.49	2.42	2.37	2.32	2.28	2.25
	8.02	**5.78**	**4.87**	**4.37**	**4.04**	**3.81**	**3.65**	**3.51**	**3.40**	**3.31**	**3.24**	**3.17**
22	4.30	3.44	3.05	2.82	2.66	2.55	2.47	2.40	2.35	2.30	2.26	2.23
	7.94	**5.72**	**4.82**	**4.41**	**3.99**	**3.76**	**3.59**	**3.45**	**3.35**	**3.26**	**3.18**	**3.12**
23	4.28	3.42	3.03	2.80	2.64	2.53	2.45	2.38	2.32	2.28	2.24	2.20
	7.88	**5.66**	**4.76**	**4.26**	**3.94**	**3.71**	**3.54**	**3.41**	**3.30**	**3.21**	**3.14**	**3.07**
24	4.26	3.40	3.01	2.78	2.62	2.51	2.43	2.36	2.30	2.26	2.22	2.18
	7.82	**5.61**	**4.72**	**4.22**	**3.90**	**3.67**	**3.50**	**3.36**	**3.25**	**3.17**	**3.09**	**3.03**
25	4.24	3.38	2.99	2.76	2.60	2.49	2.41	2.34	2.28	2.24	2.20	2.16
	7.77	**5.57**	**4.68**	**4.18**	**3.86**	**3.63**	**3.46**	**3.32**	**3.21**	**3.13**	**3.05**	**2.99**
26	4.22	3.37	2.89	2.74	2.59	2.47	2.39	2.32	2.27	2.22	2.18	2.15
	7.72	**5.53**	**4.64**	**4.14**	**3.82**	**3.59**	**3.42**	**3.29**	**3.17**	**3.09**	**3.02**	**2.96**
27	4.21	3.35	2.96	2.73	2.57	2.46	2.37	2.30	2.25	2.20	2.18	2.13
	7.68	**5.49**	**4.60**	**4.11**	**3.79**	**3.56**	**3.39**	**3.26**	**3.14**	**3.06**	**2.98**	**2.93**
28	4.20	3.34	2.95	2.71	2.56	2.44	2.36	2.29	2.24	2.19	2.15	2.12
	7.64	**5.45**	**4.57**	**4.07**	**3.76**	**3.53**	**3.36**	**3.23**	**3.11**	**3.03**	**2.95**	**2.90**
29	4.18	3.33	2.93	2.70	2.54	2.43	2.35	2.28	2.22	2.18	2.14	2.10
	7.60	**5.52**	**4.54**	**4.04**	**3.73**	**3.50**	**3.33**	**3.20**	**3.08**	**3.00**	**2.92**	**2.87**

Table J (continued)

Degrees of Freedom for Denominator	Degrees of Freedom for Numerator											
	14	**16**	**20**	**24**	**30**	**40**	**50**	**75**	**100**	**200**	**500**	**∞**
12	2.64	2.60	2.54	2.50	2.46	2.42	2.40	2.36	2.35	2.32	2.31	2.30
	4.05	**3.98**	**3.86**	**3.78**	**3.70**	**3.61**	**3.56**	**3.49**	**3.46**	**3.41**	**3.38**	**3.36**
13	2.55	2.51	2.46	2.42	2.38	2.34	2.32	2.28	2.26	2.24	2.22	2.21
	3.85	**3.78**	**3.67**	**3.59**	**3.51**	**3.42**	**3.37**	**3.30**	**3.27**	**3.21**	**3.18**	**3.16**
14	2.48	2.44	2.39	2.35	2.31	2.27	2.24	2.21	2.19	2.16	2.14	2.13
	3.70	**3.62**	**3.51**	**3.43**	**3.34**	**3.26**	**3.21**	**3.14**	**3.11**	**3.06**	**3.02**	**3.00**
15	2.43	2.39	2.33	2.29	2.25	2.21	2.18	2.15	2.12	2.10	2.08	2.07
	3.56	**3.48**	**3.36**	**3.29**	**3.20**	**3.12**	**3.07**	**3.00**	**2.97**	**2.92**	**2.89**	**2.87**
16	2.37	2.33	2.28	2.24	2.20	2.16	2.13	2.09	2.07	2.04	2.02	2.01
	3.45	**3.37**	**3.25**	**3.18**	**3.10**	**3.01**	**2.96**	**2.89**	**2.86**	**2.80**	**2.77**	**2.75**
17	2.33	2.29	2.23	2.19	2.15	2.11	2.08	2.04	2.02	1.99	1.97	1.97
	3.35	**3.27**	**3.16**	**3.08**	**3.00**	**2.92**	**2.86**	**2.79**	**2.76**	**2.70**	**2.67**	**2.65**
18	2.29	2.25	2.19	2.15	2.11	2.07	2.04	2.00	1.98	1.95	1.93	1.92
	3.27	**3.19**	**3.07**	**3.00**	**2.91**	**2.83**	**2.78**	**2.71**	**2.68**	**2.62**	**2.59**	**2.57**
19	2.26	2.21	2.15	2.11	2.07	2.02	2.00	1.96	1.94	1.91	1.90	1.88
	3.19	**3.12**	**3.00**	**2.92**	**2.84**	**2.76**	**2.70**	**2.63**	**2.60**	**2.54**	**2.51**	**2.49**
20	2.23	2.18	2.12	2.08	2.04	1.99	1.96	1.92	1.90	1.87	1.85	1.84
	3.13	**3.05**	**2.94**	**2.86**	**2.77**	**2.69**	**2.63**	**2.56**	**2.53**	**2.47**	**2.44**	**2.42**
21	2.20	2.15	2.09	2.05	2.00	1.96	1.93	1.89	1.87	1.84	1.82	1.81
	3.07	**2.99**	**2.88**	**2.80**	**2.72**	**2.63**	**2.58**	**2.51**	**2.47**	**2.42**	**2.38**	**2.36**
22	2.18	2.13	2.07	2.03	1.98	1.93	1.91	1.87	1.84	1.81	1.80	1.78
	3.02	**2.94**	**2.83**	**2.75**	**2.67**	**2.58**	**2.53**	**2.46**	**2.42**	**2.37**	**2.33**	**2.31**
23	2.14	2.10	2.04	2.00	1.96	1.91	1.88	1.84	1.82	1.79	1.77	1.76
	2.97	**2.89**	**2.78**	**2.70**	**2.62**	**2.53**	**2.48**	**2.41**	**2.37**	**2.32**	**2.28**	**2.26**
24	2.13	2.09	2.02	1.98	1.94	1.89	1.86	1.82	1.80	1.76	1.74	1.73
	2.93	**2.85**	**2.74**	**2.66**	**2.58**	**2.49**	**2.44**	**2.36**	**2.33**	**2.27**	**2.23**	**2.21**
25	2.11	2.06	2.00	1.96	1.92	1.87	1.84	1.80	1.77	1.74	1.72	1.71
	2.89	**2.81**	**2.70**	**2.62**	**2.54**	**2.45**	**2.40**	**2.32**	**2.29**	**2.23**	**2.19**	**2.17**
26	2.10	2.05	1.99	1.95	1.90	1.85	1.82	1.78	1.76	1.72	1.70	1.69
	2.86	**2.77**	**2.66**	**2.58**	**2.50**	**2.41**	**2.36**	**2.28**	**2.25**	**2.19**	**2.15**	**2.13**
27	2.08	2.03	1.97	1.93	1.88	1.84	1.80	1.76	1.74	1.71	1.68	1.67
	2.83	**2.74**	**2.63**	**2.55**	**2.47**	**2.38**	**2.33**	**2.25**	**2.21**	**2.16**	**2.12**	**2.10**
28	2.06	2.02	1.96	1.91	1.87	1.81	1.78	1.75	1.72	1.69	1.67	1.65
	2.80	**2.71**	**2.60**	**2.52**	**2.44**	**2.35**	**2.30**	**2.22**	**2.18**	**2.13**	**2.09**	**2.06**
29	2.05	2.00	1.94	1.90	1.85	1.80	1.77	1.73	1.71	1.68	1.65	1.64
	2.77	**2.68**	**2.57**	**2.49**	**2.41**	**2.32**	**2.27**	**2.19**	**2.15**	**2.10**	**2.06**	**2.03**

Table J (continued)

Degrees of Freedom for Denominator	Degrees of Freedom for Numerator											
	1	2	3	4	5	6	7	8	9	10	11	12
30	4.17	3.32	2.92	2.69	2.53	2.43	2.34	2.27	2.21	2.16	2.12	2.09
	7.56	**5.39**	**4.51**	**4.02**	**3.70**	**3.47**	**3.30**	**3.17**	**3.06**	**2.98**	**2.90**	**2.84**
32	4.15	3.30	2.90	2.67	2.51	2.40	2.32	2.25	2.19	2.14	2.10	2.07
	7.50	**5.34**	**4.46**	**3.97**	**3.66**	**3.42**	**3.25**	**3.12**	**3.01**	**2.94**	**2.86**	**2.80**
34	4.13	3.28	2.88	2.65	2.49	2.38	2.30	2.23	2.17	2.12	2.08	2.05
	7.44	**5.29**	**4.42**	**3.93**	**3.61**	**3.38**	**3.21**	**3.08**	**2.97**	**2.89**	**2.82**	**2.76**
36	4.11	3.26	2.86	2.63	2.48	2.36	2.28	2.21	2.15	2.10	2.06	2.03
	7.39	**5.25**	**4.38**	**3.89**	**3.58**	**3.35**	**3.18**	**3.04**	**2.94**	**2.86**	**2.78**	**2.72**
38	4.10	3.25	2.85	2.62	2.46	2.35	2.26	2.19	2.14	2.09	2.05	2.02
	7.35	**5.21**	**4.34**	**3.86**	**3.54**	**3.32**	**3.15**	**3.02**	**2.91**	**2.82**	**2.75**	**2.69**
40	4.08	3.23	2.84	2.61	2.45	2.34	2.25	2.18	2.12	2.08	2.04	2.00
	7.31	**5.18**	**4.31**	**3.83**	**3.51**	**3.29**	**3.12**	**2.99**	**2.88**	**2.80**	**2.73**	**2.66**
42	4.07	3.22	2.83	2.59	2.44	2.32	2.24	2.17	2.11	2.06	2.02	1.99
	7.27	**5.15**	**4.29**	**3.80**	**3.49**	**3.26**	**3.10**	**2.96**	**2.86**	**2.77**	**2.70**	**2.64**
44	4.06	3.21	2.82	2.58	2.43	2.31	2.23	2.16	2.10	2.05	2.01	1.98
	7.24	**5.12**	**4.26**	**3.78**	**3.46**	**3.24**	**3.07**	**2.94**	**2.84**	**2.75**	**2.68**	**2.62**
46	4.05	3.20	2.81	2.57	2.42	2.30	2.22	2.14	2.09	2.04	2.00	1.97
	7.21	**5.10**	**4.24**	**3.76**	**3.44**	**3.22**	**3.05**	**2.92**	**2.82**	**2.73**	**2.66**	**2.60**
48	4.04	3.19	2.80	2.56	2.41	2.30	2.21	2.14	2.08	2.03	1.99	1.96
	7.19	**5.08**	**4.22**	**3.74**	**3.42**	**3.20**	**3.04**	**2.90**	**2.80**	**2.71**	**2.64**	**2.58**
50	4.03	3.18	2.79	2.56	2.40	2.29	2.20	2.13	2.07	2.20	1.98	1.95
	7.17	**5.06**	**4.20**	**3.72**	**3.41**	**3.18**	**3.02**	**2.88**	**2.78**	**2.70**	**2.62**	**2.56**
55	4.02	3.17	2.78	2.54	2.38	2.27	2.18	2.11	2.05	2.00	1.97	1.93
	7.12	**5.01**	**4.16**	**3.68**	**3.37**	**3.15**	**2.98**	**2.85**	**2.75**	**2.66**	**2.59**	**2.53**
60	4.00	3.15	2.76	2.52	2.37	2.25	2.17	2.10	2.04	1.99	1.95	1.92
	7.08	**4.98**	**4.13**	**3.65**	**3.34**	**3.12**	**2.95**	**2.82**	**2.72**	**2.63**	**2.56**	**2.50**
65	3.99	3.14	2.75	2.51	2.36	2.24	2.15	2.08	2.02	1.98	1.94	1.90
	7.04	**4.95**	**4.10**	**3.62**	**3.31**	**3.09**	**2.93**	**2.79**	**2.70**	**2.61**	**2.54**	**2.47**
70	3.98	3.13	2.74	2.50	2.35	2.32	2.14	2.07	2.01	1.97	1.93	1.80
	7.01	**4.92**	**4.08**	**3.60**	**3.29**	**3.07**	**2.91**	**2.77**	**2.67**	**2.59**	**2.51**	**2.45**
80	3.96	3.11	2.72	2.48	2.33	2.21	2.12	2.05	1.99	1.95	1.91	1.88
	6.95	**4.88**	**4.04**	**3.56**	**3.25**	**3.04**	**2.87**	**2.74**	**2.64**	**2.55**	**2.48**	**2.41**
100	3.94	3.09	2.70	2.46	2.30	2.19	2.10	2.03	1.97	1.92	1.88	1.85
	6.90	**4.82**	**3.98**	**3.51**	**3.20**	**2.99**	**2.82**	**2.69**	**2.59**	**2.51**	**2.43**	**2.36**
125	3.92	3.07	2.68	2.44	2.29	2.17	2.08	2.01	1.95	1.90	1.86	1.83
	6.84	**4.78**	**3.94**	**3.47**	**3.17**	**2.95**	**2.79**	**2.65**	**2.56**	**2.47**	**2.40**	**2.33**

Table J (continued)

Degrees of Freedom for Denominator	Degrees of Freedom for Numerator											
	14	16	20	24	30	40	50	75	100	200	500	∞
30	2.04	1.99	1.93	1.89	1.84	1.79	1.76	1.72	1.69	1.66	1.64	1.62
	2.74	**2.66**	**2.55**	**2.47**	**2.38**	**2.29**	**2.24**	**2.16**	**2.13**	**2.07**	**2.03**	**2.01**
32	2.02	1.97	1.91	1.86	1.82	1.76	1.74	1.69	1.67	1.64	1.61	1.59
	2.70	**2.62**	**2.51**	**2.42**	**2.34**	**2.25**	**2.20**	**2.12**	**2.08**	**2.02**	**1.98**	**1.96**
34	2.00	1.95	1.89	1.84	1.80	1.74	1.71	1.67	1.64	1.61	1.59	1.57
	2.66	**2.58**	**2.47**	**2.38**	**2.30**	**2.21**	**2.15**	**2.08**	**2.04**	**1.98**	**1.94**	**1.91**
36	1.89	1.93	1.87	1.82	1.78	1.72	1.69	1.65	1.62	1.59	1.56	1.55
	2.62	**2.54**	**2.43**	**2.35**	**2.26**	**2.17**	**2.12**	**2.04**	**2.00**	**1.94**	**1.90**	**1.87**
38	1.96	1.92	1.85	1.80	1.76	1.71	1.67	1.63	1.60	1.57	1.54	1.53
	2.59	**2.51**	**2.40**	**2.32**	**2.22**	**2.14**	**2.08**	**2.00**	**1.97**	**1.90**	**1.86**	**1.84**
40	1.95	1.90	1.84	1.79	1.74	1.69	1.66	1.61	1.59	1.55	1.53	1.51
	2.56	**2.49**	**2.37**	**2.29**	**2.20**	**2.11**	**2.05**	**1.97**	**1.94**	**1.88**	**1.84**	**1.81**
42	1.94	1.89	1.82	1.78	1.73	1.68	1.64	1.60	1.57	1.54	1.51	1.49
	2.54	**2.46**	**2.35**	**2.26**	**2.17**	**2.08**	**2.02**	**1.94**	**1.91**	**1.85**	**1.80**	**1.78**
44	1.92	1.88	1.81	1.76	1.72	1.66	1.63	1.58	1.56	1.52	1.50	1.48
	2.52	**2.44**	**2.32**	**2.24**	**2.15**	**2.06**	**2.00**	**1.92**	**1.88**	**1.82**	**1.78**	**1.75**
46	1.91	1.87	1.80	1.75	1.71	1.65	1.62	1.57	1.54	1.51	1.48	1.46
	2.50	**2.42**	**2.30**	**2.22**	**2.13**	**2.04**	**1.98**	**1.90**	**1.86**	**1.80**	**1.76**	**1.72**
48	1.90	1.86	1.79	1.74	1.70	1.64	1.61	1.56	1.53	1.50	1.47	1.45
	2.43	**2.40**	**2.28**	**2.20**	**2.11**	**2.02**	**1.96**	**1.88**	**1.84**	**1.78**	**1.73**	**1.70**
50	1.90	1.85	1.78	1.74	1.69	1.63	1.60	1.55	1.52	1.48	1.46	1.44
	2.46	**2.39**	**2.26**	**2.18**	**2.10**	**2.00**	**1.94**	**1.86**	**1.82**	**1.76**	**1.71**	**1.68**
55	1.88	1.83	1.76	1.72	1.67	1.61	1.58	1.52	1.50	1.46	1.43	1.41
	2.43	**2.35**	**2.23**	**2.15**	**2.06**	**1.96**	**1.90**	**1.82**	**1.78**	**1.71**	**1.66**	**1.64**
60	1.86	1.81	1.75	1.70	1.65	1.59	1.56	1.50	1.48	1.44	1.41	1.39
	2.40	**2.32**	**2.20**	**2.12**	**2.03**	**1.93**	**1.87**	**1.79**	**1.74**	**1.68**	**1.63**	**1.60**
65	1.85	1.80	1.73	1.68	1.63	1.57	1.54	1.49	1.46	1.42	1.39	1.37
	2.37	**2.30**	**2.18**	**2.09**	**2.00**	**1.90**	**1.84**	**1.76**	**1.71**	**1.64**	**1.60**	**1.56**
70	1.84	1.79	1.72	1.67	1.62	1.56	1.53	1.47	1.45	1.40	1.37	1.35
	2.35	**2.28**	**2.15**	**2.07**	**1.98**	**1.88**	**1.82**	**1.74**	**1.69**	**1.63**	**1.56**	**1.53**
80	1.82	1.77	1.70	1.65	1.60	1.54	1.51	1.45	1.42	1.38	1.35	1.32
	2.32	**2.24**	**2.11**	**2.03**	**1.94**	**1.84**	**1.78**	**1.70**	**1.65**	**1.57**	**1.52**	**1.49**
100	1.79	1.75	1.68	1.63	1.57	1.51	1.48	1.42	1.39	1.34	1.30	1.28
	2.26	**2.19**	**2.06**	**1.98**	**1.89**	**1.79**	**1.73**	**1.64**	**1.59**	**1.51**	**1.46**	**1.43**
125	1.77	1.72	1.65	1.60	1.55	1.49	1.45	1.39	1.36	1.31	1.27	1.25
	2.23	**2.15**	**2.03**	**1.94**	**1.85**	**1.75**	**1.68**	**1.59**	**1.54**	**1.46**	**1.40**	**1.37**

***Table J* (continued)**

Degrees of Freedom for Denominator	Degrees of Freedom for Numerator											
	1	**2**	**3**	**4**	**5**	**6**	**7**	**8**	**9**	**10**	**11**	**12**
150	3.91	3.06	2.67	2.43	2.27	2.16	2.07	2.00	1.94	1.89	1.85	1.82
	6.81	**4.75**	**3.91**	**3.44**	**3.13**	**2.92**	**2.76**	**2.62**	**2.53**	**2.44**	**2.37**	**2.30**
200	3.89	3.04	2.65	2.41	2.26	2.14	2.05	1.98	1.92	1.87	1.83	1.80
	6.76	**4.71**	**3.88**	**3.41**	**3.11**	**2.90**	**2.73**	**2.60**	**2.50**	**2.41**	**2.34**	**2.28**
400	3.86	3.02	2.62	2.39	2.23	2.12	2.03	1.96	1.90	1.85	1.81	1.78
	6.70	**4.66**	**3.83**	**3.36**	**3.06**	**2.85**	**2.69**	**2.55**	**2.46**	**2.37**	**2.29**	**2.23**
1,000	3.85	3.00	2.61	2.38	2.22	2.10	2.02	1.95	1.89	1.84	1.80	1.76
	6.66	**4.62**	**3.80**	**3.34**	**3.04**	**2.82**	**2.66**	**2.53**	**2.43**	**2.34**	**2.26**	**2.20**
∞	3.84	2.99	2.60	2.37	2.21	2.09	2.01	1.94	1.88	1.83	1.79	1.75
	6.64	**4.60**	**3.78**	**3.32**	**3.02**	**2.80**	**2.64**	**2.51**	**2.41**	**2.32**	**2.24**	**2.18**

Table J (continued)

Degrees of Freedom for Denominator	Degrees of Freedom for Numerator											
	14	**16**	**20**	**24**	**30**	**40**	**50**	**75**	**100**	**200**	**500**	**∞**
150	1.76	1.71	1.64	1.59	1.54	1.47	1.44	1.37	1.34	1.29	1.25	1.22
	2.20	**2.12**	**2.00**	**1.91**	**1.83**	**1.72**	**1.66**	**1.56**	**1.51**	**1.43**	**1.37**	**1.33**
200	1.74	1.69	1.62	1.57	1.52	1.45	1.42	1.35	1.32	1.26	1.22	1.19
	2.17	**2.09**	**1.97**	**1.88**	**1.79**	**1.69**	**1.62**	**1.53**	**1.48**	**1.39**	**1.33**	**1.28**
400	1.72	1.67	1.60	1.54	1.49	1.42	1.38	1.32	1.28	1.22	1.16	1.13
	2.12	**2.04**	**1.92**	**1.84**	**1.74**	**1.64**	**1.57**	**1.47**	**1.42**	**1.32**	**1.24**	**1.19**
1,000	1.70	1.65	1.58	1.53	1.47	1.41	1.36	1.30	1.26	1.19	1.13	1.08
	2.09	**2.01**	**1.89**	**1.81**	**1.71**	**1.61**	**1.54**	**1.44**	**1.38**	**1.28**	**1.19**	**1.11**
∞	1.69	1.64	1.57	1.52	1.46	1.40	1.35	1.28	1.24	1.17	1.11	1.00
	2.07	**1.99**	**1.87**	**1.79**	**1.69**	**1.59**	**1.52**	**1.41**	**1.36**	**1.25**	**1.15**	**1.00**

Source: Reprinted by permission from *Statistical Methods* by George W. Snedecor and William G. Cochran, 7th ed. © 1980 by Iowa State University Press, Ames, Iowa 50010.

Table K
Constants for Computing Control Chart Limits

Number of Observations in Sample n	d_2	d_3	c_4	c_5
2	1.128	0.853	0.798	0.603
3	1.693	0.888	0.886	0.463
4	2.059	0.880	0.921	0.389
5	2.326	0.864	0.940	0.341
6	2.534	0.848	0.952	0.308
7	2.704	0.833	0.959	0.282
8	2.847	0.820	0.965	0.262
9	2.970	0.808	0.969	0.246
10	3.078	0.797	0.973	0.232
11	3.137	0.787	0.975	0.220
12	3.258	0.778	0.978	0.210
13	3.336	0.770	0.979	0.202
14	3.407	0.762	0.981	0.194
15	3.472	0.755	0.982	0.187
16	3.532	0.749	0.984	0.181
17	3.588	0.743	0.985	0.175
18	3.640	0.738	0.985	0.170
19	3.689	0.733	0.986	0.166
20	3.735	0.729	0.987	0.161
21	3.778	0.724	0.988	0.157
22	3.819	0.720	0.988	0.153
23	3.858	0.716	0.989	0.150
24	3.895	0.712	0.989	0.147
25	3.931	0.709	0.990	0.144

Values of c_4, d_2, d_3, are from ASTM-STP 15D by kind permission of the American Society for Testing and Materials.

APPENDIX B

Bibliography

Probability

Feller, William. *An Introduction to Probability Theory and Its Applications*, Vol. 1, 3rd ed. New York: John Wiley & Sons, 1968.

Fisz, Marek. *Probability Theory and Mathematical Statistics*. New York: John Wiley & Sons, 1963.

Laplace, Pierre Simon, Marquis de. *A Philosophical Essay on Probabilities*. New York: Dover Publications, 1951.

Parzen, Emmanuel. *Modern Probability Theory and Its Applications*. New York: John Wiley & Sons, 1960.

Regression and Correlation Analysis

Draper, N. R., and H. Smith. *Applied Regression Analysis*, 2nd ed. New York: John Wiley & Sons, 1981.

Ezekial, Mordecai, and Karl A. Fox. *Methods of Correlation and Regression Analysis*, 3rd ed. New York: John Wiley & Sons, 1959.

Neter, John, William Wasserman, and Michael H. Kutner. *Applied Linear Regression Models*, 3rd ed. Homewood, Ill.: Richard D. Irwin, 1990.

Analysis of Variance and Design of Experiments

Box, George E. P., William G. Hunter, J. Stuart Hunter, *Statistics for Experimenters*. New York: John Wiley & Sons, 1978.

Cochran, William G., and Gertrude M. Cox. *Experimental Designs*, 2nd ed. New York: John Wiley & Sons, 1957.

Cox, David R. *Planning of Experiments*. New York: John Wiley & Sons, 1958.

Guenther, W. C. *Analysis of Variance*. Englewood Cliffs, N.J.: Prentice-Hall, 1964.

Mendenhall, William. *An Introduction to Linear Models and the Design and Analysis of Experiments*. Belmont, Calif.: Wadsworth, 1968.

Neter, John, William Wasserman, and Michael H. Kutner. *Applied Linear Regression Models*, 3rd ed. Homewood, Ill.: Richard D. Irwin, 1990.

Scheffe, Henry. *The Analysis of Variance*. New York: John Wiley & Sons, 1959.

Statistical Quality Control

American Society of Testing Materials. *A.S.T.M. Manual on Quality Control of Materials*. Philadelphia, 1951.

Cowden, Dudley J. *Statistical Methods in Quality Control*. Englewood Cliffs, N.J.: Prentice-Hall, 1957.

Deming, W. E. *Some Theory of Sampling*. New York: Dover, 1984.

Dodge, H. F., and H. G. Romig. *Sampling Inspection Tables—Single and Double Sampling*, 2nd ed. New York: John Wiley & Sons, 1959.

Enrick, N. L. *Quality Control and Reliability*, 5th ed. New York: The Industrial Press, 1966.

Grant, Eugene L., and Richard S. Leavenworth. *Statistical Quality Control*, 4th ed. New York: McGraw-Hill, 1972.

Juran, J. M. (ed.). *Quality Control Handbook*, 2nd ed. New York: McGraw-Hill, 1962.

Wadsworth, Harrison, M., Kenneth S. Stephens, and B. Blanton Godfrey. *Modern Methods for Quality Control and Improvement*. New York: John Wiley & Sons, 1986.

Statistical Tables

Beyer, William H. (ed.). *Handbook of Tables for Probability and Statistics*, 2nd ed. Cleveland, Ohio: The Chemical Rubber Co., 1968.

Burington, Richard S., and Donald C. May. *Handbook of Probability and Statistics with Tables,* 2nd ed. New York: McGraw-Hill, 1970.

Dodge, H. F., and H. G. Romig. *Sampling Inspection Tables — Single and Double Sampling*, 2nd ed. New York: John Wiley & Sons, 1959.

Fisher, Ronald A., and F. Yates. *Statistical Tables for Biological, Agricultural and Medical Research*, 6th ed. London: Longman Group, 1978.

Military Standard 105D. *Sampling Procedures and Tables for Inspection by Attributes*. Washington, D.C.: U.S. Government Printing Office, 1963.

Military Standard 414. *Sampling Procedures and Tables by Variables for Percent Defective*. Washington, D.C.: U.S. Government Printing Office, 1957.

Military Standard 1235 (Ord). *Single- and Multi-Level Continuous Sampling Procedures and Tables for Inspection by Attributes*. Washington, D.C.: U.S. Government Printing Office, 1962.

Pearson, E. S., and H. O. Hartley. *Biometrika Tables for Statisticians*, 3rd ed. Cambridge, England: Cambridge University Press, 1966.

The Rand Corporation. *A Million Random Digits with 100,000 Normal Deviates.* New York: The Free Press, 1955.

ANSWERS

Answers to Selected Problems

1-6 **(a)** ordinal **(b)** nominal **(c)** interval **(d)** ratio **(e)** interval

1-8 **(a)** Ignoring order of selection, the following possibilities apply.

SUI	SIO	SOD	UIO	UOD	IOD
SUO	SID	SOB	UID	UOB	IOB
SUD	SIB	SDB	UIB	UDB	IDB
SUB					ODB

(b) 16/20 **(c)** 4/20

1-10 **(a)** deductive statistics **(b)** inductive statistics **(c)** inductive statistics

1-14

61 Richter	47 Neel
74 Bloembergen	64 Mott
98 Charpak	98 (skip)
54 Esaki	17 Purcell
15 Walton	00 Taylor
	53 Giaever

1-19 **(a)** deductive **(b)** inductive **(c)** deductive **(d)** inductive

2-1 **(a)**

Unit Cost (dollars/pound)	Frequency
10.00—under 11.00	1
11.00—under 12.00	8
12.00—under 13.00	16
13.00—under 14.00	9
14.00—under 15.00	6
15.00—under 16.00	4
16.00—under 17.00	3
17.00—under 18.00	2
18.00—under 19.00	1
	50

2-4 **(a)**

Leaf (last digit)

Stem	0.|4 7 3 8 3 6 7 5
	1.|1 0 8 5 8 2 5 2 6 2 8
	2.|9 1 9 9 2 8 2 6 2 7 6 9 0 3 6 9 4 9 9 0 7 9 4
	3.|7 2 9 0 2 5 6 3 8 3 6 9 1 1 7 7 4 4 3 6 8 6 4 3 6 1 2 6 4
	4.|0 0 0 0 2 8 4 1 5 0
	5.|4 4 9 0 8

(b)

Midpoint	Count	
0.50	8	********
1.50	11	***********
2.50	23	***********************
3.50	29	*****************************
4.50	10	**********
5.50	5	*****

2-10 **(a)**

Gasoline Mileage (miles/gallon)	**Frequency**
10.0—under 15.0	1
15.0—under 20.0	2
20.0—under 25.0	4
25.0—under 30.0	12
30.0—under 35.0	17
25.0—under 40.0	14
	50

2-14 **(a)**

Time (milliseconds)	**Cumulative Relative Frequency**	**Relative Frequency**
0.0—under 5.0	.441	.441
5.0—under 10.0	.243	.684
10.0—under 15.0	.162	.846
15.0—under 20.0	.090	.936
20.0—under 25.0	.041	.977
25.0—under 30.0	.017	.994
30.0—under 35.0	.006	1.000
	1.000	

2-16

Depth of Well (feet)	(a) Relative Frequency	(b) Frequency	(c) Cumulative Frequency
0—under 1,000	.09	63	63
1,000—under 2,000	.26	182	245
2,000—under 3,000	.37	259	504
3,000—under 4,000	.16	112	616
4,000—under 5,000	.07	49	665
5,000—under 6,000	.03	21	686
6,000—under 7,000	.01	7	693
7,000—under 8,000	.01	7	700
	1.00	700	

2-18 **(a)** $\overline{X} = 3.1$ **(b)** $m = 3$ **(c)** Mode $= 2$ **(d)** $Q_{.25} = 2$ **(e)** $Q_{.75} = 4$

2-20 **(a)**

Leaf (Last Digit)

```
       0.|3 3 4 5 6 7 7 8
       1.|0 1 2 2 2 5 5 6 8 8 8
Stem   2.|0 0 1 2 2 2 3 4 4 6 6 6 7 7 8 9 9 9 9 9 9 9
       3.|0 1 1 1 2 2 2 3 3 3 3 4 4 4 4 5 6 6 6 6 6 6 7 7 7 8 8 9 9
       4.|0 0 0 0 0 1 2 4 5 8
       5.|0 4 4 8 9
```

(b) 3.05 **(c)** 2.9

(d) **(1)** 0.94 **(2)** 2.075 **(3)** 3.7 **(4)** 4.26

2-23 **(a)** 31.208 **(b)** 33.55 **(c)** −2.342 negatively skewed

2-25 **(a)** 29.827 **(b)** 29.98

2-26 **(a)** 29.86

(b) **(1)** 27.1 **(2)** 28.3 **(3)** 30.2 **(4)** 31.4 **(5)** 32.3

(c) 30.2 **(d)** **(1)** .033 above **(2)** .2 above

2-29 **(a)** 231.17 **(b)** 230.5 **(c)** Each value is modal.
(d) 220.25 **(e)** 238.75

2-30 **(a)** 13

(b) $\overline{X} = 10.6$ $s^2 = 20.71$ $s = 4.55$

(c) **(1)** 5.75 **(2)** 11.00 **(3)** 15

(d) 9.25

2-32 **(a)** 41 **(b)** $\overline{X} = 231.17$ $s^2 = 194.16$ $s = 13.934$

(c) **(1)** 220.25 **(2)** 230.5 **(3)** 238.75

(d) 18.50

2-33 **(a)** 2.33 **(b)** $m = 11.00$ SK $= -.26$

2-36

Class Interval	Frequency f	Midpoint X	fX	$f(X-\bar{X})^2$
0.0—under 5.0	152	2.5	380.0	4,783.76
5.0—under 10.0	84	7.5	630.0	31.26
10.0—under 15.0	56	12.5	700.0	1,079.24
15.0—under 20.0	31	17.5	542.5	2,733.34
20.0—under 25.0	14	22.5	315.0	2,899.01
25.0—under 30.0	6	27.5	165.0	2,255.83
30.0—under 35.0	2	32.5	65.0	1,189.74
	345		2,797.5	14,972.18

$\bar{X} = 8.11$ $s^2 = 43.52$ $s = 6.6$

2-40 **(a)** in the interval (1.5, 19.7) **(b)** in the interval (0, 24.25)

2-42 **(a)** **(1)** .38 **(2)** .16 **(3)** .35 **(4)** .53 **(5)** .42
(b) category (2) **(c)** category (4)

2-45 **(a)** .337 **(b)** .452 **(c)** .428 **(d)** .175 **(e)** .127

2-48 **(a)** 3.4 **(b)** 3.5 **(c)** 4 **(d)** 10
(e) 6.57 **(f)** 2.56 **(g)** .75 **(h)** −.12

2-57 **(a)**

	Day 1	2	3	4	5
(1)	33.00	34.00	35.00	37.00	38.00
(2)	12.83	6.96	4.38	1.633	.816
(3)	23.00	27.50	31.50	35.50	37.25
(4)	29.50	34.00	34.00	37.00	38.00
(5)	46.50	40.50	39.00	38.50	38.75

(b)

	Day 1	2	3	4	5
(1)	.39	.20	.13	.04	.02
(2)	.82	0	.68	0	0

3-1 **(a)** $\bar{X} = 16.47$ out of control **(c)** $\bar{X} = 16.12$ in control
(b) $\bar{X} = 15.75$ in control **(d)** $\bar{X} = 15.97$ in control

3-2 **(a)** $s = .320$ in control **(c)** $s = .697$ in control
(b) $s = .288$ in control **(d)** $s = .589$ in control

3-6 **(a)** **(1)** LCL = .097, UCL = .103 **(2)** LCL = 0, UCL = .0106

(b) **(1)** LCL = .0047, UCL = .0053 **(2)** LCL = .0002, UCL = .0018

(c) **(1)** LCL = 5.525, UCL = 5.715 **(2)** LCL = 0, UCL = .297

(d) **(1)** LCL = 1.368, UCL = 1.552 **(2)** LCL = .017, UCL = .423

3-11 **(a)**

Sample	$\bar{X}$	R
1	.2572	.021
2	.2380	.011
3	.2508	.010
4	.2618	.016
5	.2490	.012
6	.2494	.011
7	.2498	.025
8	.2516	.014
9	.2502	.013
10	.2528	.013
	2.5106	.146

$\bar{\bar{X}} = .25106$ $\bar{R} = .0146$

LCL = .24264 UCL = .25948 Samples 2 and 4 are out-of-control.

(b) LCL = 0 UCL = .031 All samples are in control.

(c) $\bar{\bar{X}} = .25135$ $\bar{R} = .01488$

(1) LCL = .24277 UCL = .25993

(2) LCL = 0 UCL = .0315

3-12 **(a)** LCL = .01, UCL = .19 **(c)** LCL = .036, UCL = .104

(b) LCL = .029, UCL = .071 **(d)** LCL = .02, UCL = .14

4-1 **(b)** $b = .442$ $a = 7.94$ $\hat{Y}(X) = 7.94 + .442X$

(c) $s_Y = 72.55$

(d) $s_{Y \cdot X} = 12.49$

(e) **(1)** 96.34 **(2)** 140.54 **(3)** 184.74 **(4)** 228.94

4-4 **(a)** $b = 8.7116$ $a = \$4{,}016.76$ $\hat{Y}(X) = \$4{,}016.76 + 8.7116X$

(b) **(1)** \$4,016.76 **(2)** \$8.7116

(c) **(1)** \$21,440 **(2)** \$10.72

4-6 **(a)** $b = 10{,}650{,}000$ $a = -4{,}410$

$\hat{Y}(X) = -4{,}410 + 10{,}650{,}000X$ or $\sigma = -4{,}410 + 10{,}650{,}000\epsilon$

(c) $\epsilon = 10{,}650{,}000$

4-11 $r = -.973$

4-14 **(a)** Y = mean flow (cfps) X_1 = rainfall (inches) X_2 = releases (cfps)

$\hat{Y} = 161{,}813 + 4{,}024X_1 - .3532X_2$

(b) 16,275

(c) **(1)** 206,973 **(3)** 203,933

(2) 203,441 (4) 214,529

4-18 **(a)** $\hat{Y} = 597{,}162 + 5{,}276X_1 + .0692X_2 - 8{,}046X_3$

(b) **(1)** 226,853.8 **(3)** 238,524.1

(2) 193,121.9 **(4)** 226,853.8

4-20 X_4 = irrigation withdrawals (cfps)

(a) $\hat{Y} = 736{,}310.6 + 3{,}502.2X_1 + .0419X_2 - 9{,}338.4X_3 - .1948X_4$

(b) **(1)** 235,635.6 **(3)** 240,177.2

(2) 204,336.3 **(4)** 232,714.2

4-26 **(a)** Y = processing time (minutes) X_1 = RAM (k bytes) X_2 = input (k bytes)

$\hat{Y} = 3.75 + .054X_1 + .614X_2$

(c) 2.65

4-27 **(a)** X_3 = output (k lines) **(b)**

$\hat{Y} = 1.18 + .128X_1 + 1.234X_3$ $\hat{Y} = 2.12 + .709X_2 + .680X_3$

$S_{Y\cdot 13} = 1.62$ $S_{Y\cdot 23} = 2.43$

(c) The regression in (a) involves a substantially smaller standard error. This means that it explains more of the variation in Y and should provide better predictions of yield than the regression in (b).

4-28 $\hat{Y} = .98 + .099X_1 + .243X_2 + 1.049X_3$

5-5 $\ln \hat{Y}(X) = 1.316 + 1.24 \ln X$ $\hat{Y}(X) = e^{1.316+1.24\ln X} = 3.73X^{1.24}$

5-8 **(a)** $\ln \hat{Y}(1{,}000/X) = -7.321 + 8.349(1{,}000/X)$

$\hat{Y}(X) = e^{-7.321+8.349(1{,}000/X)} = .0006615(4{,}226)^{1{,}000/X}$

5-14 $\hat{Y}(X) = 13.0 + 5{,}494X - 427{,}500X^2$

5-15 $\hat{Y}(X) = -7.48 + 21{,}024.46X - 3{,}381{,}430X^2 + 173{,}234{,}902X^3$

5-17 **(a)** $\hat{Y} = 38.76 - .091X_1 + 9.36X_2$

(b) $X_2 = 1$ $\hat{Y} = 38.76 - .091X_1 + 9.36(1) = 48.12 - .091X_1$

(c) $X_2 = 0$ $\hat{Y} = 38.76 - .091X_1$

5-22 $\hat{Y} = 4.1 + 16.8X_1 - .09X_2 - 1.78X_3 + 8.8X_4 + 2.5X_3X_4$

5-27 **(a)** $\hat{Y} = 1{,}812.74 + 227.30X_1$ $s_{Y\cdot X_1} = 1{,}580.01$

(b) **(1)** 7,495.2 **(2)** 8.631.7

5-28 **(a)** $\hat{Y} = 4{,}497 + 221.80X_1 - 63.27X_2$ $S_{Y\cdot 12} = 1{,}484.00$

(b) **(1)** 7,511.6 **(3)** 7,987.9

(2) 8,304.2 **(4)** 7,671.6

5-29 **(a)** $\hat{Y} = 11{,}702.14 + 234.73X_1 - 39.02X_2 - 124.68X_3$ $S_{Y\cdot 123} = 1{,}478.41$

(b) **(1)** 8,529.1 **(3)** 9,312.6

(2) 9,507.7 **(4)** 7,870.8

5-30 **(a)** $\hat{Y} = -3{,}481.34 + 225.71X_1 - 20.07X_2 - 35.52X_3 + 83.35X_4$

$S_{Y\cdot 1234} = 1{,}033.59$

(b) **(1)** 6,728.5 **(2)** 8,590.2

(3) 8,489.8 **(4)** 8,867.7

6-3 **(a)** **(1)** 23/50 **(2)** 11/50 **(3)** 4/50 **(4)** 5/50 **(5)** 3/50 **(6)** 4/50
(b) **(1)** 43/50 **(2)** 7/50 **(3)** 32/50 **(4)** 15/50

6-5 **(a)** 6/10 **(b)** 7/15 **(c)** 24/42 **(d)** 49/86 **(e)** 86/153

6-13 **(a)** **(1)** .970299 **(2)** .029403 **(3)** .000297 **(4)** .000001
(b) **(1)** $.029403 + .000297 = .029700$
(2) $.000297 + .000001 = .000298$
(3) $.970299 + .029403 + .000297 = .999999$

6-15 **(a)** **(1)** $D_SW_ST_S$ **(2)** $D_SW_ST_U$ **(3)** $D_SW_UT_S$ **(4)** $D_SW_UT_U$
(5) $D_UW_ST_S$ **(6)** $D_UW_ST_U$ **(7)** $D_UW_UT_S$ **(8)** $D_UW_UT_U$
(b) **(1)** .684 **(2)** .036 **(3)** .076 **(4)** .004 **(5)** .171 **(6)** .009
(7) .019 **(8)** .001

6-21 **(a)** **(1)** 23/50 **(2)** 23/43
(b) **(1)** 4/50 **(2)** 4/7
(c) **(1)** 32/50 **(2)** 32/43
(d) **(1)** 15/50 **(2)** 15/43

6-26 **(a)** **(1)** .27 **(2)** .21 **(3)** .49 **(4)** .03
(b) **(1)** .48 **(2)** .52
(c) **(1)** .563 **(2)** .942

6-29 **(b)** **(1)** .857375 **(3)** .007125 **(5)** .142625
(2) .135375 **(4)** .000125 **(6)** .999875

6-33 **(a)** .3024 **(b)** .9976

6-37 **(a)** .9985 **(b)** .9969 **(c)** .9577

6-42 **(a)** .150 **(b)** .657 **(c)** .0064 **(d)** .840

6-49 **(a)** .80909 **(b)** .90000 **(c)** .8990 **(d)** .80909 **(e)** .90000 **(f)** .8990

7-4 **(b)** GPA

x	**Pr**$[X=x]$
2.75	.010
3.00	.125
3.25	.355
3.50	.375
3.75	.135
	1.000

7-8 **(a)** .4 **(b)** .4 **(c)** .2

7-11 **(a)** 8 **(b)** 3 **(c)** 1 **(d)** 12

7-14 $E(X) = .99$ $Var(X) = .8699$ $SD(X) = .933$

7-20 **(a)** **(1)** 6 **(2)** 12 **(3)** 3.4641

(b) **(1)** .2 **(2)** .04 **(3)** .2
(c) **(1)** $45/28 = 1.61$ **(2)** .0742 **(3)** .272
(d) **(1)** 1 **(2)** $1/6 = .1667$ **(3)** .408

7-24 **(a)** .1285 **(b)** .0875 **(c)** .0216 **(d)** .6730 **(e)** .0023 **(f)** .0000 (rounded)

7-27 **(a)** .0794 **(b)** .3233 **(c)** .5802 **(d)** .8817 **(e)** .9981

7-34 **(a)** **(1)** .9245 **(2)** .0071
(b) **(1)** .8779 **(2)** .0032
(c) **(1)** .6442 **(2)** .0071
(d) **(1)** .0090 **(2)** .0187

7-37 **(a)** **(1)** .0113 **(2)** .3001 **(3)** .7826 **(4)** .0008
(b) **(1)** 33.82″ **(2)** 34.29″ **(3)** 36.80″ **(4)** 38.06″

7-42 **(a)** .9876 **(b)** .7888 **(c)** .4714 **(d)** .2434

7-48 **(a)** 2 **(b)** $\sqrt{5}$) **(c)** 10

8-2 **(a)** **(1)** .0031 **(2)** .0361 **(3)** .1755 **(4)** .0378
(b) **(1)** .1008 **(2)** .1755 **(3)** .0378 **(4)** .0001

8-6 **(a)** **(1)** .0361 **(2)** .1563 **(3)** .1606
(b) **(1)** .3679 **(2)** .1353 **(3)** .0498

8-14 **(a)** .9502 **(b)** .0025 **(c)** .4581 **(d)** .2231

8-17 **(a)** .5934 **(b)** .0111 **(c)** .0001 **(d)** .0399

8-21 **(a)** **(1)** .6065 **(2)** .3679 **(3)** .1353 **(4)** .0067
(b) 23 feet

8-25 **(a)** .2642 **(b)** .5940 **(c)** .9084

8-30 **(a)** **(1)** .6199 **(2)** .5774 **(3)** .5512
(b) **(1)** .01913 **(2)** .01099 **(3)** .00794

8-35 **(a)** .0369 **(b)** .1114 **(c)** .4297

8-38 .0215

8-41 .9400

8-49 **(a)** Using the exponential distribution: .2493
(b) Using the normal distribution with mean $\mu = 20$ hours and $\sigma = 4$ hours: .8882

8-53 **(a)** .2476 **(b)** .6106

9-1 **(a)** 5 lbs. **(b)** \$2 **(c)** .004 m **(d)** .83 kHz

9-8 **(a)** **(1)** .838 **(2)** .4514
(b) **(1)** .53 **(2)** .6528
(c) **(1)** .375 **(2)** .8164
(d) **(1)** .265 **(2)** .9412

9-10 **(a)** .0062 **(b)** .0000 **(c)** .7888 **(d)** .7056

9-14 **(a)** **(1)** .8664 **(2)** .9876 **(3)** .9972 **(4)** .99994

(b) The increased probability for a truthful estimate results in a wider, less-precise estimate.

9-18 **(a)** .9050 **(b)** .7888 **(c)** .6826

9-22 **(a)** .7689 **(b)** .0826

9-25 **(a)** .05 **(b)** .75 **(c)** .995 **(d)** .0025 **(e)** .3995

9-27 **(a)** –2.439 leave alone **(c)** –4.054 replace
(b) .962 leave alone **(d)** –1.220 leave alone

9-29 **(a)** .8078 **(b)** .5987 **(c)** .4129

9-39 **(a)** $\chi^2 = 58.58$ $z = -2.87$ .0021
(b) $\chi^2 = 131.805$ $z = 2.33$ .0099
(c) $\chi^2 = 114.817$ $z = 1.12$ .8686
(d) $\chi^2 = 91.531$ $z = -.53$
$\chi^2 = 106.762$ $z = .55$
.4107

9-47 **(a)** $\chi^2_{.01} = 42.980$ $s = \sqrt{\frac{(1{,}500)^2(42{,}900)}{25-1}} = \$2{,}007.33$

(b) $\sigma_{\overline{X}} = \frac{\$2{,}007.33}{\sqrt{25}} = \$401.47$ $= .7888$

10-8 $54.48 \le \mu \le 60.32$ hours

10-11 **(a)** $\$95.51 \le \mu \le \$104.99$ **(c)** $\$99.48 \le \mu \le \$102.02$
(b) $\$96.91 \le \mu \le \$102.59$ **(d)** $\$99.60 \le \mu \le \$101.40$

10-19 $\overline{X} = 7.31$ $s = 8.522$ $2.77 \le \mu \le 11.85$ hours

10-20

	n	Mean	Stand. Dev.	95% Confidence Interval
(a) Uranium	21	7.476	1.531	$6.779 \le \mu \le 8.173$
(b) Thorium	21	8.900	1.981	$7.998 \le \mu \le 9.802$

10-27 $P = 34/200 = .17$ $.109 \le \pi \le .231$

10-28 **(a)** $.002 \le \pi \le .078$ **(b)** .0184

10-33 **(a)** $10.53 \le \sigma \le 16.09$ **(b)** $10.00 \le \sigma \le 17.27$ **(c)** $9.13 \le \sigma \le 19.88$

10-37 $-22.0 \le \mu_A - \mu_B \le 192.8$ days

10-38 $-160.4 \le \mu_A - \mu_B \le -41.6$ hours **x**

10-39 $23.2 \le \mu_A - \mu_B \le 147.4$ days

10-42 $n_A = 35$ $\overline{X}_A = 12.81$ $s_A = 1.148$
$n_B = 32$ $\overline{X}_B = 11.69$ $s_B = 1.035$
$s_D = .267$ $df = 65$ $.59 \le \mu_A - \mu_B \le 1.65$ hours

10-58 $\overline{X} = 13.34$ $n = 2.9818$
(a) $11.69 \le \sigma \le 14.99$ minutes **(b)** $2.15 \le \sigma \le 4.82$ minutes

10-62 $n_A = 15$ $\bar{X}_A = 1{,}016$ $n_A = 757$

$n_B = 20$ $\bar{X}_B = 787$ $s_B = 691$

$s_D = 249.2$ $df = 29$

$-281 \leq \mu_A - \mu_B \leq 739$ hours

11-1 **(a)** $H_0 : \mu \leq 37 (= \mu_0)$ lb/ton $H_1 : \mu > 37$

(b) **(1)** .1151 **(2)** .2119

11-5 $\alpha = .0162$ (smaller) $\beta = .0764$ (larger)

11-9 **(a)** **(1)** .00016 **(2)** refute

(b) **(1)** 0 **(2)** support

(c) **(1)** 0 **(2)** refute

(d) **(1)** .0139 **(2)** refute

11-13 *Accept* H_0 if $t \geq -2.624$ and *Reject* H_0 if $t < -2.624$. $t = -8.743$. Reject.

11-14 **(a)** *Accept* H_0 if $\bar{X} \leq 5{,}031.3$ and *Reject* H_0 if $\bar{X} > 5{,}031.3$. **(b)** reject

11-18 The computed value is $t = 3.395$, which corresponds to prob = .019. This refutes H_0.

11-21 **(a)** $48.94 \leq \mu \leq 58.80$ **(b)** accept

11-24 **(a)** STEP 1: $H_0 : \mu \geq 15$ hours (New test stand is no improvement.)

$H_1 : \mu < 15$ hours (New test stand is an improvement.)

STEP 2: The population standard deviation is not known, so that the Student t will serve as the test statistic.

STEP 3: The significance level is $\alpha = .01$, so that the critical value is $t_{.01} = 2.408$. The test is lower-tailed, since a low level for the mean, and hence t, will refute H_0. The following acceptance and rejection regions apply:

Accept H_0 if $t \geq -2.408$ (Retain present test stand.)

Reject H_0 if $t < -2.408$ (Replace present unit with new test stand.)

(b) The computed value is $t = -3.12$. Reject H_0 and replace.

11-28 Using $\alpha = .01$, the critical normal deviate is $z_{.01} = 2.33$. The computed value is $z = -1.53$. Accept H_0 and do not recommend the new procedure.

11-30 **(a)** .9573 **(b)** .8340 **(c)** .7088 **(d)** .5636.

11-36 **(a)**

r	Pr[$R = r$]	Pr[$R \leq r$]
0	.5766	.5766
1	.3516	.9282
2	.0670	.9952
3	.0047	1.0000
4	.0001	1.0000
5	.0000	1.0000
	1.0000	

(b) $R^* = 2$ *Accept* the shipment if $R \leq 2$ or *Reject* the shipment if $R > 2$

(c) Same as (b).

11-40 The null hypothesis is $H_0 : \mu_A \leq \mu_B$ (policy B is at least as expensive). This is an upper-tailed test. The computed value for the test statistic is $t = 1.738$. Using df $= 15$, the critical value is $t_{.05} = 1.753$. The null hypothesis must be *accepted,* and regular preventive maintenance (A) must be retained.

11-41 $H_0 : \mu_A \geq \mu_B$ (a lower-tailed test). Using

$$\bar{d} = -1.085 \qquad s_d = 2.67$$

the test statistic is $t = -1.817$. Since this exceeds the critical value $-t_{.01} = -2.539$, the null hypothesis must be *accepted,* and the traditional process should be retained.

11-44 $H_0 : \mu_A = \mu_B$. The computed value is $t = -.895$. Using degrees of freedom $20 + 25 - 2 = 43$, interpolation provides $-t_{.025} = -2.018$. The instructor *accepted* the null hypothesis.

11-47 accept

11-51 reject

11-55 **(a)** **(1)** mean **(2)** 51.64 **(3)** upper-tailed

(b) **(1)** t **(2)** -1.663 **(3)** lower-tailed

(c) **(1)** t **(2)** $-2.093, 2.093$ **(3)** two-sided

(d) **(1)** P **(2)** .2656 **(3)** upper-tailed

(e) **(1)** P **(2)** .3014 **(3)** lower-tailed

(f) **(1)** mean **(2)** .545 **(3)** upper-tailed

(g) **(1)** t **(2)** -1.711 **(3)** lower-tailed

(h) **(1)** mean **(2)** 24.75, 25.25 **(3)** two-sided

11-58 $H_0 : \mu = 5$ mg

(a) $\overline{X}_1^* = 4.95$ mg and $\overline{X}_2^* = 5.05$ mg. No corrective action is taken if the computed mean lies between these limits (and H_0 is accepted). Otherwise, corrections are made (and H_0 is rejected).

(b) **(1)** .5000 **(2)** .8413 **(3)** .1587 **(4)** .9330

11-66 **(a)** $-.469 \leq \mu_A - \mu_B \leq .395$ **(b)** $t = -.18$. Accept.

12-1 **(a)**

	Inert Material Concentration X	Theoretical Time $\mu_{Y \cdot X}$	Random Normal Deviate z	Error Term ϵ	Simulated Actual Time Y
(1)	50	2.950	.05	.005	2.955
(2)	50	2.950	−1.23	−.123	2.827
(3)	100	2.900	−.87	−.087	2.813
(4)	100	2.900	1.56	.156	3.056
(5)	200	2.800	.97	.097	2.897
(6)	250	2.750	.43	.043	2.793
(7)	300	2.700	−.98	−.098	2.602
(8)	350	2.650	−.55	−.055	2.595
(9)	400	2.600	.17	.017	2.617
(10)	500	2.500	1.08	.108	2.608

12-7 .972 97.2%

12-9 .950 95.0%

12-11 **(a)** 111.30, 125.58 **(b)** 121.47, 177.29

12-16 **(a)** \$11,994.36, \$13,462.10 **(b)** \$7.9049, \$9.5183

12-18 7,328,000, 13,972,000

12-22 **(a)** 29.626, 30.585 **(b)** 28.475, 31.736

12-25 **(a)** 298.70, 340.14 pounds **(b)** 250.57, 388.27 pounds

12-30 **(a)** **(1)** (37,036, 45,144)
(2) (41,487, 50,712)
(3) (40,245, 52,272)
(4) (39,039, 53,631)
(b) **(1)** (25,796, 56,384)
(2) (30,648, 61,551)
(3) (30,333, 62,185)
(4) (29,882, 62,788)
(c) Yes.

12-34 **(b)** $\ln \hat{Y}(X) = 7113 - .0588X$ (for amputations)
(c) .018 amputations per million hours

12-36 **(a)** $r^2 = .911$ $r = -.95$
(b) $-.01208, -.00806$ *rejected*
(c) *rejected*

12-37 **(a)** **(1)** 29.0 **(2)** 28.6 **(3)** 28.1
(b) **(1)** 28.89, 29.11 **(2)** 28.51, 28.69 **(3)** 28.00, 28.20
(c) **(1)** 28.63, 29.37 **(2)** 28.24, 28.96 **(3)** 27.74, 28.46

13-5 **(b)**

Variation	Degrees of Freedom	Sum of Squares	Mean Square	F
Treatments	3	28.275	9.425	2.27
Error	36	149.700	4.158	
Total	39	177.975		

(c) Since the computed value for F falls below the critical value of $F_{.01} = 4.38$, the null hypothesis of identical population mean yields should be *accepted.*

13-8 **(a)** Using $5 - 1 = 4$ degrees of freedom for the numerator and $23 - 5 = 18$ degrees of freedom for the denominator, the critical value is $F_{.01} = 4.58$.

(b)

Variation	Degrees of Freedom	Sum of Squares	Mean Square	F
Treatments	4	262.65	65.66	33.43
Error	18	35.35	1.96	
Total	22	298.00		

(c) Since the computed value for F falls above the critical value, the null hypothesis of identical population mean percentages trapped under the various methods should be *rejected.*

13-11 (a) 1.88, 2.72 years (b) −.10,1.10 years (c) *No.*

13-21 (a)

Sample Means

	(1) Simple	(2) Low	(3) Medium	(4) High	Factor A Mean
(1) Short	.110	.125	.295	.300	.20750
(2) Middle	.155	.235	1.425	4.800	1.65375
(3) Long	.400	.320	1.020	1.370	.77750
Factor B Mean	.22167	.22667	.91333	2.15667	.87958

ANOVA Table

Variation	Degrees of Freedom	Sum of Squares	Mean Square	F
Explained by Factor A	2	8.4916	4.2458	3.634
Explained by Factor B	3	14.9474	4.9825	4.264
Explained by Interactions	6	15.0616	2.5103	2.148
Error or Unexplained	12	14.0210	1.1684	
Total	23	52.5217		

(b) (1) $F_{.05} = 3.00$ accepted (2) $F_{.05} = 3.89$ accepted (3) $F_{.05} = 3.49$ rejected

13-27

Sample Means

	(1) $e = 5$	(2) $e = 15$	(3) $e = 25$	(4) $e = 50$	Factor A Mean
(1) $N' = 10$	4.00	4.48	4.86	6.48	4.955
(2) $N' = 50$	2.75	2.15	1.95	2.20	2.2625
(3) $N' = 100$	2.70	1.83	1.61	2.00	2.035
Factor B Mean	3.15	2.82	2.8067	3.56	3.08417

ANOVA Table

Variation	Degrees of Freedom	Sum of Squares	Mean Square	F
Explained by Factor A	2	105.518	52.76	4,769
Explained by Factor B	3	5.663	1.89	171
Explained by Interactions	6	16.788	2.80	253
Error or Unexplained	48	.531	.01	
Total	59	128.500		

The null hypotheses of identical means must be *rejected for both* factors. Also, the null hypothesis of zero interactions must be *rejected.*

13-34 **(a)**

Variation	Degrees of Freedom	Sum of Squares	Mean Square	F
Treatments	3	28.275	9.425	14.99
Blocks	9	132.725	14.747	
Error	27	16.975	.629	
Total	39	177.975		

(b) For the one-way analysis of variance, $SSE = 149.70$. For the two-way analysis of variance, $SSE = 16.975$, a dramatic reduction. *Yes,* the inclusion of the blocking variable does provide a promising reduction in unexplained variation.

(c) Since the computed value of $F = 14.99$ exceeds the critical value $F_{.01} = 4.60$ the null hypothesis of identical means must be *rejected.*

13-39 **(a)**

Variation	Degrees of Freedom	Sum of Squares	Mean Square	F
Treatments	3	30.870	10.290	12.30
Error	25	20.923	.837	
Total	28	51.793		

(b) *Yes,* since the computed value of $F = 12.30$ exceeds the critical value of $F_{.01} = 4.68$, the null hypothesis of identical means must be *rejected.*

13-40 **(a)** $-4.00 \leq \mu_1 - \mu_2 \leq -1.36$ months (significant)

$-2.68 \leq \mu_1 - \mu_3 \leq .16$ months

$-1.88 \leq \mu_1 - \mu_4 \leq .76$ months

$.04 \leq \mu_2 - \mu_3 \leq 2.80$ months (significant)

$.85 \leq \mu_2 - \mu_4 \leq 3.39$ months (significant)

$-.68 \leq \mu_3 - \mu_4 \leq 2.08$ months

(b, c) All pairs involving μ_2 have significant differences, with paint (2) yielding significantly longer mean lifetime than each of the other types.

14-2 **(a)** $2 \times 2 \times 2 \times 2 = 16$ **(c)** $2 \times 4 \times 3 = 24$
(b) $3 \times 2 \times 2 \times 2 = 24$ **(d)** $3 \times 4 \times 2 = 24$

14-4 **(b)** The following individual 99% confidence intervals apply.

Effect	99% Confidence Interval	
Main		
Engine Type E	5.25 ± 3.93	(significant)
Suspension S	-3.75 ± 3.93	
Pass. Loading L	-4.75 ± 3.93	(significant)
Interactions		
$E \times S$	-2.25 ± 3.93	
$E \times L$	1.75 ± 3.93	
$S \times L$	2.75 ± 3.93	
$E \times S \times L$	-1.75 ± 3.93	

(c) Only engine type (E) and passenger loading (L) have significant main effects, with the greater horsepower improving gasoline mileage from 1.32 to 9.18 mpg and the lower passenger loading improving it by about the same. There are no significant interactions.

14-7 **(c)** The following individual 99% confidence intervals apply.

Effect	99% Confidence Interval	
Main		
Temperature T	-2.5 ± 1.7	(significant)
Tank Pressure P	-2.5 ± 1.7	(significant)
Settling Time S	5.5 ± 1.7	(significant)
Interactions		
$T \times P$	$.5 \pm 1.7$	
$T \times S$	$-.5 \pm 1.7$	
$P \times S$	1.5 ± 1.7	
$T \times P \times S$	$.5 \pm 1.7$	

(d) The significant effects are the main ones for settling time (for which an increase from 30 to 60 minutes will increase yield about 5.5 grams per liter), temperature, and tank pressure.

14-10 **(b)** The effect standard error is .26. The Student t statistic is $t_{.005} = 4.032$. The following individual 99% confidence intervals are computed.

Effect	99% Confidence Interval	
Main		
Brand B	-1.63 ± 1.05	(significant)
Location L	1.13 ± 1.05	(significant)
Drying Conditions D	2.13 ± 1.05	(significant)
Number of Coats C	$-.38 \pm 1.05$	
Interactions		
$B \times L$	$.13 \pm 1.05$	
$B \times D$	$.63 \pm 1.05$	
$B \times C$	$-.88 \pm 1.05$	
$L \times D$	$.38 \pm 1.05$	
$L \times C$	$.38 \pm 1.05$	
$D \times C$	$-.63 \pm 1.05$	

(c) The drying condition main effect is statistically significant. Humid condition will increase drying time by 1.08 to 3.18 hours. Switching from brand X to brand Y will reduce drying time by .58 to 2.68 hours. Moving from the shade to the sun will increase drying time by .08 to 2.18 hours.

14-14 The confidence level is arbitrary. The Student t critical value is $t_{.005} = 4.032$ for individual 99% confidence intervals. The following apply.

Effect	99% Confidence Interval	
Main		
Wheel Diameter W	$.4 \pm 4.8$	
Axle Length A	$.9 \pm 4.8$	
Wheel Base B	1.6 ± 4.8	
Load L	-8.4 ± 4.8	(significant)
Interactions		
$W \times A$	$-.6 \pm 4.8$	
$W \times B$	$.1 \pm 4.8$	
$W \times L$	1.6 ± 4.8	
$A \times B$	$.1 \pm 4.8$	
$A \times L$	$.1 \pm 4.8$	
$B \times L$	-2.1 ± 4.8	

Only the load main effect is statistically significant. The failure time advantage of keeping the load at 2 tons will increase time to failure somewhere between 3.6 and 13.2 hours.

LAPIN INDEX

Normal Distribution

Suppose a test of student scores is normally distributed w̄ a mean 60 and stand. dev. of 10. Now the question is what proportion of scores are below 80. Then:

$$z = \frac{x-\mu}{\sigma} = \frac{80-60}{10} = 2$$, from table D, 0.9772 scores are below 80

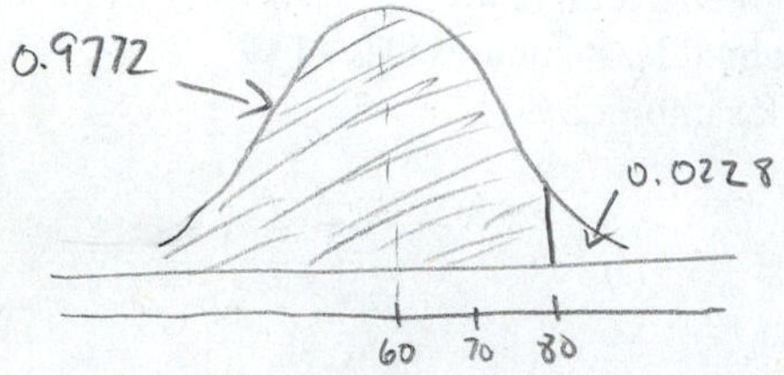

below 70, $$z = \frac{x-\mu}{\sigma} = \frac{70-60}{10} = 1$$

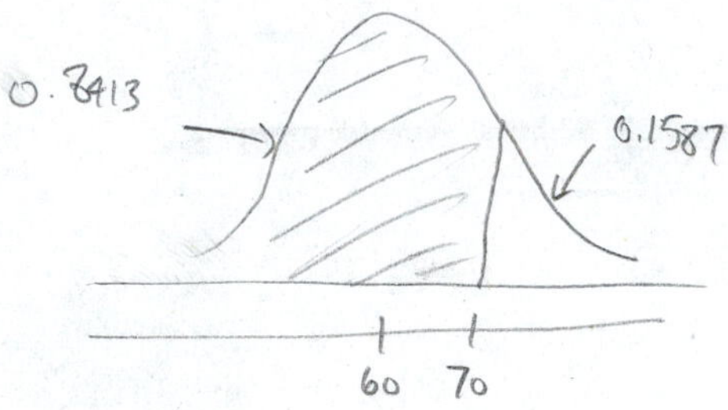

or

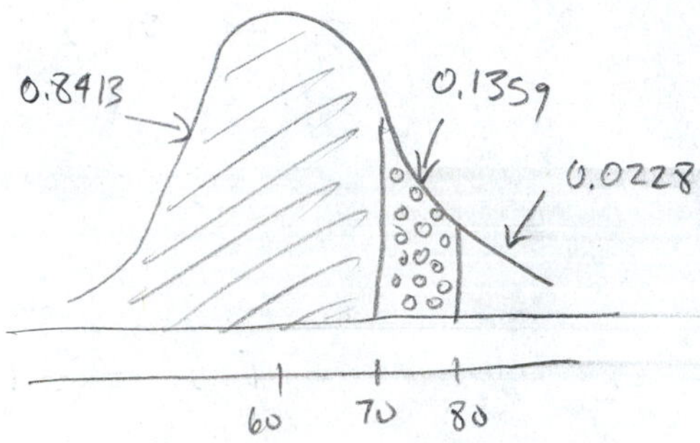

Standard Normal Distribution ⇒ 7.4, use of z to find error
⇒ z values on pg. 536
⇒ $z = \frac{(x-\mu)}{\sigma}$ σ = Standard. Dev
μ = mean
x = value.

7-35) e) $Pr[1.5 \le x \le 1.75] = Pr[x \le 1.75] - Pr[x \le 1.5] = \phi(1.75) - \phi(1.5)$
$= 0.9599 - 0.9332 = 0.267$

g) $Pr[x \le -2.80]$ or $Pr[x \ge 0.65] = Pr[x \le -2.80] + Pr[x \ge 0.65]$
$= \phi(-2.80) + (1 - \phi 0.65) = 0.0026 + (1 - 0.7922)$
$= 0.2604$

| more pg. 227 |

8-1 → Poisson Distribution (Table C), 8-2 → Exponential Distribution, 8-3 → Gamma Distr.

8-1 8-1) $P(x;\lambda;t) = P[x \le x] = \Sigma p(k;\lambda;t)$
→ pg(233)

λt is # of occurences going to happen over period of t.

8-2) Exponential Distribution
$f(t) = Pr[T \le t] = 1 - e^{-\lambda t}$

= Probability that the time between 2 successive occurences of incidents is less than or equal to t.

And the density function is $f(t) = \lambda e^{-\lambda t}$ for $t \ge 0$
$= 0$ otherwise

λ = mean process rate (mean time between 2 successive events)

8-14) Given: $\lambda = 2$ incoming calls/s, $\mu = 5$ calls answered/sec

$avg = \mu - \lambda = 3$, $\therefore \lambda = 3$

a) Pr[a calls waiting < 1sec] = Pr[a call to be answered comes in < 1sec] $= 1 - e^{-\lambda t} = 1 - e^{-3 \cdot 1}$
$= 0.9502$

b) Pr[a call waits > 2sec] $= 1 - Pr[x \le 2] = 1 - (1 - e^{-2 \cdot 3}) = e^{-6} = 0.0025$

c) $Pr[0.2 \le x \le 0.8] = (1 - e^{-3 \cdot 0.2}) - (1 - e^{-3 \cdot 0.8}) = e^{-0.6} - e^{-2.4} = 0.548812 - 0.090718$
$= 0.4581$

CHAPTER 9 Table D

9-10) Here, standard normal deviate $z = \frac{(0.062 - 0.060)}{0.0008} = 2.5$

$\sigma_{\bar{x}} = \frac{\sigma}{\sqrt{n}} = \frac{0.008}{\sqrt{100}} = 0.0008$

$Pr[x > 0.062] = 1 - Pr[x \le 0.062]$
$= 1 - \phi(2.5)$
$= 0.0062$

b) $z = \frac{(0.0505 - 0.060)}{0.0008} = -11.88$

$Pr[x \le 0.0505] = \phi(-11.88) = 0$

d) $z_1 = \frac{(0.0585 - 0.060)}{0.0008} = -1.88$, $z_2 = \frac{(0.0605 - 0.060)}{0.0008} = +0.63$

$Pr[0.0585 \le x \le 0.0605] = Pr[x \le 0.0605] - Pr[x \le 0.0585]$
$= \phi(0.63) - \phi(-1.88)$
$= 0.7056$

9-12) Since the population is normally distributed, the sample may also be taken as normally distributed

a) Sample stand. dev. ⇒ $\sigma_{\bar{x}} = \frac{\sigma}{\sqrt{n}} = \frac{2}{\sqrt{25}} = 0.4$

since X (setting time) deviates from the mean hour by 1 hour

$x = (\mu \pm 1)$

$z_1 = \left[\frac{((\mu+1) - \mu))}{0.4}\right] = 2.5$ $z_2 = \frac{((\mu-1) - \mu)}{0.4} = -2.5$

$Pr[(\mu-1) \le x \le (\mu+1)] = \phi(2.5) - \phi(-2.5) = 0.9876$

b) $\sigma_{\bar{x}} = \frac{\sigma}{\sqrt{n}} = \frac{4}{\sqrt{25}} = 0.8$ c) $\sigma_{\bar{x}} = \frac{8}{\sqrt{25}} = 1.6$ ...

9-22) Population size $N = 500$, sample size $n = 100$, st. dev. of pop. $= 0.003$

Thus, $\sigma_x = \left(\frac{\sigma}{n}\right)\sqrt{\frac{N-n}{N-1}} = 0.00027$

a) $\mu = 0.5003$ The shipment will be accepted if $0.4995 \le x \le 0.5005$

$$Z_1 = \frac{0.4995 - 0.5003}{0.00027} = -2.96 \qquad Z_2 = \frac{0.5005 - 0.5003}{0.00027} = 0.74$$

$$Pr[0.4995 \le x \le 0.5005] = Pr[x \le .5005] - Pr[x \le 0.4995]$$
$$= \phi 0.74 - \phi(-2.96)$$
$$= 0.7689$$

b) $\mu = 0.4999$

Shipment rejected if $0.5005 \le x \le 0.4995$

$$Z_1 = \frac{0.4995 - 0.4999}{0.00027} = -1.48 \qquad Z_2 = \frac{0.5005 - 0.4999}{0.00027} = 2.22$$

$$Pr[x \le 0.4995] + Pr[x \ge 0.5005] = \phi(-1.48) + 1 - \phi(2.22) = 0.0826$$

9-23) For t distribution, the upper tail area or probability is denoted by α where $\alpha = Pr[t > t_\alpha]$. The critical value t_α corresponding to any α and degree of freedom df ($= n-1$ where n is the sample size) can be read from Table G

a) $\alpha = 0.05$, $df = 11$; $t_\alpha = 1.796$

b) $\alpha = 0.10$, $df = 6$; $t_\alpha = 1.440$

c) $\alpha = 0.005$, $df = 25$; $t_\alpha = 2.787$

d) $\alpha = 0.0005$, $df = 30$ $t_\alpha = 3.646$

0.00003

CHAPTER 10 (10.1–10.4)

Interval estimator of Mean (Knowns)

$\bar{X} \pm z_{\alpha/2} \frac{\sigma}{\sqrt{n}}$

Lower Confidence Limit $\Rightarrow \bar{X} - z_{\alpha/2} \frac{\sigma}{\sqrt{n}}$

Upper $\Rightarrow \bar{X} + z_{\alpha/2} \frac{\sigma}{\sqrt{n}}$

$\sigma_{\bar{x}} = \frac{\sigma}{\sqrt{n}}$

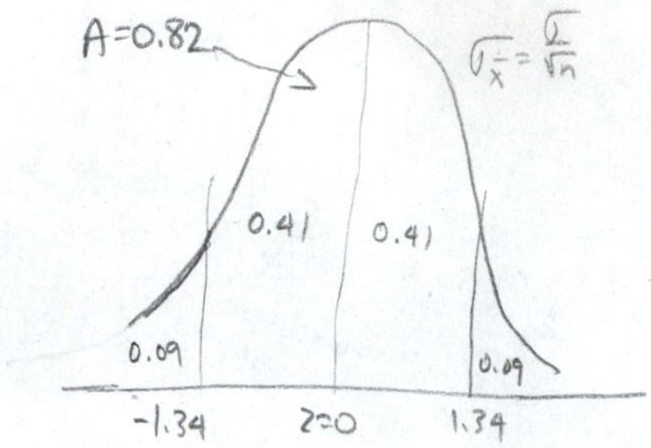

SELECTING SAMPLE SIZE

$$n = \left(\frac{z_{\alpha/2} \sigma}{d} \right)^2$$

SMALL POPULATION

$$\bar{X} \pm z_{\alpha/2} \frac{\sigma}{\sqrt{n}} \left(\sqrt{\frac{N-n}{N-1}} \right)$$

INTERVAL ESTIMATOR OF PROPORTION (10.3)

LARGE POPULATION : $\pi = P \pm z_{\alpha/2} \sqrt{\frac{P(1-P)}{n}}$ $\sqrt{\frac{N-n}{N-1}}$ ← SMALL

REQUIRED SAMPLE SIZE FOR ESTIMATING PROPORTION

$$n = \frac{z^2_{\alpha/2} \pi(1-\pi)}{d^2}$$

Confidence Interval of the variance (10.4)

$$\frac{(n-1)s^2}{\chi^2_{\alpha/2}} \leq \sigma^2 \leq \frac{(n-1)s^2}{\chi^2_{1-\alpha/2}}$$

Confidence Interval $\Rightarrow \bar{\mu}_A - \bar{\mu}_B = \bar{\mu}_D$ $\quad \sigma_D^2 = \frac{\sigma_A^2}{n_A} + \frac{\sigma_B^2}{n_B}$

Random Variable $\Rightarrow \bar{X}_A - \bar{X}_B = D$

$\mu_A - \mu_B = D \pm z_{\alpha/2} s_D$

pg. 319, 330

CHAPTER 11

Conclusion in terms of Alternate Hypothesis | σ pop. variance

Two-Tail	One Tail (right)	One Tail (Left)
$H_0: \mu = 100$	$H_0: p = 0.10$	$H_0: p = 0.10$
$H_A: \mu \neq 100$	$H_A: p > 0.10$	$H_A: p < 0.10$

Test statistic $\Rightarrow$ $z = \dfrac{\bar{x} - \mu}{\sigma/\sqrt{n}\sqrt{\frac{N-n}{N-1}}}$ $\bar{x}$ - sample mean μ - pop. mean stated in H_0 σ = pop. std. dev. n = sample size

if test statistic lies in rejection region, the null hypothesis is rejected, else the null hypothesis is not rejected

Type I $\Rightarrow$ reject null hypothesis when its true
Type II $\Rightarrow$ accept H_0 when its false

Two tail, reject H_0 if $|z| > z_{\alpha/2}$, $t_{\alpha/2, n-1}$
right tail, reject H_0 if $z > z_\alpha$ $t_{\alpha, n-1}$
left tail reject H_0 if $z < -z_\alpha$ $t_{\alpha, n-1}$

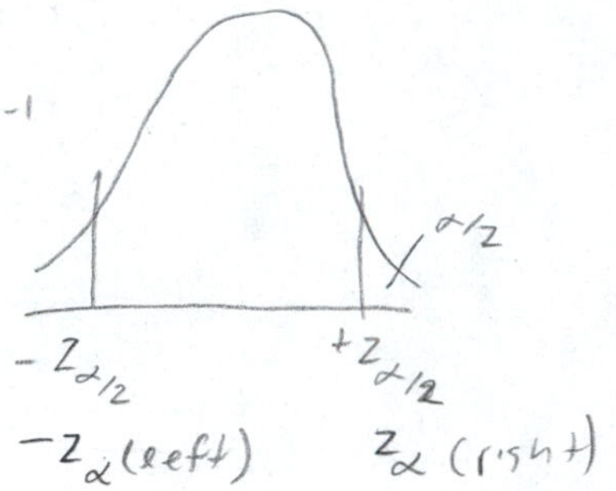

H_0 rejected: conclude: there is enough statistical evidence to infer that the alternative hypothesis is true

H_0 not rejected: not enough statistical evidence to infer that the alternative hypothesis is true

p = sample proportion
n = sample size

expected # successes $= np$
std. dev of " $= \sqrt{np(1-p)}$
std. dev of proportion $= \sqrt{\dfrac{p(1-p)}{n}}$

Test hypothesis about population proportion

$$z = \frac{p - \pi_{H_0}}{\sqrt{\pi_{H_0}(1 - \pi_{H_0})/n}}$$